General Relativity and Cosmology

General Relativity and Cosmology

Edited by
Benjamin Phillips

Willford Press
www.willfordpress.com

Published by Willford Press,
118-35 Queens Blvd., Suite 400,
Forest Hills, NY 11375, USA

ISBN:978-1-64728-355-1

Cataloging-in-Publication Data

General relativity and cosmology / edited by Benjamin Phillips.
p. cm.
Includes bibliographical references and index.
ISBN 978-1-64728-355-1
1. General relativity (Physics). 2. Cosmology. 3. Relativity (Physics). 4. Physics. I. Phillips, Benjamin.
QC173.6 .G46 2021
530.11--dc23

For information on all Willford Press publications
visit our website at www.willfordpress.com

WILLFORD PRESS

Contents

Preface

This book has been a concerted effort by a group of academicians, researchers and scientists, who have contributed their research works for the realization of the book. This book has materialized in the wake of emerging advancements and innovations in this field. Therefore, the need of the hour was to compile all the required researches and disseminate the knowledge to a broad spectrum of people comprising of students, researchers and specialists of the field.

General relativity is the theory of gravitation which provides a complete description of gravity as a geometric property of space and time. The understanding of how matter and radiation warps the geometry of space and time is governed by the Einstein field equations. This theory departs significantly from classical physics especially in relation to the geometry of space, the passage of time, the propagation of light and the motion of bodies in free fall. This is evident in its treatment of gravitational time delay, gravitational time dilation, the gravitational redshift of light and gravitational lensing. The study of the origin and evolution of the universe, starting from the Big Bang to the present and the description of its ultimate fate in the future is under the domain of cosmology. It further studies the large scale dynamics and structures of the universe. There have been significant advances in our understanding of the universe, due to advances in the observations of the microwave background, gravitational lensing and distant supernovae. The detection of gravitational waves in recent times has further strengthened the theories of the Big Bang and cosmic inflation. This book contains some path-breaking studies in the field of cosmology. It provides comprehensive insights into general relativity and its ramifications relative to the understanding of the universe and its dynamics. Scientists and students actively engaged in these areas will find this book full of crucial and unexplored concepts.

At the end of the preface, I would like to thank the authors for their brilliant chapters and the publisher for guiding us all-through the making of the book till its final stage. Also, I would like to thank my family for providing the support and encouragement throughout my academic career and research projects.

Editor

1

Absorption of electromagnetic and gravitational waves by Kerr black holes

Luiz C.S. Leite [a,b], Sam R. Dolan [b,*], Luís C.B. Crispino [a]

[a] Faculdade de Física, Universidade Federal do Pará, 66075-110, Belém, Pará, Brazil
[b] Consortium for Fundamental Physics, School of Mathematics and Statistics, University of Sheffield, Hicks Building, Hounsfield Road, Sheffield S3 7RH, United Kingdom

ARTICLE INFO

Editor: M. Trodden

ABSTRACT

We calculate the absorption cross section for planar waves incident upon Kerr black holes, and present a unified picture for scalar, electromagnetic and gravitational waves. We highlight the spin-helicity effect that arises from a coupling between the rotation of the black hole and the helicity of a circularly-polarized wave. For the case of on-axis incidence, we introduce an extended 'sinc approximation' to quantify the spin-helicity effect in the strong-field regime.

1. Introduction

Black holes, once dismissed as a mathematical artifact of Einstein's theory of general relativity (GR), have come to play a central role in modern astronomy and theoretical physics [1,2]. In astronomy, black holes provide a solution: in galaxy formation scenarios, in active galactic nuclei and in core-collapse supernovae, for instance. In theoretical physics, black holes pose a challenge: as spacetime curvature grows without bound in GR, the classical theory breaks down. Yet, novel quantum gravity effects apparently remain shrouded by a horizon endowed with generic thermodynamic properties [3].

Two recent advances in interferometry have opened new data channels on astrophysical black holes. In September 2015, LIGO detected the first gravitational-wave signal: a characteristic 'chirp' from a black hole binary merger [4]. Hundreds more chirps are anticipated over the next decade [5]. In April 2017, the Event Horizon Telescope (EHT) [6] – a global array of radio telescopes linked by very long baseline interferometry – observed the supermassive black hole candidates Sgr. A* and M87* at a resolution three orders of magnitude beyond that of the Hubble telescope [7]. Ultimately, the EHT will seek to study the black hole shadow itself [8–10], using techniques to surpass the diffraction limit [11].

These experimental advances motivate study of the interaction of electromagnetic waves (EWs) and gravitational waves (GWs) with black holes [12–14]. EWs and GWs propagating on curved spacetimes in vacuum share some traits. For example, both possess two independent (transverse) polarizations that are parallel-transported along null geodesics in the ray-optics limit. Yet there are key physical differences. GWs are tenuous, in the sense that they are not significantly attenuated or rescattered by matter sources. GWs are typically long-wavelength and polarized, because rotating quadrupoles (for example, binary systems or asymmetric neutron stars) predominantly emit circular-polarized waves at twice the rotational frequency [15]. For example, $\lambda \sim 10^{-3}$ m for EHT observations, whereas $\lambda \sim 10^{7}$ m for GW150914.

In this Letter we examine the absorption of a monochromatic planar wave of frequency ω incident upon a Kerr black hole of mass M and angular momentum J in vacuum. We calculate the absorption cross section σ_{abs}, i.e., the cross-sectional area of the black hole shadow [8–10] beyond the ray-optics approximation. For the first time, we present unifying results for scalar ($s = 0$), electromagnetic ($s = 1$) and gravitational ($s = 2$) waves. Our results highlight the influence of two key phenomena: superradiance and the spin-helicity effect, described below.

The absorption scenario, illustrated in Fig. 1, is encapsulated by several dimensionless parameters: the ratio of the gravitational length to the (reduced) wavelength $GM\omega/c^3$; the dimensionless black hole spin $a^* \equiv a/M$ where $a = Jc^2/GM$ $(0 \leq a^* < 1)$; the spin of the field $s = 0, 1, 2$; the angle of incidence with respect to the black hole axis γ; and the helicity of the circular polarization ± 1. We adopt natural units such that $G = c = 1$.

* Corresponding author.
E-mail address: s.dolan@sheffield.ac.uk (S.R. Dolan).

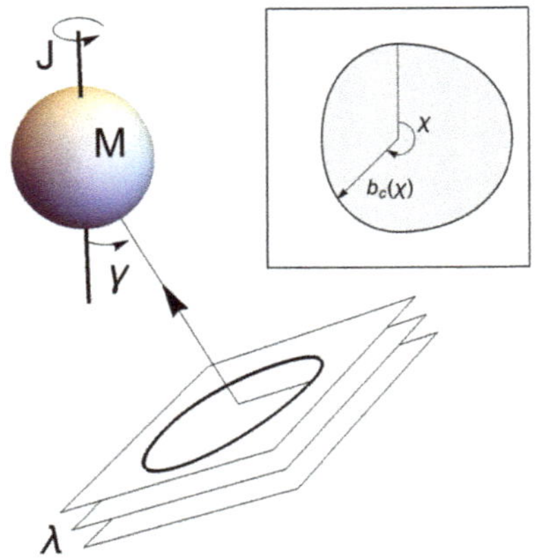

Fig. 1. A planar wave of frequency $\omega = 2\pi c/\lambda$ incident upon a rotating black hole of mass M and angular momentum J at an angle γ. *Inset:* the locus $b_c(\chi)$ of the black hole shadow on the wavefront.

2. Concepts

2.1. Black hole shadows

An observer studying a black hole in vacuum with a pinhole camera will see a dark region on the image plane defined by the set of null-geodesic rays entering the pinhole which, when traced backwards in time, pass into the black hole. The boundary of the shadow is determined by those rays which asymptote towards an (unstable) photon orbit, defining an angular radius $\alpha(\chi)$ in terms of the projection angle χ. Alternatively, a shadow can be defined on a planar surface in terms of an impact parameter $b(\chi)$, using those rays orthogonal the surface, as shown in Fig. 1. Far from the black hole, there is an approximately linear relationship $b(\chi) = r_0\alpha(\chi) + O\left(\frac{GM}{c^2 r_0}\right)$; the two approaches are closely related. Here we extend the latter approach to consider monochromatic waves of a finite wavelength.

In the geometrical-optics limit ($\lambda \to 0$), an observer at radial coordinate r_0 sees a shadow of angular radius α where [16]

$$\sin^2\alpha = \frac{27}{4}\frac{(\rho-1)}{\rho^3}, \qquad \rho \equiv \frac{r_0 c^2}{GM}. \tag{1}$$

For Sgr A*, $\alpha \approx 25$ μarcsec, with $r_0 \approx 8.3$ kpc and $M \approx 4.1 \times 10^6\, M_\odot$ [17]. In Kerr spacetime, α is a function of angle χ relative to the (projected) spin axis.

Here we seek to study Kerr shadows beyond the geometrical-optics regime. We shall focus on the difference between $\sigma_{abs}(\omega)$, the absorption cross section at fixed frequency ω, and the σ_{geo}, the geometric cross section defined by

$$\sigma_{geo} = \tfrac{1}{2}\int_0^{2\pi} b_c^2(\chi)d\chi. \tag{2}$$

2.2. Superradiance and spin-helicity

Superradiance is a radiation-enhancement mechanism by which a black hole may shed mass and angular momentum and yet still increase its horizon area, and thus its entropy [18]. As a consequence, σ_{abs} may become negative at low frequencies, through stimulated emission. The effect is strongly enhanced by spin s.

The spin-helicity effect is a coupling between a rotating source, such as a Kerr black hole, and the helicity of a polarized wave of finite wavelength λ [19]. A rotating spacetime distinguishes and separates waves of opposite helicity [20–22]. In the weak-field, rays are deflected through an angle $\zeta\Theta_E$, with $\Theta_E \equiv \frac{4GM}{c^2 b}$ the Einstein angle and $\zeta = 1 + \ldots$ an asymptotic series in which the spin-helicity effect is anticipated at $O\left(\frac{J\lambda}{Mcb^2}\right)$ [19]. In the strong-field, we anticipate that waves with a counter-rotating circular polarization are preferentially absorbed ($\sigma^-_{abs} > \sigma^+_{abs}$).

3. Method

3.1. Waves on the Kerr spacetime

The Kerr spacetime is described in Boyer–Lindquist coordinates $\{t, r, \theta, \phi\}$ by the line element

$$ds^2 = -\frac{1}{\Sigma}(\Sigma - 2Mr)\,dt^2 - \frac{4Mar\sin^2\theta}{\Sigma}dtd\phi + \frac{\Sigma}{\Delta}dr^2 + \Sigma d\theta^2 + \frac{(r^2+a^2)^2\sin^2\theta - \Delta a^2\sin^4\theta}{\Sigma}d\phi^2, \tag{3}$$

where $\Sigma \equiv r^2 + a^2\cos^2\theta$, and $\Delta \equiv r^2 - 2Mr + a^2$. We focus on the $a^2 < M^2$ case of a rotating BH with two distinct horizons: an internal (Cauchy) horizon located at $r_- = M - \sqrt{M^2 - a^2}$ and an external (event) horizon at $r_+ = M + \sqrt{M^2 - a^2}$.

In the vicinity of a Kerr black hole, perturbing fields are described by a single master equation, first obtained by Teukolsky [23] using the Newman–Penrose formalism. In vacuum the master equation takes the form

$$\left[\frac{(r^2+a^2)^2}{\Delta} - a^2\sin^2\theta\right]\frac{\partial^2\psi}{\partial t^2} + \frac{4Mar}{\Delta}\frac{\partial^2\psi}{\partial t\partial\phi} + \left[\frac{a^2}{\Delta} - \frac{1}{\sin^2\theta}\right]\frac{\partial^2\psi}{\partial\phi^2} - \Delta^{-\mathfrak{s}}\frac{\partial}{\partial r}\left(\Delta^{\mathfrak{s}+1}\frac{\partial\phi}{\partial r}\right) - \frac{1}{\sin\theta}\frac{\partial}{\partial\theta}\left(\sin\theta\frac{\partial\psi}{\partial\theta}\right) + (\mathfrak{s}^2\cot^2\theta - \mathfrak{s})\psi - 2\mathfrak{s}\left[\frac{a(r-M)}{\Delta} + \frac{i\cos\theta}{\sin^2\theta}\right]\frac{\partial\psi}{\partial\phi} - 2\mathfrak{s}\left[\frac{M(r^2-a^2)}{\Delta} - r - ia\cos\theta\right]\frac{\partial\psi}{\partial t} = 0, \tag{4}$$

where $\mathfrak{s}$ is the spin-weight of the field. We use $\mathfrak{s} = -s$ throughout, where $s = 0, 1, 2$ for scalar, electromagnetic and gravitational fields, respectively. One can separate variables in Eq. (4) using the standard ansatz

$$\psi_{\mathfrak{s}lm\omega}(t, r, \theta, \phi) = R_{\mathfrak{s}lm\omega}(r)S_{\mathfrak{s}lm\omega}(\theta)e^{-i(\omega t - m\phi)}, \tag{5}$$

to obtain angular and radial equations,

$$\frac{1}{\sin\theta}\frac{d}{d\theta}\left(\sin\theta\frac{dS_{\mathfrak{s}lm\omega}}{d\theta}\right) + U_{\mathfrak{s}lm\omega}(\theta)S_{\mathfrak{s}lm\omega} = 0, \tag{6}$$

$$\Delta^{-\mathfrak{s}}\frac{d}{dr}\left(\Delta^{\mathfrak{s}+1}\frac{dR_{\mathfrak{s}lm\omega}}{dr}\right) + V_{\mathfrak{s}lm\omega}(r)R_{\mathfrak{s}lm\omega} = 0, \tag{7}$$

where

$$U_{\mathfrak{s}lm\omega} \equiv \lambda_{\mathfrak{s}lm\omega} + 2am\omega - 2a\omega\mathfrak{s}\cos\theta - \frac{(m + a\cos\theta)^2}{\sin^2\theta} + \mathfrak{s}, \qquad V_{\mathfrak{s}lm\omega} \equiv \frac{1}{\Delta}\left[K^2 - 2(r-M)K\right] - \lambda_{\mathfrak{s}lm\omega} + 4i\omega\mathfrak{s}r, \tag{8}$$

and $K \equiv (r^2 + a^2)\omega - am$. The angular functions $S_{\mathfrak{s}lm\omega}(\theta)$ are known as spin-weighted spheroidal harmonics, and have as limiting cases the spheroidal harmonics ($\mathfrak{s} = 0$) and the spin-weighted spherical harmonics ($a\omega = 0$).

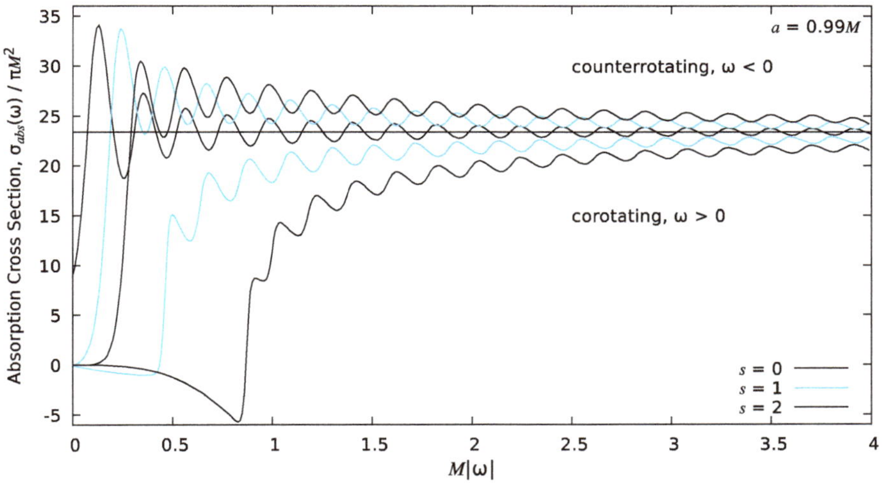

Fig. 2. The absorption cross section σ_{abs} for massless bosonic fields incident on a rapidly-rotating Kerr BH ($a = 0.99M$, $\gamma = 0$). For circularly-polarized fields ($s > 0$), the co-rotating ($\omega > 0$) and counter-rotating ($\omega < 0$) helicities are absorbed differently, due a coupling between the field helicity and the BH rotation.

We seek solutions of Eq. (7) that are purely ingoing at the event horizon, satisfying the following boundary conditions:

$$R_{\mathfrak{s}lm\omega} \sim \begin{cases} \mathcal{T}_{\mathfrak{s}lm\omega} e^{-i(\omega - m\Omega_h) r_\star} \Delta^{-\mathfrak{s}}, & r \to r_+, \\ \mathcal{I}_{\mathfrak{s}lm\omega} r^{-1} e^{-i\omega r_\star} + \mathcal{R}_{\mathfrak{s}lm\omega} r^{-(2\mathfrak{s}+1)} e^{i\omega r_\star}, & r \to +\infty, \end{cases} \tag{9}$$

where $\Omega_h \equiv \frac{a}{2Mr_+}$ is the angular frequency of the black hole horizon. Here $r_\star$ is the tortoise coordinate $r_\star \equiv \int dr \frac{(r^2+a^2)}{\Delta}$ such that $r_\star \to +\infty$ when $r \to +\infty$ and $r_\star \to -\infty$ when $r \to r_+$.

3.2. The absorption cross section

For an asymptotic incident plane wave traveling in the direction $\hat{n} = \sin\gamma\,\hat{x} + \cos\gamma\,\hat{z}$ the absorption cross section σ_{abs} is given by [24]

$$\sigma_{abs} = \frac{4\pi^2}{\omega^2} \sum_{l=|\mathfrak{s}|}^{+\infty} \sum_{m=-l}^{+l} \left| S_{\mathfrak{s}lm\omega}(\gamma) \right|^2 \Gamma_{\mathfrak{s}lm\omega}. \tag{10}$$

The transmission factor $\Gamma_{\mathfrak{s}lm\omega}$ is the ratio of the energy passing into to the hole to that encroaching from infinity, $\frac{dE_{\text{hole}}}{dE_{\text{in}}}$ [18]. It takes the same sign as $\omega(\omega - m\Omega_h)$, so it is *negative* for low-frequency co-rotating modes. Using energy balance, $dE_{\text{hole}} = dE_{\text{in}} - dE_{\text{out}}$, one obtains [24]

$$\Gamma_{0lm\omega} = 1 - \left| \frac{\mathcal{R}_{0lm\omega}}{\mathcal{I}_{0lm\omega}} \right|^2, \tag{11a}$$

$$\Gamma_{-1lm\omega} = 1 - \frac{B^2_{lm\omega}}{16\omega^4} \left| \frac{\mathcal{R}_{-1lm\omega}}{\mathcal{I}_{-1lm\omega}} \right|^2, \tag{11b}$$

$$\Gamma_{-2lm\omega} = 1 - \frac{\text{Re}^2(C) + 144M^2\omega^2}{256\omega^8} \left| \frac{\mathcal{R}_{-2lm\omega}}{\mathcal{I}_{-2lm\omega}} \right|^2, \tag{11c}$$

for the scalar ($\mathfrak{s} = 0$), electromagnetic ($\mathfrak{s} = -1$), and gravitational ($\mathfrak{s} = -2$) cases, respectively. Here $B^2_{lm\omega} \equiv \lambda^2_{-1lm\omega} + 4am\omega - 4a^2\omega^2$, $\text{Re}^2(C) = [(\lambda_{-2lm\omega} + 2)^2 + 4am\omega - 4(a\omega)^2](\lambda^2_{-2lm\omega} + 36am\omega - 36a^2\omega^2) + (2\lambda_{-2lm\omega} + 3)(96a^2\omega^2 - 48am\omega) - 144a^2\omega^2$, and $\mathcal{I}_{\mathfrak{s}lm\omega}$, $\mathcal{R}_{\mathfrak{s}lm\omega}$ are the coefficients appearing in the ingoing solutions of Eq. (9).

3.3. Numerical method

In order to determine the absorption cross section via Eq. (10) we first computed the spin-weighted spheroidal harmonics $S_{\mathfrak{s}lm\omega}$ and the transmission factors $\Gamma_{\mathfrak{s}lm\omega}$ by solving Eqs. (6) and (7) with numerical methods.

We obtained the spin-weighted spheroidal harmonics $S_{\mathfrak{s}lm\omega}$ and its corresponding eigenvalues $\lambda_{\mathfrak{s}lm\omega}$ using the *spectral eigenvalue method* as described in Ref. [13,25]. We have tested the angular eigenvalues $\lambda_{\mathfrak{s}lm\omega}$ obtained via the spectral eigenvalue method against the low-$a\omega$ formula provided in Ref. [26], obtaining a satisfying concordance.

The transmission factors were obtained as follows: in the scalar case ($\mathfrak{s} = 0$), we rewrote the radial equation into a Schrödinger-like form and numerically integrated it using the scheme detailed in Ref. [14]; in the electromagnetic ($\mathfrak{s} = -1$) and gravitational ($\mathfrak{s} = -2$) cases, we rewrote the radial Teukolsky equation using the *Detweiler* [27] and *Sasaki–Nakamura* [28] transformations, respectively. We numerically integrated the Detweiler and Sasaki–Nakamura equations from $r = r_h$ to $r = r_\infty$, where $r_h \sim 1.001r_+$ and $r_\infty \sim 10^3 r_+$ are within the near-horizon and the far-field regimes, respectively. At $r = r_\infty$, we extract the values of the ingoing and outgoing coefficients via (9) and compute the transmission factors via (11). To assure the reliability of our results, we have checked them using independent codes [13].

4. Results

4.1. Absorption cross sections

Fig. 2 shows the absorption cross section σ_{abs} for planar waves in all massless bosonic fields ($s = 0$, 1 and 2) impinging upon a rapidly-rotating Kerr BH ($a^* = 0.99$) parallel to the rotation axis ($\gamma = 0$). At long wavelengths, the incident wave stimulates superradiant emission from the black hole [29], with transmission turning negative for modes satisfying $\omega(\omega - m\Omega_h) < 0$. For on-axis incidence $\gamma = 0$, only the $m = -\mathfrak{s}$ modes contribute to the mode sum (10). Thus, σ_{abs} is negative for polarized fields ($s > 0$), but not for the scalar field ($s = 0$). The superradiant effect occurs principally in the $l = m = -\mathfrak{s}$ mode, and is much stronger for gravitational waves than for electromagnetic waves.

The absorption cross section for the co- and counter-rotating helicities are quite distinct, with the latter ($\omega < 0$) more strongly absorbed than the former ($\omega > 0$). This is a clear manifestation of

Table 1
The impact parameter b_c, orbital frequency Ω_c and Lyapunov exponent Λ_c for circular polar null geodesics, to 4 decimal places. See Eq. (13).

a^*	0	0.5	0.8	0.99	1
b_c/M	$\sqrt{27}$	5.1205	4.9849	4.8383	4.8284
$\Omega_c M$	$\frac{1}{\sqrt{27}}$	0.1958	0.2019	0.2089	0.2094
$\Lambda_c M$	$\frac{1}{\sqrt{27}}$	0.1884	0.1788	0.1633	0.1620

the spin-helicity effect for electromagnetic and gravitational waves. In the limit $M|\omega| \to \infty$, the difference falls off at $O(M|\omega|)^{-1}$ and σ_{abs} approaches the geodesic capture cross section σ_{geo}. We now attempt to quantify this effect.

4.2. *High frequency model*

Fig. 2 exhibits regular oscillations in $\sigma_{\text{abs}}(\omega)$ arising from successive l modes in Eq. (10). For scalar fields it was previously shown [30,14] that such oscillations are linked to the Regge pole spectrum of the black hole, whose asymptotic properties are set by the angular frequency Ω_c and Lyapunov exponent Λ_c of the circular photon orbits of the spacetime. At high frequencies for $\gamma = 0$, we find that σ_{abs} is well described by the *sinc approximation* [31, 30,14],

$$\sigma_{\text{abs}} \approx \sigma_{\text{sinc}} \equiv \mathcal{C}_s + \varepsilon \mathcal{A}_s \sin(\mathcal{B}_s/\varepsilon), \tag{12}$$

where $\varepsilon \equiv (M|\omega|)^{-1}$ and $\{\mathcal{A}_s, \mathcal{B}_s, \mathcal{C}_s\}$ are spin-dependent terms to be described more fully below.

A sinc approximation of this form was first developed by Sanchez [31] in 1977, for scalar fields on the Schwarzschild spacetime. For the Kerr spacetime with a scalar field incident along the axis ($\gamma = 0$), it was shown in Ref. [14] (based on the method of Ref. [30]) that Eq. (12) that remains valid with

$$\mathcal{A}_0 = -\frac{4\pi \Lambda_c e^{-\pi \Lambda_c/\Omega_c}}{\Omega_c^2}, \quad \mathcal{B}_0 = \frac{2\pi}{M\Omega_c}, \tag{13}$$

and $\mathcal{C}_0 = \sigma_{\text{geo}} = \pi b_c^2$. Sample values for b_c, Ω_c and Λ_c are given in Table 1. The method for obtaining these values is covered in Ref. [14].

For $s > 0$, we now propose an extended model to include terms at $O(\varepsilon)$:

$$\mathcal{A}_{s>0} = \mathcal{A}_0, \tag{14a}$$

$$\mathcal{B}_{s>0} = \mathcal{B}_0\left[1 + \varepsilon\left(\bar{b}_s \pm s a^* \Delta b_s\right) + O(\varepsilon^2)\right], \tag{14b}$$

$$\mathcal{C}_{s>0} = \mathcal{C}_0\left[1 + \varepsilon\left(\bar{c}_s \pm s a^* \Delta c_s\right) + O(\varepsilon^2)\right]. \tag{14c}$$

The coefficients Δb_s and Δc_s encapsulate the effect of the spin-helicity interaction, with $+$ in Eq. (14) for the co-rotating helicity, and $-$ for the counter-rotating helicity. To find the coefficients we fitted the model to our numerical data σ_{abs} across the domain $M|\omega| \in [2.5, 4]$ for $0 \le a^* \le 0.99$. Fig. 3 shows that the model (12)–(14) fits the data well across the domain in ω.

We may draw several inferences from the best-fit parameter values shown in Fig. 3(c). First, that $\Delta b_1 = \Delta b_2$ and $\Delta c_1 = \Delta c_2$ to within the fitting error. This implies that the spin-helicity effect for gravitational waves is twice as large as for electromagnetic waves, as expected. Second, that $\mathcal{C}_s \Delta b_s < 0$, so counter-rotating helicities are preferentially absorbed. Third, that $\Delta c_s \to \Delta b_s$ as $a^* \to 0$, which was not anticipated *a priori*. Fourth, that $\mathcal{B}_0 \Delta b_s s a^*$, the spin-helicity part of the phase term in the sinc approximation (12), varies monotonically from 0 in the Schwarzschild case up to approximately $s\pi$ in the extremal limit ($a \to M$). Evidence of this phase shift can be seen in Fig. 3(a).

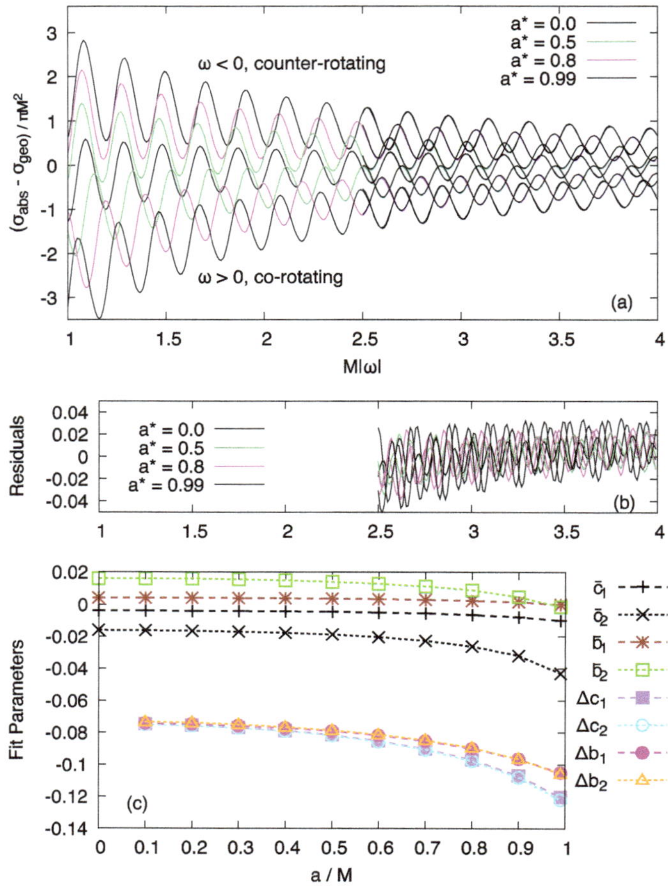

Fig. 3. (a) Fitting the sinc approximation model (12)–(14) to numerical data for $a^* \in \{0, 0.5, 0.8, 0.99\}$ across the domain $M|\omega| \in [2.5, 4]$. (b) The residuals of the fit, $|\sigma_{\text{abs}} - \sigma_{\text{sinc}}|/\pi M^2$. (c) The best-fit values for the parameters $\{\bar{c}_s, \bar{b}_s, \Delta c_s, \Delta b_s\}$ in Eq. (14).

5. Final remarks

We have calculated the absorption cross section for scalar, electromagnetic, and gravitational massless plane waves impinging upon a Kerr BH along its rotation axis. For the first time, we have presented a unified picture of the absorption spectrum for all the bosonic fields. We showed that superradiance leads to stimulated emission, rather than absorption, at low frequencies for co-rotating circular polarizations; and that counter-rotating polarizations are more heavily absorbed in general. We have proposed and tested an extended version of the sinc approximation, to encapsulate the spin-helicity effect at short wavelengths, where its effect falls off with λ/M.

An open question is whether the spin-helicity effect shown here can be quantitatively described using *spinoptics* [20–22]. That is, can a modified geometric-optics approximation, incorporating next-to-leading order helicity-dependent corrections in the eikonal equations, successfully reproduce the $1/M|\omega|$ terms in Eqs. (14)? Future work in this direction could prove illuminating.

Acknowledgements

The authors would like to thank Conselho Nacional de Desenvolvimento Científico e Tecnológico (CNPq) and Coordenação de Aperfeiçoamento de Pessoal de Nível Superior (CAPES), in Brazil, for partial financial support. S.D. acknowledges financial support from the Engineering and Physical Sciences Research Council

(EPSRC) under Grant No. EP/M025802/1 and from the Science and Technology Facilities Council (STFC) under Grant No. ST/L000520/1.

References

[1] K.S. Thorne, R.H. Price, D.A. MacDonald, Black Holes: The Membrane Paradigm, Yale University Press, 1986.

[2] V.P. Frolov, I.D. Novikov (Eds.), Black Hole Physics: Basic Concepts and New Developments, 1998.

[3] R.M. Wald, The thermodynamics of black holes, Living Rev. Relativ. 4 (2001) 6, http://dx.doi.org/10.12942/lrr-2001-6, arXiv:gr-qc/9912119.

[4] B.P. Abbott, et al., Observation of gravitational waves from a binary black hole merger, Phys. Rev. Lett. 116 (6) (2016) 061102, http://dx.doi.org/10.1103/PhysRevLett.116.061102, arXiv:1602.03837.

[5] B.P. Abbott, et al., Binary black hole mergers in the first advanced LIGO observing run, Phys. Rev. X 6 (4) (2016) 041015, http://dx.doi.org/10.1103/PhysRevX.6.041015, arXiv:1606.04856.

[6] A. Ricarte, J. Dexter, The Event Horizon Telescope: exploring strong gravity and accretion physics, Mon. Not. R. Astron. Soc. 446 (2015) 1973–1987, http://dx.doi.org/10.1093/mnras/stu2128, arXiv:1410.2899.

[7] D. Castelvecchi, How to hunt for a black hole with a telescope the size of Earth, Nature 543 (2017) 478–480.

[8] H. Falcke, F. Melia, E. Agol, Viewing the shadow of the black hole at the galactic center, Astrophys. J. 528 (2000) L13, http://dx.doi.org/10.1086/312423, arXiv:astro-ph/9912263.

[9] R.-S. Lu, A.E. Broderick, F. Baron, J.D. Monnier, V.L. Fish, S.S. Doeleman, V. Pankratius, Imaging the supermassive black hole shadow and jet base of M87 with the Event Horizon Telescope, Astrophys. J. 788 (2014) 120, http://dx.doi.org/10.1088/0004-637X/788/2/120, arXiv:1404.7095.

[10] T. Johannsen, Testing the no-hair theorem with observations of black holes in the electromagnetic spectrum, Class. Quantum Gravity 33 (12) (2016) 124001, http://dx.doi.org/10.1088/0264-9381/33/12/124001, arXiv:1602.07694.

[11] K. Akiyama, et al., Imaging the Schwarzschild-radius-scale structure of M87 with the Event Horizon Telescope using sparse modeling, Astrophys. J. 838 (1) (2017) 1, http://dx.doi.org/10.3847/1538-4357/aa6305, arXiv:1702.07361.

[12] K. Glampedakis, N. Andersson, Scattering of scalar waves by rotating black holes, Class. Quantum Gravity 18 (2001) 1939–1966, http://dx.doi.org/10.1088/0264-9381/18/10/309, arXiv:gr-qc/0102100.

[13] S.R. Dolan, Scattering and absorption of gravitational plane waves by rotating black holes, Class. Quantum Gravity 25 (2008) 235002, http://dx.doi.org/10.1088/0264-9381/25/23/235002, arXiv:0801.3805.

[14] C.F.B. Macedo, L.C.S. Leite, E.S. Oliveira, S.R. Dolan, L.C.B. Crispino, Phys. Rev. D 88 (2013) 064033, http://dx.doi.org/10.1103/PhysRevD.88.064033.

[15] B.P. Abbott, et al., The basic physics of the binary black hole merger GW150914, Ann. Phys. 529 (2017) 0209, http://dx.doi.org/10.1002/andp.201600209, arXiv:1608.01940.

[16] J.L. Synge, The escape of photons from gravitationally intense stars, Mon. Not. R. Astron. Soc. 131 (3) (1966) 463–466.

[17] A. Grenzebach, V. Perlick, C. Lämmerzahl, Photon regions and shadows of Kerr–Newman–NUT black holes with a cosmological constant, Phys. Rev. D 89 (12) (2014) 124004, http://dx.doi.org/10.1103/PhysRevD.89.124004, arXiv:1403.5234.

[18] R. Brito, V. Cardoso, P. Pani Superradiance, Lect. Notes Phys. 906 (2015) 1–237, http://dx.doi.org/10.1007/978-3-319-19000-6, arXiv:1501.06570.

[19] B. Mashhoon, Electromagnetic scattering from a black hole and the glory effect, Phys. Rev. D 10 (1974) 1059–1063, http://dx.doi.org/10.1103/PhysRevD.10.1059.

[20] V.P. Frolov, A.A. Shoom, Spinoptics in a stationary spacetime, Phys. Rev. D 84 (2011) 044026, http://dx.doi.org/10.1103/PhysRevD.84.044026, arXiv:1105.5629.

[21] V.P. Frolov, A.A. Shoom, Scattering of circularly polarized light by a rotating black hole, Phys. Rev. D 86 (2012) 024010, http://dx.doi.org/10.1103/PhysRevD.86.024010, arXiv:1205.4479.

[22] C.-M. Yoo, Notes on spinoptics in a stationary spacetime, Phys. Rev. D 86 (2012) 084005, http://dx.doi.org/10.1103/PhysRevD.86.084005, arXiv:1207.6833.

[23] S.A. Teukolsky, Rotating black holes: separable wave equations for gravitational and electromagnetic perturbations, Phys. Rev. Lett. 29 (16) (1972) 1114.

[24] J.A.H. Futterman, F.A. Handler, R.A. Matzner, Scattering from Black Holes, Cambridge University Press, 1988.

[25] G.B. Cook, M. Zalutskiy, Gravitational perturbations of the Kerr geometry: high-accuracy study, Phys. Rev. D 90 (12) (2014) 124021.

[26] E. Berti, V. Cardoso, M. Casals, Eigenvalues and eigenfunctions of spin-weighted spheroidal harmonics in four and higher dimensions, Phys. Rev. D 73 (2) (2006) 024013.

[27] S. Detweiler, On the Equations Governing the Electromagnetic Perturbations of the Kerr Black Hole, Proceedings of the Royal Society of London A: Mathematical, Physical and Engineering Sciences, vol. 349, The Royal Society, 1976, pp. 217–230.

[28] M. Sasaki, T. Nakamura, Gravitational radiation from a Kerr black hole. I. Formulation and a method for numerical analysis, Prog. Theor. Phys. 67 (6) (1982) 1788–1809.

[29] J.G. Rosa, Superradiance in the sky, Phys. Rev. D 95 (6) (2017) 064017, http://dx.doi.org/10.1103/PhysRevD.95.064017, arXiv:1612.01826.

[30] Y. Decanini, G. Esposito-Farese, A. Folacci, Universality of high-energy absorption cross sections for black holes, Phys. Rev. D 83 (2011) 044032, http://dx.doi.org/10.1103/PhysRevD.83.044032, arXiv:1101.0781.

[31] N.G. Sanchez, Absorption and emission spectra of a Schwarzschild black hole, Phys. Rev. D 18 (1978) 1030, http://dx.doi.org/10.1103/PhysRevD.18.1030.

2

Bouncing black holes in quantum gravity and the Fermi gamma-ray excess

Aurélien Barrau [a,*], Boris Bolliet [a], Marrit Schutten [a,b], Francesca Vidotto [b]

[a] Laboratoire de Physique Subatomique et de Cosmologie, Université Grenoble-Alpes, CNRS-IN2P3, 53, avenue des Martyrs, 38026 Grenoble cedex, France
[b] Radboud University, Institute for Mathematics, Astrophysics and Particle Physics, Mailbox 79, P.O. Box 9010, 6500 GL Nijmegen, The Netherlands

ARTICLE INFO

Editor: H. Peiris

ABSTRACT

Non-perturbative quantum-gravity effects can change the fate of black holes and make them bounce in a time scale shorter than the Hawking evaporation time. In this article, we show that this hypothesis can account for the GeV excess observed from the galactic center by the *Fermi* satellite. By carefully taking into account the secondary component due to the decay of unstable hadrons, we show that the model is fully self-consistent. This phenomenon presents a specific redshift-dependence that could allow to distinguish it from other astrophysical phenomena possibly contributing to the GeV excess.

1. Introduction

The Planck scale is currently out of reach from any direct local experiment by a factor of approximately 10^{15}. It is therefore hard to test quantum gravity. Many efforts have however been devoted to quantum gravity phenomenology in the last decade (see, *e.g.*, [1–3] and references therein for some general arguments) and it is not unreasonable to expect measurable consequences. Most efforts in the recent years have focused on the early Universe or on modified dispersion relations impacting the propagation of gamma-rays on huge distances. In this article, we focus on a recent result associated with black holes physics, first exposed in [4]. The main idea is grounded in a robust result of loop quantum cosmology: quantum gravity might manifest itself in the form of an effective pressure that counterbalances the classically attractive gravity when matter reaches the Planck density [5]. For a black hole, this means that matter's collapse could stop before the central singularity forms. The classical singularity is replaced in the quantum theory by a phase of maximum density – a "Planck star" [4]. The absence of the central singularity allows for the dynamical trapping horizon (shrinking of light surfaces) to be converted in an anti-trapping horizon (expanding of light surfaces), that releases matter and eventually disappears. This is a non-perturbative quantum-gravity process that tunnels a classical black hole into a classical white hole. Because of the gravitational redshift, the process is almost instantaneous in proper time but appears as very long if measured by an external distant observer.

The viability of the model is supported by the existence of a classical metric satisfying the Einstein equations outside the space-time region where matter collapses into a black hole and then emerges from a white hole[1] [7]. This can be achieved without violating causality nor the semiclassical approximation, as quantum effects piles up outside the horizon over a very long time.

The time quantum effects take to pile up outside the horizon determines the lifetime of the black hole, and its phenomenology. This was first investigated in [8] for a long lifetime (comparable but shorter than the Hawking evaporation time). Further studies in [9] and [10] were developed considering a wider range of possible lifetimes and the integrated signal coming from a diffuse emission.

The tunneling process connects two classically disconnected solutions. Einstein equations should therefore be violated during the evolution, but the model allows for a violation that takes place only over a finite region. This is where full quantum gravity dominates.[2] This process seems to be quite generically allowed for a wide range of viable quantum theories of gravity. Interestingly, in covariant loop quantum gravity (LQG) it is possible to perform the

* Corresponding author.
E-mail addresses: Aurelien.Barrau@cern.ch (A. Barrau), Bolliet@lpsc.in2p3.fr (B. Bolliet), M.Shutten@students.ru.nl (M. Schutten), F.Vidotto@science.ru.nl (F. Vidotto).

[1] A modifications was suggested in [6] where the scenario was made asymmetric, with a black hole phase longer than the white hole one. Such a modification overcomes complications coming from a possible instability in the white-hole phase.

[2] A possibility could be to study an effective metric associated with this finite region, as originally done by Hayward [11]. See [12] for recent results in this direction, recently extended to rotating metrics [13].

calculation of the tunneling amplitudes [14] that provides an estimation of the black-hole lifetime.

In this work, we address the puzzle posed by the observation by the *Fermi* telescope of a GeV photon excess, coming from the galactic center. Different explanations – including standard astrophysical sources – have been considered to explain it. Here we investigate whether bouncing (primordial) black holes could explain this specific excess and if this hypothesis has specific features that could allow to distinguish it from more conventional explanations.

In the first part, we briefly explain what are the parameters of the model and their possible values. In the second part, we present the way we have calculated and modeled the gamma-ray emission from bouncing black holes. In the third part we show the fit to the GeV *Fermi* excess we are interested in. In the fourth part, we suggest ways to discriminate our model from other possible explanations and normalize the mass spectrum. Some prospects are then discussed in the conclusion.

2. Parameters of the model

A precise astrophysical model for the emission from a bouncing black hole is not available, but heuristic arguments lead to consider two different emission mechanisms [9]. One, designated as the *low-energy* component, is grounded in a simple and conservative dimensional analysis. The mean energy of the emitted signal is assumed to be such that the corresponding wavelength matches the size of the horizon. This is a reasonable expectation, agreeing with the Hawking spectrum. The other one, designated as the *high-energy* component, has a smaller wavelength and depends on the conditions at which the black hole formed. In the model, the matter forming the black hole reemerges rapidly in the white-hole phase. The gravitational blueshift felt by radiation in the contracting phase is precisely compensated by the very same amount of redshift in the expanding phase.

If the considered model is correct, the bounce should take place for all kinds of black holes, but observable effects become experimentally accessible only for primordial black holes (PBHs), *i.e.* black holes that formed in the early universe with a potentially wide mass spectrum. In particular, they can form with masses smaller than the Solar mass so that their bouncing time can be of the order of the age of the Universe (more massive black holes would require much more than the Hubble time to bounce and nothing would be visible). Studying the phenomenology of bouncing black holes, we are interested only by primordial black holes. Many different processes that can lead to the formation of black holes in the early Universe were suggested, see, *e.g.*, [15] for a recent review. In the simplest models, PBHs form by collapse of over-dense regions. Given the mass of a black hole, its formation time is then (approximately) known and so is the spectrum of the radiation that collapsed to form it – and that will emerge from the bounce in the *high-energy* component of the signal considered here.

The most important parameter of the model is the bouncing time of black holes. It can be written as [7]

$$\tau = 4kM^2, \tag{1}$$

in Planck units, where M is the mass of the black hole and k is a free parameter. This is a key-point: the bounce time scales as M^2 whereas the Hawking evaporation requires a time of order M^3. The parameter k is bounded from below at the value $k_{min} = 0.05$ which ensures that the quantum effects do pile up enough to appear outside of the black hole horizon so that the bounce can take place. It is also bounded from above at a value $k_{max}(M)$ which translates the fact that the bouncing time needs to be smaller than the Hawking time,[3] otherwise the black hole would disappear before bouncing and the evaporation could not be considered anymore as a small correction associated with a dissipative process, as assumed in the model.

A signal detected today comes from black holes that have lived for a time equal to the Hubble time t_H. Fixing the lifetime to t_H, Eq. (1) gives the corresponding mass of the bouncing black hole, that determines the energy of the emitted radiation. We ask the following question: is there an allowed value of k such that this emission can explain the GeV excess observed by the *Fermi* telescope? We note immediately that the GeV energy scale is far below any possible contribution coming from the high-energy component of our model: even for the smaller possible value of k the emitted energy is of order a TeV. On the other hand, the low-energy component can indeed match the observed signal. Our analysis therefore focuses on this component. To have an emitted energy of the order of 1 GeV, that is of order $10^{-19}E_{Pl}$, the size of the black hole should be of the order of $10^{19}l_{Pl}$ and its mass of the oder $M \sim 10^{19}M_{Pl}$. The Hubble time is $t_H \sim 10^{60}t_{Pl}$. Requiring the Hubble time to be equal to the bouncing time leads to $k \sim 10^{22}$. How does this compare with the Hawking time? The Hawking time is roughly $t_{Haw} \sim 10^3M^3$, that is of the order of $10^{60}t_{Pl}$ for the mass we are interested in. This is of the same order of magnitude than the bouncing time.[4] This is therefore a quite interesting situation from the theoretical point of view in the sense that the required value of the parameter is not random or arbitrary in the (very large) allowed interval but a near-extremal one.

To summarize, the *high-energy* component of the signal emitted by bouncing black holes cannot explain the *Fermi* excess but the *low-energy* component might do so if the free parameter k is chosen near its highest possible value.

3. Modeling of the gamma-ray emission

Whatever the details of the emission mechanism, as soon as fundamental particles are emitted at energies higher than the QCD confinement scale, quarks and gluons are emitted and do fragmentate into subsequent hadrons. For a bouncing black hole emitting quanta with energies greater than, say, 100 MeV, it is required to consider not only the primary (*i.e.* direct) emission of gamma-rays but also the secondary component, due to the decay of unstable hadrons produced by fragmentation. This has been studied with analytical approximations for evaporating black holes in [17,18]. In this work we use a full Monte Carlo analysis based the "Lund" PYTHIA code (with some scaling approximations in the low energy range) [19] to determine the normalized differential fragmentation functions $dg(\epsilon, E)/d\epsilon$, where E is the quark energy and ϵ is the photon energy. This takes into account a large number of physical aspects, including hard and soft interactions, parton distributions, initial- and final-state parton showers, multiple interactions, fragmentation and decay.

For all energies, we have found that the obtained spectra can be well fitted by a function

$$f(E,\epsilon) = \frac{a\epsilon^b}{\pi\gamma}\left[\frac{\gamma^2}{(\epsilon-\epsilon_0)^2+\gamma^2}\right]e^{-\left(\frac{4\epsilon}{E}\right)^3}, \tag{2}$$

[3] More precisely, the bounce time is constrained to be smaller than "Page time" at which the black holes would have lost half of its mass by Hawking evaporation because this time signs the entrance in the full quantum gravity regime [16].

[4] In our study we disregard the mass loss due to Hawking evaporation. In fact, even if the bouncing time considered here is comparable with the Hawking one, Hawking radiation decreases the mass of the black hole only by a small amount without changing its order of magnitude.

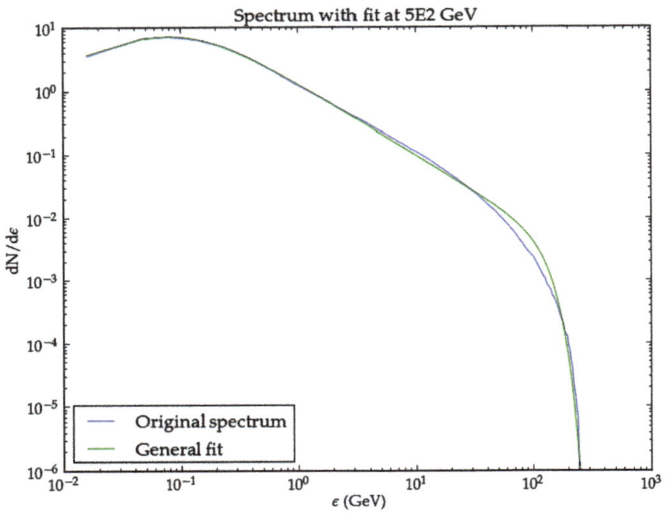

Fig. 1. Spectrum of gamma-rays generated by 5×10^2 GeV jets. The green histogram corresponds to the output of the simulation and the blue curve to the analytical fit. (For interpretation of the references to color in this figure legend, the reader is referred to the web version of this article.)

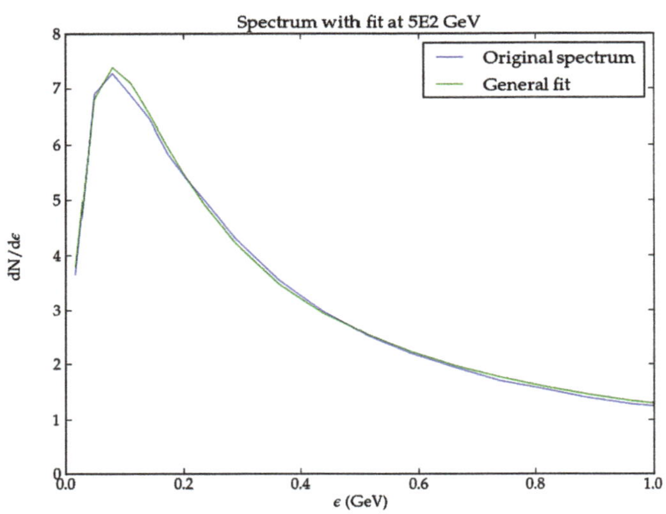

Fig. 2. Zoom on the low-energy part of the spectrum of gamma-rays generated by 5×10^2 GeV jets. The green histogram corresponds to the output of the simulation and the blue curve to the analytical fit. (For interpretation of the references to color in this figure legend, the reader is referred to the web version of this article.)

with $a = 50.7$, $b = 0.847$, $\gamma = 0.0876$ and $\epsilon_0 = 0.0418$ if the energies are given in GeV. The low-energy peak of the spectrum is well approximated by a Cauchy function. It is then roughly a power law, followed by an exponential cutoff around the initial jet energy (see Figs. 1 and 2).

As soon as the jet reaches an energy much higher that the associated quark mass, the result does not depend substantially on the quark type. Depending on the mean energy E of the primary component, the number of types of emitted quarks – that is with $m < E$ – is accounted for. The normalization is chosen to be consistent with the primary emission.

For the *low-energy* component, the shape of the primary signal is not completely determined by the model. We have used a Gaussian function, centered on the energy estimated in the previous Section, with a relative width taken as the second free parameter of the model. Its exact value depends on the details of the astrophysical phenomena occurring during the bounce and this is far beyond the scope of this study. The full signal can be written as

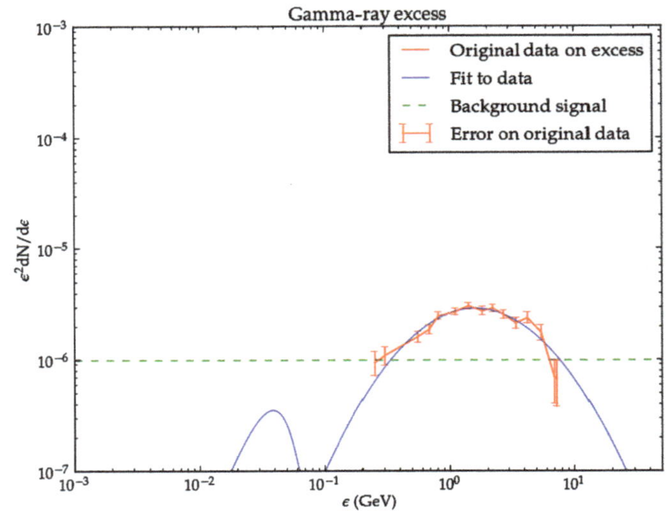

Fig. 3. Best fit to the *Fermi* excess with bouncing black holes. (For interpretation of the references to color in this figure legend, the reader is referred to the web version of this article.)

$$Ae^{-\frac{(\epsilon-E)^2}{2\sigma^2}} + 3N\sqrt{2\pi}A\sigma f(E,\epsilon), \tag{3}$$

where N is the number of species of quarks with $m < E$.

For the *high-energy* component, which is irrelevant for this study but potentially interesting for other works, the same strategy can be followed. The primary component is then a Planck law and the full signal can be written as

$$A\frac{\epsilon^2}{e^{E/T}-1} + 36AT^3\zeta(3)f(E,\epsilon). \tag{4}$$

Interestingly, this formula can also be used to model the full spectrum of an evaporating black hole since the Hawking spectrum is also very close to a Planck law.

4. Fitting Fermi data

The *Fermi* Gamma-ray Space Telescope is a space observatory being used for gamma-ray astronomy observations from low Earth orbit. Its main instruments are the Large Area Telescope (LAT), intended to perform an all-sky survey studying astrophysical and cosmological phenomena, and the Gamma-ray Burst Monitor (GBM), used to study transients.

An excess in the *Fermi*-LAT data has been reported within the inner 10 arcmin of the Galactic center (see, *e.g.*, [20–22]) and up to larger galactic latitudes (see, *e.g.*, [23–26]). A huge number of works have been published on possible explanations. Our opinion is that an astrophysical origin, notably associated with millisecond pulsars, is the most convincing one (see, *e.g.*, [27]). It is however not fully satisfactory and dark-matter like hypotheses are worth being considered (see, *e.g.*, [23]). Here we investigate whether this signal can be due to bouncing black holes.

We stress that the explanation we suggest is specifically associated with the quantum gravity scenario considered in this work. The time integrated spectrum of black holes evaporating by the usual Hawking process is scaling as E^{-3} and there is no way it can account for the *Fermi* excess. As explained before, two parameters are required to fully determine the low-energy component of bouncing black holes: their bouncing time and the width of the primary Gaussian. The best fit (with a near-extremal bouncing time) is shown in Fig. 3. The agreement with data is good, with a χ^2 per degree of freedom of 1.05. Notice that what is plotted here is not the differential spectrum but the spectral energy density

$(\epsilon^2 dN/d\epsilon)$, as used for most experimental publications. The key point we want to stress is that although the number of secondary gamma-rays is higher than the number of primary gamma-rays, their spectral energy density is much lower. This is of utmost importance for this study: as the background has a basically constant spectral energy density, this means that the anomaly can be accounted for without any spurious excess in the 10–100 MeV range, where is situated the peak of the secondary component. This peak remains much below the background and the signal can be explained with no contradiction with the data.

This also shows why the *high-energy* component cannot be used to explain the excess. The energy of its primary component is in all cases too high and its secondary component would not have a high enough spectral energy density.

5. Discrimination with dark matter and mass spectrum

The model presented in this work is unquestionably quite exotic when compared with astrophysical hypotheses. But the important point is than it can, in principle, be distinguished both from astrophysical explanations and from other "beyond the Standard Model" scenarios. The reason for that is a peculiar redshift dependence. When looking at a galaxy at redshift z, the measured energy of the signal emitted either by decaying WIMPS or by astrophysical objects will be $E/(1+z)$ if the rest-frame energy is E. But this is not true for the bouncing black holes signal. The reason for this is that black holes that have bounced far away and are observed now must have a shorter bouncing time and therefore a smaller mass. Their emission energy – in the low energy channel we are considering in this article – is therefore higher and this partly compensates for the redshift effect. Following [9], we can write down the observed wavelength of the signal from a host galaxy at redshift z, taking into account both the expansion of the universe and the change of bouncing time, as:

$$\lambda_{obs}^{BH} \sim \frac{2Gm}{c^2}(1+z) \times \sqrt{\frac{H_0^{-1}}{6k\Omega_\Lambda^{1/2}} \sinh^{-1}\left[\left(\frac{\Omega_\Lambda}{\Omega_M}\right)^{1/2} (z+1)^{-3/2}\right]}, \tag{5}$$

where we have reinserted the Newton constant G and the speed of light c; H_0, Ω_Λ and Ω_M being the Hubble constant, the cosmological constant, and the matter density. On the other hand, for standard sources, the measured wavelength is just related to the observed wavelength by

$$\lambda_{obs}^{other} = (1+z)\lambda_{emitted}^{other}\,. \tag{6}$$

The redshift dependence specific of our model makes it possibly testable against other proposals. Obviously, detecting such a signal from far away galaxies is challenging but we hope this work might motivate some experimental prospects for the next generation of gamma-ray satellites. On Fig. 4, we have displayed the evolution of the wavelength, normalized to the rest-frame wavelength, as a function of the redshift for both a conventional source (upper curve) and the model considered in this work (lower curve). By "conventional" we mean here basically all other models we are aware of, including astrophysical sources and the decay of supersymmetric particles. Obviously it is easy to distinguish between both cases: in the hypothesis of bouncing black holes, the wavelength does not vary much because black holes bouncing far away are smaller and therefore emit higher-energy photons.

Interestingly, there might be another specific observational signature for this model. In addition to specific signals coming from identified galaxies, one should also expect a diffuse background. As

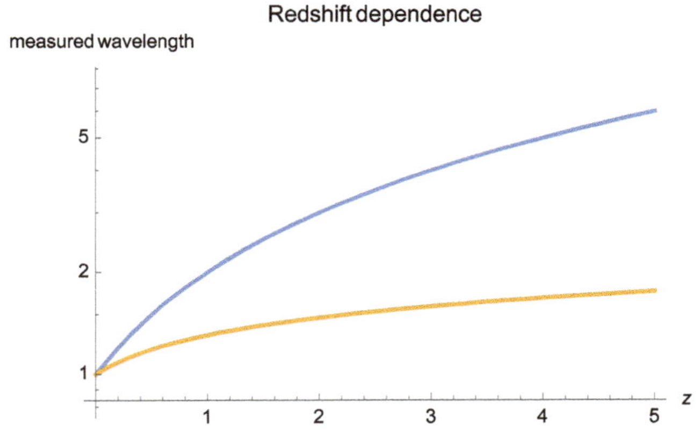

Fig. 4. Measured wavelength, normalized to the rest-frame wavelength, as a function of the redshift. The upper curve is for a conventional signal and the lower curve is for the model considered in this article.

we have demonstrated in [10], for the *low energy* component of the bouncing signal, considered here, the integrated emission exhibits an interesting feature. The integrated spectrum, defined as

$$\frac{dN_{mes}}{dEdtdS} = \int \Phi_{ind}((1+z)E, R) \cdot n(R) \cdot A(E) \cdot f(E,R)dR, \tag{7}$$

(where $\Phi_{ind}(E,R)$ denotes the individual flux emitted by a single bouncing black hole at distance R and at energy E, $A(E)$ is the angular acceptance of the detector multiplied by its efficiency, $f(E,R)$ is the absorption function, and $n(R)$ is the number of black holes bouncing at distance R per unit time and volume) was indeed shown to be nearly the *same* than the individual spectrum but with a slight distortion on the left tail [10]. This is another signature for this specific model.

The order of magnitude of the number of bouncing black holes in the galactic-center region required to account for the observed flux is around 100 per second. The associated mass is negligible when compared to the expected dark matter density, even when integrated over a long time interval. If the mass spectrum of primordial black holes was known, which is not the case, in principle it would be possible to fix the total mass associated with bouncing black holes. As a reasonable toy model, let us assume that the mass spectrum is given by

$$\frac{d^2N}{dMdV} = pM^{-\alpha}. \tag{8}$$

If the number of exploding black holes required to explain the data on a time interval $d\tau$ is N_{exp}, one can estimate the associated mass variation

$$dM = \frac{d\tau}{8kM}. \tag{9}$$

Calling M_0 the mass corresponding to a black hole exploding today, one then gets

$$N_{exp} = \int_{M_0}^{M_0+dM} pM^{-\alpha}dM\,. \tag{10}$$

This allows, in principle, to determine p and therefore to normalize the spectrum.

6. Conclusion

Black holes could bounce once they have reached the "Planck star" stage. This can be seen as a tunneling into an expanding explosive phase. The process appears generic in quantum gravity. In this article, we have shown that this phenomenon could explain the GeV excess measured by the *Fermi* satellite. This would open the fascinating possibility to observe (non perturbative) quantum gravity processes at energies 19 orders of magnitude below the Planck scale. Interestingly, the explanation we suggest is fully self-consistent in the sense that the hadronic "noise" due to decaying pions remains much below the observed background. Unquestionably, there are other – less exotic – ways to explain the *Fermi* excess. But the important point we have made is that this model has a specific redshift dependence which, in principle, can lead to a clear signature for future experiments. On the theoretical side, the important next step would be to fix the free parameter of the model from the full theory so that the energy of the signal is fixed from first principle and not anymore tuned to fit the data (see [28] for a recent step in this direction). Another interesting possible improvement would be to take into account the distribution of actual bouncing times for individual black holes around the mean time τ fixed by the theory.

Acknowledgements

B.B. is supported by a grant from ENS-Lyon. F.V. is supported by a Veni grant from the Netherlands Organisation for Scientific Research (NWO).

References

[1] S. Liberati, L. Maccione, J. Phys. Conf. Ser. 314 (2011) 012007, arXiv:1105.6234.
[2] S. Hossenfelder, L. Smolin, Phys. Can. 66 (2010) 99, arXiv:0911.2761.
[3] G. Amelino-Camelia, Living Rev. Relativ. 16 (2013) 5, arXiv:0806.0339.
[4] C. Rovelli, F. Vidotto, Int. J. Mod. Phys. D 23 (2014) 1442026, arXiv:1401.6562.
[5] A. Ashtekar, T. Pawlowski, P. Singh, Phys. Rev. Lett. 96 (141301) (2006), arXiv: gr-qc/0602086.
[6] T. De Lorenzo, A. Perez, arXiv:1512.04566, 2015.
[7] H.M. Haggard, C. Rovelli, Phys. Rev. D 92 (2015) 104020, arXiv:1407.0989.
[8] A. Barrau, C. Rovelli, Phys. Lett. B 739 (2014) 405, arXiv:1404.5821.
[9] A. Barrau, C. Rovelli, F. Vidotto, Phys. Rev. D 90 (2014) 127503, arXiv:1409.4031.
[10] A. Barrau, B. Bolliet, F. Vidotto, C. Weimer, J. Cosmol. Astropart. Phys. 1602 (2016) 022, arXiv:1507.05424.
[11] S.A. Hayward, Phys. Rev. Lett. 96 (031103) (2006), arXiv:gr-qc/0506126.
[12] T.D. Lorenzo, C. Pacilio, C. Rovelli, S. Speziale, Gen. Relativ. Gravit. 47 (2015), arXiv:1412.6015, 2014.
[13] T.D. Lorenzo, A. Giusti, S. Speziale, Gen. Relativ. Gravit. 48 (March 2016), arXiv: 1510.08828, 2015.
[14] M. Christodoulou, C. Rovelli, S. Speziale, I. Vilensky, arXiv:1605.05268, 2016.
[15] A.M. Green, arXiv:1403.1198, 2014.
[16] A. Almheiri, D. Marolf, J. Polchinski, J. Sully, J. High Energy Phys. 02 (2013) 062, arXiv:1207.3123.
[17] J.H. MacGibbon, B.R. Webber, Phys. Rev. D 41 (1990) 3052.
[18] J.H. MacGibbon, Phys. Rev. D 44 (1991) 376.
[19] T. Sjöstrand, S. Ask, J.R. Christiansen, R. Corke, N. Desai, P. Ilten, S. Mrenna, S. Prestel, C.O. Rasmussen, P.Z. Skands, Comput. Phys. Commun. 191 (2015) 159, arXiv:1410.3012.
[20] D. Hooper, L. Goodenough, Phys. Lett. B 697 (2011) 412, arXiv:1010.2752.
[21] K.N. Abazajian, M. Kaplinghat, Phys. Rev. D 86 (2012) 083511, Erratum: Phys. Rev. D 87 (2013) 129902, arXiv:1207.6047.
[22] C. Gordon, O. Macias, Phys. Rev. D 88 (2013) 083521, Erratum: Phys. Rev. D 89 (4) (2014) 049901, arXiv:1306.5725.
[23] T. Daylan, D.P. Finkbeiner, D. Hooper, T. Linden, S.K.N. Portillo, N.L. Rodd, T.R. Slatyer, Phys. Dark Universe 12 (2016) 1, arXiv:1402.6703.
[24] F. Calore, I. Cholis, C. Weniger, J. Cosmol. Astropart. Phys. 1503 (2015) 038, arXiv:1409.0042.
[25] D. Hooper, T.R. Slatyer, Phys. Dark Universe 2 (2013) 118, arXiv:1302.6589.
[26] W.-C. Huang, A. Urbano, W. Xue, arXiv:1307.6862, 2013.
[27] R. Bartels, S. Krishnamurthy, C. Weniger, Phys. Rev. Lett. 116 (2016) 051102, arXiv:1506.05104.
[28] M. Christodoulou, C. Rovelli, S. Speziale, I. Vilensky, arXiv:1605.05268, 2016.

3

Configurational entropy of charged AdS black holes

Chong Oh Lee

Department of Physics, Kunsan National University, Kunsan 573-701, Republic of Korea

ARTICLE INFO

Editor: M. Cvetič

ABSTRACT

When we consider charged AdS black holes in higher dimensional spacetime and a molecule number density along coexistence curves is numerically extended to higher dimensional cases. It is found that a number density difference of a small and large black holes decrease as a total dimension grows up. In particular, we find that a configurational entropy is a concave function of a reduced temperature and reaches a maximum value at a critical (second-order phase transition) point. Furthermore, the bigger a total dimension becomes, the more concave function in a configurational entropy while the more convex function in a reduced pressure.

1. Introduction

It has been found that there is a first-order phase transition in the Schwarzschild–AdS black hole from analogy between black hole and standard thermodynamic system [1]. It has been suggested that the critical behavior of charged AdS black holes remind Van der Waals liquid-gas phase transition [2,3]. Thermodynamics of black holes in the extended space has been recently investigated by treating the cosmological constant as the thermodynamic pressure [4]. It has been found that the thermodynamic pressure of charged AdS black holes is proportional to their thermodynamic volume [5]. The $P - V$ criticality of charged AdS black holes in the extended space has been investigated [6,7]

A present-day concept of configurational entropy has been suggested in Ref. [8] in search for the informational entropy in the context of communication theory. It was recently obtained in Ref. [9] through investigation of measure of ordering in field configuration space for spatially localized energy solutions of nonlinear models and used to study instability of a variety of objects [10–17].

The number density of black hole molecules was recently introduced in Ref. [18] for investigation of measure of the microscopic degrees of freedom of black holes. It was extensively studied for $f(R)$ AdS black holes and Gauss–Bonnet AdS black holes [19] and for the generalization of charged AdS black hole specific volume and number density [20].

The paper is organized as follows: in the next section we investigate configurational entropy in charged AdS black holes in higher dimensional spacetime. In the four-dimensional case, we explicitly present the configurational entropy. In the five-dimensional/six-dimensional case, its numerical result is given. In the last subsection we discuss their thermodynamic properties. In the last section we give our conclusion.

E-mail address: cohlee@gmail.com.

2. Configurational entropy

In statistical thermodynamics the most general formula between the entropy and the set of probabilities of their microscopic states is given as the Boltzmann–Gibbs entropy S_{BG}

$$S_{BG} = -k_B \sum p_i \ln p_i, \tag{2.1}$$

with $\sum p_i = 1$ where k_B is the Boltzmann constant, and p_i is the probability of a microstate. Each microstate has equal probability as the following

$$p_i = \frac{1}{W}, \tag{2.2}$$

where W is the number of microstates. Then the Boltzmann–Gibbs entropy S_{BG} (2.1) reduces to

$$S_{BG} = k_B \ln W, \tag{2.3}$$

where W is treated as the number of possible configurations at the given energy, and the Boltzmann–Gibbs entropy S_{BG} (2.3) becomes the configurational entropy in the microcanonical ensemble. Especially, supposing there are two different molecules with the total number of molecules N_0, then the number of one type of molecule is N_1 and the number of another type of molecule N_1. The configurational entropy S is written as

$$S = k_B \ln W = k_B \ln\left(\frac{N_0!}{N_1!N_2!}\right), \tag{2.4}$$

which leads to

$$S = k_B(N_0 \ln N_0 - N_1 \ln N_1 - N_2 \ln N_2), \tag{2.5}$$

by employing sterling's approximation $\ln N! \approx N \ln N$. Since the effective number density n for the AdS black hole is given as [6,7]

$$n = \frac{N}{V} = \frac{1}{2l_P^2 r_h} \tag{2.6}$$

where V is the thermodynamic volume and l_P is Planck length

$$l_P = \sqrt{\hbar G/c^3}, \tag{2.7}$$

the configurational entropy of charged AdS black holes per unit volume s_{conf} reduces to

$$s_{conf} = -\left[n_1 \ln\left(\frac{n_1}{n_1+n_2}\right) + n_2 \ln\left(\frac{n_2}{n_1+n_2}\right)\right] \tag{2.8}$$

by using geometric units $G = c = k_B = \hbar = 1$.

The charged AdS black hole metric in higher dimensional spacetime is given as

$$ds^2 = -f(r)dt^2 + \frac{dr^2}{dr^2} + r^2 d\Omega_{d-2}^2, \tag{2.9}$$

with

$$f(r) = 1 - \frac{m}{r^{d-3}} + \frac{q^2}{r^{2(d-3)}} + \frac{r^2}{l^2}, \tag{2.10}$$

where parameter m relates to the ADM mass M, which is identified with enthalpy H $(M \equiv H = U + PV)$ [4]

$$M \equiv H = \frac{\pi^{\frac{d-1}{2}}(d-2)m}{8\pi\Gamma(\frac{d-1}{2})}, \tag{2.11}$$

and parameter q relates to the black hole charge Q [2,3]

$$Q = \frac{\pi^{\frac{d-1}{2}}\sqrt{2(d-2)(d-3)}q}{4\pi\Gamma(\frac{d-1}{2})}. \tag{2.12}$$

One may treat the cosmological constant Λ as the thermodynamic pressure P

$$P = -\frac{\Lambda}{8\pi} = \frac{(d-1)(d-2)}{8\pi l^2}, \tag{2.13}$$

and its conjugate quantity as the thermodynamic volume [5].

$$V = \frac{2\pi^{\frac{d-1}{2}} r_h^{d-1}}{\Gamma(\frac{d-1}{2})}. \tag{2.14}$$

The d-dimensional black hole temperature T can read

$$T = \frac{1}{4\pi r_h}\left[(d-3) + \frac{16\pi P}{d-2} r_h^2 - \frac{(d-3)q^2}{r_h^{2(d-3)}}\right], \tag{2.15}$$

which leads to the thermodynamic pressure P

$$P = \frac{(d-2)T}{4r_h} - \frac{(d-2)(d-3)}{16\pi r_h^2} + \frac{(d-2)(d-3)q^2}{16\pi r_h^{2(d-2)}}. \tag{2.16}$$

Employing the Legendre transform of enthalpy $\tilde{G} = H - TS$, the Gibbs free energy $\tilde{G}$ is given as

$$\tilde{G} = \frac{\pi^{\frac{d-1}{2}}}{8\pi\Gamma(\frac{d-1}{2})}\left[r_h^{d-3} - \frac{16\pi P r_h^{d-1}}{(d-1)(d-2)} + \frac{(2d-5)q^2}{r_h^{d-3}}\right]. \tag{2.17}$$

The specific volume v of the black hole fluid is identified with the horizon radius of the black hole through comparing with the Van der Waals equation [6,7]

$$v = \frac{4l_P^{d-2}}{d-2}, \tag{2.18}$$

and the equation of state is given as

$$P = \frac{T}{v} - \frac{d-3}{\pi(d-2)v^2} + \frac{4^{2d-5}(d-3)q^2}{4\pi(d-2)^{2d-5}v^{2(d-2)}}. \tag{2.19}$$

The critical point is obtained by solving the following two equations

$$\frac{\partial P}{\partial v} = 0, \quad \frac{\partial^2 P}{\partial v^2} = 0, \tag{2.20}$$

which leads to

$$P_c = \frac{(d-3)^2}{\pi(d-2)^2 v_c^2}, \tag{2.21}$$

$$T_c = \frac{4(d-3)^2}{\pi(d-2)(2d-5)v_c}, \tag{2.22}$$

$$v_c = \frac{4}{d-2}\left[(d-2)(2d-5)q^2\right]^{\frac{1}{2(d-3)}}, \tag{2.23}$$

$$\tilde{G}_c = \frac{\pi^{\frac{d-1}{2}}\sqrt{(d-2)(2d-5)}q}{2\pi\Gamma(\frac{d-1}{2})(d-1)}. \tag{2.24}$$

Employing the reduced physical parameters as

$$p = \frac{P}{P_c}, \quad \tau = \frac{T}{T_c}, \quad \nu = \frac{v}{v_c}, \quad G = \frac{\tilde{G}}{\tilde{G}_c}, \tag{2.25}$$

the Gibbs free energy (2.17) is written as

$$G = \frac{1}{4}\left[(d-1)\nu^{d-3} - \frac{(d-3)^2 p\nu^{d-1}}{d-2} + \frac{d-1}{(d-2)\nu^{d-3}}\right], \tag{2.26}$$

and the equation of state is

$$p = \frac{4(d-2)\tau}{(2d-5)\nu} - \frac{d-2}{(d-3)\nu^2} + \frac{1}{(d-3)(2d-5)\nu^{2(d-2)}}, \tag{2.27}$$

which leads to

$$\tau = \frac{(2d-5)p\nu}{4(d-2)} - \frac{2d-5}{4(d-3)\nu} - \frac{1}{4(d-2)(d-3)\nu^{2d-5}}. \tag{2.28}$$

Since the first-order phase transition occurs between the small and large black hole along the coexistence curve except the critical point $\tau = \tau_c$, the two states have the same Gibbs free energy, and Eqs. (2.26) and (2.28) are written as

$$\begin{aligned} G_1 &= \frac{1}{4}\left[(d-1)\nu_1^{d-3} - \frac{(d-3)^2 p\nu_1^{d-1}}{d-2} + \frac{d-1}{(d-2)\nu_1^{d-3}}\right], \\ &= \frac{1}{4}\left[(d-1)\nu_2^{d-3} - \frac{(d-3)^2 p\nu_2^{d-1}}{d-2} + \frac{d-1}{(d-2)\nu_2^{d-3}}\right] \\ &= G_2, \end{aligned} \tag{2.29}$$

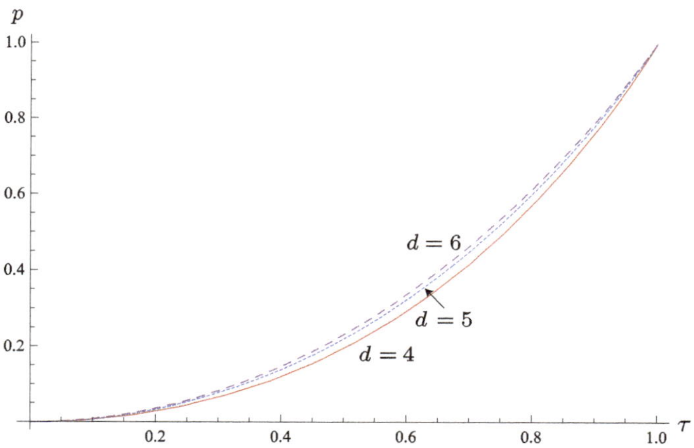

Fig. 1. Plot of the reduced pressure p as the function of the reduced temperature τ (red solid curve for $d=4$, blue dotted curve for $d=5$, and purple dashed curve for $d=6$, respectively).

$$\tau = \frac{(2d-5)pv_1}{4(d-2)} - \frac{2d-5}{4(d-3)v_1} - \frac{1}{4(d-2)(d-3)v_1^{2d-5}}, \\ = \frac{(2d-5)pv_2}{4(d-2)} - \frac{2d-5}{4(d-3)v_2} - \frac{1}{4(d-2)(d-3)v_2^{2d-5}}. \tag{2.30}$$

2.1. *The four-dimensional case*

In the case of $d=4$, the above Eqs. (2.29) and (2.30) reduce to [18,19]

$$\frac{pv_1^4 - 6v_1^2 - 3}{8v_1} = \frac{pv_2^4 - 6v_2^2 - 3}{8v_2}, \tag{2.31}$$

$$\frac{3pv_1^4 + 6v_1^2 - 1}{8v_1^3} = \frac{3pv_2^4 + 6v_2^2 - 1}{8v_2^3}, \tag{2.32}$$

$$2\tau = \frac{3pv_1^4 + 6v_1^2 - 1}{8v_1^3} + \frac{3pv_2^4 + 6v_2^2 - 1}{8v_2^3}. \tag{2.33}$$

For convenience, we now employ the parameters as the following

$$x = v_1 + v_2, \qquad y = v_1 v_2, \tag{2.34}$$

and the above Eqs. (2.31)~(2.33) are written as

$$-px^2y + py^2 + 6y - 3 = 0, \tag{2.35}$$

$$2py^3 + x^2 - 6y^2 - y = 0, \tag{2.36}$$

$$-3pxy^3 + x^3 - 6xy^2 - 3xy + 16\tau y^3 = 0. \tag{2.37}$$

These equations can be analytically solved, and the corresponding reduced pressure p is obtained as

$$p = \frac{2^{\frac{4}{3}}\tau^2(-\tau + \sqrt{\tau^2 - 2})^{\frac{2}{3}}}{[2^{\frac{1}{3}} + (-\tau + \sqrt{\tau^2 - 2})^{\frac{2}{3}}]^2}, \tag{2.38}$$

which is shown as red solid curve in Fig. 1.

Introducing the number density of black hole molecules $n = 1/v$, we have

$$\frac{n_1 - n_2}{n_c} = \frac{v_2 - v_1}{v_1 v_2} = \frac{x^2 - 4y}{y} = \sqrt{6 - 6\sqrt{p}} \\ = \sqrt{6 - \frac{6 \times 2^{\frac{2}{3}}\tau(-\tau + \sqrt{\tau^2 - 2})^{\frac{1}{3}}}{2^{\frac{1}{3}} + (-\tau + \sqrt{\tau^2 - 2})^{\frac{2}{3}}}}, \tag{2.39}$$

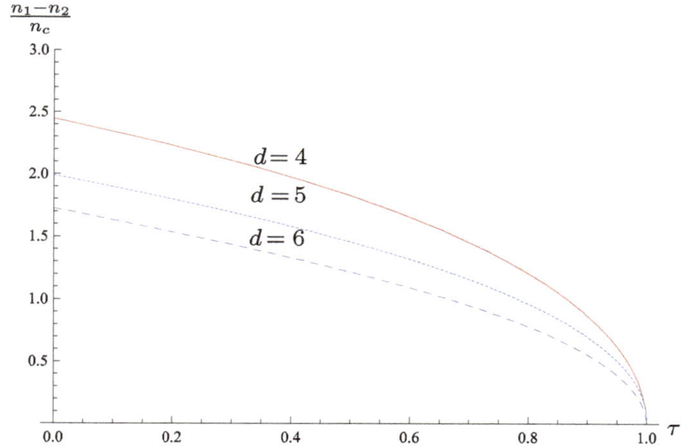

Fig. 2. Plot of the number density difference of the small and large black holes $(n_1 - n_2)/n_c$ as the function of the reduced temperature τ (red solid curve for $d=4$, blue dotted curve for $d=5$, and purple dashed curve for $d=6$, respectively).

which is shown as red solid curve in Fig. 2. Here, n_1 and n_2 are explicitly calculated as

$$n_1 = \frac{1}{v_1} = \frac{\sqrt{x^2 - 4y} + x}{2y} = \frac{\left(\sqrt{3 - \sqrt{p}} + \sqrt{3 - 3\sqrt{p}}\right)\sqrt{p}}{\sqrt{2}}, \\ = \frac{1}{\sqrt{2}}\left(\sqrt{3 - \frac{2^{\frac{2}{3}}\tau(-\tau + \sqrt{\tau^2 - 2})^{\frac{1}{3}}}{2^{\frac{1}{3}} + (-\tau + \sqrt{\tau^2 - 2})^{\frac{2}{3}}}} + \sqrt{3 - \frac{3 \times 2^{\frac{2}{3}}\tau(-\tau + \sqrt{\tau^2 - 2})^{\frac{1}{3}}}{2^{\frac{1}{3}} + (-\tau + \sqrt{\tau^2 - 2})^{\frac{2}{3}}}}\right) \\ \times \frac{2^{\frac{2}{3}}\tau(-\tau + \sqrt{\tau^2 - 2})^{\frac{1}{3}}}{2^{\frac{1}{3}} + (-\tau + \sqrt{\tau^2 - 2})^{\frac{2}{3}}}, \tag{2.40}$$

$$n_2 = \frac{1}{v_2} = \frac{\sqrt{x^2 - 4y} - x}{2y} = \frac{\left(\sqrt{3 - \sqrt{p}} - \sqrt{3 - 3\sqrt{p}}\right)\sqrt{p}}{\sqrt{2}}, \\ = \frac{1}{\sqrt{2}}\left(\sqrt{3 - \frac{2^{\frac{2}{3}}\tau(-\tau + \sqrt{\tau^2 - 2})^{\frac{1}{3}}}{2^{\frac{1}{3}} + (-\tau + \sqrt{\tau^2 - 2})^{\frac{2}{3}}}} - \sqrt{3 - \frac{3 \times 2^{\frac{2}{3}}\tau(-\tau + \sqrt{\tau^2 - 2})^{\frac{1}{3}}}{2^{\frac{1}{3}} + (-\tau + \sqrt{\tau^2 - 2})^{\frac{2}{3}}}}\right) \\ \times \frac{2^{\frac{2}{3}}\tau(-\tau + \sqrt{\tau^2 - 2})^{\frac{1}{3}}}{2^{\frac{1}{3}} + (-\tau + \sqrt{\tau^2 - 2})^{\frac{2}{3}}}. \tag{2.41}$$

Substituting with the configurational entropy (2.8), we get

$$s_{conf} = -\left[\frac{\left(\sqrt{3 - \sqrt{p}} + \sqrt{3 - 3\sqrt{p}}\right)\sqrt{p}}{\sqrt{2}} \right. \\ \times \ln\left\{\frac{1}{2}\left(1 + \frac{\sqrt{3 - 3\sqrt{p}}}{\sqrt{3 - \sqrt{p}}}\right)\right\}$$

$$
\begin{aligned}
&+\frac{\left(\sqrt{3-\sqrt{p}}-\sqrt{3-3\sqrt{p}}\right)\sqrt{p}}{\sqrt{2}}\\
&\quad\times\ln\left\{\frac{1}{2}\left(1-\frac{\sqrt{3-3\sqrt{p}}}{\sqrt{3-\sqrt{p}}}\right)\right\}\Bigg]\\
&=-\left[\frac{1}{\sqrt{2}}\left(\sqrt{3-\frac{2^{\frac{2}{3}}\tau(-\tau+\sqrt{\tau^2-2})^{\frac{1}{3}}}{2^{\frac{1}{3}}+(-\tau+\sqrt{\tau^2-2})^{\frac{2}{3}}}}\right.\right.\\
&\quad\left.+\sqrt{3-\frac{3\times 2^{\frac{2}{3}}\tau(-\tau+\sqrt{\tau^2-2})^{\frac{1}{3}}}{2^{\frac{1}{3}}+(-\tau+\sqrt{\tau^2-2})^{\frac{2}{3}}}}\right)\\
&\quad\times\frac{2^{\frac{2}{3}}\tau(-\tau+\sqrt{\tau^2-2})^{\frac{1}{3}}}{2^{\frac{1}{3}}+(-\tau+\sqrt{\tau^2-2})^{\frac{2}{3}}}\\
&\quad\times\ln\left\{\frac{1}{2}\left(1+\frac{\sqrt{3-3\times\frac{2^{\frac{2}{3}}\tau(-\tau+\sqrt{\tau^2-2})^{\frac{1}{3}}}{2^{\frac{1}{3}}+(-\tau+\sqrt{\tau^2-2})^{\frac{2}{3}}}}}{\sqrt{3-\frac{2^{\frac{2}{3}}\tau(-\tau+\sqrt{\tau^2-2})^{\frac{1}{3}}}{2^{\frac{1}{3}}+(-\tau+\sqrt{\tau^2-2})^{\frac{2}{3}}}}}\right)\right\}\\
&\quad+\frac{1}{\sqrt{2}}\left(\sqrt{3-\frac{2^{\frac{2}{3}}\tau(-\tau+\sqrt{\tau^2-2})^{\frac{1}{3}}}{2^{\frac{1}{3}}+(-\tau+\sqrt{\tau^2-2})^{\frac{2}{3}}}}\right.\\
&\quad\left.-\sqrt{3-\frac{3\times 2^{\frac{2}{3}}\tau(-\tau+\sqrt{\tau^2-2})^{\frac{1}{3}}}{2^{\frac{1}{3}}+(-\tau+\sqrt{\tau^2-2})^{\frac{2}{3}}}}\right)\\
&\quad\times\frac{2^{\frac{2}{3}}\tau(-\tau+\sqrt{\tau^2-2})^{\frac{1}{3}}}{2^{\frac{1}{3}}+(-\tau+\sqrt{\tau^2-2})^{\frac{2}{3}}}\\
&\quad\left.\times\ln\left\{\frac{1}{2}\left(1-\frac{\sqrt{3-3\times\frac{2^{\frac{2}{3}}\tau(-\tau+\sqrt{\tau^2-2})^{\frac{1}{3}}}{2^{\frac{1}{3}}+(-\tau+\sqrt{\tau^2-2})^{\frac{2}{3}}}}}{\sqrt{3-\frac{2^{\frac{2}{3}}\tau(-\tau+\sqrt{\tau^2-2})^{\frac{1}{3}}}{2^{\frac{1}{3}}+(-\tau+\sqrt{\tau^2-2})^{\frac{2}{3}}}}}\right)\right\}\right.,
\end{aligned}
\tag{2.42}
$$

which is shown as red solid curve in Fig. 3.

2.2. The five-dimensional case

As discussed in the previous section we will apply a similar analysis to the five-dimensional case.

$$\frac{p\nu_1^6-3\nu_1^4-1}{3\nu_1^2}=\frac{p\nu_2^6-3\nu_2^4-1}{3\nu_2^2}, \tag{2.43}$$

$$\frac{10p\nu_1^6+15\nu_1^4-1}{24\nu_1^5}=\frac{10p\nu_2^6+15\nu_2^4-1}{24\nu_2^5}, \tag{2.44}$$

$$2\tau=\frac{10p\nu_1^6+15\nu_1^4-1}{24\nu_1^5}+\frac{10p\nu_2^6+15\nu_2^4-1}{24\nu_2^5}, \tag{2.45}$$

which leads to

$$-px^2y^2+2py^3+3y^2-1=0 \tag{2.46}$$

$$10py^5+x^4-3x^2y-15y^4+y^2=0 \tag{2.47}$$

$$10pxy^5-x^5+5x^3y+15xy^4-5xy^2+48\tau y^5=0, \tag{2.48}$$

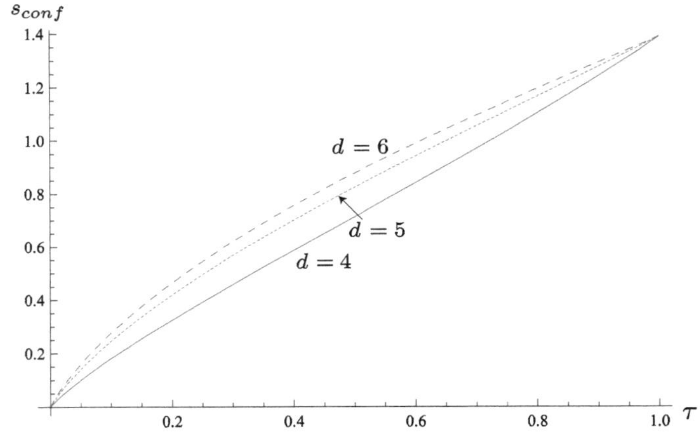

Fig. 3. Plot of the reduced configurational entropy s_{conf} as the function of the reduced temperature τ (red solid curve for $d=4$, blue dotted curve for $d=5$, and purple dashed curve for $d=6$, respectively).

which is a complicated high-order polynomial equation and it is difficult to analyze exactly. However some numerical investigation can be employed and the reduced pressure p as the function of the reduced temperature τ is numerically calculated as blue dotted curve in Fig. 1. The number density difference of the small, and large black holes $(n_1-n_2)/n_c$ as the function of the reduced temperature τ is numerically obtained as blue dotted curve in Fig. 2, and the configurational entropy s_{conf} as the function of the reduced temperature τ is numerically given as blue dotted curve in Fig. 3.

2.3. The six-dimensional case

We now apply similar numerical investigations to the six-dimensional case.

$$\frac{9p\nu_1^8-20\nu_1^6-5}{16\nu_1^3}=\frac{9p\nu_2^8-20\nu_2^6-5}{16\nu_2^3}, \tag{2.49}$$

$$\frac{21p\nu_1^8+28\nu_1^6-1}{48\nu_1^7}=\frac{21p\nu_2^8+28\nu_2^6-1}{48\nu_2^7}, \tag{2.50}$$

$$2\tau=\frac{21p\nu_1^8+28\nu_1^6-1}{48\nu_1^7}+\frac{21p\nu_2^8+28\nu_2^6-1}{48\nu_2^7}, \tag{2.51}$$

which leads to

$$-9px^4y^3+27px^2y^4-9py^5+20x^2y^3-5x^2-20y^4+5y=0, \tag{2.52}$$

$$21py^7+x^6-5x^4y+6x^2y^2-28y^6-y^3=0, \tag{2.53}$$

$$-21pxy^7+x^7-7x^5y+14x^3y^2-28xy^6-7xy^3+96\tau y^7=0, \tag{2.54}$$

which is numerically solved and the reduced pressure p as the function of the reduced temperature τ, the number density difference of the small, and large black holes $(n_1-n_2)/n_c$ as the function of the reduced temperature τ, and the configurational entropy s_{conf} as the function of the reduced temperature τ are given as purple dashed curve in Fig. 1, Fig. 2, and Fig. 3, respectively.

2.4. The thermodynamic properties

We now discuss the thermodynamic properties in charged AdS black holes in higher dimensional spacetime.

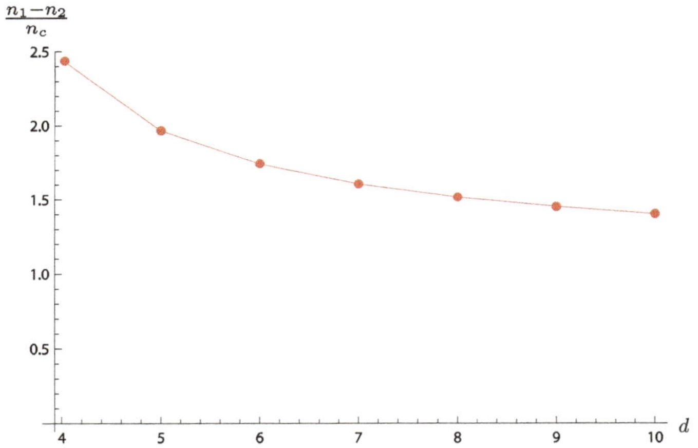

Fig. 4. Plot of the maximum number density difference of the small and large black holes $(n_1 - n_2)/n_c$ as the function of the total dimension d.

As shown in Fig. 1, and Fig. 2, as the reduced temperature τ grows up the reduced pressure p monotonically increases while the number density difference of the small, and large black holes $(n_1 - n_2)/n_c$ decreases. As shown in Fig. 3, the bigger the total dimension d becomes, the more concave function in the configurational entropy s_{conf}. Especially, as the reduced temperature τ grows up, s_{cong} monotonically increases, and reaches the maximum value at the critical (second-order phase transition) point. The maximum number density difference of the small and large black holes $(n_1 - n_2)/n_c$ at $\tau = 0$ is obtained as

$$\frac{n_1 - n_2}{n_c} = \left[(d-2)(2d-5)\right]^{\frac{1}{2(d-3)}}. \tag{2.55}$$

As shown in Fig. 4, it decreases as the total dimension d increases.

3. Conclusion

We considered higher dimensional charged AdS black holes and investigated the number density difference of the small and large black holes $(n_1 - n_2)/n_c$, and the reduced configurational entropy s_{conf} in the context of the molecule number density. We explicitly obtained the general form of maximum value of $(n_1 - n_2)/n_c$ at $\tau = 0$, and found that its maximum value decreases as the total dimension d increases. Especially, the configurational entropy s_{conf} monotonically increases as the reduced temperature τ grows up. It finally reaches a maximum value at a critical (second-order phase transition) point. This result is natural since in any system containing two different types of molecules, when they have the same number of molecules, the number of microstates W reaches maximum value. Furthermore, such result is consistent with that of the Van der Waals system. It was shown that the critical behaviour of charged AdS black holes coincides with those of the Van der Waals system [6]. In particular, when the second-order phase transition between liquid and gas occurs at the critical point, the distinction between the liquid and gas phases of the Van der Waals fluid is almost non-existent near the critical point and the molecules in the liquid and gas states are almost identical. Then, the number of microstates W becomes maximum.

References

[1] S.W. Hawking, D.N. Page, Commun. Math. Phys. 87 (1983) 577, http://dx.doi.org/10.1007/BF01208266.
[2] A. Chamblin, R. Emparan, C.V. Johnson, R.C. Myers, Phys. Rev. D 60 (1999) 064018, http://dx.doi.org/10.1103/PhysRevD.60.064018, arXiv:hep-th/9902170.
[3] A. Chamblin, R. Emparan, C.V. Johnson, R.C. Myers, Phys. Rev. D 60 (1999) 104026, http://dx.doi.org/10.1103/PhysRevD.60.104026, arXiv:hep-th/9904197.
[4] D. Kastor, S. Ray, J. Traschen, Class. Quantum Gravity 26 (2009) 195011, http://dx.doi.org/10.1088/0264-9381/26/19/195011, arXiv:0904.2765 [hep-th].
[5] M. Cvetic, G.W. Gibbons, D. Kubiznak, C.N. Pope, Phys. Rev. D 84 (2011) 024037, http://dx.doi.org/10.1103/PhysRevD.84.024037, arXiv:1012.2888 [hep-th].
[6] D. Kubiznak, R.B. Mann, J. High Energy Phys. 1207 (2012) 033, http://dx.doi.org/10.1007/JHEP07(2012)033, arXiv:1205.0559 [hep-th].
[7] S. Gunasekaran, R.B. Mann, D. Kubiznak, J. High Energy Phys. 1211 (2012) 110, http://dx.doi.org/10.1007/JHEP11(2012)110, arXiv:1208.6251 [hep-th].
[8] C.E. Shannon, Bell Syst. Tech. J. 27 (1948) 379, Bell Syst. Tech. J. 27 (1948) 623.
[9] M. Gleiser, N. Stamatopoulos, Phys. Lett. B 713 (2012) 304, http://dx.doi.org/10.1016/j.physletb.2012.05.064, arXiv:1111.5597 [hep-th].
[10] M. Gleiser, N. Stamatopoulos, Phys. Rev. D 86 (2012) 045004, http://dx.doi.org/10.1103/PhysRevD.86.045004, arXiv:1205.3061 [hep-th].
[11] M. Gleiser, D. Sowinski, Phys. Lett. B 727 (2013) 272, http://dx.doi.org/10.1016/j.physletb.2013.10.005, arXiv:1307.0530 [hep-th].
[12] M. Gleiser, N. Graham, Phys. Rev. D 89 (8) (2014) 083502, http://dx.doi.org/10.1103/PhysRevD.89.083502, arXiv:1401.6225 [astro-ph.CO].
[13] M. Gleiser, N. Jiang, Phys. Rev. D 92 (4) (2015) 044046, http://dx.doi.org/10.1103/PhysRevD.92.044046, arXiv:1506.05722 [gr-qc].
[14] A.E. Bernardini, R. da Rocha, Phys. Lett. B 762 (2016) 107, http://dx.doi.org/10.1016/j.physletb.2016.09.023, arXiv:1605.00294 [hep-th].
[15] A.E. Bernardini, N.R.F. Braga, R. da Rocha, Phys. Lett. B 765 (2017) 81, http://dx.doi.org/10.1016/j.physletb.2016.12.007, arXiv:1609.01258 [hep-th].
[16] R. Casadio, R. da Rocha, Phys. Lett. B 763 (2016) 434, http://dx.doi.org/10.1016/j.physletb.2016.10.072, arXiv:1610.01572 [hep-th].
[17] N.R.F. Braga, R. da Rocha, Phys. Lett. B 767 (2017) 386, http://dx.doi.org/10.1016/j.physletb.2017.02.031, arXiv:1612.03289 [hep-th].
[18] S.W. Wei, Y.X. Liu, Phys. Rev. Lett. 115 (11) (2015) 111302, http://dx.doi.org/10.1103/PhysRevLett.115.111302, Phys. Rev. Lett. 116 (16) (2016) 169903, http://dx.doi.org/10.1103/PhysRevLett.116.169903 (Erratum), arXiv:1502.00386 [gr-qc].
[19] J.X. Mo, G.Q. Li, Phys. Rev. D 92 (2) (2015) 024055, http://dx.doi.org/10.1103/PhysRevD.92.024055, arXiv:1604.07931 [gr-qc].
[20] Z. Wang, M. He, C. Fang, D. Sun, J. Deng, Gen. Relativ. Gravit. 49 (4) (2017) 53, http://dx.doi.org/10.1007/s10714-017-2216-9, arXiv:1703.01387 [gr-qc].

4

Cutoff in the Lyman-α forest power spectrum: Warm IGM or warm dark matter?

Antonella Garzilli [a,*], Alexey Boyarsky [a], Oleg Ruchayskiy [b]

[a] *Lorentz Institute, Leiden University, Niels Bohrweg 2, Leiden, NL-2333 CA, The Netherlands*
[b] *Discovery Center, Niels Bohr Institute, Blegdamsvej 17, DK-2100 Copenhagen, Denmark*

ARTICLE INFO

Editor: S. Dodelson

Keywords:
Cosmology, Warm dark matter, Large scale structure of Universe
Methods, Numerical, Observational
Quasars, Absorption lines

ABSTRACT

We re-analyse high redshift and high resolution Lyman-α forest spectra considered in [1], seeking to constrain the properties of warm dark matter particles. Compared to this previous work, we consider a wider range of thermal histories of the intergalactic medium. We find that both warm and cold dark matter models can explain the cut-off observed in the flux power spectra of high-resolution observations equally well. This implies, however, very different thermal histories and underlying reionization models. We discuss how to remove this degeneracy.

1. Introduction

Dark matter is a central ingredient of the current standard cosmological model. It drives the formation of structures, and explains the masses of galaxies and galaxy clusters. If dark matter is made of particles, these yet-unseen particles should have been created in the early Universe long before the recombination epoch. If such particles were relativistic at early times, they would stream out from overdense regions, smoothing out primordial density fluctuations. The signature of such *warm dark matter* (WDM) scenario would be the suppression of the matter power spectrum at scales below their free-streaming horizon. From cosmological data at large scales (CMB and galaxy surveys) we know that such a suppression should be sought at comoving scales well below a Mpc.

The Lyman-α forest has been used for measuring the matter power spectrum at such scales [2–4]. In previous works only upper bounds had been reported on the mass of the thermal relic [5–10]. However, while in the SDSS spectra there is no cut-off in the transmitted flux power spectrum, there is a cut-off in the high resolution spectra, for example [4,11,7]. Recently [1] has observed the cut-off of the flux power spectrum at scales $k \sim 0.03$ s/km and redshifts $z = 4.2$–5.4.

However, the Lyman-α forest method measures not the distribution of dark matter itself, but only the neutral hydrogen density as a proxy for the overall matter density. The process of reionization heats the hydrogen and prevents it from clustering at small scales at the redshifts in question [12]. Therefore, the observed hydrogen distribution eventually stops to follow the DM distribution. Indeed, it was demonstrated in [1] that within ΛCDM cosmology there exists a suitable thermal history of intergalactic medium (IGM) that is consistent with the observed cutoff. This does not mean, however, that this scenario is realized in nature.

In this Letter we investigate this issue in depth. We ask whether *the cutoff in the flux power spectrum can be attributed to the suppression of small scales with warm dark matter* and what this means for the thermal history of IGM. To this end we reanalyze the data used in [1]. We use *the same* suite of hydrodynamical simulations of the IGM evolution with cold and warm dark matter models as in [1] and demonstrate that the data is described equally well by the model, where flux power spectrum suppression is mainly due to WDM.

2. Data and model

The data set is constituted by 25 high-resolution quasar spectra, in the redshift interval $4.48 \leq z_{QSO} \leq 6.42$. The spectra were taken with the Keck High Resolution Echelle Spectrometer (HIRES) and the Magellan Inamory Kyocera Echelle (MIKE) spectrograph on the Magellan clay telescope. The QSO spectra are divided into four

* Corresponding author.
E-mail address: garzilli@lorentz.leidenuniv.nl (A. Garzilli).

redshift bins centered on: $z = 4.2, 4.6, 5.0, 5.4$. The resulting range of wave-numbers probed by this dataset is $k = 0.005$–0.08 s/km.

At these redshifts, the IGM is thought to be in a highly ionized state, being photo-ionized and photo-heated by early sources. Both the WDM cosmology and the IGM temperature affect the amount of flux power spectrum at small scales through three distinct physical mechanisms: (1) a suppression in the initial matter power spectrum; (2) Jeans broadening; and (3) Doppler broadening of the absorption lines [12–17]. The first mechanism is cosmological, the latter two are astrophysical. The Doppler broadening is a one dimensional smoothing effect that originates from observing the hot IGM along a line of sight. The Maxwellian distribution of velocities in the gas then leads to the broadening effect. The Jeans broadening smooths the three-dimensional underlying gas distribution relative to the dark matter.

The level of ionization is captured by *the effective optical depth*, $\tau_{\rm eff}$, that is computed from the mean flux, $\langle F \rangle$, through the relation $\langle F(z) \rangle = \exp(-\tau_{\rm eff}(z))$. Because the IGM spans a wide range of density, describing the IGM temperature may be complicated in principle. But, assuming that the IGM is heated by photoheating, the temperature of the IGM follows a simple power-law *temperature-density relation* [18]:

$$T(\delta) = T_0(z)\big(1+\delta\big)^{\gamma(z)-1}, \tag{1}$$

where $\delta = \delta\rho_{\rm m}/\bar{\rho}_{\rm m}$ is the matter overdensity and $T_0(z)$, $\gamma(z)$ are unknown functions of redshift. The results of Ref. [1] are based on single power-law parametrizations, $T_0(z)$ and $\gamma(z)$. In this letter we let the parameters of the IGM thermal state vary independently in each redshift bin, with a total of 8 parameters describing the IGM thermal state ($T_0(z_i)$ and $\gamma(z_i)$ in 4 distinct redshift intervals).[1]

We want to point out that T_0 and γ are not varied in post-processing. The original work of [1] considered 9 simulation runs with distinct thermal histories for each cosmology considered. The different thermal histories are realized by changing the photo-heating function in the simulations. The resulting values of T_0 and γ are approximately distributed on a regular grid. In [1] the effect of Jeans smoothing is accounted by considering two additional simulation runs, where the time at which the ultraviolet background is switched on, $z_{\rm reion}$, is varied. We caution the reader that the resulting constraints on $z_{\rm reion}$ must not be intended as a measurement of the time of reionization, because this depends on the details of the implementation of the ultraviolet background. Instead, varying $z_{\rm reion}$ must be considered as a way to account for the unknown level of Jeans smoothing. Finally, as in [1], we allow the effective optical depth vary independently in each redshift bin, $\tau_{\rm eff}[z_i]$.

It should be noted that this interpolation scheme between simulations with different temperatures may also vary the amount of Jeans broadening (also known as the "filtering scale"). While the degeneracy between the WDM cosmologies and the Doppler smoothing has been extensively considered in the literature, the degeneracy between Jeans smoothing and WDM cosmology has not been considered in depth so far. In particular this has not been done for the suite of simulations in the original work [1] on which we base our analysis. We leave the study of the degeneracy between the Jeans smoothing and WDM for future work.

The results also depend on the cosmological parameters n_s, Ω_M, σ_8, H_0. However the small scale Lyman-α data by itself does not sufficiently constrain the cosmological parameters. Therefore, in the final likelihood function for these parameters we used best fit Planck values [19] with Gaussian priors (as in [1]), $\Omega_M = 0.315 \pm 0.017$, $\sigma_8 = 0.829 \pm 0.013$, $n_s = 0.9603 \pm 0.0073$.

[1] Ref. [1] also performed such a "binned analysis", see the detailed comparison below.

Table 1
Parameter estimation from Bayesian analysis. We show the 1-σ and 2-σ confidence intervals. We only show the parameters that are constrained at 1 or 2-σ level.

parameter	mean	1-σ	2-σ
H_0 [km/s/Mpc]	63	< 67	-
$m_{\rm WDM}$ [keV]	3.9	[143, 2.3]	> 2.1
$T_0(z=4.2)$ [10^3 K]	10.6	[9.4, 11.8]	[8.3, 12.9]
$T_0(z=4.6)$ [10^3 K]	9.8	[8.6, 11.1]	[7.5, 12.2]
$T_0(z=5.0)$ [10^3 K]	4.0	[2.0, 5.6]	< 6.9
$T_0(z=5.4)$ [10^3 K]	3.8	< 4.5	< 8.2
$\tau_{\rm eff}(z=4.2)$	1.12	[1.05, 1.19]	[1.00, 1.25]
$\tau_{\rm eff}(z=4.6)$	1.30	[1.21, 1.39]	[1.15, 1.47]
$\tau_{\rm eff}(z=5.0)$	1.88	[1.74, 2.00]	[1.64, 2.13]
$\tau_{\rm eff}(z=5.4)$	2.91	[2.69, 3.10]	[2.54, 3.31]
$\gamma(z=4.2)$	1.3	> 1.1	-
$\gamma(z=5.4)$	1.3	> 1.1	-

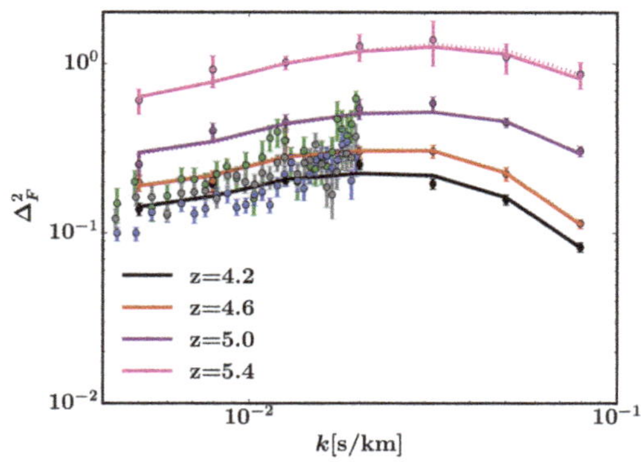

Fig. 1. Measured flux power spectrum in dimensionless units, $\Delta_F^2(k) = P_F(k) \times k/\pi$, compared with the theoretical model with the best-fitting values of the astrophysical and cosmological parameters for WDM and CDM cosmologies. The solid refer the best-fitting values for WDM cosmology. The dotted lines refer to the best-fitting case for CDM cosmology. These best-fitting models largely overlap, except at the highest redshift and on the smallest scales. The blue, gray and green points are SDSS-III/BOSS DR9 data for $z = 4.0$, $z = 4.2$ and $z = 4.4$ from [20]. (For interpretation of the references to color in this figure legend, the reader is referred to the web version of this article.)

3. Results

In Table 1 we give the result of the parameter estimation. Fig. 1 shows the theoretical flux power spectrum for the mean values of the parameters, compared with the MIKE and HIRES data used in this analysis. In order to clarify the effect of different thermal histories on our constraints, we show the effect of changing the thermal parameters (T_0 and γ) and ionization parameters ($\tau_{\rm eff}$) and the mass of the thermal relic ($1/m_{\rm wdm}$) in Fig. 2, analogous to Figs. 5 and 6 of [1].

In Fig. 3 we show the 2D confidence regions between $m_{\rm wdm}$, and $T_0 \equiv T(\delta = 0)$ (marginalizing over the other parameters). We see that at redshifts $z = 4.2$, 4.6 there is no degeneracy and an IGM temperature $T_0 \sim 10^4$ K is needed to explain the observed flux power spectrum independently of $m_{\rm wdm}$. If dark matter is "too warm" ($m_{\rm wdm} < 1.5$ keV) it produces too sharp of a cut-off in the power spectrum and is inconsistent with the data.

At the $z = 5.0$ bin the situation is different. For the masses $m_{\rm wdm} \sim 2.2$–3.3 keV even very low temperatures $T_0 \lesssim 2500$ K are consistent with the data. In this case the cutoff in the flux power spectrum is explained by WDM rather than by the temperature. The situation is analogous at $z = 5.4$. Table 1 summarizes the parameter estimation.

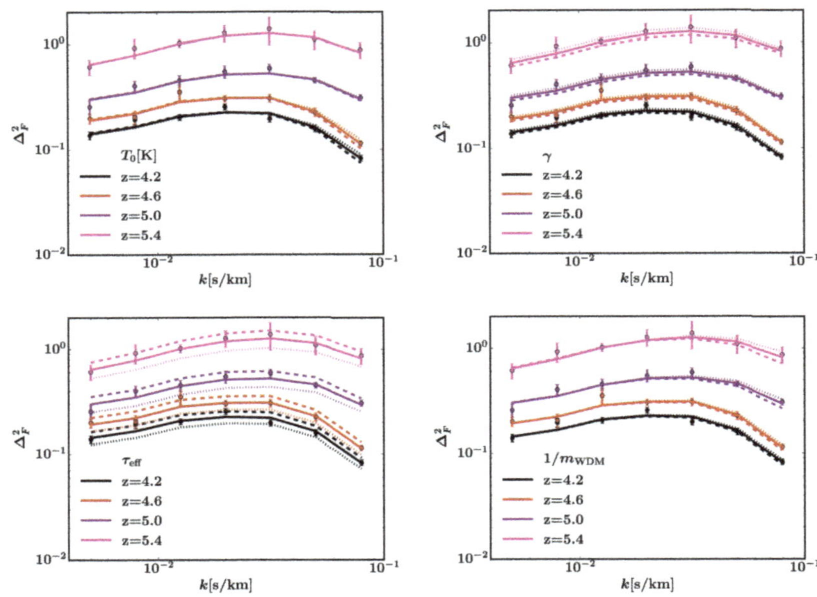

Fig. 2. Effect of the IGM parameters and $m_{\rm wdm}$ on the flux power spectrum in dimensionless units, $\Delta_F^2(k) = P_F(k) \times k/\pi$. In the top-left (top-right, bottom-left, bottom-right) panel we show the effect of varying T_0 (γ, $\tau_{\rm eff}$, $1/m_{\rm wdm}$) by $\pm 10\%$ with respect to the best-fitting values for WDM cosmology. The solid line corresponds to the best-fitting case for WDM cosmology, the dashed (dotted) line corresponds to the relevant parameter increased (decreased) by 10%. (For interpretation of the references to color in this figure legend, the reader is referred to the web version of this article.)

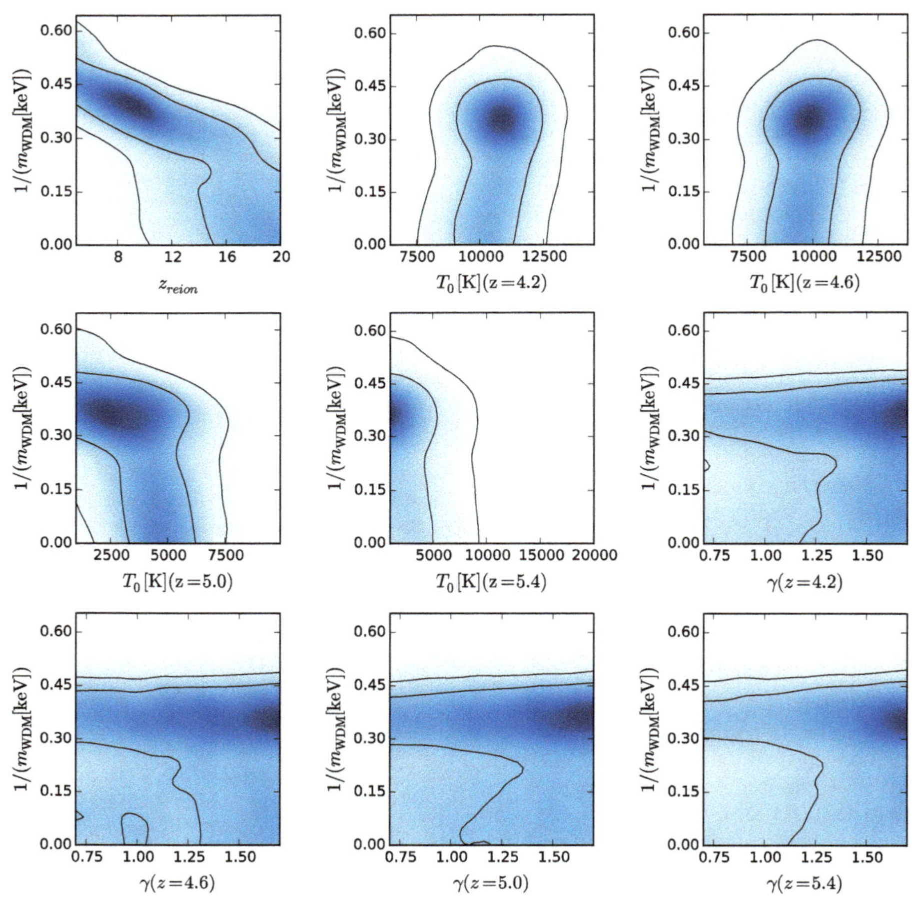

Fig. 3. Confidence regions between $m_{\rm wdm}$, and T_0 and γ at all redshift, and $z_{\rm reion}$. We show $1/m_{\rm wdm}$ instead of $m_{\rm wdm}$ for visualization purposes. $m_{\rm wdm}$ is degenerate with $z_{\rm reion}$, that is the redshift at which the ultraviolet background has been switched on in the simulations, and T_0 at the redshift $z = 5.0$. $m_{\rm wdm}$ is not degenerate with the T_0 for the other redshift intervals. There is no obvious degeneracy with γ. Regarding $m_{\rm wdm}$ and T_0, at the redshifts $z = 4.2$, 4.6 there is no degeneracy and $T_0 \sim 10^4$ K is needed to explain the observed flux power spectrum, independently of $m_{\rm wdm}$. At $z = 5.0$ even very low temperatures $T_0 \lesssim 2500$ K are consistent with the data, and the cutoff in the flux power spectrum is explained by WDM rather than by the temperature. At $z = 5.4$ the analysis prefers low values of $T_0 \sim 5 \times 10^3$ K, independently of $m_{\rm wdm}$.

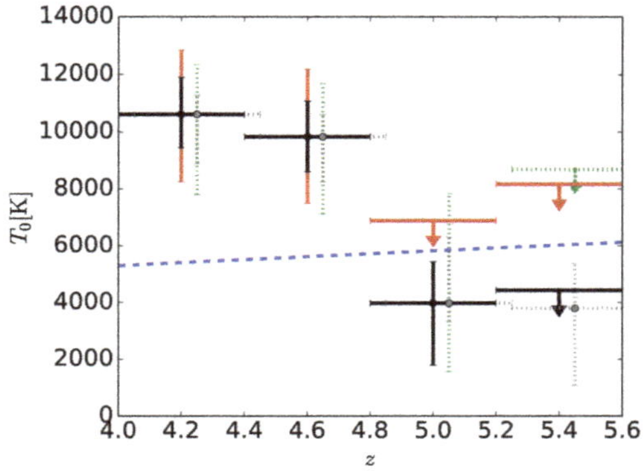

Fig. 4. The evolution of the IGM mean temperature, T_0, in redshift. Black vertical bars are 1-σ confidence limits; red vertical bars are 2-σ confidence limits. Filled dots are the parameter mean; the arrows mark the upper limits. The horizontal bars span the redshift interval of Lyman α absorbers considered for each measurement of the flux power spectrum. The solid (dotted) lines refer to the constraints on temperature for WDM (CDM) cosmology (the constraints in CDM have been shifted by $z = 0.05$ for improving the readability of the figure). At $z = 5.0$ there is a 1-σ level detection and only an upper limit at 2-σ level in WDM cosmology, instead there are both 1 and 2-σ detections for CDM cosmology. At $z = 5.4$, there are only upper limits at 1 and 2-σ levels for WDM cosmology and 1-σ detection and 2-σ upper limit for CDM cosmology. Hence, the constraints on the temperature are substantially equivalent in the two cosmologies. The blue dashed line is the asymptotic IGM mean temperature in the case of early hydrogen and first helium reionization from a stellar ionizing spectrum with slope $\alpha = 2$, being the ionizing spectrum $J_\nu \propto \nu^{-\alpha}$. (For interpretation of the references to color in this figure legend, the reader is referred to the web version of this article.)

Another important property of Fig. 3 is that even assuming CDM cosmology, the temperature T_0 is a non-monotonic function of redshift and should be colder than ~ 8000 K at $z = 5.0$–5.4, see Fig. 4.[2]

The resulting χ^2 for the Bayesian analysis is ~ 25, for 30 degrees of freedom (49 data points − 19 free parameters). This is in agreement with the fact that the covariance matrix is uncertain and that has been multiplied by a factor that boosts the resulting error bars by 30%, with respect to the error bars computed by bootstrapping. This is done in the original analysis in order to account for presumed sample variance effect that affect other statistics like the transmitted flux PDF. The sample variance effect may affect the transmitted flux power spectrum, even if a detailed computation has not been performed.

For completeness we have also performed frequentist analysis for the same χ^2 considered in the Bayesian analysis. As shown in Fig. 5 the two analyses are in broad agreement with each other.

We would like to stress that our results depend crucially on allowing for a non-monotonic redshift dependence of $T_0(z)$. In [1] it was shown that assuming a power-law (*monotonic*) redshift dependence for $T_0(z)$ and $\gamma(z)$, one predicts higher temperatures of IGM for the same data. In this case the CDM cosmology is preferred over WDM, leading to the 2σ lower bound $m_{\rm wdm} \geq 3.3$ keV [1]. The "binned analysis" of [1] gave results similar to those, reported here. The authors of [1] however rejected these results, considering a temperature jump at $z = 5$–5.4 to be "unphysical" and arguing that the low χ^2 is a sign of overfitting.

In our opinion the present analysis implies that more data is needed to study such a scenario, as it currently does not allow to make any definitive conclusion and in particular does not allow to rule it out. Moreover, as mentioned above, the error bars in [1] were inflated by 30% and therefore we consider the reduced $\chi^2 = 25/30 \approx 0.83$ to be consistent with 1. We see that 2σ

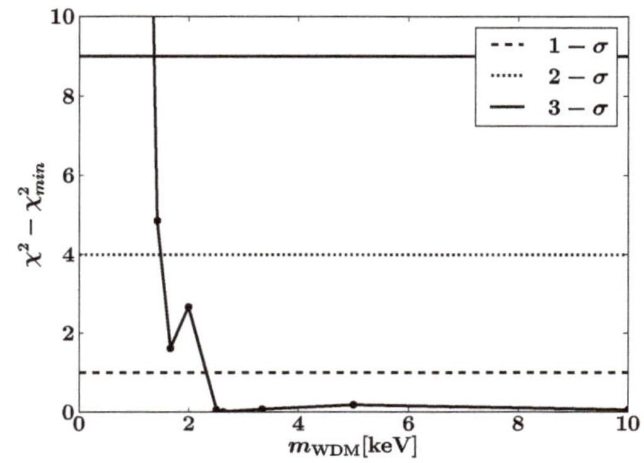

Fig. 5. The results of the frequentist analysis: the $\chi^2 - \chi^2_{min}$ versus the WDM mass, $m_{\rm WDM}$. There are two minima of the χ^2 curve, CDM and $m_{\rm wdm} = 2.7$ keV.

lower bound on the WDM mass relaxes down to $m_{\rm wdm} \geq 2.1$ keV (consistently with the results of binned analysis of [1]). Moreover, the non-monotonic thermal history makes the WDM with $m_{\rm wdm} = 2$–3 keV an equally good fit as CDM. The best fit values of T_0 can be inaccurate as they lie below the lowest simulation point in T_0 grid. Therefore more simulations are needed to settle this question. This is currently work in progress. In the absence of such additional studies the proposed non-monotonic thermal history cannot be ruled out based on the existing Lyman-α data.

For the interpretation of these results it is crucial to overview what is known about the thermal state of the IGM both theoretically and observationally. We argue below that the measured thermal history is in agreement with current models of galaxy formation and reionization.

4. State of the IGM at $z \sim 5$

The IGM temperature can be *determined* from the broadening of the Lyman-α absorption lines in QSO spectra [21–31,16]. Alternatively, it has been proposed to determine the IGM temperature by measuring the level of the transmitted flux [32–34,30], however there is no agreement between the two methods yet, see [35].

All the measurement of the IGM temperature in the literature assumed CDM cosmology. Because of the existing degeneracy between the IGM temperature and WDM, the assumption of the WDM cosmology could change the deduced values of the IGM temperature. Nevertheless, in the absence of such measurements, we compare our estimates for the IGM temperature with the measurements based on the CDM assumption.

The IGM temperature at $z < 5$ is constrained relatively well to be at the level $T_0 \gtrsim (8\text{–}10) \times 10^3$ K [25,22,27,28]. At $z = 6.0$ there is a single measurement, [29], that restricts T_0 to the range $5000 < T_0 < 10000$ K (68% CL) (see e.g. [1] for discussion). The simplest interpretation of these data (also adopted in [1]) is that the temperature is growing monotonically with redshift. Instead, given the large error bars of the measurements, and taking into account adiabating cooling one may expect a drop of temperature at $z \sim 5$ with a subsequent rise to $\sim 10^4$ K at $z \sim 4.6$ in agreement with other measurements from [25,22,27,28]. This increase in IGM temperature can be explained with an early start of HeII reionization predicted by some models of reionization by quasars, [36] (see recent discussion of such "two-component" reionization models in [37]).

In such a scenario, the temperature at $5 < z < 6$ depends on how long the first stage of reionization lasted and what the temperature of IGM was at $z \gtrsim 6$. As mentioned above, the measurement [29] at $z \sim 6$ has large uncertainties. Theoretically, $T_0(z = 6)$ depends on how early the first stage of hydrogen (and HeI) reion-

[2] The temperature values that we have estimated at high redshift could be inaccurate, because the lowest temperature in the simulation grid was 5400 K.

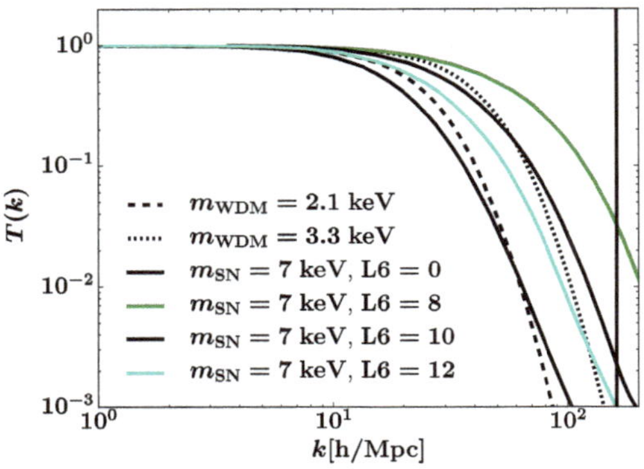

Fig. 6. Comparison between the linear transfer functions, $T(k)$, of thermal relic (WDM) and sterile neutrinos (SN). The dashed (dotted) black line is the linear transfer function for $m_{\rm WDM} = 2.1$ keV ($m_{\rm WDM} = 3.3$ keV) as computed in [10]. The colored (green, red, cyan) lines are realistic linear transfer functions for some of the sterile neutrino models with $m_{\rm SN}^{\rm NRP} = 7$ keV. The linear transfer functions with $L_6 = 10$ and 12 (red and cyan lines) are partially warmer that the lower bound of [1] (the dotted black line), but still satisfy the constraints from this letter (the dashed black line) until the maximum k-mode used in the reference numerical simulations. The linear transfer function with $L_6 = 8$ (green line) is colder than the bound of [1]. The linear transfer function with $L_6 = 0$ (blue line) violates the constraint from this letter. The solid vertical line is the maximum k-mode used in the reference simulations. (For interpretation of the references to color in this figure legend, the reader is referred to the web version of this article.)

izations has ended, and what sources drove it (cf. [38,39]). It has been speculated that hydrogen is reionized by the metal-free (Population III) stars, whose spectral hardness predicts high values of the temperature. However, the properties of Population III stars are purely speculative - we do not know how long they lasted and whether they were indeed the sources of reionization. For example, reionization could be due to a more metal rich population of stars with softer stellar spectra [40], leading to a lower values of IGM temperature at $z \sim 6$. To settle this question, an independent constraint on the ultraviolet background at high redshift would be needed, however there are no such measurements to-date. The lower limit of [29] is $T_0(z=6) \approx 5 \times 10^3$ K or even slightly below, fully consistent with the low values at $z = 5.0$–5.4 (Table 1) reached via adiabatic cooling.

We note that an indirect argument in favour of the IGM temperatures at high redshifts being $\sim 10^4$ K, is the "missing satellite problem" - high temperature would prevent gas from collapsing into dark matter halos with a mass below $\sim 10^7 M_\odot$, thus suppressing the formation of small galaxies (see e.g. [41–44]), explaining in particular the small number of satellites of the Milky Way. However, in WDM cosmologies the matter power spectrum is suppressed at the smallest scales, thus solving the missing satellite problem even if the gas was sufficiently cooler.

Finally, we use our results to explore the constraints on sterile neutrino dark matter [45], resonantly produced in the presence of lepton asymmetry [46–48]. This is a non-thermal warm dark matter, whose primordial phase-space density distribution resembles a mixture of cold + warm dark matter components [49,50], demonstrating a shallower cut-off. In Fig. 6 we compare the linear transfer function (the square root of the ratio of the modified linear matter power spectrum to that of cold dark matter, $T(k) = \sqrt{P_{\rm WDM}(k)/P_{\rm CDM}(k)}$) of thermal relic WDM with a mass $m_{\rm wdm} = 2.1$ keV (lower bound from this work) and a $m_{\rm wdm} = 3.3$ keV [1] with those of resonantly produced sterile neutrinos with the mass 7 keV (motivated by the recent reports of an unidentified spectral line at the energy $E \sim 3.5$ keV in the stacked X-ray spectra of Andromeda galaxy, Perseus galaxy clusters, stacked galaxy clusters and the Galactic Center of the Milky Way [51–53]). We show that depending on the value of the lepton asymmetry, $L_6 \equiv 10^6(n_{\nu_e} - n_{\bar{\nu}_e})/s$ (see [50] for details) the linear power can be colder than that of thermal relics with $m_{\rm wdm} = 2.1$ keV (Fig. 6), thus being fully admissible by the data.[3] Notice that the non-resonant sterile neutrino dark matter with a 7 keV mass would be excluded at more than 3σ level by previous constraints from the SDSS [7,6].

5. Conclusion and future work

We demonstrated that the cut-off in the flux power spectrum, observed in the high resolution Lyman-α forest data may either be due to free-streaming of dark matter particles or be explained by the temperature of the intergalactic medium. Taking into account measurements at redshifts $z \sim 6$ and at $z < 5$ we see that if dark matter is *warm*, this requires non-monotonic dependence on the IGM temperature on z with the local minimum at $z \sim 5.0$–5.4. Even cold dark matter slightly prefers a non-monotonic $T_0(z)$.[4] Improving our knowledge of the IGM temperature at $z \sim 5$–6 will therefore either result in very strong Lyman-α bounds on DM free-streaming, essentially excluding its influence on observable small-scale structures, or (if temperature will be found to be well below 5000 K) would lead to the discovery of WDM.

A method that would allow to measure the IGM temperature at the redshifts of interest was presented in [16]. It is based on the following idea: for high resolution spectra it is not necessary to study average deviation from the QSO continuum per redshift bins (as it is done in lower resolution case) but it is possible to identify individual absorption lines and to measure their broadening. The thermal Doppler effect broadens the natural lorentzian line profile of the Lyman-α transition proportionally to the square root of the temperature, and one would like to use this information to determine the temperature of the IGM *directly*. However, there are other effects that contribute to the line width - the physical extent and the clustering of the underlying filaments. The method of [16] potentially allows to disentangle these effects. In view of our results it is important to attempt to apply this method to observational data. This is a method that has been tested with simulations at redshift ~ 3, and it still has to be seen if it works at redshift 5.

Acknowledgements

The authors are grateful to Matteo Viel, for sharing with them the code used in [1], and making this analysis possible. The authors thank James Bolton, Joop Schaye and Tom Theuns for useful discussions on the IGM temperature at high redshift. AG thanks Mark Lovell for sharing his knowledge about generating SN power spectra. She also thanks Bin Hu, Samuel Leach, Matteo Martinelli and Jesus Torrado for useful discussions on MCMC method. The authors thank the anonymous referee for his helpful comments, that improved the manuscript by large extent. The research was supported in part by the European Research Council under the European Union's Seventh Framework Programme (FP7/2007–2013)/ERC grant agreements 278594-GasAroundGalaxies.

[3] Our computations of the phase-space distribution functions for sterile neutrinos are based on [48] and the linear power spectrum is obtained with the modified CAMB code developed in [49]. We do not expect the most recent computations [54, 55] to affect our results.

[4] As stated in [56], a model of fluctuating UVB, with spatially constant mean free path for the hydrogen-ionizing photons (similar to the one used in [1] and in this work) may not be adequate to explain the observed scatter in the optical depth at the redshifts $5.1 \le z \le 5.7$. Proper modeling of patchy reionization may affect the conclusions about the IGM state. We leave this for future work.

References

[1] M. Viel, G.D. Becker, J.S. Bolton, M.G. Haehnelt, Warm dark matter as a solution to the small scale crisis: new constraints from high redshift Lyman-α forest data, Phys. Rev. D 88 (2013) 043502, http://dx.doi.org/10.1103/PhysRevD.88.043502, arXiv:1306.2314.

[2] R.A.C. Croft, D.H. Weinberg, N. Katz, L. Hernquist, Recovery of the power spectrum of mass fluctuations from observations of the Lyman alpha forest, Astrophys. J. 495 (1998) 44, http://dx.doi.org/10.1086/305289, arXiv:astro-ph/9708018.

[3] P. McDonald, J. Miralda-Escude, M. Rauch, W.L.W. Sargent, T.A. Barlow, R. Cen, J.P. Ostriker, The observed probability distribution function, power spectrum, and correlation function of the transmitted flux in the Lyman-alpha forest, Astrophys. J. 543 (2000) 1–23, http://dx.doi.org/10.1086/317079, arXiv:astro-ph/9911196.

[4] R.A.C. Croft, D.H. Weinberg, M. Bolte, S. Burles, L. Hernquist, N. Katz, D. Kirkman, D. Tytler, Towards a precise measurement of matter clustering: Lyman alpha forest data at redshifts 2–4, Astrophys. J. 581 (2002) 20–52, http://dx.doi.org/10.1086/344099, arXiv:astro-ph/0012324.

[5] S.H. Hansen, J. Lesgourgues, S. Pastor, J. Silk, Constraining the window on sterile neutrinos as warm dark matter, Mon. Not. R. Astron. Soc. 333 (2002) 544–546, http://dx.doi.org/10.1046/j.1365-8711.2002.05410.x, arXiv:astro-ph/0106108.

[6] A. Boyarsky, J. Lesgourgues, O. Ruchayskiy, M. Viel, Lyman-alpha constraints on warm and on warm-plus-cold dark matter models, J. Cosmol. Astropart. Phys. 0905 (2009) 012, http://dx.doi.org/10.1088/1475-7516/2009/05/012, arXiv:0812.0010.

[7] U. Seljak, A. Makarov, P. McDonald, H. Trac, Can sterile neutrinos be the dark matter?, Phys. Rev. Lett. 97 (2006) 191303, http://dx.doi.org/10.1103/PhysRevLett.97.191303, arXiv:astro-ph/0602430.

[8] M. Viel, G.D. Becker, J.S. Bolton, M.G. Haehnelt, M. Rauch, W.L.W. Sargent, How cold is cold dark matter? Small scales constraints from the flux power spectrum of the high-redshift Lyman-alpha forest, Phys. Rev. Lett. 100 (2008) 041304, http://dx.doi.org/10.1103/PhysRevLett.100.041304, arXiv:0709.0131.

[9] M. Viel, J. Lesgourgues, M.G. Haehnelt, S. Matarrese, A. Riotto, Can sterile neutrinos be ruled out as warm dark matter candidates?, Phys. Rev. Lett. 97 (2006) 071301, http://dx.doi.org/10.1103/PhysRevLett.97.071301, arXiv:astro-ph/0605706.

[10] M. Viel, J. Lesgourgues, M.G. Haehnelt, S. Matarrese, A. Riotto, Constraining warm dark matter candidates including sterile neutrinos and light gravitinos with WMAP and the Lyman-alpha forest, Phys. Rev. D 71 (2005) 063534, http://dx.doi.org/10.1103/PhysRevD.71.063534, arXiv:astro-ph/0501562.

[11] T.S. Kim, M. Viel, M.G. Haehnelt, R.F. Carswell, S. Cristiani, The power spectrum of the flux distribution in the Lyman-alpha forest of a large sample of UVES QSO absorption spectra (LUQAS), Mon. Not. R. Astron. Soc. 347 (2004) 355, http://dx.doi.org/10.1111/j.1365-2966.2004.07221.x, arXiv:astro-ph/0308103.

[12] N.Y. Gnedin, L. Hui, Probing the universe with the Lyman alpha forest: 1. Hydrodynamics of the low density IGM, Mon. Not. R. Astron. Soc. 296 (1998) 44–55, http://dx.doi.org/10.1046/j.1365-8711.1998.01249.x.

[13] T. Theuns, J. Schaye, M.G. Haehnelt, Broadening of QSO Lyα forest absorbers, Mon. Not. R. Astron. Soc. 315 (2000) 600–610, http://dx.doi.org/10.1046/j.1365-8711.2000.03423.x.

[14] V. Desjacques, A. Nusser, Joint modeling of the probability distribution and power spectrum of the Ly-alpha forest: comparison with observations at $z = 3$, Mon. Not. R. Astron. Soc. 361 (2005) 1257–1272, http://dx.doi.org/10.1111/j.1365-2966.2005.09254.x, arXiv:astro-ph/0410618.

[15] M.S. Peeples, D.H. Weinberg, R. Dave, M.A. Fardal, N. Katz, Pressure support vs. thermal broadening in the Lyman-alpha forest I: effects of the equation of state on longitudinal structure, Mon. Not. R. Astron. Soc. 404 (2010) 1281–1294, http://dx.doi.org/10.1111/j.1365-2966.2010.16383.x, arXiv:0910.0256.

[16] A. Garzilli, T. Theuns, J. Schaye, The broadening of Lyman-α forest absorption lines, Mon. Not. R. Astron. Soc. 450 (2) (2015) 1465–1476, http://dx.doi.org/10.1093/mnras/stv394, arXiv:1502.05715.

[17] G. Kulkarni, J.F. Hennawi, J. Oñorbe, A. Rorai, V. Springel, Characterizing the pressure smoothing scale of the intergalactic medium, Astrophys. J. 812 (2015) 30, http://dx.doi.org/10.1088/0004-637X/812/1/30, arXiv:1504.00366.

[18] L. Hui, N.Y. Gnedin, Equation of state of the photoionized intergalactic medium, Mon. Not. R. Astron. Soc. 292 (1997) 27, http://dx.doi.org/10.1093/mnras/292.1.27, arXiv:astro-ph/9612232.

[19] Planck Collaboration, P.A.R. Ade, N. Aghanim, C. Armitage-Caplan, M. Arnaud, M. Ashdown, F. Atrio-Barandela, J. Aumont, C. Baccigalupi, A.J. Banday, et al., Planck 2013 results. XVI. Cosmological parameters, Astron. Astrophys. 571 (2014) A16, http://dx.doi.org/10.1051/0004-6361/201321591, arXiv:1303.5076.

[20] N. Palanque-Delabrouille, C. Yèche, A. Borde, J.-M. Le Goff, G. Rossi, M. Viel, É. Aubourg, S. Bailey, J. Bautista, M. Blomqvist, A. Bolton, J.S. Bolton, N.G. Busca, B. Carithers, R.A.C. Croft, K.S. Dawson, T. Delubac, A. Font-Ribera, S. Ho, D. Kirkby, K.-G. Lee, D. Margala, J. Miralda-Escudé, D. Muna, A.D. Myers, P. Noterdaeme, I. Pâris, P. Petitjean, M.M. Pieri, J. Rich, E. Rollinde, N.P. Ross, D.J. Schlegel, D.P. Schneider, A. Slosar, D.H. Weinberg, The one-dimensional Lyα forest power spectrum from BOSS, Astron. Astrophys. 559 (2013) A85, http://dx.doi.org/10.1051/0004-6361/201322130, arXiv:1306.5896.

[21] T. Theuns, S. Zaroubi, A wavelet analysis of QSO spectra, Mon. Not. R. Astron. Soc. 317 (2000) 989, http://dx.doi.org/10.1046/j.1365-8711.2000.03729.x, arXiv:astro-ph/0002172.

[22] P. McDonald, J. Miralda-Escude, M. Rauch, W.L.W. Sargent, T.A. Barlow, R. Cen, A measurement of the temperature-density relation in the intergalactic medium using a new Lyman-alpha absorption line fitting method, Astrophys. J. 562 (2001) 52–75, http://dx.doi.org/10.1086/323426, arXiv:astro-ph/0005553, Astrophys. J. 598 (2003) 712 (Erratum).

[23] M. Zaldarriaga, L. Hui, M. Tegmark, Constraints from the Lyman alpha forest power spectrum, Astrophys. J. 557 (2001) 519–526, http://dx.doi.org/10.1086/321652, arXiv:astro-ph/0011559.

[24] M. Viel, M.G. Haehnelt, Cosmological and astrophysical parameters from the SDSS flux power spectrum and hydrodynamical simulations of the Lyman-alpha forest, Mon. Not. R. Astron. Soc. 365 (2006) 231–244, http://dx.doi.org/10.1111/j.1365-2966.2005.09703.x, arXiv:astro-ph/0508177.

[25] J. Schaye, T. Theuns, M. Rauch, G. Efstathiou, W.L.W. Sargent, The thermal history of the intergalactic medium, Mon. Not. R. Astron. Soc. 318 (2000) 817, http://dx.doi.org/10.1046/j.1365-8711.2000.03815.x, arXiv:astro-ph/9912432.

[26] M. Ricotti, N.Y. Gnedin, J.M. Shull, The evolution of the effective equation of state of the IGM, Astrophys. J. 534 (2000) 41–56, http://dx.doi.org/10.1086/308733, arXiv:astro-ph/9906413.

[27] A. Lidz, C.A. Faucher-Giguere, A. Dall'Aglio, M. McQuinn, C. Fechner, M. Zaldarriaga, L. Hernquist, S. Dutta, A measurement of small scale structure in the $2.2 \leq z \leq 4.2$ Lyman-alpha forest, Astrophys. J. 718 (2010) 199–231, http://dx.doi.org/10.1088/0004-637X/718/1/199, arXiv:0909.5210.

[28] G.D. Becker, J.S. Bolton, M.G. Haehnelt, W.L.W. Sargent, Detection of extended He II reionization in the temperature evolution of the intergalactic medium, Mon. Not. R. Astron. Soc. 410 (2011) 1096, http://dx.doi.org/10.1111/j.1365-2966.2010.17507.x, arXiv:1008.2622.

[29] J.S. Bolton, G.D. Becker, S. Raskutti, J.S.B. Wyithe, M.G. Haehnelt, W.L.W. Sargent, Improved measurements of the intergalactic medium temperature around quasars: possible evidence for the initial stages of He II reionization at $z \sim 6$, Mon. Not. R. Astron. Soc. 419 (2012) 2880–2892, http://dx.doi.org/10.1111/j.1365-2966.2011.19929.x.

[30] A. Garzilli, J.S. Bolton, T.S. Kim, S. Leach, M. Viel, The intergalactic medium thermal history at redshift $z = 1.7$–3.2 from the Lyman alpha forest: a comparison of measurements using wavelets and the flux distribution, Mon. Not. R. Astron. Soc. 424 (2012) 1723, http://dx.doi.org/10.1111/j.1365-2966.2012.21223.x, arXiv:1202.3577.

[31] G.C. Rudie, C.C. Steidel, M. Pettini, The temperature-density relation in the intergalactic medium at Redshift $< z >= 2.4$, Astrophys. J. 757 (2012) L30, http://dx.doi.org/10.1088/2041-8205/757/2/L30, arXiv:1209.0005.

[32] J.S. Bolton, M. Viel, T.S. Kim, M.G. Haehnelt, R.F. Carswell, Possible evidence for an inverted temperature-density relation in the intergalactic medium from the flux distribution of the Lyman-alpha forest, Mon. Not. R. Astron. Soc. 386 (2008) 1131–1144, http://dx.doi.org/10.1111/j.1365-2966.2008.13114.x, arXiv:0711.2064.

[33] M. Viel, J.S. Bolton, M.G. Haehnelt, Cosmological and astrophysical constraints from the Lyman-alpha forest flux probability distribution function, Mon. Not. R. Astron. Soc. 399 (2009) L39–L43, http://dx.doi.org/10.1111/j.1745-3933.2009.00720.x, arXiv:0907.2927.

[34] F. Calura, E. Tescari, V. D'Odorico, M. Viel, S. Cristiani, T.S. Kim, J.S. Bolton, The Lyman alpha forest flux probability distribution at $z > 3$, Mon. Not. R. Astron. Soc. 422 (2012) 3019, http://dx.doi.org/10.1111/j.1365-2966.2012.20811.x, arXiv:1201.5121.

[35] E. Rollinde, T. Theuns, J. Schaye, I. Pâris, P. Petitjean, Sample variance and Lyman α forest transmission statistics, Mon. Not. R. Astron. Soc. 428 (2013) 540–550, http://dx.doi.org/10.1093/mnras/sts057.

[36] M. McQuinn, A. Lidz, M. Zaldarriaga, L. Hernquist, P.F. Hopkins, S. Dutta, C.-A. Faucher-Giguère, He II reionization and its effect on the intergalactic medium, Astrophys. J. 694 (2009) 842–866, http://dx.doi.org/10.1088/0004-637X/694/2/842.

[37] P. Madau, F. Haardt, Cosmic reionization after Planck: could quasars do it all?, arXiv:1507.07678.

[38] F. Haardt, P. Madau, Radiative transfer in a clumpy universe. II. The ultraviolet extragalactic background, Astrophys. J. 461 (1996) 20, http://dx.doi.org/10.1086/177035.

[39] L. Hui, Z. Haiman, The thermal memory of reionization history, Astrophys. J. 596 (2003) 9–18, http://dx.doi.org/10.1086/377229.

[40] B. Ciardi, A. Ferrara, The first cosmic structures and their effects, Space Sci. Rev. 116 (2005) 625–705, http://dx.doi.org/10.1007/s11214-005-3592-0.

[41] A.J. Benson, C.G. Lacey, C.M. Baugh, S. Cole, C.S. Frenk, The effects of photoionization on galaxy formation. 1. Model and results at $z = 0$, Mon. Not. R. Astron. Soc. 333 (2002) 156, http://dx.doi.org/10.1046/j.1365-8711.2002.05387.x, arXiv:astro-ph/0108217.

[42] A.J. Benson, C.S. Frenk, C.G. Lacey, C.M. Baugh, S. Cole, The effects of photoionization on galaxy formation. 2. Satellites in the local group, Mon. Not. R. Astron. Soc. 333 (2002) 177, http://dx.doi.org/10.1046/j.1365-8711.2002.05388.x, arXiv:astro-ph/0108218.

[43] A.V. Maccio', X. Kang, F. Fontanot, R.S. Somerville, S.E. Koposov, et al., On the origin and properties of Ultrafaint Milky Way Satellites in a LCDM Universe, Mon. Not. R. Astron. Soc. 402 (2009) 1995–2008, http://dx.doi.org/10.1111/j.1365-2966.2009.16031.x.

[44] T. Sawala, C.S. Frenk, A. Fattahi, J.F. Navarro, R.G. Bower, R.A. Crain, C. Dalla Vecchia, M. Furlong, A. Jenkins, I.G. McCarthy, Y. Qu, M. Schaller, J. Schaye, T. Theuns, Bent by baryons: the low mass galaxy-halo relation, ArXiv e-prints.

[45] S. Dodelson, L.M. Widrow, Sterile-neutrinos as dark matter, Phys. Rev. Lett. 72 (1994) 17–20, http://dx.doi.org/10.1103/PhysRevLett.72.17, arXiv:hep-ph/9303287.

[46] X.-D. Shi, G.M. Fuller, A new dark matter candidate: nonthermal sterile neutrinos, Phys. Rev. Lett. 82 (1999) 2832–2835, http://dx.doi.org/10.1103/PhysRevLett.82.2832, arXiv:astro-ph/9810076.

[47] K. Abazajian, G.M. Fuller, M. Patel, Sterile neutrino hot, warm, and cold dark matter, Phys. Rev. D 64 (2001) 023501, http://dx.doi.org/10.1103/PhysRevD.64.023501.

[48] M. Laine, M. Shaposhnikov, Sterile neutrino dark matter as a consequence of nuMSM-induced lepton asymmetry, J. Cosmol. Astropart. Phys. 0806 (2008) 031, http://dx.doi.org/10.1088/1475-7516/2008/06/031, arXiv:0804.4543.

[49] A. Boyarsky, J. Lesgourgues, O. Ruchayskiy, M. Viel, Realistic sterile neutrino dark matter with keV mass does not contradict cosmological bounds, Phys. Rev. Lett. 102 (2009) 201304, http://dx.doi.org/10.1103/PhysRevLett.102.201304, arXiv:0812.3256.

[50] A. Boyarsky, O. Ruchayskiy, M. Shaposhnikov, The role of sterile neutrinos in cosmology and astrophysics, Annu. Rev. Nucl. Part. Sci. 59 (2009) 191–214, http://dx.doi.org/10.1146/annurev.nucl.010909.083654, arXiv:0901.0011.

[51] E. Bulbul, M. Markevitch, A. Foster, R.K. Smith, M. Loewenstein, S.W. Randall, Detection of an unidentified emission line in the stacked X-ray spectrum of galaxy clusters, Astrophys. J. 789 (2014) 13, http://dx.doi.org/10.1088/0004-637X/789/1/13, arXiv:1402.2301.

[52] A. Boyarsky, O. Ruchayskiy, D. Iakubovskyi, J. Franse, Unidentified line in X-ray spectra of the Andromeda galaxy and Perseus galaxy cluster, Phys. Rev. Lett. 113 (2014) 251301, http://dx.doi.org/10.1103/PhysRevLett.113.251301, arXiv:1402.4119.

[53] A. Boyarsky, J. Franse, D. Iakubovskyi, O. Ruchayskiy, Checking the dark matter origin of 3.53 keV line with the Milky Way center, Phys. Rev. Lett. 115 (16) (2015) 161301, http://dx.doi.org/10.1103/PhysRevLett.115.161301, arXiv:1408.2503.

[54] J. Ghiglieri, M. Laine, Improved determination of sterile neutrino dark matter spectrum, arXiv:1506.06752.

[55] T. Venumadhav, F.-Y. Cyr-Racine, K.N. Abazajian, C.M. Hirata, Sterile neutrino dark matter: a tale of weak interactions in the strong coupling epoch, arXiv:1507.06655.

[56] G.D. Becker, J.S. Bolton, P. Madau, M. Pettini, E.V. Ryan-Weber, B.P. Venemans, Evidence of patchy hydrogen reionization from an extreme Lyα trough below redshift six, Mon. Not. R. Astron. Soc. 447 (2015) 3402, http://dx.doi.org/10.1093/mnras/stu2646, arXiv:1407.4850.

5

Darkness without dark matter and energy – generalized unimodular gravity

A.O. Barvinsky [a,b], A.Yu. Kamenshchik [c,d,*]

[a] Theory Department, Lebedev Physics Institute, Leninsky Prospect 53, Moscow 119991, Russia
[b] Tomsk State University, Department of Physics, Lenin Ave. 36, Tomsk 634050, Russia
[c] Dipartimento di Fisica e Astronomia, Università di Bologna and INFN, Via Irnerio 46, 40126 Bologna, Italy
[d] L.D. Landau Institute for Theoretical Physics of the Russian Academy of Sciences, Kosygin str. 2, 119334 Moscow, Russia

ARTICLE INFO

Editor: M. Trodden

Keywords:
Unimodular gravity
Cosmology
Dark matter

ABSTRACT

We suggest a Lorentz non-invariant generalization of the unimodular gravity theory, which is classically equivalent to general relativity with a locally inert (devoid of local degrees of freedom) perfect fluid having an equation of state with a constant parameter w. For the range of w near -1 this dark fluid can play the role of dark energy, while for $w = 0$ this dark dust admits spatial inhomogeneities and can be interpreted as dark matter. We discuss possible implications of this model in the cosmological initial conditions problem. In particular, this is the extension of known microcanonical density matrix predictions for the initial quantum state of the closed cosmology to the case of spatially open Universe, based on the imitation of the spatial curvature by the dark fluid density. We also briefly discuss quantization of this model necessarily involving the method of gauge systems with reducible constraints and the effect of this method on the treatment of recently! suggested mechanism of vacuum energy sequestering.

1. Introduction

Dark matter and dark energy phenomena form a dark side of modern precision cosmology and, therefore, represent an unprecedentedly rich playground for various modifications of general relativity (GR). Perhaps, conceptually the most interesting versions of these modifications are the ones which do not involve special types of gravitating matter and originate from the purely metric sector of the theory, like local $f(R)$-gravity or nonlocal cosmology models [1,2]. Usually such modifications are equivalent to adding or removing some local degrees of freedom. Even more interesting is the case when a nontrivial modification occurs without changing the balance of local physical variables – darkness arises without dark energy or dark matter constituents. Known examples of such a concept include, in particular, the unimodular (UM) gravity [3–5], the theory of vacuum energy sequestering [6,7], QCD holonomy mechanism of dark energy [8] and others. Unimodular gravity differs from the Einstein GR by the requirement that at the kinematical level the full set of metric coefficients is subject to the restriction of the unit determinant of the metric tensor. Rather anti-intuitive conclusion that this theory has the same number of local degrees of freedom as GR [9] can be explained by the fact that reduction in the number of independent field variables is compensated by the reduction of the local gauge invariance group, and the main effect of the unimodular modification is the origin of one global degree of freedom playing the role of the cosmological constant.

Extension of the physical sector of the theory by a partial violation of gauge invariance is a well-known and rather popular phenomenon. In particular, reduction from Lorentz symmetry to anisotropic scaling invariance in Lifshitz models is very productive in condensed matter theory context [10], while a similar modification in Horava gravity models [11] opens prospects for renormalizable unitarity preserving gravity theories. Other examples can be found in [12,13]. Here we will consider the synthesis of Lorentz violation with the concept of unimodular gravity [3–5]. This generalized UM gravity incorporates Lorentz violation in the definition of the reduced configuration space of metric coefficients – instead of the requirement of a unit metric determinant this theory is based on the metric field satisfying the following constraint

$$N = N(\gamma), \quad \gamma \equiv \det \gamma_{ij}, \tag{1}$$

* Corresponding author.
E-mail addresses: barvin@td.lpi.ru (A.O. Barvinsky), kamenshchik@bo.infn.it (A.Yu. Kamenshchik).

where $N=(-g^{00})^{-1/2}$ is the lapse function and $N(\gamma)$ is some function of γ – the determinant of the spatial metric γ_{ij} in the ADM $(3+1)$-decomposition of metric coefficients $g_{\mu\nu}$,

$$g_{\mu\nu}dx^\mu dx^\nu=(N_i N^i-N^2)\,dt^2+2N_i\,dt\,dx^i+\gamma_{ij}dx^i\,dx^j. \qquad (2)$$

Here $x^\mu=(t,x^i)$, $\mu=0,1,2,3$, $i=1,2,3$ and $N_i=g_{0i}$ is the corresponding shift function.

The motivation for such a generalization of the unimodular gravity is as follows. To begin with, the class of metrics subject to (1) includes the original unimodular theory corresponding to $N(\gamma)=1/\sqrt{\gamma}$. The right hand side of (1) is invariant under spatial rotations, so that this is a minimal breakdown of Lorentz symmetry from $O(1,3)$ to $O(3)$. Another reason to consider it is an interesting fact that at the classical level such a theory effectively incorporates a special type of matter source – dark fluid with a nonlinear (general barotropic) equation of state. Thus it goes beyond a conventional unimodular gravity by generating the perfect fluid characterized not by just vacuum energy with $p=-\varepsilon$, but by a nontrivial pressure as well. Finally, for a simple class of power-like functions $N(\gamma)$ in (1) it generates an equation of state $p=w\varepsilon$ with a constant w and, moreover, in the comoving reference frame of this fluid renders the density and pressure constant both in space and time.[1] Thus, similarly to the original unimodular gravity it can incorporate as a spacetime constant of motion the analogue of dark energy which is free from clustering but has a constant polytropic parameter w different from -1. In the particular case of a pressureless dust with $w=0$, corresponding to $N(\gamma)=\text{const}$, the density of this dust is characterized by a single function of spatial coordinates entirely fixed by the initial conditions, which can be interpreted as a model of inhomogeneous distribution of dark matter similar to the mechanism of mimetic model [14].

Here we analyze this model at the classical level and show that on shell (without extra matter sources) it is equivalent to general relativity with this special type of perfect fluid. Its "darkness" can be intuitively interpreted as the absence of *local* degrees of freedom of this fluid, and its effective manifestation can in principle be either the dark energy or dark matter. Rigorous counting its degrees of freedom, which is important for the quantization of this model, requires the analysis of its local gauge invariance. Usual diffeomorphism invariance is obviously broken by the restriction (1) on metric coefficients, which leads to a preferred spacetime foliation by spacelike hypersurfaces. However, there exist reduced diffeomorphisms which leave the theory locally gauge invariant and turn out to be a generalization of volume preserving diffeomorphisms of the unimodular gravity. We briefly discuss them and show that their origin naturally leads to the theory with reducible (linearly dependent) generators. At the quantum level it is subject to Batalin–Vilkovisky technique [16] which allows one to quantize the theory without explicitly disentangling its physical sector.

We conclude the paper by the discussion of how this model can be used within the initial conditions problem in cosmology. Dark fluid of generalized UM gravity can be used to imitate the effect of spatial curvature. This might extend the predictions of the cosmological density matrix construction [17], which are valid only in the spatially closed model, to the phenomenologically more preferable open model with flat space foliation. Another potential application could be the mechanism of sequestering the back reaction effect of quantum vacuum energy recently suggested as a possible solution of Planckian hierarchy and cosmological constant problems [6,7]. Remarkably, the method of careful treatment of the global physical mode responsible for the locally inert dark fluid is the same as that of the sequestering mechanism – the canonical version of the BV method [16], which might clarify acausality puzzles of this mechanism and extend it to noncompact spacetimes.

[1] Since this model violates general coordinate invariance this property of density and pressure becomes frame dependent.

2. Dark fluid and its generalized unimodular invariance

The simplest way to handle the constraint (1) on metric coefficients is not to explicitly substitute it in the Einstein action, but rather incorporate it into the action with the Lagrange multiplier λ,

$$S=\int d^4x\left\{\frac{M_P^2}{2}\,g^{1/2}R(g)-\lambda\left(\frac{1}{\sqrt{-g^{00}}}-N(\gamma)\right)\right\}. \qquad (3)$$

Varying this action with respect to λ and $g_{\mu\nu}$ one obtains the restriction (1) on the metric and the Einstein equation with the perfect fluid matter stress tensor

$$R^{\mu\nu}-\frac{1}{2}\,g^{\mu\nu}R=\frac{1}{M_P^2}\,T^{\mu\nu}, \qquad (4)$$

$$\begin{aligned}T^{\mu\nu}&\equiv-\frac{2}{g^{1/2}}\frac{\delta}{\delta g_{\mu\nu}}\int d^4x\,\lambda\left(\frac{1}{\sqrt{-g^{00}}}-N(\gamma)\right)\\&=\varepsilon\,u^\mu u^\nu+p\big(g^{\mu\nu}+u^\mu u^\nu\big), \qquad (5)\end{aligned}$$

where the four-velocity $u^\mu=-g^{\mu0}N$ is a future pointing vector normal to spacelike hypersurfaces of the ADM foliation (2), and its energy density and pressure read

$$\varepsilon=\frac{\lambda}{2\sqrt{\gamma}},\quad p=\frac{\lambda}{\sqrt{\gamma}}\left(\frac{\gamma}{N}\frac{dN}{d\gamma}\right). \qquad (6)$$

Thus, this dark fluid satisfies the equation of state $p=w\varepsilon$ with a generally nonconstant parameter $w=w(\gamma)$ given by

$$w=2\frac{\gamma}{N}\frac{dN}{d\gamma}=2\frac{d\ln N}{d\ln\gamma}. \qquad (7)$$

Similarly to the UM gravity [3] the generalized unimodularity condition (1) is not invariant under generic diffeomorphisms of the metric – Lie derivatives with respect to the 4-dimensional vector field ξ^μ which in the $(3+1)$-decomposition can be written down as a column,

$$\delta_\xi g^{\mu\nu}=-\nabla^\mu\xi^\nu-\nabla^\nu\xi^\mu,\quad \xi^\mu=\begin{bmatrix}\xi^0\\ \xi^i\end{bmatrix}. \qquad (8)$$

However, this condition remains invariant under reduced diffeomorphisms with respect to the subset of vector fields $\boldsymbol{\xi}^\mu$ satisfying the equation

$$\begin{aligned}\delta_{\boldsymbol{\xi}}\big(N-N(\gamma)\big)\Big|_{N=N(\gamma)}&=N\big[\,\partial_t\boldsymbol{\xi}^0-(1+w)\,N^i\partial_i\boldsymbol{\xi}^0-w\,\partial_i\boldsymbol{\xi}^i\,\big]\\&=0, \qquad (9)\end{aligned}$$

which in the UM gravity case, $w=-1$, obviously reduces to the equation on parameters of volume preserving diffeomorphisms $\partial_\mu\boldsymbol{\xi}^\mu=0$ [3].

With the decomposition of ξ^i into the longitudinal and transverse parts,[2]

$$\boldsymbol{\xi}^i=\sqrt{\gamma}\,(\gamma^{ij}\partial_j\varphi+\xi^i_\perp),\quad \partial_i(\sqrt{\gamma}\,\xi^i_\perp)=0, \qquad (10)$$

[2] Since general diffeomorphism invariance is broken, the transformation properties of φ and $\xi^i_\perp$ are no longer of a scalar and vector type, and the $\sqrt{\gamma}$-factor is added merely for reasons of convenience.

the equation (9) can be solved with respect to φ in terms of the spatially nonlocal Green's function of the Laplacian operator Δ weighted by the function w,

$$\varphi = \frac{1}{w\Delta} D_t \boldsymbol{\xi}^0, \quad \Delta = \partial_i \gamma^{ij} \sqrt{\gamma}\, \partial_j, \quad D_t = \partial_t - (1+w)\, N^i \partial_i. \tag{11}$$

The gauge parameter $\boldsymbol{\xi}^\alpha$ can be represented in terms of a projector $\boldsymbol{\Pi}^\alpha_\beta$ acting on a generic diffeomorphism parameter ξ^β

$$\boldsymbol{\xi}^\alpha = \boldsymbol{\Pi}^\alpha_\beta \xi^\beta, \quad \boldsymbol{\Pi}^\alpha_\beta = \begin{bmatrix} 1 & 0 \\ \sqrt{\gamma}\, \partial^i \frac{1}{w\Delta} D_t & \sqrt{\gamma}\left(\delta^i_j - \partial^i \frac{1}{\Delta} \partial_j\right) \end{bmatrix}, \tag{12}$$

so that the generators

$$\boldsymbol{R}^{\mu\nu}{}_\beta = -2\nabla^{(\mu} \boldsymbol{\Pi}^{\nu)}_\beta \tag{13}$$

of the gauge invariance transformations of the action (3) are not linearly independent. They are annihilated by the zero vector Z_0^β of the projector $\boldsymbol{\Pi}^\alpha_\beta$,

$$\boldsymbol{R}^{\mu\nu}{}_\beta Z_0^\beta = 0, \quad Z_0^\beta = \begin{bmatrix} 0 \\ \sqrt{\gamma}\, \partial^i \end{bmatrix} \tag{14}$$

Thus, this is the gauge theory with reducible generators, which should be subject to the BV technique of [16]. It is important that the generators (13) are nonlocal, and this would present certain difficulties in the framework of the *Lagrangian* quantization which is strongly based on the locality of gauge generators and structure constants. However, this nonlocality is in space rather than in time, so that time locality of the formalism is preserved and, therefore, guarantees applicability of the *canonical* quantization to be implemented in the future [18].[3]

3. Dynamics of dark fluid in the comoving frame

The dynamics of the Lagrange multiplier λ and the corresponding density and pressure is determined from the conservation law for the stress tensor (5)

$$\nabla^\mu T_{\mu\nu} = \nabla^\mu [(\varepsilon + p) u_\mu u_\nu] + \nabla_\nu p = 0, \tag{15}$$

where in the definition of covariant derivatives we interpret ε and p as scalars, that is $\nabla_\nu p = \partial_\nu p$ and $\nabla_\nu \varepsilon = \partial_\nu \varepsilon$, to match with the definition of covariant derivatives acting on Einstein tensor in the l.h.s. of Einstein equation. Since the theory is not invariant with respect to general coordinate transformations the density and pressure are not scalars, and their properties are frame dependent. Three independent diffeomorphisms preserving the condition (1) derived above are sufficient to make a transform to the distinguished comoving frame of the dark fluid. In this frame $u^i \sim g^{0i} = 0$, and the temporal component of (15), $\nabla^\mu T_{\mu\nu} u^\nu \equiv -u^\mu \nabla_\mu \varepsilon - (\varepsilon + p)\nabla_\mu u^\mu = 0$, gives

$$0 = \frac{\dot\varepsilon}{\varepsilon} + (1+w)\frac{\dot\gamma}{2\gamma} = \partial_t \left(\ln\varepsilon + \frac{1}{2}\ln\gamma + \ln N \right), \tag{16}$$

where we took into account that $\nabla_\mu u^\mu = \dot\gamma / 2N\gamma$ and $w\dot\gamma/2\gamma = \partial_t \ln N$. Therefore

$$\varepsilon N \sqrt{\gamma} = S(\mathbf{x}), \tag{17}$$

where $S(\mathbf{x})$ is a time integration constant – some function of spatial coordinates.

Space components of the conservation law (15) give in the same gauge

$$0 = \nabla^\mu T_{\mu i} = \partial_i (w\varepsilon) + \frac{\partial_i N}{N}(1+w)\varepsilon, \tag{18}$$

where we took into account that $u^\mu \nabla_\mu u_i = \partial_i N/N$. For the case of $w \neq 0$, dividing this equation by $w\varepsilon$ we immediately have $\partial_i(\ln w + \ln\varepsilon + \ln\gamma/2 + \ln N) = 0$ in virtue of the relation $\partial_i \ln N/w = \partial_i \ln\gamma/2$, so that

$$w\,\varepsilon N \sqrt{\gamma} = T(t). \tag{19}$$

Combining (17) and (19) together, one finds

$$w = \frac{T(t)}{S(\mathbf{x})}, \tag{20}$$

which means that for a class of models with a constant nonvanishing w both functions also degenerate to constants in space and time,

$$N = \mathrm{const}\,\gamma^{w/2}, \quad \varepsilon = \frac{\mathrm{const}}{\gamma^{(w+1)/2}}, \quad w = \mathrm{const} \neq 0. \tag{21}$$

For the case of the dust with zero w and a constant lapse (originally considered in [19]) only the first term of Eq. (18) remains, so that one nontrivial function of spatial coordinates $S(\mathbf{x})$ still survives

$$N = \mathrm{const}, \quad \varepsilon = \frac{S(\mathbf{x})}{N\sqrt{\gamma}} \equiv \frac{\tilde S(\mathbf{x})}{\sqrt{\gamma}}, \quad w = 0. \tag{22}$$

In fact, these two cases of *dark energy* with a constant w close to -1 and *dark dust* seem to saturate physically reasonable cosmological setups in the generalized UM theory. This follows from a simple observation that a nontrivial function $S(\mathbf{x})$ is obviously a part of initial conditions, but the parameter w is determined by a *kinematical* restriction (1) of the configuration space of the theory and should not depend on its particular initial conditions like (20) unless it is some universal constant.[4]

This can easily be illustrated by a simple example which shows that the attempt to model a fairly generic equation of state $p = p(\varepsilon)$ by an appropriate choice of function $N(\gamma)$ in (1) actually fails. Consider a popular Chaplygin gas model with $p = -A/\varepsilon$. Independently of the unimodular setup, the conservation of its stress tensor, $\dot\varepsilon + (p+\varepsilon)\dot\gamma/2\gamma = 0$, gives a well known relation between the energy density and γ [20,21], $\varepsilon = \sqrt{A + B(\mathbf{x})/\gamma}$, where $B(\mathbf{x})$ is a time integration constant – some function of spatial coordinates. Together with the equations (17) and (19) this relation yields the expression for N in terms of γ, $N = \sqrt{-S(\mathbf{x})\,T(t)}/\sqrt{\gamma}$. According to the assumptions of our generalized unimodular gravity both N and w are the functions of one variable γ, which means that both the ratio (20) and the product of $S(\mathbf{x})$ and $T(t)$ should be the functions of γ. This is possible only when $S(\mathbf{x})$ and $B(\mathbf{x})$ are constant and γ is a function of time, which means that this case, in contrast to the $w = const$ case above, is valid only for a spatially homogeneous model. Similar situation holds for other equations of state with $w \neq const$.

[3] Of course, transition from the canonical path integral to the Lagrangian one will again raise the issue of locality accompanied by the associated issues of renormalizability, etc., but this problem definitely goes beyond the present discussion of the quantization of the model.

[4] Boundary conditions can in principle be incorporated into the Lagrangian as local total derivative terms forming boundary integrals in the action, which is however not the case of (1).

4. Conclusions

Thus, we see that there exists a class of models with a broken Lorentz invariance generalizing unimodular gravity theory, which generate dark fluid with a barotropic equation of state with a constant w. Similarly to UMG the gravitational dynamics of this fluid is characterized by an independent of space and time constant which is fixed by initial conditions. The spacetime rigidity of this constant implies that this fluid does not carry local degrees of freedom, but rather describes a global variable incapable of clustering. Therefore it can play the role of dark energy, especially in view of the fact that the parameter w can occupy a continuous range of values near $w = -1$. For a special case of $w = 0$ the rigidity condition relaxes to one constant in time function of space coordinates $S(\mathbf{x})$, so that this *dark dust* can be interpreted, similarly to mimetic gravity [14], as a candidate for dark matter.

Our work, in fact, suggests a new concept in cosmology and gravity theory which can be called "darkness" designating the general mechanism based not on local degrees of freedom, but rather on global, topological ones, that could underlie the whole bunch of phenomena and their models, including dark energy, Horava gravity theory [11], quantum initial value problem [17,22], cosmological constant sequestering formalism [6,7], etc.

Breakdown of Lorentz invariance is perhaps too high a price for the generation of darkness phenomena in cosmology. However, Lorentz symmetry violation has become very popular in recent years due to the fact that the extension of Lifshitz anisotropic scaling invariance to gravity - Horava gravity models - is a way to recover unitarity in renormalizable higher derivative quantum gravity [10,11]. Moreover, breakdown of Lorentz invariance can be an inalienable feature of cosmological initial conditions. The suggestion of the initial quantum state of the Universe in the form of the microcanonical density matrix [17] implies existence of the distinguished foliation of spacetime by spatial hypersurfaces. This foliation underlies the construction of this initial state density matrix and persists in the further cosmological evolution. Therefore, there is no reason to reject violation of Lorentz symmetry at a deeper kinematical level, like in the condition (1).

The density matrix state [17] is conceptually very attractive because of the minimum set of assumptions underlying it [22] and, moreover, because of a mechanism restricting the cosmological ensemble to subplanckian domain in UV limit and avoiding the IR catastrophe, characteristic of the no-boundary wavefunction. However, it applies only to a closed Universe with a negative contribution $\Omega_K = -K/H^2a^2$ of the positive spatial curvature, $K = +1$, in the full set of cosmological density parameters, $\Omega_K + \Omega_\Lambda + \Omega_m = 1$, where a is a scale factor of the FRW metric, $H = \dot{a}/Na$ is its Hubble factor and $K = \pm 1, 0$ is the sign of the 3-metric curvature scalar respectively for closed, open or spatially flat FRW cosmology. Therefore, even though the density matrix prescription generates good hill-top initial conditions for inflation (at the *maximum* of the inflaton potential) [23], it does not include the case of a spatially flat FRW model, $K = 0$, most natural from the viewpoint of the observational status of inflationary scenario ($\Omega_K = 0.000 \pm 0.005$ according to combined Planck, lensing and BAO data [24]).

Remarkably, the generalized UM model with $w = -1/3$ can imitate the effect of positive/negative spatial curvature in the Friedmann equation with $K = \pm 1$, $\gamma \sim a^3$, provided the integration constant in the expression (21) for a dark fluid density ε is negative/positive. Under a proper normalization of the flat space scale factor a the dark fluid density becomes $\varepsilon_K = -3M_P^2 K/a^2$ and fully imitates the spatial curvature contribution $\Omega_K = \varepsilon_K/3M_P^2 H^2$ to the *flat space* Friedmann equation

$$H^2 = \frac{\varepsilon_m + \varepsilon_\Lambda + \varepsilon_K}{3M_P^2}. \tag{23}$$

This would allow one to extend the conclusions of [17] to FRW models in the flat and even hyperbolic space foliations, and this is one of the motivations for our generalized UM gravity model.

What underlies this phenomenon, which as we see can effectively change even the space topology [25], is a global degree of freedom encoded at the level of the Lagrangian formalism in the integration constant. Like in a conventional unimodular gravity the mechanism of this is based on a subtle interplay of physical and gauge degrees of freedom - in the generalized version it is technically more involved, but conceptually similar to the original unimodular case. A similar mechanism due to the interplay of conformal invariance and field reparametrization can be observed in the mimetic gravity theory [14], though the latter incorporates a new local (dust matter) degree of freedom [15], whereas in our case this is the global topological variable parameterizing the dark fluid.

It should be emphasized that our generalized model is not a gauge fixed version of general relativity. In UM gravity the cosmological constant Λ is incorporated as an integration constant of equations of motion and this makes a great conceptual difference from GR with a given Λ. A similar situation happens here, but the integration "constant" is much richer - this is the perfect fluid stress tensor without local degrees of freedom.

Here we analyzed the generalized UM gravity at the classical level. At the quantum level its global mode should either be disentangled explicitly or treated within the quantization method for constrained systems. In either case rigorous quantization requires the construction of the canonical formalism. As is known, UM gravity in this formalism [9] has instead of the GR Hamiltonian constraint the vanishing of the spatial gradient of this constraint, which eventually results in a freely chosen value of Λ as an integration constant. As will be shown in a forthcoming paper [18], a similar but more involved constraint appears here. At the Lagrangian level this is a conservation of perfect fluid stress tensor leading to the rigidity of its energy density and pressure, which can be interpreted as the absence of clustering of dark energy (or, in a particular case of a zero pressure, as dark matter).

At the quantum level, especially in the transition from the canonical to the Lagrangian quantization, the situation becomes nontrivial because linear dependence of the gauge invariance generators (14) implies reducibility of the first class constraints of the canonical formalism, which is subject to BV formalism for systems with linearly dependent generators [16]. Additional difficulty is that this reducibility is of a spatially nonlocal nature because of nonlocal generators (13).

Treatment of this problem was endeavored in [9,26] and has lead to a special procedure of averaging over 3-dimensional space - the counterpart to the analogous *spacetime* averaging in the vacuum energy sequestering mechanism of [6,7].[5] Weak point in this averaging procedure is an ad hoc choice of the integration measure. In particular, it fails to be well defined in noncompact asymptotically flat spacetimes. Moreover, physical predictions of [6,7,9,26] depend on this measure, whereas the freedom in its choice should be physically irrelevant because it reflects invariance of the BV quantization scheme under the change of the basis of gauge generators (13) or canonical constraints. Careful analysis of this problem will be a subject of our future work [18].

[5] It should be emphasized that this mechanism, which is an interesting part of solution of hierarchy and cosmological constant problems, can also be generalized in a Lorentz non-invariant way, what can be done by a covariantization analogous to the covariant formulation of UM gravity [4,9] - parametrization of the distinguished spacetime foliation by an auxiliary antisymmetric tensor or vector density. This, however, will have to be achieved by parameterizing all 4-dimensional coordinates in terms of four embedding functions [18].

This analysis should, perhaps, resolve the conundrum of nonlocality and acausality in sequestering mechanism of [6], change the conclusions on spacetime compactness in the epoch of transient cosmological expansion [7] and, thus, extend cosmological applications to spatially open models.

Acknowledgements

This work was supported by the RFBR grant No. 17-02-00651. The work of A.O.B. was also supported by the Tomsk State University Competitiveness Improvement Program.

References

[1] A.A. Starobinsky, Phys. Lett. B 91 (1980) 99;
A.A. Starobinsky, JETP Lett. 86 (2007) 157.
[2] S. Deser, R.P. Woodard, Phys. Rev. Lett. 99 (2007) 111301.
[3] J.J. van der Bij, H. van Dam, Y.J. Ng, Physica A 116 (1982) 307.
[4] M. Henneaux, C. Teitelboim, Phys. Lett. B 222 (1989) 195.
[5] W.G. Unruh, Phys. Rev. D 40 (1989) 1048;
K.V. Kuchar, Phys. Rev. D 43 (1991) 3332;
G.F.R. Ellis, H. van Elst, J. Murugan, J.P. Uzan, Class. Quantum Gravity 28 (2011) 225007.
[6] N. Kaloper, A. Padilla, Phys. Rev. Lett. 112 (2014) 091304;
N. Kaloper, A. Padilla, Phys. Rev. Lett. 118 (2017) 061303;
N. Kaloper, A. Padilla, D. Stefanyszyn, G. Zahariade, Phys. Rev. Lett. 116 (2016) 051302.
[7] N. Kaloper, A. Padilla, Phys. Rev. D 90 (2014) 084023.
[8] A.R. Zhitnitsky, Phys. Rev. D 89 (6) (2014) 063529;
A.R. Zhitnitsky, Phys. Rev. D 90 (4) (2014) 043504;
A.R. Zhitnitsky, Phys. Rev. D 92 (4) (2015) 043512.
[9] R. Bufalo, M. Oksanen, A. Tureanu, Eur. Phys. J. C 75 (2015) 477.
[10] E.M. Lifshitz, Zh. Eksp. Teor. Fiz. 11 (1941) 255–269;
E.H. Fradkin, Front. Phys. 82 (2013) 1.
[11] P. Horava, Phys. Rev. D 79 (2009) 084008;
A.O. Barvinsky, D. Blas, M. Herrero-Valea, S.M. Sibiryakov, C.F. Steinwachs, Phys. Rev. D 93 (6) (2016) 064022.
[12] R. Righi, G. Venturi, Nuovo Cimento A 43 (1978) 145;
R. Righi, G. Venturi, Nuovo Cimento A 52 (1979) 166;
R. Righi, G. Venturi, Nuovo Cimento A 52 (1979) 511;
R. Righi, G. Venturi, Lett. Nuovo Cimento 31 (1981) 487.
[13] A.M. Akhmeteli, Int. J. Quantum Inf. 9 (2011) S17;
A.M. Akhmeteli, J. Math. Phys. 52 (2011) 082303;
A.M. Akhmeteli, Eur. Phys. J. C 73 (2013) 2371.
[14] A.H. Chamseddine, V. Mukhanov, J. High Energy Phys. 11 (2013) 135.
[15] A. Barvinsky, J. Cosmol. Astropart. Phys. 1401 (2014) 014, arXiv:1311.3111.
[16] I.A. Batalin, G.A. Vilkovisky, Phys. Lett. B 102 (1981) 27;
I.A. Batalin, G.A. Vilkovisky, Phys. Lett. B 120 (1983) 166;
I.A. Batalin, G.A. Vilkovisky, Phys. Rev. D 28 (1983) 2567;
I.A. Batalin, G.A. Vilkovisky, Phys. Rev. D 30 (1984) 508 (Erratum);
I.A. Batalin, E.S. Fradkin, Lett. Nuovo Cimento 38 (1983) 393.
[17] A.O. Barvinsky, A.Yu. Kamenshchik, J. Cosmol. Astropart. Phys. 09 (2006) 014.
[18] A. Barvinsky, A.Yu. Kamenshchik, work in progress.
[19] D.E. Burlankov, Physics of Space and Time, 2014, Nizhny Novgorod (in Russian).
[20] A.Y. Kamenshchik, U. Moschella, V. Pasquier, Phys. Lett. B 511 (2001) 265.
[21] N. Bilic, G.B. Tupper, R.D. Viollier, Phys. Lett. B 535 (2002) 17.
[22] A.O. Barvinsky, Phys. Rev. Lett. 99 (2007) 071301.
[23] A.O. Barvinsky, A.Yu. Kamenshchik, D.V. Nesterov, J. Cosmol. Astropart. Phys. 01 (2016) 036;
A.O. Barvinsky, A.Yu. Kamenshchik, D.V. Nesterov, Eur. Phys. J. C 75 (12) (2015) 584.
[24] P.A.R. Ade, et al., Astron. Astrophys. 594 (2016) A13.
[25] A.Y. Kamenshchik, I.M. Khalatnikov, Int. J. Mod. Phys. D 21 (2012) 1250004.
[26] R. Bufalo, J. Klusoň, M. Oksanen, Phys. Rev. D 94 (2016) 044005.

6

Doomsdays in a modified theory of gravity: A classical and a quantum approach

Imanol Albarran [a,b], Mariam Bouhmadi-López [c,d,*], Che-Yu Chen [e,f], Pisin Chen [e,f,g]

[a] Departamento de Física, Universidade da Beira Interior, Rua Marquês D'Ávila e Bolama, 6201-001 Covilhã, Portugal
[b] Centro de Matemática e Aplicações da Universidade da Beira Interior (CMA-UBI), Rua Marquês D'Ávila e Bolama, 6201-001 Covilhã, Portugal
[c] Department of Theoretical Physics, University of the Basque Country UPV/EHU, P.O. Box 644, 48080 Bilbao, Spain
[d] IKERBASQUE, Basque Foundation for Science, 48011, Bilbao, Spain
[e] Department of Physics and Center for Theoretical Sciences, National Taiwan University, Taipei, 10617, Taiwan
[f] LeCosPA, National Taiwan University, Taipei, 10617, Taiwan
[g] Kavli Institute for Particle Astrophysics and Cosmology, SLAC National Accelerator Laboratory, Stanford University, Stanford, CA 94305, USA

ARTICLE INFO

Editor: J. Hisano

Keywords:
Quantum cosmology
Modified theories of gravity
Dark energy doomsdays
Palatini type of theories

ABSTRACT

By far cosmology is one of the most exciting subject to study, even more so with the current bulk of observations we have at hand. These observations might indicate different kinds of doomsdays, if dark energy follows certain patterns. Two of these doomsdays are the Little Rip (LR) and Little Sibling of the Big Rip (LSBR). In this work, aside from proving the unavoidability of the LR and LSBR in the Eddington-inspired-Born-Infeld (EiBI) scenario, we carry out a quantum analysis of the EiBI theory with a matter field, which, from a classical point of view would inevitably lead to a universe that ends with either LR or LSBR. Based on a modified Wheeler–DeWitt equation, we demonstrate that such fatal endings seems to be avoidable.

1. Introduction

The scrutiny of extensions on General Relativity (GR) is a well motivated topic in cosmology. Some phenomena, such as the current red accelerating expansion of the universe or gravitational singularities like the big bang, would presage extensions of GR in the infra-red as well as in the ultra-violet limits. Among these extensions, the EiBI theory [1], which is constructed on a Palatini formalism, is an appealing model in the sense that it is inspired by the Born-Infeld electrodynamics [2] and the big bang singularity can be removed through a regular stage with a finite physical curvature [1]. Various important issues of the EiBI theory have been addressed such as cosmological solutions [3–9], compact objects [10–15], cosmological perturbations [16–18], parameter constraints [19–21], and the quantization of the theory [22,23]. However, some possible drawbacks of the theory were discovered in Ref. [24]. Finally, some interesting generalizations of the theory were proposed in Refs. [25–28].

As is known, the cause of the late time accelerating expansion of the universe can be resorted to phantom dark energy, which violates the null energy condition (at least from a phenomenological point of view) while remains consistent with observations so far. Nonetheless, the phantom energy may induce more cosmological singularities in GR (curvature singularities). In particular there are three kinds of behaviors intrinsic to phantom models, which can be characterized by the behaviors of the scale factor a, the Hubble rate $H=\dot{a}/a$, and its cosmic derivatives $\dot{H}$ near the singular points: (a) The big rip singularity (BR) happens at a finite cosmic time t when $a\to\infty$, $H\to\infty$, and $\dot{H}\to\infty$ [29–38], (b) the LR happens at $t\to\infty$ when $a\to\infty$, $H\to\infty$ and $\dot{H}\to\infty$ [39–47], (c) the LSBR happens at $t\to\infty$ when $a\to\infty$, $H\to\infty$, while $\dot{H}\to$ constant [48–50]. All these three scenarios would lead to the universe to rip itself as all the structures in the universe would be destroyed no matter what kind of binding energy is involved.

Interestingly, even though the EiBI theory can cure the big bang, in Refs. [5,6] it was found that the BR and LR are unavoidable in the EiBI setup, hinting that the EiBI theory is still not complete and some quantum treatments near these singular events may be necessary. In this paper, we will extend the investigations in Ref. [22] where we showed that the BR in the EiBI phantom model is expected to be cured in the context of quantum geometrodynamics.

* Corresponding author.
E-mail addresses: imanol@ubi.pt (I. Albarran), mariam.bouhmadi@ehu.eus (M. Bouhmadi-López), b97202056@gmail.com (C.-Y. Chen), pisinchen@phys.ntu.edu.tw (P. Chen).

We will carry an analysis to encompass the rest of truly phantom dark energy abrupt events; i.e. the LR and LSBR.

2. The EiBI model: the LR and LSBR

The EiBI action proposed in [1] is (from now on, we assume $8\pi G = c = 1$)

$$\mathcal{S}_{EiBI} = \frac{2}{\kappa}\int d^4x\Big[\sqrt{|g_{\mu\nu}+\kappa R_{\mu\nu}(\Gamma)|} - \lambda\sqrt{-g}\Big] + S_m(g), \quad (1)$$

where $|g_{\mu\nu}+\kappa R_{\mu\nu}|$ is the determinant of the tensor $g_{\mu\nu}+\kappa R_{\mu\nu}$. The parameter κ, which characterizes the theory, is assumed to be positive to avoid the imaginary effective sound speed instabilities usually associated with a negative κ [20] and λ is related to the effective cosmological constant. S_m is the matter Lagrangian. The field equations are obtained by varying (1) with respect to $g_{\mu\nu}$ and the connection Γ. In a flat, homogeneous and isotropic (FLRW) universe filled with a perfect fluid whose energy density and pressure are ρ and p, respectively, the Friedmann equations of the physical metric $g_{\mu\nu}$ and of the auxiliary metric compatible with Γ are [6]

$$\kappa H^2 = \frac{8}{3}\Big[\bar\rho + 3\bar p - 2 + 2\sqrt{(1+\bar\rho)(1-\bar p)^3}\,\Big] \times \frac{(1+\bar\rho)(1-\bar p)^2}{[(1-\bar p)(4+\bar\rho-3\bar p)+3\frac{d\bar p}{d\bar\rho}(1+\bar\rho)(\bar\rho+\bar p)]^2}, \quad (2)$$

and

$$\kappa H_q^2 = \kappa\Big(\frac{1}{b}\frac{db}{d\tilde t}\Big)^2 = \frac{1}{3} + \frac{\bar\rho+3\bar p-2}{6\sqrt{(1+\bar\rho)(1-\bar p)^3}}, \quad (3)$$

where $\bar\rho \equiv \kappa\rho$ and $\bar p \equiv \kappa p$.[1] On the above equations a and b are the scale factor of the physical and auxiliary metrics, respectively. $\tilde t$ is a rescaled time such that the auxiliary metric can be written in a FLRW form.

In GR, the LR and LSBR can be driven (separately) by two phantom energy models as follows [44,48]

$$p_{LR} = -\rho_{LR} - A_{LR}\sqrt{\rho_{LR}}\,, \qquad p_{LSBR} = -\rho_{LSBR} - A_{LSBR}\,,$$

where A_{LR} and A_{LSBR} are positive constants. Therefore,

$$\frac{\rho_{LR}}{\rho_0} = \Big(\frac{3A_{LR}}{2\sqrt{\rho_0}}\ln(a/a_0)+1\Big)^2,$$
$$\rho_{LSBR} = 3A_{LSBR}\ln(a/a_0) + \rho_0\,, \quad (4)$$

where we take $\rho_{LR} = \rho_{LSBR} = \rho_0$ when $a = a_0$ [44,48]. The abrupt events happen at an infinite future where a and ρ diverge. Inserting these phantom energy contents into the EiBI model, i.e., Eqs. (2) and (3), and considering the large a limit (for ρ given in Eqs. (4)), we have

$$\kappa H^2 \approx \frac{\bar\rho}{3} \to \infty\,, \qquad \kappa H_q^2 \approx \frac{1}{3}\,, \quad (5)$$

and

$$\dot H \approx \begin{cases} \dfrac{A_{LR}}{2}\sqrt{\rho_{LR}}\,, & \text{LR} \\ \dfrac{A_{LSBR}}{2}\,, & \text{LSBR} \end{cases} \quad (6)$$

for these two phantom energy models. Therefore, the LR and LSBR of the physical metric are unavoidable within the EiBI model while the auxiliary metric behaves as a de-Sitter phase at late time.

3. The EiBI quantum geometrodynamics: the LR and LSBR minisuperspace model

The deduction of the WDW equation of the EiBI model is based on the construction of a classical Hamiltonian that is promoted to a quantum operator. As shown in [22], this can be achieved more straightforwardly by considering an alternative action which is dynamically equivalent to the EiBI action (1):

$$\mathcal{S}_a = \lambda\int d^4x\sqrt{-q}\Big[R(q) - \frac{2\lambda}{\kappa} + \frac{1}{\kappa}\Big(q^{\alpha\beta}g_{\alpha\beta} - 2\sqrt{\frac{g}{q}}\Big)\Big] + S_m(g). \quad (7)$$

In Ref. [8] it has been shown that the field equations obtained by varying the action (7) with respect to $g_{\mu\nu}$ and the auxiliary metric $q_{\mu\nu}$ are the same to those derived from the action (1). Starting from action (7) and inserting the FLRW ansatz, the Lagrangian of this model in which matter field is described by a perfect fluid can be written as (see Ref. [22])

$$\mathcal{L} = \lambda Mb^3\Big[-\frac{6\dot b^2}{M^2b^2} - \frac{2\lambda}{\kappa} + \frac{1}{\kappa}(X^2+3Y^2-2XY^3)\Big] - 2\rho Mb^3XY^3, \quad (8)$$

where $X \equiv N/M$ and $Y \equiv a/b$. N and M are the lapse functions of $g_{\mu\nu}$ and $q_{\mu\nu}$, respectively. Note that ρ is a function of a, i.e., $\rho = \rho(bY)$ and it is given in Eqs. (4).

3.1. *The classical analysis of the Hamiltonian system*

The system described by the Lagrangian $\mathcal{L}$ is a constrained system. The conjugate momenta can be obtained as follows:

$$p_b \equiv \frac{\partial\mathcal{L}}{\partial\dot b} = -\frac{12\lambda b\dot b}{M}, \quad (9)$$
$$p_X \equiv \frac{\partial\mathcal{L}}{\partial\dot X} = 0, \quad (10)$$
$$p_Y \equiv \frac{\partial\mathcal{L}}{\partial\dot Y} = 0, \quad (11)$$
$$p_M \equiv \frac{\partial\mathcal{L}}{\partial\dot M} = 0. \quad (12)$$

Therefore, the system has three primary constraints [52,53]:

$$p_X \sim 0, \quad (13)$$
$$p_Y \sim 0, \quad (14)$$
$$p_M \sim 0, \quad (15)$$

where $\sim$ denotes the weak equality, i.e., equality on the constraint surface. The total Hamiltonian of the system can be defined by [52,53]

$$\mathcal{H}_T = \dot b p_b - \mathcal{L} + \lambda_X p_X + \lambda_Y p_Y + \lambda_M p_M, \quad (16)$$

where λ_X, λ_Y, and λ_M are Lagrangian multipliers associated with each primary constraint. According to the consistent conditions of each primary constraint, i.e., their conservation in time:

[1] Notice that we are dealing with Palatini type of models which are also known as affine models. On these types of theories (cf. the action (1)) there is a metric $g_{\mu\nu}$ and a connection Γ which does not correspond to the Christoffel symbols of the metric. However, it is always possible to define a metric compatible with that connection [51] and this is the metric that we are referring to as the auxiliary metric. The same applies to the action (7) where we denote the auxiliary metric as $q_{\mu\nu}$ and the physical metric $g_{\mu\nu}$. This is the standard and usual nomenclature in Palatini/affine theories.

$[p_X, \mathcal{H}_T] \sim 0$, $[p_Y, \mathcal{H}_T] \sim 0$, and $[p_M, \mathcal{H}_T] \sim 0$, one further obtains three secondary constraints[2] [52,53]:

$$C_X \equiv \lambda X - Y^3(\lambda + \kappa\rho) \sim 0, \tag{17}$$

$$C_Y \equiv 3\lambda - 3XY(\lambda + \kappa\rho) - XY^2 b\kappa\rho' \sim 0, \tag{18}$$

$$C_M \equiv \frac{p_b^2}{24\lambda b} - \frac{2\lambda^2 b^3}{\kappa} + \frac{\lambda}{\kappa} b^3 X^2 + \frac{3\lambda}{\kappa} b^3 Y^2 - \frac{2XY^3 b^3}{\kappa}(\lambda + \kappa\rho) \sim 0. \tag{19}$$

The prime denotes the derivative with respect to $a = bY$. Furthermore, it can be shown that the total Hamiltonian is a constraint of the system:

$$\mathcal{H}_T = -MC_M + \lambda_X p_X + \lambda_Y p_Y + \lambda_M p_M \sim 0. \tag{20}$$

Because the Poisson brackets of the total Hamiltonian with all the constraints should vanish weakly by definition, $\mathcal{H}_T$ is a first class constraint and we will use it to construct the modified WDW equation.

This system has six independent constraints: p_X, p_Y, p_M, C_X, C_Y, and C_M. After calculating their Poisson brackets with each other, we find that except for p_M, which is a first class constraint, the other five constraints are second class [52,53]. The existence of the first class constraint p_M implies a gauge degree of freedom in the system and one can add a gauge fixing condition into the system to make the constraint second class. An appropriate choice of the gauge fixing condition is $M =$ constant and after fixing the gauge, the conservation in time of this gauge fixing condition, i.e., $[M, \mathcal{H}_T] = 0$, implies $\lambda_M = 0$.

3.2. Quantization of the system

To construct the WDW equation, we impose the first class constraint $\mathcal{H}_T$ as a restriction on the Hilbert space where the wave function of the universe $|\Psi\rangle$ is defined, $\hat{\mathcal{H}}_T|\Psi\rangle = 0$. The hat denotes the operator. The remaining constraints $\chi_i = \{M, p_M, p_X, p_Y, C_X, C_Y\}$ are all second class and we need to consider the Dirac brackets to construct the commutation relations and promote the phase space functions to operators [53]. Note that C_M can be used to construct the first class constraint $\mathcal{H}_T$, i.e., Eq. (20), so it is excluded from the set χ_i.

The Dirac bracket of two phase space functions F and G are defined by [53]

$$[F, G]_D \equiv [F, G] - [F, \chi_i]\Delta_{ij}[\chi_j, G], \tag{21}$$

where Δ_{ij} is the matrix satisfying

$$\Delta_{ij}[\chi_j, \chi_k] = \delta_{ik}. \tag{22}$$

The existence of the matrix Δ_{ij} is proven in Dirac's lecture [53].

According to Ref. [53], the second class constraints can be treated as zero operators after promoting them to quantum operators as long as the Dirac brackets are used to construct the commutation relations:

$$[\hat{F}, \hat{G}] = i\hbar[F, G]_{D,\ (F=\hat{F},\ G=\hat{G})}. \tag{23}$$

[2] We remind that the Poisson bracket is defined as

$$[F, G] = \frac{\partial F}{\partial q_i}\frac{\partial G}{\partial p_i} - \frac{\partial F}{\partial p_i}\frac{\partial G}{\partial q_i},$$

where q_i are the variables and p_i their conjugate momenta. Notice that the repeating suffices denote the summation.

This is due to the fact that the Dirac brackets of the constraints χ_i with any phase space function vanish strongly (they vanish without inserting any constraint). After some calculations, the Dirac brackets between the fundamental variables take the forms

$$\begin{aligned} &[b, p_b]_D = [b, p_b] = 1, \\ &[b, X]_D = 0, \\ &[b, Y]_D = 0, \\ &[X, Y]_D = 0, \\ &[X, p_b]_D = f_1(X, Y, b) = f_1(b), \\ &[Y, p_b]_D = f_2(X, Y, b) = f_2(b), \end{aligned} \tag{24}$$

where f_1 and f_2 are two non-vanishing functions. Notice that f_1 and f_2 can be written as functions of b because it is legitimate to insert the constraints C_X and C_Y to replace X and Y with b when calculating the Dirac brackets.

On the XYb basis, if we define

$$\begin{aligned} \langle XYb|\hat{b}|\Psi\rangle &= b\langle XYb|\Psi\rangle, \\ \langle XYb|\hat{X}|\Psi\rangle &= X\langle XYb|\Psi\rangle, \\ \langle XYb|\hat{Y}|\Psi\rangle &= Y\langle XYb|\Psi\rangle, \\ \langle XYb|\hat{p_b}|\Psi\rangle &= -i\hbar\frac{\partial}{\partial b}\langle XYb|\Psi\rangle \\ &\quad - f_1\frac{\partial}{\partial X}\langle XYb|\Psi\rangle - f_2\frac{\partial}{\partial Y}\langle XYb|\Psi\rangle, \end{aligned} \tag{25}$$

it can be shown that the resulting commutation relations satisfy Eqs. (23) and (24). Furthermore, the momentum operator $\hat{p_b}$ can be written as

$$\langle \xi\zeta b|\hat{p_b}|\Psi\rangle = -i\hbar\frac{\partial}{\partial b}\langle \xi\zeta b|\Psi\rangle, \tag{26}$$

after an appropriate redefinition of the wave functions: $\langle XYb| \to \langle \xi(X, Y, b), \zeta(X, Y, b), b|$. Therefore, in the new $\xi\zeta b$ basis, the modified WDW equation $\langle \xi\zeta b|\hat{\mathcal{H}}_T|\Psi\rangle = 0$ can be written as

$$\frac{-1}{24\lambda}\langle \xi\zeta b|\frac{\hat{p_b}^2}{b}|\Psi\rangle + V(b)\langle \xi\zeta b|\Psi\rangle = 0, \tag{27}$$

where the term containing $\hat{p_b}^2$ is determined by Eq. (26) and its explicit form depends on the factor orderings. Note that the eigenvalues X and Y can be written as functions of b according to the constraints C_X and C_Y, hence it leads to the potential $V(b)$ as follows

$$V(b) = \frac{2\lambda^2 b^3}{\kappa} + \frac{\lambda}{\kappa} b^3 X(b)^2 - \frac{3\lambda}{\kappa} b^3 Y(b)^2. \tag{28}$$

3.3. Wheeler–DeWitt equation: factor ordering 1

In order to prove that our results are independent of the factor ordering, we make two choices of it. First, we consider $\langle \xi\zeta b|b^3\hat{\mathcal{H}}_T|\Psi\rangle = 0$ and choose the following factor ordering:

$$b^2\hat{p_b}^2 = -\hbar^2\left(b\frac{\partial}{\partial b}\right)\left(b\frac{\partial}{\partial b}\right) = -\hbar^2\left(\frac{\partial}{\partial x}\right)\left(\frac{\partial}{\partial x}\right), \tag{29}$$

where $x = \ln(\sqrt{\lambda}b)$. Near the LR singular event, the energy density ρ behaves as $\rho \propto (\ln a)^2$. On that regime, the dependence between the auxiliary scale factor b and a is $b \propto a\ln a$. On the other hand, near the LSBR event the energy density behaves as $\rho \propto \ln a$ and b behaves as $b \propto a\sqrt{\ln a}$. For both cases, the WDW equation can be written as

$$\left(\frac{d^2}{dx^2}+\frac{48}{\kappa\hbar^2}e^{6x}\right)\Psi(x)=0, \tag{30}$$

when x and a go to infinity. Note that we have replaced the partial derivatives with ordinary derivatives and $\Psi(x)\equiv\langle\xi\zeta b|\Psi\rangle$. The wave function reads [55]

$$\Psi(x)=C_1J_0(A_1e^{3x})+C_2Y_0(A_1e^{3x}), \tag{31}$$

and consequently when $x\to\infty$, its asymptotic behavior reads [55]

$$\Psi(x)\approx\sqrt{\frac{2}{\pi A_1}}e^{-3x/2}\Big[C_1\cos\left(A_1e^{3x}-\frac{\pi}{4}\right)+C_2\sin\left(A_1e^{3x}-\frac{\pi}{4}\right)\Big], \tag{32}$$

where

$$A_1\equiv\frac{4}{\sqrt{3\kappa\hbar^2}}. \tag{33}$$

Here $J_\nu(x)$ and $Y_\nu(x)$ are Bessel function of the first kind and second kind, respectively. It can be seen that the wave function vanishes when a and x go to infinity.

3.4. Wheeler–DeWitt equation: factor ordering 2

From the WDW equation (27), we can as well derive a quantum Hamiltonian by choosing another factor ordering

$$\frac{\hat{p_b}^2}{b}=-\hbar^2\left(\frac{1}{\sqrt{b}}\frac{\partial}{\partial b}\right)\left(\frac{1}{\sqrt{b}}\frac{\partial}{\partial b}\right). \tag{34}$$

Before proceeding further, we highlight that this quantization is based on the Laplace–Beltrami operator which is the Laplacian operator in minisuperspace [54]. This operator depends on the number of degrees of freedom involved. For the case of a single degree of freedom, it can be written as in Eq. (34).

Under this factor ordering and after introducing a new variable $y\equiv(\sqrt{\lambda}b)^{3/2}$, in both cases (LR and LSBR) the WDW equation can be written as

$$\left(\frac{d^2}{dy^2}+\frac{64}{3\kappa\hbar^2}y^2\right)\Psi(y)=0, \tag{35}$$

when a and y approach infinity. The solution of the previous equation reads [55]

$$\Psi(y)=C_1\sqrt{y}J_{1/4}(A_1y^2)+C_2\sqrt{y}Y_{1/4}(A_1y^2), \tag{36}$$

and when $y\to\infty$, therefore, [55]

$$\Psi(y)\approx\sqrt{\frac{2}{\pi A_1 y}}\Big[C_1\cos\left(A_1y^2-\frac{3\pi}{8}\right)+C_2\sin\left(A_1y^2-\frac{3\pi}{8}\right)\Big]. \tag{37}$$

Consequently, the wave functions approach zero when a goes to infinity. According to the DeWitt criterium for singularity avoidance [56], the LR and LSBR is expected to be avoided independently of the factor orderings considered in this work.

3.5. Expected values

We have shown that the DeWitt criterium of singularity avoidance is fulfilled hinting that the universe would escape the LR and LSBR in the EiBI model once the quantum effects are important. We next estimate the expected value of the scale factor of the universe a by estimating the expected value of b. The reason we have to resort to the expected value of b rather than a is that in the classical theory [8] that we have quantized the dynamics is only endowed to the scale factor b. We remind at this regard that when approaching the LR and LSBR, $b\propto a\ln a$ and $b\propto a\sqrt{\ln a}$, respectively, at least within the classical framework. Therefore if the expected value of b, which we will denote as **b**, is finite, then we expect that the expected value of a; i.e. **a** would be finite as well. Therefore, none of the cosmological and geometrical divergences present at the LR and LSBR would take place.

We next present a rough estimation for an upper limit of **b** for the two quantization procedures presented on the previous subsection.

- Factor ordering I:
  The expected value of b at late-time can be estimated as follows:

$$\mathbf{b}=\int_{x_1}^{\infty}\Psi^*(x)\frac{e^x}{\sqrt{\lambda}}\Psi(x)\,dx, \tag{38}$$

  where x_1 is large enough to ensure the validity of the approximated potential in (30), i.e., $\delta\to0$. In this limit, we can use the asymptotic behavior for the wave function cf. Eq. (32). Then, it can be shown that the approximated value of **b** is bounded as

$$\int_{x_1}^{\infty}\Psi^*(x)\frac{e^x}{\sqrt{\lambda}}\Psi(x)\,dx<\frac{|C_1|^2+|C_2|^2}{\pi A_1\sqrt{\lambda}}e^{-2x_1}. \tag{39}$$

  Therefore, we can conclude that **b** has an upper finite limit. Consequently, the LR and LSBR are avoided.
- Factor ordering II:
  In this case the expected value of b can be written as

$$\mathbf{b}=\int_{y_1}^{\infty}\Psi^*(y)\frac{y^{\frac{2}{3}}}{\sqrt{\lambda}}\Psi(y)\,f(y)\,dy, \tag{40}$$

  where y_1 is large enough to ensure the validity of the approximated potential in (35), i.e., $\eta\to0$. In addition, we have introduced a phenomenological weight $f(y)$ such that the norm of the wave function is well defined and finite for large y [57–59]. In fact, we could as well choose $f(y)=y^{-\alpha}$ with $2/3<\alpha$. After some simple algebra, we obtain

$$\mathbf{b}<\frac{2\left(|C_1|^2+|C_2|^2\right)}{\pi A_1\sqrt{\lambda}}\int_{y_1}^{\infty}y^{-\frac{1}{3}}f(y)\,. \tag{41}$$

  Consequently, we get

$$\mathbf{b}<\frac{2\left(|C_1|^2+|C_2|^2\right)}{\pi A_1\sqrt{\lambda}(\alpha-2/3)}y_1^{\frac{2}{3}-\alpha}. \tag{42}$$

  Once again, we reach the conclusion that **b** is finite. Therefore, the LR and LSBR are avoided.

4. Conclusions

Singularities seem inevitable in most theories of gravity. It is therefore natural to ask whether by including quantum effects would the singularities be removed. In the case of the EiBI scenario, while the big bang singularity can be removed, the intrinsic phantom dark energy doomsday remains inevitable [6]. We solved the modified Wheeler–DeWitt equation of the EiBI model for a homogeneous and isotropic universe whose matter content corresponds to two kinds of perfect fluid. Those fluids within a classical universe would unavoidably induce LR or LSBR. We show that within the quantum approach we invoked, the DeWitt criterion is fulfilled and it leads toward the potential avoidance of the LR and LSBR. Our conclusion appears unaffected by the choice of factor ordering.

Acknowledgements

The work of IA was supported by a Santander-Totta fellowship "Bolsas de Investigação Faculdade de Ciências (UBI) – Santander Totta". The work of MBL is supported by the Basque Foundation of Science Ikerbasque. She also wishes to acknowledge the partial support from the Basque Government Grant No. IT956-16 (Spain) and FONDOS FEDER under grant FIS2014-57956-P (Spanish government). This research work is supported partially by the Portuguese grand UID/MAT/00212/2013. CYC and PC are supported by Taiwan National Science Council under Project No. NSC 97-2112-M-002-026-MY3 and by Leung Center for Cosmology and Particle Astrophysics, National Taiwan University.

References

[1] M. Bañados, P.G. Ferreira, Phys. Rev. Lett. 105 (2010) 011101, Erratum: Phys. Rev. Lett. 113 (11) (2014) 119901.
[2] M. Born, L. Infeld, Proc. R. Soc. Lond. A 144 (1934) 425.
[3] J.H.C. Scargill, M. Bañados, P.G. Ferreira, Phys. Rev. D 86 (2012) 103533.
[4] P.P. Avelino, R.Z. Ferreira, Phys. Rev. D 86 (2012) 041501.
[5] M. Bouhmadi-López, C.Y. Chen, P. Chen, Eur. Phys. J. C 74 (2014) 2802.
[6] M. Bouhmadi-López, C.Y. Chen, P. Chen, Eur. Phys. J. C 75 (2015) 90.
[7] M. Bouhmadi-López, C.Y. Chen, P. Chen, Phys. Rev. D 90 (2014) 123518.
[8] T. Delsate, J. Steinhoff, Phys. Rev. Lett. 109 (2012) 021101.
[9] I. Cho, H.C. Kim, T. Moon, Phys. Rev. Lett. 111 (2013) 071301.
[10] P. Pani, V. Cardoso, T. Delsate, Phys. Rev. Lett. 107 (2011) 031101.
[11] P. Pani, T. Delsate, V. Cardoso, Phys. Rev. D 85 (2012) 084020.
[12] T. Harko, F.S.N. Lobo, M.K. Mak, S.V. Sushkov, Phys. Rev. D 88 (2013) 044032.
[13] Y.H. Sham, L.M. Lin, P.T. Leung, Astrophys. J. 781 (2014) 66.
[14] S.W. Wei, K. Yang, Y.X. Liu, Eur. Phys. J. C 75 (2015) 253, Erratum: Eur. Phys. J. C 75 (2015) 331.
[15] G.J. Olmo, D. Rubiera-Garcia, H. Sanchis-Alepuz, Eur. Phys. J. C 74 (2014) 2804.
[16] C. Escamilla-Rivera, M. Bañados, P.G. Ferreira, Phys. Rev. D 85 (2012) 087302.
[17] K. Yang, X.L. Du, Y.X. Liu, Phys. Rev. D 88 (2013) 124037.
[18] X.L. Du, K. Yang, X.H. Meng, Y.X. Liu, Phys. Rev. D 90 (2014) 044054.
[19] J. Casanellas, P. Pani, I. Lopes, V. Cardoso, Astrophys. J. 745 (2012) 15.
[20] P.P. Avelino, Phys. Rev. D 85 (2012) 104053.
[21] P.P. Avelino, J. Cosmol. Astropart. Phys. 1211 (2012) 022.
[22] M. Bouhmadi-López, C.Y. Chen, J. Cosmol. Astropart. Phys. 1611 (11) (2016) 023.
[23] F. Arroja, C.Y. Chen, P. Chen, D.h. Yeom, arXiv:1612.00674 [gr-qc].
[24] P. Pani, T.P. Sotiriou, Phys. Rev. Lett. 109 (2012) 251102.
[25] A.N. Makarenko, S. Odintsov, G.J. Olmo, Phys. Rev. D 90 (2014) 024066.
[26] S.D. Odintsov, G.J. Olmo, D. Rubiera-Garcia, Phys. Rev. D 90 (2014) 044003.
[27] J. Beltrán Jiménez, L. Heisenberg, G.J. Olmo, J. Cosmol. Astropart. Phys. 1411 (2014) 004.
[28] C.Y. Chen, M. Bouhmadi-López, P. Chen, Eur. Phys. J. C 76 (2016) 40.
[29] A.A. Starobinsky, Gravit. Cosmol. 6 (2000) 157.
[30] R.R. Caldwell, Phys. Lett. B 545 (2002) 23.
[31] R.R. Caldwell, M. Kamionkowski, N.N. Weinberg, Phys. Rev. Lett. 91 (2003) 071301.
[32] S.M. Carroll, M. Hoffman, M. Trodden, Phys. Rev. D 68 (2003) 023509.
[33] L.P. Chimento, R. Lazkoz, Phys. Rev. Lett. 91 (2003) 211301.
[34] M.P. Dąbrowski, T. Stachowiak, M. Szydłowski, Phys. Rev. D 68 (2003) 103519.
[35] P.F. González-Díaz, Phys. Lett. B 586 (2004) 1.
[36] P.F. González-Díaz, Phys. Rev. D 69 (2004) 063522.
[37] M. Bouhmadi-López, Y. Tavakoli, P. Vargas Moniz, J. Cosmol. Astropart. Phys. 1004 (2010) 016.
[38] I. Albarran, M. Bouhmadi-López, J. Cosmol. Astropart. Phys. 1508 (08) (2015) 051.
[39] T. Ruzmaikina, A.A. Ruzmaikin, Sov. Phys. JETP 30 (1970) 372.
[40] S.'i. Nojiri, S.D. Odintsov, S. Tsujikawa, Phys. Rev. D 71 (2005) 063004.
[41] S.'i. Nojiri, S.D. Odintsov, Phys. Rev. D 72 (2005) 023003.
[42] H. Štefančić, Phys. Rev. D 71 (2005) 084024.
[43] M. Bouhmadi-López, Nucl. Phys. B 797 (2008) 78.
[44] P.H. Frampton, K.J. Ludwick, R.J. Scherrer, Phys. Rev. D 84 (2011) 063003.
[45] I. Brevik, E. Elizalde, S.'i. Nojiri, S.D. Odintsov, Phys. Rev. D 84 (2011) 103508.
[46] M. Bouhmadi-López, P. Chen, Y.W. Liu, Eur. Phys. J. C 73 (2013) 2546.
[47] I. Albarran, M. Bouhmadi-López, C. Kiefer, J. Marto, P. Vargas Moniz, Phys. Rev. D 94 (6) (2016) 063536.
[48] M. Bouhmadi-López, A. Errahmani, P. Martín-Moruno, T. Ouali, Y. Tavakoli, Int. J. Mod. Phys. D 24 (10) (2015) 1550078.
[49] I. Albarran, M. Bouhmadi-López, F. Cabral, P. Martín-Moruno, J. Cosmol. Astropart. Phys. 1511 (11) (2015) 044.
[50] J. Morais, M. Bouhmadi-López, K. Sravan Kumar, J. Marto, Y. Tavakoli, Phys. Dark Universe 15 (2017) 7.
[51] R.M. Wald, General Relativity, University of Chicago Press, Chicago, 1984, (I,A).
[52] M. Henneaux, C. Teitelboim, Quantization of Gauge Systems, Princeton University Press, 1992.
[53] P.A.M. Dirac, Lectures on Quantum Mechanics, Yeshiva University, New York, 1964.
[54] C. Kiefer, Quantum Gravity, third edition, Oxford University Press, Oxford, 2012.
[55] M. Abramowitz, I. Stegun, Handbook on Mathematical Functions, Dover, 1980.
[56] B.S. DeWitt, Phys. Rev. 160 (1967) 1113.
[57] A.O. Barvinsky, Phys. Rep. 230 (1993) 237.
[58] A.Y. Kamenshchik, S. Manti, Phys. Rev. D 85 (2012) 123518.
[59] A.O. Barvinsky, A.Y. Kamenshchik, Phys. Rev. D 89 (4) (2014) 043526.

7

Effect of the chameleon scalar field on brane cosmological evolution

Y. Bisabr, F. Ahmadi *

Department of Physics, Shahid Rajaee Teacher Training University, Lavizan, Tehran 16788, Iran

ARTICLE INFO

Editor: H. Peiris

ABSTRACT

We have investigated a brane world model in which the gravitational field in the bulk is described both by a metric tensor and a minimally coupled scalar field. This scalar field is taken to be a chameleon with an appropriate potential function. The scalar field interacts with matter and there is an energy transfer between the two components. We find a late-time asymptotic solution which exhibits late-time accelerating expansion. We also show that the Universe recently crosses the phantom barrier without recourse to any exotic matter. We provide some thermodynamic arguments which constrain both the direction of energy transfer and dynamics of the extra dimension.

1. Introduction

General Relativity has brilliant successes in explaining gravitational phenomena in Solar System. It is also a powerful tool to explain theoretically many observational facts about the Universe such as expansion of the universe, light element abundances and gravitational waves. Despite all the successes, there are also some unresolved problems such as inflation, the cosmological constant problem and the problems associated with the dark sector, i.e., dark matter and dark energy. These problems have motivated people to seek for some modifications of the theory. Among many possibilities, there are models that deal with extra dimensions. Most of these models propose that our four-dimensional world is a hypersurface (or brane) embedded in a higher dimensional space–time (or bulk). The gravitational field propagates into the bulk while matter systems or standard fields are confined to live in the brane. The most well-known model in this context is the model proposed by Randall and Sundrum (RS). In the so-called RSI model [1], they proposed a mechanism to solve the hierarchy problem with use of two branes, while in the RSII model [2] they considered a single brane with a positive tension. In the latter model, the extra-dimension is compactified and a four-dimensional Newtonian gravity is recovered at low energies. The cosmological evolution of such a brane world scenario has been extensively investigated and modifications of the gravitational equations have been studied [3,4].

The basic idea in the brane world models can be extended to scalar-tensor brane models in which gravity in the bulk is described by a five-dimensional spacetime metric together with a scalar field (see for instance [5]). There are different motivations for introducing a bulk scalar field in brane world scenarios. This scalar field may be used to formulate a low-energy effective theory [6] or to address the gauge hierarchy problem [7]. One of the important motivations to introduce such a bulk scalar field is to stabilize the distance between the two branes in the RSI model [8]. Another explored possibility is formulating inflation without an inflaton field on the brane [9]. It is also shown that decaying of the bulk scalar field can lead to entropy production [10].

There is also a recent tendency in the Literature [11] to interpret the scalar field as a chameleon field [12]. In these chameleon brane world models the scalar field interacts with the matter system via the metric tensor and it is assumed that it can be heavy enough in the environment of the laboratory tests so that the local gravity constraints are satisfied. Meanwhile, it can be light enough in the low-density environment to be considered as cosmologically viable. In the present work we will investigate such a gravitational model with the assumption that the scalar field has a minimal coupling with gravity in the bulk. We will focus on the late-time behavior of the Universe and show that even though the scalar field is normal in the sense that its energy–momentum tensor satisfies weak energy condition, it causes the Universe to cross the phantom boundary.

The present work is organized as follows: In section 2, we introduce the chameleon brane world model and derive the field equations. In section 3, we write the field equations for a five-dimensional metric and then induce them on the brane with ap-

* Corresponding author.
E-mail addresses: y-bisabr@srttu.edu (Y. Bisabr), fahmadi@srttu.edu (F. Ahmadi).

propriate boundary conditions. In section 4, we will consider some cosmological aspects of the model. We first show that due to interaction of the scalar field and matter evolution of both corresponding energy densities are modified. One immediate implication of such a modification is an energy transfer between the two components. In our analysis the chameleon actually appears as a normal field satisfying weak energy condition. By finding a late-time asymptotic solution for the field equations we will show that the Universe suffers a late-time accelerating expansion and a recent cross-over from a decelerated to an accelerated phase. In section 5, we provide some thermodynamic arguments for the interaction process. In section 6, we draw our conclusions.

2. The model

We consider the following action[1]

$$S=\frac{1}{2}\int d^5x\sqrt{-g}\left[R-g^{AB}\nabla_A\phi\nabla_B\phi-2V(\phi)\right] + \int d^4xL_m(\psi_m\bar{h}_{\mu\nu}) \tag{1}$$

where the first term is the five-dimensional gravity in the presence of a minimally coupled scalar field ϕ. The second term is the action of some matter fields on the brane which is taken to be coupled to the scalar field via $\bar{h}_{\mu\nu}=A^2(\phi)h_{\mu\nu}$ with $h_{\mu\nu}$ and $\bar{h}_{\mu\nu}{}^2$ being four-dimensional metrics on the brane.

Varying the action with respect to the metric g^{AB}, gives

$$G_{AB}=[T_{AB}|_{bulk}+T_{AB}|_{brane}] \tag{2}$$

where

$$T_{AB}|_{bulk}=\nabla_A\phi\nabla_B\phi-\frac{1}{2}g_{AB}\nabla_C\phi\nabla^C\phi-g_{AB}V(\phi) \tag{3}$$

and

$$T_{AB}|_{brane}=\delta^{\mu}_{A}\delta^{\nu}_{B}\tau_{\mu\nu}\frac{\delta(y)}{b} \tag{4}$$

Here we take $g_{AB}dz^Adz^B=h_{\mu\nu}dx^\mu dx^\nu+b^2(t,y)dy^2$ and $\tau_{\mu\nu}=A^2(\phi)\tau^m_{\mu\nu}$ with $\tau^m_{\mu\nu}=\frac{-2}{\sqrt{-h}}\frac{\delta L_m}{\delta h^{\mu\nu}}$. We will consider $\tau_{\mu\nu}$ as the stress-tensor of a perfect fluid with energy density ρ_b and pressure P_b. Variation of the action with respect to ϕ, leads to

$$\Box\phi-\frac{dV}{d\phi}=-\beta(\phi)T|_{brane} \tag{5}$$

where $\beta(\phi)=\frac{d\ln A(\phi)}{d\phi}$. By applying Bianchi identities to (2), we obtain

$$\nabla_AT^{AB}|_{brane}=-\nabla_AT^{AB}|_{bulk}=\beta(\phi)T|_{brane}\nabla^B\phi \tag{6}$$

3. The brane-world paradigm

We use the five-dimensional metric

$$\begin{aligned} dS^2&=h_{\mu\nu}dx^\mu dx^\nu+b^2(t,y)dy^2\\ &=-\tilde{n}^2(t,y)dt^2+\tilde{a}^2(t,y)[\frac{dr^2}{(1-kr^2)}+r^2(d\theta^2+\sin^2\theta d\phi^2)]\\ &\quad+\tilde{b}^2(t,y)dy^2 \end{aligned} \tag{7}$$

[1] We work in the unit system in which $k_5=1$.

[2] Latin indices denote 5-dimensional components $A, B, ... = 0, .., 4$ while Greek indices run over four-dimensional brane $\mu, \nu, ... = 0, ..., 3$ and y is the coordinate transverse to the brane.

with $k=0,+1,-1$. The metric coefficients are subjected to the conditions

$$\tilde{n}(t,y)|_{brane}=1,\qquad \tilde{a}(t,y)|_{brane}=a(t),\qquad \tilde{b}(t,y)|_{brane}=b(t) \tag{8}$$

with $a(t)$ being the scale factor. To write the bulk field equations in compact form, we define [4]

$$F(t,y)\equiv\frac{(\tilde{a}'\tilde{a})^2}{\tilde{b}^2}-\frac{(\dot{\tilde{a}}\tilde{a})^2}{\tilde{n}^2}-k\tilde{a}^2 \tag{9}$$

where a prime denotes a derivative with respect to y. The (0,0) and (5,5) components of the field equations become

$$F'=\frac{2\tilde{a}'\tilde{a}^3}{3}T^0_0|_{bulk} \tag{10}$$

$$\dot{F}=\frac{2\dot{\tilde{a}}\tilde{a}^3}{3}T^5_5|_{bulk} \tag{11}$$

If we take

$$T^A_B|_{bulk}=diag[-\rho_\phi,P_\phi,P_\phi,P_\phi,P_T] \tag{12}$$

and assume that ϕ and therefore $T^0_0|_{bulk}=-\rho_\phi$ are independent of y, then we can integrate (10) which gives

$$F-\frac{1}{6}\tilde{a}^4T^0_0|_{bulk}+C_1=F+\frac{1}{6}\tilde{a}^4\rho_\phi+C_1=0 \tag{13}$$

where C_1 is a constant of integration. Since ϕ is only time-dependent, we have

$$T^0_0|_{bulk}=\frac{1}{2}\nabla_0\phi\nabla^0\phi-V(\phi) \tag{14}$$

$$T^5_5|_{bulk}=-\frac{1}{2}\nabla_0\phi\nabla^0\phi-V(\phi) \tag{15}$$

This results in $T^0_0|_{bulk}-T^5_5|_{bulk}=\nabla_0\phi\nabla^0\phi$. From time-derivative of (10) and derivative of (11) with respect to y, one then finds

$$\frac{d}{dt}T^0_0|_{bulk}=-\frac{(\frac{d\tilde{a}^4}{dt})(\nabla_0\phi\nabla^0\phi)}{\tilde{a}^4} \tag{16}$$

Using this equation and (13), we arrive at

$$\dot{F}=\frac{2}{3}\dot{\tilde{a}}\tilde{a}^3T^5_5|_{bulk}-\frac{dC_1}{dt} \tag{17}$$

Comparing this with (11), indicates that C_1 is time-independent. From (9) and (13), we can write

$$(\frac{\dot{\tilde{a}}}{\tilde{n}\tilde{a}})^2=\frac{1}{6}\rho_\phi+(\frac{\tilde{a}'}{\tilde{b}\tilde{a}})^2-\frac{K}{\tilde{a}^2}+\frac{C_1}{\tilde{a}^4} \tag{18}$$

For inducing the field equations on the brane, one usually uses the junction conditions. They simply relate the jumps of derivative of the metric across the brane to the stress-energy tensor inside the brane. To do this, we first note that homogeneity and isotropy imply that

$$T^A_B|_{brane}=\frac{\delta(y)}{b}diag[-\rho_b,P_b,P_b,P_b,0] \tag{19}$$

and

$$T^\mu_\nu|_{brane}(x^\alpha,0)=\lim_{\epsilon\to0}\int_{-\frac{\epsilon}{2}}^{\frac{\epsilon}{2}}T^\mu_\nu|_{brane}bdy=\tau^\mu_\nu(x^\alpha) \tag{20}$$

where the energy density ρ_b and pressure P_b are only functions of time. The junction condition is

$$\frac{[\tilde{a}']}{ab} = -\frac{1}{3}\rho_b \tag{21}$$

where $[Q] = Q(0^+) - Q(0^-)$ denotes the jump of function Q across $y = 0$. Assuming the symmetry $y \leftrightarrow -y$, the generalized Friedmann equation becomes

$$H^2 = \frac{1}{6}\rho_\phi + \frac{1}{36}\rho_b^2 + \frac{C_1}{a^4} - \frac{k}{a^2} \tag{22}$$

where $H \equiv \frac{\dot{a}}{a}$ is the Hubble parameter.

4. Cosmological implications

In this section we will consider some cosmological implications of the above model.

4.1. The conservation equation on the brane

We would like to consider a class of solutions of the field equations (2) under the assumption that the metric coefficients in (7) are separable functions of their arguments [13]. In this class, we have

$$\tilde{n}(t, y) = n(y), \qquad \tilde{a}(t, y) = a(t)Y(y), \qquad \tilde{b}(t, y) = b(t) \tag{23}$$

together with $Y(y)|_{brane} = Y(0) = 1$ and $n(y)|_{brane} = n(0) = 1$. From $G_{05} = 0$, it follows that

$$(\frac{n'}{n}) = (1 - s)(\frac{Y'}{Y}), \qquad \frac{\dot{b}}{b} = s\frac{\dot{a}}{a} \tag{24}$$

where s is an arbitrary constant. This leads to a relation between $a(t)$ and $b(t)$, namely $b(t) = C_2 a^s$ with C_2 being a constant of integration.

There is a constraint on the parameter s coming from arguments related to temporal variation of the gravitational coupling. These arguments lead to $(\frac{\dot{G}}{G}) = -sH$ [1] [2].[3] On the other hand, observations on the time variation of G give $\frac{\dot{G}}{G} = gH$, with g being bounded by $|g| \leq 0.1$ [15]. Thus the absolute value of s is constrained to be $|s| \leq 0.1$.

One can use the equation (20) to write (6) on the brane $(y = 0)$

$$\dot{\rho}_\phi + 3H(\omega_\phi + 1)\rho_\phi + \frac{\dot{b}}{b}\dot{\phi}^2 = Q \tag{25}$$

$$\dot{\rho}_b + 3H(\omega_b + 1 + \frac{s}{3})\rho_b = -Q \tag{26}$$

where $Q = \beta(\phi)(3\omega_b - 1)\dot{\phi}\rho_b$. The solution of the latter is

$$\rho_b = \rho_{0b} a^{-3(\omega_b + 1 + \frac{s}{3})} e^{(1-3\omega_b)\int \beta d\phi} \tag{27}$$

with ρ_{0b} being an integration constant. This solution indicates that the evolution of the matter density is modified due to interaction with ϕ. This expression can be also written as [16]

$$\rho_b = \rho_{0b} a^{-3(\omega_b + 1 + \frac{s}{3}) + \epsilon} \tag{28}$$

with ϵ being defined by

$$\epsilon = \frac{(1 - 3\omega_b)\int \beta d\phi}{\ln a} \tag{29}$$

Before going further, we would like to show that contrary to the usual dark energy fields the scalar field ϕ satisfies the weak energy condition. To do this, we first use the relation (23) to write (5) on the brane

$$\ddot{\phi} + (3 + s)H\dot{\phi} + \frac{dV}{d\phi} = \beta(\phi)(3\omega_b - 1)\rho_b \tag{30}$$

Moreover, equation (16) on the brane gives

$$\ddot{\phi} + \frac{dV}{d\phi} = -4H\dot{\phi} \tag{31}$$

Combining these two equations leads to

$$(s - 1)H\dot{\phi} = \beta(\phi)(3\omega_b - 1)\rho_b \tag{32}$$

We then write (25) in the following form

$$\dot{\rho}_\phi + 3\frac{\dot{a}}{a}(\omega_\phi + 1)\rho_\phi + \frac{\dot{a}}{a}\dot{\phi}^2 = 0, \tag{33}$$

where (23) and (32) have been used. From the definition $\rho_\phi = \frac{1}{2}\dot{\phi}^2 + V(\phi)$, we have

$$\dot{\rho}_\phi = (\ddot{\phi} + \frac{dV}{d\phi})\dot{\phi} \tag{34}$$

Now combining (33), (34) and using (31) gives

$$\dot{\phi}^2 = (\omega_\phi + 1)\rho_\phi \tag{35}$$

which means that $(\omega_\phi + 1) > 0$ and the scalar field ϕ satisfies the weak energy condition.

4.2. Late-time behavior

We are interested in late-time behavior of the Universe. To deal with this issue we look for late-time asymptotic solutions of the field equations. When $t \to \infty$ (or $a \to \infty$), equations (22) and (30) reduce to

$$H^2 \approx \frac{1}{6}\rho_\phi = \frac{1}{6}(\frac{1}{2}\dot{\phi}^2 + V(\phi)) \tag{36}$$

$$a^{-(s+3)}\frac{d}{dt}(\dot{\phi}a^{(s+3)}) \approx -\frac{dV(\phi)}{d\phi} \tag{37}$$

As a usual rule for solving this set of equations, one usually gives the potential function $V(\phi)$ as an input and then finds the functions $a(t)$ and $\phi(t)$. However, we would like to follow a different strategy. We will take $\dot{\phi} = a^n$, with n being a constant parameter, as an input and find $V(\phi)$ and $a(t)$ so that the equations (36) and (37) are satisfied. The solutions are

$$a(t) = C^{-\frac{1}{n}} t^{-\frac{1}{n}} \tag{38}$$

$$V(\phi) = V_0 e^{-2C\phi} \tag{39}$$

where $C = \frac{(-n)}{2\sqrt{3}}(\frac{-(s+3)}{n})^{\frac{1}{2}}$ and $V_0 = \frac{6(n+(s+3))}{n^2(s+3)}$. This set of solutions indicates that the Universe is accelerating for $-1 < n < 0$. The functions $a(t)$ and $V(\phi)$ are plotted in Fig. 1. The figure shows that $V(\phi)$ has a run-away form as it should be since ϕ is a chameleon field [12].

The Universe has not been in an accelerating phase at all the time and has suffered a transition from an early decelerating phase to a recent accelerating one. To check that whether or nor the present model can generate such a phase transition, we look at the effective equation of state parameter ω_{eff}. We first re-write (26) in the form

$$\dot{\rho}_b + 3H(\omega_{eff} + 1)\rho_b = 0 \tag{40}$$

where

[3] If spacetime has one spatial extra dimension, then there will be a relation such as $bG = G_*$ [14] where G and G_* are, respectively, four and five dimensional gravitational couplings and b is radius of the extra dimension. Then $\frac{\dot{G}}{G} = -\frac{\dot{b}}{b} = -sH$ where G_* is assumed to be a constant.

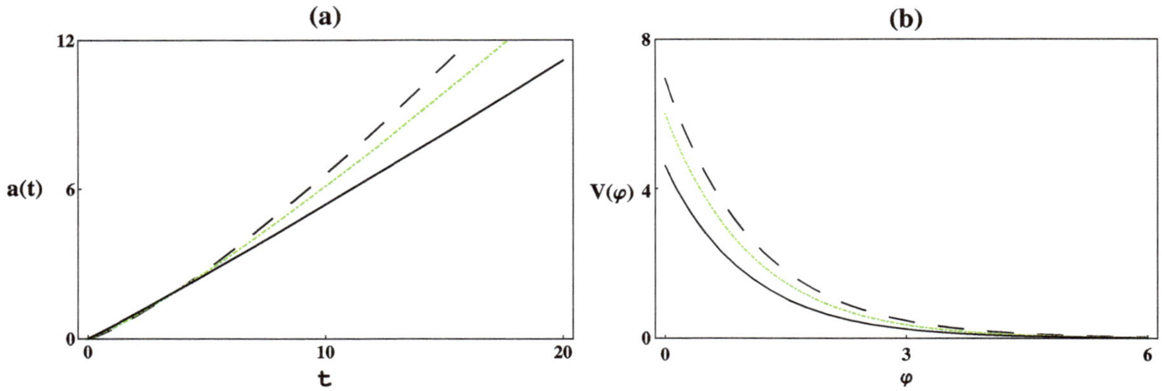

Fig. 1. The plot of $a(t)$ and $V(\phi)$ for $n=-0.8$ and $s=0.08$ (dashed), $n=-0.85$ and $s=0.09$ (dotted) and $n=-0.95$ and $s=0.1$ (solid).

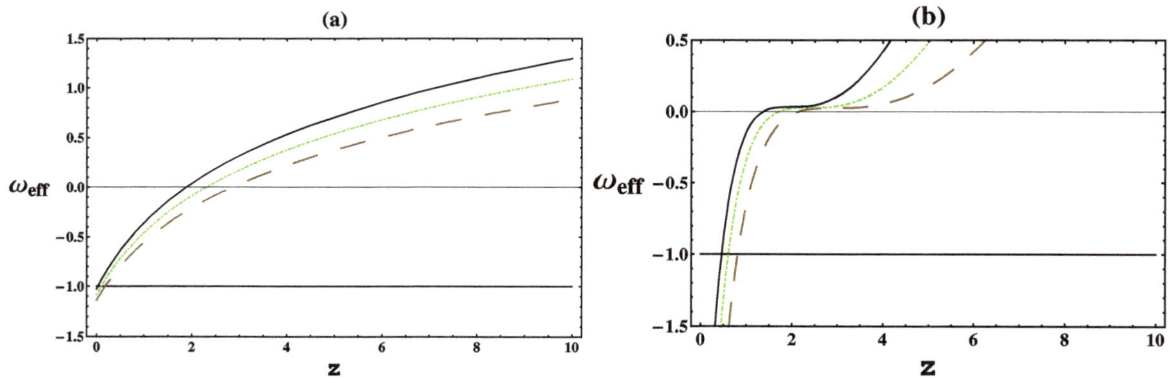

Fig. 2. The plot of ω_{eff} in terms of z for $\beta(\phi)=\phi$ panel (a) and $\beta(\phi)=\phi^3$ panel (b). The curves correspond to $n=-0.65$ and $s=0.08$ (dashed), $n=-0.7$ and $s=0.09$ (dotted) and $n=-0.75$ and $s=0.1$ (solid).

$$\omega_{eff}=\omega_b+\frac{s}{3}+\frac{Q}{3H\rho_b} \tag{41}$$

$$\omega_{eff}=\frac{s}{3}-\frac{1}{3}\beta(\phi)\frac{\dot{\phi}}{H} \quad (\omega_b=0)$$

Using the solution (38) and (39), gives $\frac{\dot{\phi}}{H}=2\sqrt{3}(\frac{-n}{(s+3)})^{\frac{1}{2}}$ which is a constant. This means that deceleration to acceleration phase transition needs $\beta(\phi)$ not to be a constant.

Among many possible choices for $\beta(\phi)$, let us choose a simple one $\beta(\phi)=\phi$ as an input coupling function. It corresponds to $A(\phi)=e^{\frac{1}{2}\phi^2}$. The resulting effective equation of state parameter is plotted in Fig. 2. As it is clear from the figure, the function ω_{eff} exhibits a recent signature flip. It also shows that the Universe recently enters the phantom region.

The choice $\beta(\phi)=\phi$ is not the only one that leads to a transition from decelerating to accelerating phase. The panel (b) of the Fig. 2 shows ω_{eff} for another choice $\beta(\phi)=\phi^3$. It should be remarked that in both cases deceleration to acceleration transition takes place when $\beta>0$ or $Q<0$. It means that in the interacting process described by (25) and (26), the direction of energy flow is so that matter is created. This seems to be consistent with the results reported in [17].

5. Thermodynamic analysis

A thermodynamic description of a homogeneous and isotropic interacting perfect fluid requires a knowledge of the particle flux $N^{\alpha}=nu^{\alpha}$ and the entropy flux $S^{\alpha}=su^{\alpha}$ where $n=N/a^3$, $s=n\sigma$ and $\sigma=S/N$ is specific entropy (per particle) of the created or annihilated particles. Since energy density of matter is given by $\rho_b=nM$ with M being the mass of each particle, the appearance of the extra term in the energy balance equation (26) means that this term can be attributed to a change of n or M. Here we assume that the mass of each matter particle remains constant and the extra term in the energy balance equation only leads to a change of the number density n. In this case, the equation (26) can be written as[4]

$$\dot{n}+3H(1+\frac{s}{3})n=n\Gamma \tag{42}$$

where $\Gamma\equiv\beta(\phi)\dot{\phi}$ is the rate of creation (or annihilation) of particles. The direction of energy transfer between matter and the scalar field depends on the sign of Γ. If $\Gamma>0$ (or $Q<0$), the energy goes inside of the matter system and matter is created. If $\Gamma<0$ (or $Q>0$) the direction of energy transfer is reversed and matter is annihilated.

From $\sigma=S/N$, we have

$$\frac{\dot{\sigma}}{\sigma}=\frac{\dot{S}}{S}-\frac{\dot{N}}{N} \tag{43}$$

With use of (42), the latter can be written as

$$\frac{\dot{S}}{S}=\frac{\dot{\sigma}}{\sigma}+(\Gamma-sH) \tag{44}$$

Since $n\propto a^{-3+(\epsilon-s)}$, the total number of particles scale as $N\propto a^{\epsilon-s}$. Thus (43) can also be written as

$$\frac{\dot{S}}{S}=\frac{\dot{\sigma}}{\sigma}+(\epsilon-s)H \tag{45}$$

[4] Throughout this section we have set $\omega_b=0$.

In an adiabatic process, when the overall energy transfer is such that the specific entropy per particle remains constant ($\dot{\sigma} = 0$) [18], the second law of thermodynamics ($\dot{S} \geq 0$) implies that $\epsilon - s \geq 0$ in an expanding Universe. In this case, when $\epsilon < 0$ the parameter s is allowed to take only negative values. Alternatively speaking, the extra dimension shrinks with expansion of the Universe (see (24)).

In the non-adiabatic case, on the other hand, the second law of thermodynamics requires that

$$\Gamma \geq sH - \frac{\dot{\sigma}}{\sigma} \tag{46}$$

which is also a constraint on the creation (or annihilation) rate and evolution of the extra dimension.

6. Conclusion

We have investigated a brane world scenario in which gravity is described by a five-dimensional metric together with a minimally coupled scalar field. The scalar field is a chameleon and interacts with the matter sector. Due to this interaction the energy associated with both the scalar field and matter system are not separately conserved. Thus evolution of matter energy density modifies and is controlled by Q. When $Q > 0$ matter is created and energy is injecting into the matter system so that the latter will dilute more slowly compared to its standard evolution $\rho_b \propto a^{-3(\omega_b+1)}$. On the other hand, when $Q < 0$ the reverse is true, namely that matter is annihilated and the direction of energy transfer is outside of the matter system (and into the scalar field) so that the rate of dilution is faster than the standard one.

The main results of our analysis are the following:

1) We have found a late-time asymptotic solution that exhibits accelerating expansion. There is also a recent transition from a decelerating phase to an accelerating one.

2) The interaction of chameleon field with matter plays an important role in this phase transition. In order that this transition takes place, the coupling function should be an evolving function (or $\beta(\phi)$ should not be a constant).

3) Our analysis also indicates that the Universe has recently entered the phantom region. We emphasize that this behavior is not attributed to any exotic matter system.

4) A thermodynamic analysis puts constraints on Γ and evolution of the extra dimension in adiabatic and non-adiabatic cases.

There are some problems that are not investigated in the present analysis such as behavior of the Universe at early times or the cosmological constant problem. They are deserved to be investigated elsewhere.

References

[1] L. Randall, R.R. Sundrum, Phys. Rev. Lett. 83 (1999) 3370, arXiv:hep-ph/9905221.

[2] L. Randall, R. Sundrum, Phys. Rev. Lett. 83 (1999) 4690, arXiv:hep-th/9906064.

[3] P. Brax, C. van de Bruck, Class. Quantum Gravity 20 (2003) R201, arXiv:hep-th/0303095;
D. Langlois, Prog. Theor. Phys. Suppl. 148 (2003) 181, arXiv:hep-th/0209261;
R. Maartens, in: J. Pascual-Sanchez, et al. (Eds.), Reference Frames and Gravitomagnetism, World Scientific, 2001, pp. 93–119, arXiv:gr-qc/0101059.

[4] P. Binetruy, C. Deffayet, D. Langlois, Nucl. Phys. B 565 (2000) 269, arXiv:hep-th/9905012;
P. Binetruy, C. Deffayet, U. Ellwanger, D. Langlois, Phys. Lett. B 477 (2000) 285, arXiv:hep-th/9910219.

[5] K.i. Maeda, D. Wands, Phys. Rev. D 62 (2000) 124009, arXiv:hep-th/0008188;
S.C. Davis, J. High Energy Phys. 0203 (2002) 058, arXiv:hep-ph/0111351;
M. Parry, S. Pichler, J. Cosmol. Astropart. Phys. 0411 (2004) 005, arXiv:hep-ph/0410025;
C. Bogdanos, A. Dimitriadis, K. Tamvakis, Class. Quantum Gravity 24 (2007) 3701, arXiv:hep-th/0611181;
M. Heydari-Fard, H.R. Sepangi, J. Cosmol. Astropart. Phys. 0901 (2009) 034, arXiv:0901.0855 [gr-qc];
Arianto, F.P. Zen, S. Feranie, I.P. Widyatmika, B.E. Gunara, Phys. Rev. D 84 (2011) 044008, arXiv:1103.1703 [gr-qc].

[6] S. Kanno, J. Soda, Phys. Rev. D 66 (2002) 083506, arXiv:hep-th/0207029;
K. Yang, Y. Liu, Y. Zhong, X. Du, S.W. Wei, Phys. Rev. D 86 (2012) 127502, arXiv:1212.2735 [hep-th].

[7] K. Yang, Y. Liu, Y. Zhong, X. Du, S. Wei, Phys. Rev. D 86 (2012) 127502, arXiv:1212.2735 [hep-th];
Q. Xie, Z. Zhao, Y. Zhong, J. Yang, X. Zhou, J. Cosmol. Astropart. Phys. 1503 (2015) 014, arXiv:1410.5911 [hep-th].

[8] W.D. Goldberger, M.B. Wise, Phys. Rev. Lett. 83 (1999) 4922, arXiv:hep-ph/9907447;
W.D. Goldberger, M.B. Wise, Phys. Rev. D 60 (1999) 107505, arXiv:hep-ph/9907218;
P. Kanti, K.A. Olive, M. Pospelov, Phys. Lett. B 481 (2000) 386, arXiv:hep-ph/0002229.

[9] Y. Himemoto, M. Sasaki, Phys. Rev. D 63 (2001) 044015, arXiv:gr-qc/0010035.

[10] J. Yokoyama, Y. Himemoto, Phys. Rev. D 64 (2001) 083511, arXiv:hep-ph/0103115.

[11] Kh. Saaidi, A. Mohammadi, Phys. Rev. D 85 (2012) 023526;
Kh. Saaidi, A. Mohammadi, T. Golanbari, H. Sheikhahmadi, B. Ratra, Phys. Rev. D 86 (2012) 045007.

[12] J. Khoury, A. Weltman, Phys. Rev. Lett. 93 (2004) 171104, arXiv:astro-ph/0309300;
J. Khoury, A. Weltman, Phys. Rev. D 69 (2004) 044026, arXiv:astro-ph/0309411.

[13] J. Ponce de Leon, Gen. Relativ. Gravit. 38 (2006) 61–81, arXiv:gr-qc/0412005;
J. Ponce de Leon, J. Cosmol. Astropart. Phys. 1003 (2010) 030, arXiv:1001.1961.

[14] N. Arkani-Hamed, S. Dimopoulos, G. Dvali, Phys. Lett. B 429 (1998) 263.

[15] J.P. Uzan, Rev. Mod. Phys. 75 (2003) 403, arXiv:hep-ph/0205340.

[16] Y. Bisabr, Phys. Rev. D 86 (2012) 127503, arXiv:1212.2709;
Y. Bisabr, Gen. Relativ. Gravit. 44 (2012) 427, arXiv:1110.3421.

[17] I. Prigogine, J. Geheniau, E. Gunzig, P. Nardone, Gen. Relativ. Gravit. 21 (1989) 767.

[18] J.A.S. Lima, Phys. Rev. D 54 (1996) 2571.

8

Embedding cosmological inflation, axion dark matter and seesaw mechanism in a 3-3-1 gauge model

J.G. Ferreira Jr., C.A. de S. Pires *, J.G. Rodrigues, P.S. Rodrigues da Silva

Departamento de Física, Universidade Federal da Paraíba, Caixa Postal 5008, 58051-970, João Pessoa, PB, Brazil

ARTICLE INFO

Editor: M. Trodden

ABSTRACT

The Peccei–Quinn symmetry is an intrinsic global symmetry of the 3-3-1 gauge models. Its spontaneous breaking mechanism engendering an invisible KSVZ-like axion links the 3-3-1 models with new physics at $\sim 10^{10}$ GeV scale. The axion that results from this mechanism is an interesting candidate for the dark matter of the universe, while its real partner may drive inflation if radiative corrections are taken into account. This is obtained by connecting the type I seesaw mechanism with the spontaneous breaking of the Peccei–Quinn symmetry. In the end of the day we have a scenario providing a common answer to the strong-CP problem, inflation, dark matter and neutrino mass.

1. Introduction

The $SU(3)_C \times SU(3)_L \times U(1)_N$ (3-3-1) gauge models for the electroweak interactions are interesting in their own right. For example, in these models generation cannot replicate unrestrictedly as in the standard model (SM), since it is not exact replica of one another and each is separately anomalous. However, when three generations are taken into account, gauge anomaly is automatically canceled [1,2], providing a reason for the existence of three families of fermions.

Moreover, the set of constraints provided by the gauge invariance of the Yukawa interactions together with those coming from the anomaly cancellation conditions are enough to fix the electric charges of the particles in the 3-3-1 model, thus providing an understanding of the pattern of electric charge quantization [3,4].

In what concerns the Peccei–Quinn (PQ) symmetry, it is an automatic symmetry of these models, thus elegantly solving the strong CP-problem [5]. However, the original versions of the 3-3-1 gauge models furnish an unrealistic axion because of its sizable couplings with the standard particles [6,7]. In order to have an invisible axion a neutral scalar singlet must be added to the conventional scalar sector [8–11].

Regarding neutrino masses, canonical seesaw mechanisms, as type I and type II, as well as the inverse seesaw mechanism are easily implemented in the framework of the 3-3-1 models [12–16].

From the phenomenological point of view, a remarkable aspect of the 3-3-1 models relies on flavor physics. Rare decays, lepton number violation and flavor changing neutral current are natural outcome of the model [17–24]. Recent collider phenomenology of these models are performed in Refs. [25–27].

Last in the sequence but not least in importance, we remember that conventional particle content of some 3-3-1 models includes a stable and neutral particle that may play the role of cold dark matter in the WIMP form [28–31]. These interesting features turn the 3-3-1 models an appealing candidates for physics beyond the SM. In this point we call the attention to the fact that the physics of the early universe, particularly inflation, has been poorly explored within these models [32,33]. Thus, in view of the recent experimental advances in probing inflation observables, it turns imperative to search for mechanisms that allow implementation of inflation in the framework of the 3-3-1 gauge models.

Concerning implementation of cosmological issues within phenomenological gauge models, as the SM, we remark that there are two distinct ways of providing a common solution to cosmological inflation, cold dark mater and neutrino masses within the SM. The first arises within the type I seesaw mechanism for small neutrino masses. By adding right handed neutrinos and at least one neutral scalar in the singlet form to the SM, besides considering spontaneous breaking of global lepton number within type I seesaw mechanism, one has that the real part of the neutral singlet may drive inflation while the imaginary part may be the dark matter of the universe [34,35].

On the other hand, the implementation of the PQ symmetry in the standard model may be accomplished by adding exotic vector

* Corresponding author.
E-mail address: cpires@fisica.ufpb.br (C.A. de S. Pires).

like quarks, right-handed neutrinos and neutral scalar singlet to its particle content. This scenario is called SMASH [36–40]. The PQ symmetry is spontaneously broken when the neutral scalar singlet develops vacuum expectation value (VEV) different from zero. In this circumstance, the imaginary part of this scalar singlet will be the invisible axion, which may play the role of dark matter, while the real part may drive inflation. Moreover, on coupling the neutral scalar singlet to the right-handed neutrino, through an Yukawa interaction, the VEV of the neutral scalar, that must lie in the range (10^{10}–10^{11}) GeV, will generate mass to heavy neutrinos that may trigger the type I seesaw mechanism yielding small masses for the standard neutrinos. The problem with this scenario is that it generates an inflaton potential of the type $\lambda\phi^4$ which is practically excluded by the current bounds from PLANCK15 [41]. A way of circumventing such a problem is, either to consider that the inflaton couples non-minimally with the scalar curvature R, or take into account radiative corrections to the inflaton potential.

The main difference among these two proposals relies on the dark matter sector. In the former case the dark matter candidate, a Majoron, gains mass from quantum gravity effect and then is classified as warm dark matter, while in the latter case the dark matter candidate is the axion and is classified as cold dark matter candidate. The axion gains mass from QCD and quantum gravity effects. Care must be taken because quantum gravity effect may destabilize the axion as dark matter candidate. We take care of this by means of large discrete symmetry.

Since the PQ symmetry is an automatic symmetry of the 3-3-1 gauge models, consequently its version involving right-handed neutrinos realizes automatically the SMASH proposal. In this case it turns imperative to check if the real partner of the axion will drive inflation. We show that this is possible when radiative corrections are taken into account. We examine also reheating phase. It is engendered by the decay of the inflaton into the conventional scalars. The model easily provides a reheating temperature of 10^9 GeV for typical values of the parameters required by canonical inflation models. In addition, standard neutrinos will gain mass through the type I seesaw mechanism and the axion is the natural dark matter candidate of the model.

The paper is divided in the following way: In Sec. 2 we revisit the 3-3-1 model that contains an invisible axion in its spectrum. Next, in Sec. 3, we develop the inflationary paradigm in such model. We finally conclude in Sec. 4.

2. The 3-3-1 model, the Peccei–Quinn symmetry and the invisible axion

The model developed here is one proposed in Ref. [42] which is a modification of the original one [1,43,44]. To realize our proposal, heavy neutrinos in the singlet form must be added to the leptonic sector of the model

$$f_L^a = \begin{pmatrix} \nu_L^a \\ e_L^a \\ (\nu_R^c)^a \end{pmatrix} \sim (1\,,\,3\,,\,-1/3)\,,\;\; e_{aR} \sim (1,1,-1)\,,$$

$$N_{aR} \sim (1,1,0) \tag{1}$$

with $a = 1, 2, 3$ representing the three known generations. We are indicating the transformation under 3-3-1 after the similarity sign, "$\sim$".

The quark sector is kept intact with one generation of left-handed fields coming in the triplet fundamental representation of $SU(3)_L$ and the other two composing an anti-triplet representation with the content

$$Q_{iL} = \begin{pmatrix} d_{iL} \\ -u_{iL} \\ d'_{iL} \end{pmatrix} \sim (3\,,\,\bar{3}\,,\,0)\,,\;\; Q_{3L} = \begin{pmatrix} u_{3L} \\ d_{3L} \\ u'_{3L} \end{pmatrix} \sim (3\,,\,3\,,\,1/3)\,, \tag{2}$$

and the right-handed fields

$$u_{iR} \sim (3,1,2/3)\,,\, d_{iR} \sim (3,1,-1/3)\,,\, d'_{iR} \sim (3,1,-1/3)$$
$$u_{3R} \sim (3,1,2/3)\,,\, d_{3R} \sim (3,1,-1/3)\,,\, u'_{3R} \sim (3,1,2/3), \tag{3}$$

where $j = 1, 2$ represent different generations. The primed quarks are the exotic ones but with the usual electric charges.

In order to generate the masses for the gauge bosons and fermions, the model requires only three Higgs scalar triplets. For our proposal here we add a neutral scalar in the singlet form such that the scalar content is composed by

$$\chi = \begin{pmatrix} \chi^0 \\ \chi^- \\ \chi'^0 \end{pmatrix} \sim (1\,,\,3\,,\,-1/3)\,,\;\; \eta = \begin{pmatrix} \eta^0 \\ \eta^- \\ \eta'^0 \end{pmatrix} \sim (1\,,\,3\,,\,-1/3),$$

$$\rho = \begin{pmatrix} \rho^+ \\ \rho^0 \\ \rho'^+ \end{pmatrix} \sim (1\,,\,3\,,\,2/3)\,,\quad \phi \sim (1,1,0). \tag{4}$$

Thus the particle content of the model in Ref. [42] is extended by the fields N_{a_R} and ϕ.

In order to keep intact the physics results of the Ref. [42], the Lagrangian of the model must be invariant by the following set of discrete symmetries $Z_{11} \otimes Z_2$ but now with Z_{11} acting as

$$\begin{aligned}
\phi &\to \omega_1^{-1}\phi\,, & f_{aL} &\to \omega_1 f_{aL}\,,\\
\rho &\to \omega_2^{-1}\rho\,, & d_{aR} &\to \omega_2 d_{aR}\,,\\
\chi &\to \omega_3^{-1}\chi\,, & u'_{3R} &\to \omega_3 u'_{3R}\,,\\
Q_{iL} &\to \omega_4^{-1}Q_{iL}\,, & d'_{iR} &\to \omega_4 d'_{iR}\,,\\
\eta &\to \omega_5^{-1}\eta\,, & u_{aR} &\to \omega_5 u_{aR}\,,\\
Q_{3L} &\to \omega_0 Q_{3L}\,, & N_R &\to \omega_5^{-1} N_R\,,\\
e_{aR} &\to \omega_3 e_{aR}\,,
\end{aligned} \tag{5}$$

where $\omega_k \equiv e^{2\pi i \frac{k}{11}}$, $\{k = 0, \pm1, ..., \pm5\}$.

The Z_2 symmetry must act as

$$(\rho\,,\chi\,,d'_R\,,u'_{3R}\,,u_R\,,d_R\,,e_R) \;\to\; -(\rho\,,\chi\,,d'_R\,,u'_{3R}\,,u_R\,,d_R\,,e_R)\,. \tag{6}$$

These discrete symmetries yield the following Yukawa couplings

$$\begin{aligned}
\mathcal{L}^Y &= G_1\bar{Q}_{3L}u'_{3R}\chi + G_2^{ij}\bar{Q}_{iL}d'_{jR}\chi^* + G_3^{3a}\bar{Q}_{3L}u_{aR}\eta + G_4^{ia}\bar{Q}_{iL}d_{aR}\eta^*\\
&+ G_5^{3a}\bar{Q}_{3L}d_{aR}\rho + G_6^{ia}\bar{Q}_{iL}u_{aR}\rho^* + g_{ab}\bar{f}_{aL}e_{bR}\rho\\
&+ h_{ab}\bar{f}_{aL}\eta N_{bR} + h'_{ab}\phi\bar{N}^C_{aR}N_{bR} + \text{H.c.}
\end{aligned} \tag{7}$$

The transformations displayed in Eqs. (5) and (6) are a little different from the original case [42]. The reason for this is to accommodate the last two terms in the Lagrangian above which are crucial for our proposal, as we will see later. The physics of the original case remains the same because the new terms involve heavy neutrinos that are standard model singlets.

The potential does not change. It is exactly the same as in the original case, i.e.,

$$
\begin{aligned}
V_H = {} & \mu_\phi^2\phi^2 + \mu_\chi^2\chi^2 + \mu_\eta^2\eta^2 + \mu_\rho^2\rho^2 + \lambda_1\chi^4 + \lambda_2\eta^4 + \lambda_3\rho^4 \\
& + \lambda_4(\chi^\dagger\chi)(\eta^\dagger\eta) + \lambda_5(\chi^\dagger\chi)(\rho^\dagger\rho) + \lambda_6(\eta^\dagger\eta)(\rho^\dagger\rho) \\
& + \lambda_7(\chi^\dagger\eta)(\eta^\dagger\chi) + \lambda_8(\chi^\dagger\rho)(\rho^\dagger\chi) + \lambda_9(\eta^\dagger\rho)(\rho^\dagger\eta) \\
& + \lambda_{10}(\phi\phi^*)^2 + \lambda_{11}(\phi\phi^*)(\chi^\dagger\chi) + \lambda_{12}(\phi\phi^*)(\rho^\dagger\rho) \\
& + \lambda_{13}(\phi\phi^*)(\eta^\dagger\eta) + \lambda_\phi\epsilon^{ijk}\eta_i\rho_j\chi_k\phi + \text{H.c.} \qquad (8)
\end{aligned}
$$

Other tiny change arises in the definition of the PQ charges. In order to have chiral quarks under $U(1)_{PQ}$, we need the following transformation

$$
\begin{aligned}
& u_{aL} \to e^{-i\alpha X_u} u_{aL}\,, \quad u_{aR} \to e^{i\alpha X_u} u_{aR}\,, \\
& u'_{3L} \to e^{-i\alpha X'_u} u'_{3L}\,, \quad u'_{3R} \to e^{i\alpha X'_u} u'_{3R}\,, \\
& d_{aL} \to e^{-i\alpha X_d} d_{aL}\,, \quad d_{aR} \to e^{i\alpha X_d} d_{aR}\,, \\
& d'_{iL} \to e^{-i\alpha X'_d} d'_{iL}\,, \quad d'_{iR} \to e^{i\alpha X'_d} d'_{iR}\,. \qquad (9)
\end{aligned}
$$

For the leptons we can define their PQ charges by

$$
\begin{aligned}
& e_{aL} \to e^{i\alpha X_e} e_{aL}\,, \quad e_{aR} \to e^{i\alpha X_{eR}} e_{aR}\,, \quad N_{aL} \to e^{i\alpha X_N} N_{aL} \\
& \nu_{aL} \to e^{i\alpha X_\nu} \nu_{aL}\,, \quad \nu_{aR} \to e^{i\alpha X_{\nu R}} \nu_{aR}\,. \qquad (10)
\end{aligned}
$$

With these assignments and taking the Yukawa interactions in Eq. (7) into account, as well as the non-Hermitean terms $\eta\rho\chi\phi$, we easily see that the PQ charges for the scalars are constrained and imply the following relations:

$$
X_d = -X_u\,, \quad X_{d'} = -X_{u'}\,, \quad X_\nu = X_{eR}\,, \quad X_e = X_{\nu R}\,. \qquad (11)
$$

We can make the further choice $X_d = X_{d'}$, leading to

$$
X_d = X_{d'} = -X_u = -X_{u'} = -X_e = X_{eR} = X_\nu = -X_{\nu R} = X_N\,, \qquad (12)
$$

implying that the PQ symmetry is chiral for the leptons, too. The scalars transform as

$$
\begin{aligned}
& \phi \to e^{-2i\alpha X_d}\phi\,, && \eta^0 \to e^{2i\alpha X_d}\eta^0 \\
& \eta^- \to \eta^-\,, && \eta'^0 \to e^{2i\alpha X_d}\eta'^0 \\
& \rho^+ \to \rho^+\,, && \rho^0 \to e^{-2i\alpha X_d}\rho^0 \\
& \rho'^+ \to \rho'^+\,, && \chi^0 \to e^{2i\alpha X_d}\chi^0 \\
& \chi^- \to \chi^-\,, && \chi'^0 \to e^{2i\alpha X_d}\chi'^0.
\end{aligned}
\qquad (13)
$$

It is now clear that the entire Lagrangian of the model is $U(1)_{PQ}$ invariant, providing a natural solution to the strong-CP problem.

To accomplish our proposal, let us consider that only χ'^0, ρ^0, η^0 and ϕ develop VEV and expand such fields in the standard way,

$$
\begin{aligned}
& \chi'^0 = \frac{1}{\sqrt{2}}(v_{\chi'} + R_{\chi'} + iI_{\chi'})\,, \quad \eta^0 = \frac{1}{\sqrt{2}}(v_\eta + R_\eta + iI_\eta)\,, \\
& \rho^0 = \frac{1}{\sqrt{2}}(v_\rho + R_\rho + iI_\rho)\,, \quad \phi = \frac{1}{\sqrt{2}}(v_\phi + R_\phi + iI_\phi)\,. \qquad (14)
\end{aligned}
$$

With such expansion, we obtain the set of constraint equations that guarantee that the potential has a minimum

$$
\begin{aligned}
& \mu_\chi^2 + \lambda_1 v_{\chi'}^2 + \frac{\lambda_4}{2}v_\eta^2 + \frac{\lambda_5}{2}v_\rho^2 + \frac{\lambda_{11}}{2}v_\phi^2 + \frac{A}{v_{\chi'}^2} = 0, \\
& \mu_\eta^2 + \lambda_2 v_\eta^2 + \frac{\lambda_4}{2}v_{\chi'}^2 + \frac{\lambda_6}{2}v_\rho^2 + \frac{\lambda_{13}}{2}v_\phi^2 + \frac{A}{v_\eta^2} = 0, \\
& \mu_\rho^2 + \lambda_3 v_\rho^2 + \frac{\lambda_5}{2}v_{\chi'}^2 + \frac{\lambda_6}{2}v_\eta^2 + \frac{\lambda_{12}}{2}v_\phi^2 + \frac{A}{v_\rho^2} = 0, \\
& \mu_\phi^2 + \lambda_{10} v_\phi^2 + \frac{\lambda_{11}}{2}v_{\chi'}^2 + \frac{\lambda_{12}}{2}v_\rho^2 + \frac{\lambda_{13}}{2}v_\eta^2 + \frac{A}{v_\phi^2} = 0\,, \qquad (15)
\end{aligned}
$$

where we have defined $A \equiv \lambda_\phi v_\eta v_\rho v_{\chi'} v_\phi$. The physical scalars are obtained by substituting these constraints into the mass matrices given by the second derivative of the potential. The axion arises from the mass matrix M_I^2 given by

$$
-\frac{A}{2}\begin{pmatrix}
\frac{1}{v_{\chi'}^2} & \frac{1}{v_\eta v_{\chi'}} & \frac{1}{v_\rho v_{\chi'}} & \frac{1}{v_{\chi'} v_\phi} \\
\frac{1}{v_\eta v_{\chi'}} & \frac{1}{v_\eta^2} & \frac{1}{v_\eta v_\rho} & \frac{1}{v_\eta v_\phi} \\
\frac{1}{v_\rho v_{\chi'}} & \frac{1}{v_\eta v_\rho} & \frac{1}{v_\rho^2} & \frac{1}{v_\eta v_\rho} \\
\frac{1}{v_{\chi'} v_\phi} & \frac{1}{v_\eta v_\phi} & \frac{1}{v_\eta v_\rho} & \frac{1}{v_\phi^2}
\end{pmatrix} \qquad (16)
$$

in the basis $(I_{\chi'}, I_\eta, I_\rho, I_\phi)$. Its diagonalization furnishes an axion given by, $a = \frac{1}{\sqrt{1+\frac{v_{\chi'}^2}{v_\phi^2}}}\left(I_\phi - \frac{v_{\chi'}}{v_\phi}I_{\chi'}\right)$. As $v_\phi \gg v_{\chi'}$ we have that $a \sim I_\phi$.

Now let us focus on the CP-even component of ϕ. It will be our inflaton candidate. It composes the following mass matrix M_R^2 given by

$$
\begin{pmatrix}
\lambda_1 v_{\chi'}^2 - \frac{A}{2v_{\chi'}^2} & \frac{\lambda_4 v_{\chi'} v_\eta}{2} + \frac{A}{2v_\eta v_{\chi'}} & \frac{\lambda_5 v_{\chi'} v_\rho}{2} + \frac{A}{2v_\rho v_{\chi'}} & \frac{A}{2v_\phi v_{\chi'}} \\
\frac{\lambda_4 v_{\chi'} v_\eta}{2} + \frac{A}{2v_\eta v_{\chi'}} & \lambda_2 v_\eta^2 - \frac{A}{2v_\eta^2} & \frac{\lambda_6 v_\eta v_\rho}{2} + \frac{A}{2v_\rho v_\eta} & \frac{A}{2v_\eta v_\phi} \\
\frac{\lambda_5 v_{\chi'} v_\rho}{2} + \frac{A}{2v_\rho v_{\chi'}} & \frac{\lambda_6 v_\eta v_\rho}{2} + \frac{A}{2v_\rho v_\eta} & \lambda_3 v_\rho^2 - \frac{A}{2v_\rho^2} & \frac{A}{2v_\rho v_\phi} \\
\frac{A}{2v_\phi v_{\chi'}} & \frac{A}{2v_\eta v_\phi} & \frac{A}{2v_\rho v_\phi} & \lambda_{10} v_\phi^2 - \frac{A}{2v_\phi^2}
\end{pmatrix} \qquad (17)
$$

in the basis $(R_{\chi'}, R_\eta, R_\rho, R_\phi)$. As $v_\phi \gg v_\rho, v_\eta, v_{\chi'}$, we have that R_ϕ decouples and its mass is predicted to be $m_{R_\phi}^2 \sim \lambda_{10} v_\phi^2$.

In this point we call the attention to the fact that the presence of large discrete symmetries has the function of stabilizing the axion against quantum gravity effects. In order to see this, perceive that the effective operators responsible for the gravitational contribution to the axion mass is of the form ϕ^n/M_{Pl}^{n-4}. A Z_N symmetry automatically suppress terms of this kind till some $n = N-1$. The main surviving term contributing to the axion mass is the one with $n = N$. It is true that with Z_{11} the axion is protected only for energy scales not bigger than $\langle\phi\rangle \simeq 10^{10}$ GeV. Nevertheless, this is not a threat for the model since we still have values for the θ angle and axion mass (gravitationally induced) [45],

$$
\begin{aligned}
& M_a^{Grav} \simeq \sqrt{\frac{\langle\phi\rangle^{N-2}}{M_{Pl}^{N-4}}} \simeq 10^{-12}\,\text{eV} \simeq 10^{-7} m_a\,, \\
& \theta_{eff} \simeq \frac{\langle\phi\rangle^N}{M_{Pl}^{N-4}\Lambda_{QCD}^4} \simeq 10^{-19}\,, \qquad (18)
\end{aligned}
$$

where we have used $M_{Pl} \simeq 10^{19}$ GeV, $\Lambda_{QCD} \simeq 300$ MeV, and $m_a \simeq 10^{-5}$ eV is the instanton induced axion mass. These values are consistent with astrophysical and cosmological bounds (see PDG [46]). If we had taken $\langle\phi\rangle \simeq 10^{11}$ GeV, the axion would still be protected under gravitation, but the θ value would be on the threshold of its bound $\theta_{eff} \lesssim 10^{-9}$. So we can have a valid solution to the strong-CP problem for Z_{11} for scales $\langle\phi\rangle \lesssim 10^{10}$ GeV in this version of 3-3-1. We finish this section remarking that the incorporation of PQ symmetry in 3-3-1 model as done here has as main purpose the explanation of the strong CP-problem and, as a byproduct, the invisible axion fulfilling the conditions to be a viable dark matter candidate.

3. Implementing inflation

Here we consider inflation in the specific framework of the 3-3-1 model presented in the previous section. Our aim is to show that the real component of the ϕ field will drive inflation with its potential satisfying the slow roll conditions while providing the current prediction for the scalar spectral index, n_s, and obeying the current bound on the scalar to tensor ratio, r.

First thing to note is that the ϕ potential involves the terms,

$$V_\phi = \mu_\phi^2 \phi^* \phi + \lambda_{10}(\phi^*\phi)^2 + \lambda_{11}(\phi^*\phi)(\chi^\dagger\chi) + \lambda_{12}(\phi^*\phi)(\rho^\dagger\rho) + \lambda_{13}(\phi^*\phi)(\eta^\dagger\eta) + \lambda_\phi \epsilon^{ijk}\eta_i\rho_j\chi_k\phi + \text{H.c.} \tag{19}$$

However, as $v_\phi >> v_{\eta,\rho,\chi}$, the terms in this potential that really matter during inflation are

$$V_\phi = \mu_\phi^2 \phi^*\phi + \lambda_{10}(\phi^*\phi)^2. \tag{20}$$

This is the chaotic inflation scenario with the inflaton being the real part of ϕ. From now on we use the notation $R_\phi \equiv \Phi$

A VEV around 10^{10} GeV for ϕ implies that the dominant term in the above potential is $\lambda_{10}\Phi^4$. However, as we know, the $\lambda_{10}\Phi^4$ chaotic inflation is not favored by recent values of r measured by PLANCK2015 [41]. Thus, in order to circumvent this problem, we take into account radiative corrections to the potential which now reads

$$V(\Phi) = V_{tree} + V_{eff}, \tag{21}$$

with $V_{tree} = \lambda_{10}\Phi^4$ and V_{eff} being the radiative corrections due to the coupling of Φ to the particle content of the 3-3-1 model. The radiative corrections are engendered by the couplings of our inflaton with the right-handed neutrinos and the scalars whose intensities are determined by the parameters $\lambda_{11}, \lambda_{12}, \lambda_{13}, \lambda_\phi$ and h'. As we will see below, successful reheating after inflation requires $\lambda_{\phi,11,12,13}$ very small. Thus the intensity of the radiative corrections is practically determined by h' which is the coupling of the inflaton, Φ, to the heavy neutrino, N_R, and is given by the last term of the Lagrangian in Eq. (7).

According to Coleman–Weinberg approach[47], we obtain the following expression to the effective potential

$$V_{eff} = \frac{1}{64\pi^2}\sum_i \left[(-1)^{2J}(2J+1)m_i^4 \ln\frac{m_i^2}{\Delta^2}\right], \tag{22}$$

where m_i is the ϕ-field dependent mass where $i = \eta, \rho, \chi, \phi, N_R$. J is the spin of the respective contribution. The intensities of the scalar contributions are dictated by the couplings $\lambda_{\phi,11,12,13}$ while the intensities of the heavy neutrino contributions are dictated by h''s. As we will see in the end of this section, efficient reheating implies $\lambda_{\phi,11,12,13} \ll h'$. Thus, for reasons of simplicity, we just consider contributions due to N_R. This means to take in Eq. (22) $m_{i=N_R} = -\sqrt{2}h'\Phi$. In this circumstance, for our proposal here, and for simplicity reasons, it is just sufficient to consider one family of heavy neutrinos. After all this the potential that really matters during the inflationary period is given by

$$V(\Phi) \approx \lambda'\left(\Phi^4 + a'\Phi^4 \ln\frac{\Phi}{\Delta}\right), \tag{23}$$

where $\lambda' = \frac{\lambda_{10}}{4}$ and $a' = \frac{a+160\lambda'^2}{32\pi^2\lambda'} \approx \frac{a}{32\pi^2\lambda'}$. Δ is a renormalization scale. This approximation is justified because the amplitude of curvature perturbation demands a small λ_{10}. The term a carries the radiative contribution and in our case it is given by $a = -16h'^4$. The negative sign is a characteristic feature of fermion contributions. We would like to point out that, according to our assumptions, the loop dominant contribution that generates the expression above comes from box diagram composed by four heavy neutrinos running in the box. This is the reason why the loop correction given above gets proportional to h'^4. If we go further with the calculation, the two loop diagram is a correction of the box diagram and get proportional to h'^6 which is smaller than one loop contribution for $h' < 1$. Thus we do not need to worry about higher order corrections than one-loop.

We can now treat the issue of inflation, which occurs as long as the slow roll approximation is satisfied ($\epsilon \ll 1$, $\eta \ll 1$, $\zeta^2 \ll 1$). Throughout this section we follow the approach given in Refs. [34,35]. The slow roll parameters are given by [48]

$$\epsilon(\Phi) = \frac{m_P^2}{16\pi}\left(\frac{V'}{V}\right)^2, \qquad \eta(\Phi) = \frac{m_P^2}{8\pi}\left(\frac{V''}{V}\right),$$
$$\zeta^2(\Phi) = \frac{m_P^4}{64\pi^2}\left(\frac{V'''V''}{V^2}\right), \tag{24}$$

where $m_P = 1.22 \times 10^{19}$ GeV.

The spectral index n_S, the scalar to tensor ratio r and the running of spectral index $\alpha \equiv \frac{dn_S}{d\ln k}$ are defined as [49]

$$n_S = 1 - 6\epsilon + 2\eta, \qquad r = 16\epsilon,$$
$$\alpha = 16\epsilon\eta - 24\epsilon^2 - 2\zeta^2. \tag{25}$$

For a wave number $k = 0.05$ Mpc^{-1}, the Planck results indicate $n_S = 0.9644 \pm 0.0049$ and $r < 0.149$ [41].

The number of e-folds is given by

$$N = \frac{-8\pi}{m_P^2}\int_{\Phi_i}^{\Phi_f}\frac{V}{V'}d\Phi, \tag{26}$$

where Φ_f marks the end of inflation and is defined by $(\epsilon, \eta, \zeta^2)$=1. To find Φ_i we set $N = 50$, 60 and 70 and solve Eq. (26) for Φ_i.

Another important parameter is the amplitude of curvature perturbation

$$\Delta_R^2 = \frac{8V}{3m_P^4\epsilon}. \tag{27}$$

Planck measurement of this parameter gives $\Delta_R^2 = 2.215 \times 10^{-9}$ for a wave number $k = 0.05$ Mpc^{-1}. We use this experimental value of Δ_R^2 to fix λ'.

Let us discuss our results beginning with Fig. 1. There we show the behavior of the scalar to tensor ratio, r, related to a' for some values of Δ. First of all, so as to have an idea of the values of Φ_i and Φ_f, for the case of $\Delta = 3m_P$, and considering the setup presented above, we have that inflation ends with $\Phi_f \sim 10^{18}$ GeV and, for the particular case of 60 e-folds, we get the initial value $\Phi_i \sim 4 \times 10^{19}$ GeV. Note that as a' goes to zero all the curves converge to a point around $r = 0.26$. This is the expected value for r provided by Φ^4 chaotic inflation. Thus, in our case, the current bounds on r requires $a' \neq 0$. This means that radiative corrections turn to be absolutely necessary in our analysis. We also stress that the scalar to tensor ratio demands trans-Planckian regime for Δ because the sub-Planckian case faces problems in the integration on the e-fold number to reach the value 60 unless a' goes to zero, again recovering the Φ^4 chaotic inflation. Even for the trans-Planckian case, on assuming $a' < 0$, the current values of r do not allow Δ to exceed the regime of $\sim 6m_P$. In other words, our inflation model requires sizable radiative corrections in order to obey

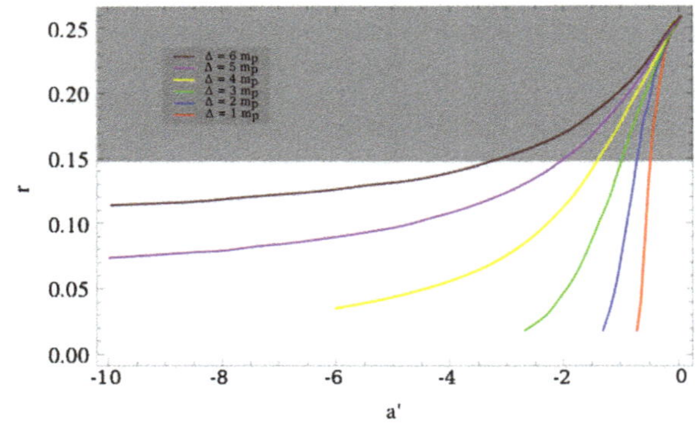

Fig. 1. r vs a' for several values of Δ. The region in gray is excluded by Planck.

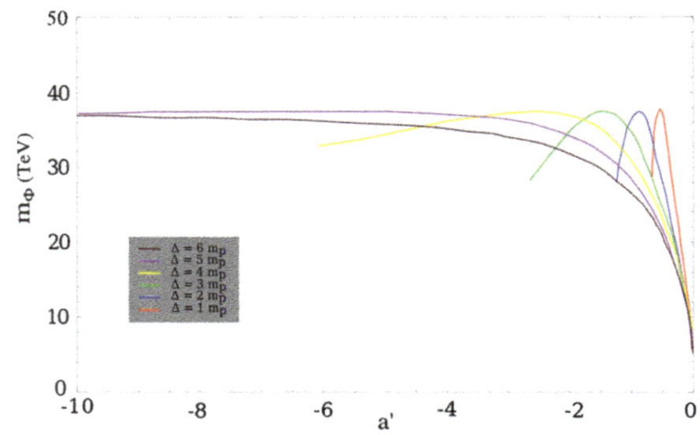

Fig. 3. m_ϕ vs a' for several values of Δ.

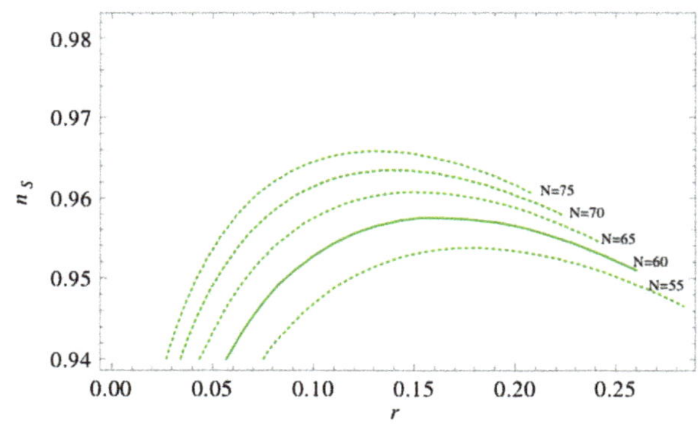

Fig. 2. n_S vs r for $\Delta = 3m_P$.

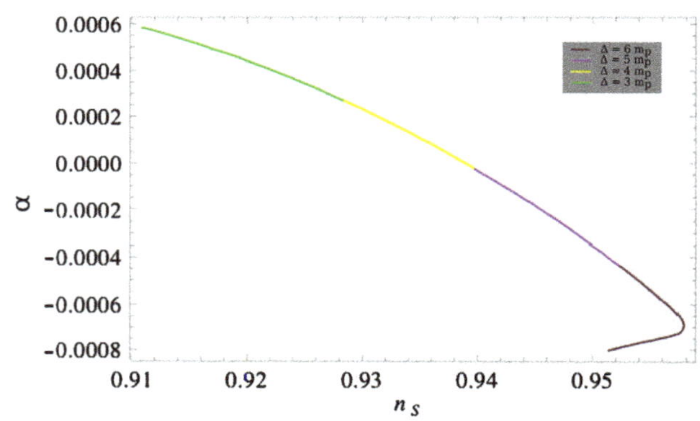

Fig. 4. α vs n_S for several values of Δ.

the current value of n_S and the bound on r. All this run into $a' \neq 0$ and Δ around few m_P.

In Fig. 2 we present our results for n_S and r in a plot confronting n_s with r for $\Delta = 3m_P$ and a' obeying the values corresponding to the green curve in Fig. 1 for several e-fold values. As we can see in that plot, the model predictions for n_S and r are in perfect agreement with the experimental bounds provided by PLANCK2015. This result is valid for any other choice of the values for the parameter Δ presented in Fig. 1.

Another interesting outcome we have obtained concerns the inflaton mass. Its expression at tree level is extracted from the diagonalization of the mass matrix M_R^2 in Eq. (17). As reheating demands very tiny λ_ϕ and $v_\phi \gg v_\rho, v_\eta, v_{\chi'}$, then the $(M_R^2)_{44}$ element of that matrix decouples incurring into the following expression for the inflaton mass at tree level, $m_\Phi \sim \sqrt{2\lambda_{10}}v_\phi$. When radiative corrections are plugged in, this expression receives a correction that depends on the parameters a' and Δ. In Fig. 3 we plot the behavior of the inflaton mass m_Φ with a' for some values of Δ. Even if v_ϕ is around 10^{10} GeV, but as the coupling λ_{10} is very small, as required by reheating phase, the inflaton gains a small mass when compared to the conventional chaotic inflation case. According to the prediction of our model, the inflaton may develop mass until few tens of TeV. This has implications to the reheating phase, as discussed below.

For sake of completeness, in Fig. 4 we plot the running index α versus n_S for some values of Δ. There we have a relatively small α value for all points as it has to be in chaotic inflation.

We finish this section by discussing reheating [50]. First of all notice that our inflaton couples to the heavy neutrinos through the Yukawa coupling in Eq. (7), and to scalars according to the last four terms in the potential in Eq. (19). Because $v_\phi \sim 10^{10}$ GeV, the inflaton develops mass around tens of TeV, as shown in Fig. 3. This order of magnitude for the inflaton mass forbids that it decays into a pair of heavy neutrinos because, as we will see below, in the model developed here heavy neutrinos gain mass around $\sim 10^7$ GeV. Thus reheating will be solely due to the decay of the inflaton into a pair of scalars.

Perceive that the 3-3-1 model in question has a complex scalar sector that turns quite impossible to obtain a well-behaved mixing matrix that connects the symmetrical scalars to the physical ones. This is particularly sure with the CP-even neutral scalars. This is because its mass matrix, given in Eq. (17), is a 4×4 one. Its diagonalization requires numerical approach. For our proposal here it is just enough to parametrizes the couplings among the inflaton and a pair of Higgs, provided by those last four terms in the potential in Eq. (19), by the general form: $\frac{\lambda}{8}v_\phi \Phi hh$ where λ represents any combination of the couplings $\lambda_{\phi,11,12,13}$ while h represents any combination of the eigenstates R_η, R_ρ and $R_{\chi'}$, or simply the heavier one, that is practically $R_{\chi'}$. This does not matter too much because the reheating phase puts constraint into the couplings λ's. We assume that one of the couplings $\lambda_{\phi,11,12,13}$ will be the dominant one and proceed with calculation.

In view of this, the expression for the decay width of the channel $\Phi \to hh$ is

$$\Gamma(\Phi \to hh) \sim \frac{\lambda^2 v_\phi^2}{32\pi m_\Phi}. \tag{28}$$

In this case reheating temperature is estimated to be [51]

$$T_R \sim 0.1\sqrt{\Gamma(\Phi \to hh) m_P}. \tag{29}$$

For $v_\phi = 10^{10}$ GeV and $m_\Phi \sim 10$ TeV, a reheating temperature around 10^9 GeV requires $\lambda \sim 10^{-6}$. This means that the couplings $\lambda_{\phi,11,12,13}$ in the potential must develop values at most around this order of magnitude which are typical values in chaotic inflation models [51]. In summary, although the inflaton has an unusual small mass, the model is efficient in reheating the universe.

4. Some remarks and conclusions

When η^0 and ϕ develop VEVs, the last two terms in the Lagrangian in Eq. (7) yields Dirac and Majorana mass terms for ν_L and N_R,

$$\mathcal{L} \supset M_D \bar{\nu}_L N_R + M \bar{N}_R^C N_R + \text{H.c.}, \tag{30}$$

where $M_D = h\frac{v_\eta}{\sqrt{2}}$ and $M = \frac{h' v_\phi}{\sqrt{2}}$. These terms provide the following mass matrix for the six massive neutrinos,

$$M_\nu = \begin{pmatrix} 0 & M_D \\ M_D^T & M \end{pmatrix}. \tag{31}$$

This is the mass matrix for the type I seesaw mechanism whose diagonalization, for $M \gg M_D$, leads to [52–56]

$$m_{\nu_L} \simeq \frac{M_D^2}{M} \quad \text{and} \quad M_R \simeq M. \tag{32}$$

We are interested in estimating the order of magnitude of the masses, only. As $a' \approx \frac{a}{32\pi^2\lambda'}$ and $a = -16h'^4$, for a' around 10 and $\lambda' \sim 10^{-14}$, as required by $\Delta_R^2 = 2.215 \times 10^{-9}$, we get $h' \sim 10^{-3}$, which results in $M_R \sim 10^7$ GeV. So as to obtain standard neutrinos at eV scale in agreement with solar and atmospheric neutrino oscillation, we just need $M_D \sim (10^{-1} - 10^{-2})$ GeV. This is obtained for h in the range $\sim (10^{-3} - 10^{-4})$ for $v_\eta \sim 10^2$ GeV. Such range of values for h are of the same order of the Yukawa couplings in the standard model.

Axion dark matter is considered as an attractive alternative to thermal WIMP dark matter. Our axion is invisible and receives mass through chiral anomaly, $m_a^2 \sim \frac{\Lambda_{QCD}^4}{f_{pq}^2}$, which is about 10^{-3} eV for $f_{pq} \sim 10^{10}$ GeV and $\Lambda_{QCD} \sim 10^{-1}$ GeV, turning our axion a natural candidate for cold dark matter. As PQ symmetry is broken during inflation, our axion will be produced in the early universe through the misalignment mechanism and its relic abundance is cast in Refs. [57,58].

Just few words about heavy neutrinos with masses around 10^7 GeV. These neutrinos interact with charged scalars, as allowed by the Yukawa coupling $h\bar{f}_L \eta N_R$, and may give rise to baryogenesis through leptogenesis. Because of the complexity and importance of such subject, we treat it separately elsewhere. However, for a previous treatment of this issue in a similar situation, but different scenario, we refer the reader to the Refs. [59,60].

In summary, several papers have proposed extensions of the standard model that provide a common origin to the understanding of the strong CP-problem, dark matter, inflation, and small neutrino masses. In this paper we argued that such proposal is elegantly realized in the framework of a 3-3-1 gauge model. In it the strong CP-problem is solved with the PQ symmetry whose associated axion is invisible and may constitute the dark matter of the universe. Inflation is driven by the real part of the neutral scalar singlet that contains the axion. Successful inflation was obtained by considering radiative corrections to the inflaton potential. The model has an unusual inflaton with mass of tens of TeV. Reheating is achieved through the decay of the inflaton into scalars, and neutrinos gain small mass through the type I seesaw mechanism.

We end by saying that we do not expect that the 3-3-1 model be the final theory valid in the range from TeV up to Planck scale. It is more probable that it is an effective theory of a more fundamental one that prevails in the high energy scale, as for example grand unification theories. If this is the case, it is reasonable to expect that the predictions done here be preserved in the context of such a final theory.

Acknowledgements

This work was supported by Conselho Nacional de Pesquisa e Desenvolvimento Científico – CNPq (C.A.S.P, P.S.R.S.) and Coordenação de Aperfeiçoamento de Pessoal de Nível Superior – CAPES (J.G.R.).

References

[1] M. Singer, J.W.F. Valle, J. Schechter, Phys. Rev. D 22 (1980) 738.
[2] P.H. Frampton, Phys. Rev. Lett. 69 (1992) 2889.
[3] C.A. de Sousa Pires, O.P. Ravinez, Phys. Rev. D 58 (1998) 035008, Phys. Rev. D 58 (1998) 35008, arXiv:hep-ph/9803409.
[4] C.A. de Sousa Pires, Phys. Rev. D 60 (1999) 075013, arXiv:hep-ph/9902406.
[5] P.B. Pal, Phys. Rev. D 52 (1995) 1659, arXiv:hep-ph/9411406.
[6] S. Weinberg, Phys. Rev. Lett. 40 (1978) 223.
[7] F. Wilczek, Phys. Rev. Lett. 40 (1978) 279.
[8] J.E. Kim, Phys. Rev. Lett. 43 (1979) 103.
[9] M.A. Shifman, A.I. Vainshtein, V.I. Zakharov, Nucl. Phys. B 166 (1980) 493.
[10] M. Dine, W. Fischler, M. Srednicki, Phys. Lett. B 104 (1981) 199.
[11] A.G. Dias, V. Pleitez, Phys. Rev. D 69 (2004) 077702, arXiv:hep-ph/0308037.
[12] J.C. Montero, C.A. De, S. Pires, V. Pleitez, Phys. Rev. D 65 (2002) 095001, arXiv:hep-ph/0112246.
[13] A.G. Dias, C.A. de, S. Pires, P.S. Rodrigues da Silva, A. Sampieri, Phys. Rev. D 86 (2012) 035007, arXiv:1206.2590 [hep-ph].
[14] P.V. Dong, H.N. Long, Phys. Rev. D 77 (2008) 057302, arXiv:0801.4196 [hep-ph].
[15] S.M. Boucenna, J.W.F. Valle, A. Vicente, Phys. Rev. D 92 (2015) 053001, arXiv:1502.07546 [hep-ph].
[16] M. Reig, J.W.F. Valle, C.A. Vaquera-Araujo, Phys. Rev. D 94 (2016) 033012, arXiv:1606.08499 [hep-ph].
[17] J.A. Rodriguez, M. Sher, Phys. Rev. D 70 (2004) 117702, arXiv:hep-ph/0407248.
[18] C. Promberger, S. Schatt, F. Schwab, Phys. Rev. D 75 (2007) 115007, arXiv:hep-ph/0702169.
[19] C. Promberger, S. Schatt, F. Schwab, S. Uhlig, Phys. Rev. D 77 (2008) 115022, arXiv:0802.0949 [hep-ph].
[20] A.C.B. Machado, J.C. Montero, V. Pleitez, Phys. Rev. D 88 (2013) 113002, arXiv:1305.1921 [hep-ph].
[21] A.J. Buras, F. De Fazio, J. Girrbach, J. High Energy Phys. 02 (2014) 112, arXiv:1311.6729 [hep-ph].
[22] A.J. Buras, F. De Fazio, J. Girrbach-Noe, J. High Energy Phys. 08 (2014) 039, arXiv:1405.3850 [hep-ph].
[23] R.M. Fonseca, M. Hirsch, Phys. Rev. D 94 (2016) 115003, arXiv:1607.06328 [hep-ph].
[24] F.S. Queiroz, C. Siqueira, J.W.F. Valle, Phys. Lett. B 763 (2016) 269, arXiv:1608.07295 [hep-ph].
[25] C. Salazar, R.H. Benavides, W.A. Ponce, E. Rojas, J. High Energy Phys. 07 (2015) 096, arXiv:1503.03519 [hep-ph].
[26] M. Lindner, M. Platscher, F.S. Queiroz, arXiv:1610.06587 [hep-ph], 2016.
[27] Q.-H. Cao, D.-M. Zhang, arXiv:1611.09337 [hep-ph], 2016.
[28] C.A. de, S. Pires, P.S. Rodrigues da Silva, J. Cosmol. Astropart. Phys. 0712 (2007) 012, arXiv:0710.2104 [hep-ph].
[29] J.K. Mizukoshi, C.A. de, S. Pires, F.S. Queiroz, P.S. Rodrigues da Silva, Phys. Rev. D 83 (2011) 065024, arXiv:1010.4097 [hep-ph].
[30] P.S. Rodrigues da Silva, arXiv:1412.8633 [hep-ph], 2014.
[31] P.V. Dong, C.S. Kim, D.V. Soa, N.T. Thuy, Phys. Rev. D 91 (2015) 115019, arXiv:1501.04385 [hep-ph].
[32] D.T. Huong, H.N. Long, Phys. At. Nucl. 73 (2010) 791, arXiv:0807.2346 [hep-ph].
[33] H.N. Long, http://dx.doi.org/10.3844/pisp.2016.1.14, 2015, arXiv:1501.01852 [hep-ph].
[34] V.N. Senoguz, Q. Shafi, Phys. Lett. B 668 (2008) 6, arXiv:0806.2798 [hep-ph].
[35] S.M. Boucenna, S. Morisi, Q. Shafi, J.W.F. Valle, Phys. Rev. D 90 (2014) 055023, arXiv:1404.3198 [hep-ph].
[36] A.G. Dias, A.C.B. Machado, C.C. Nishi, A. Ringwald, P. Vaudrevange, J. High Energy Phys. 06 (2014) 037, arXiv:1403.5760 [hep-ph].
[37] G. Barenboim, W.-I. Park, Phys. Lett. B 756 (2016) 317, arXiv:1508.00011 [hep-ph].
[38] G. Ballesteros, J. Redondo, A. Ringwald, C. Tamarit, arXiv:1608.05414 [hep-ph], 2016.
[39] G. Ballesteros, J. Redondo, A. Ringwald, C. Tamarit, arXiv:1610.01639 [hep-ph], 2016.
[40] A. Salvio, Phys. Lett. B 743 (2015) 428, arXiv:1501.03781 [hep-ph].
[41] P.A.R. Ade, et al., Planck, Astron. Astrophys. 594 (2016) A20, arXiv:1502.02114 [astro-ph.CO].
[42] A.G. Dias, C.A. de, S. Pires, P.S. Rodrigues da Silva, Phys. Rev. D 68 (2003) 115009, arXiv:hep-ph/0309058.
[43] J.C. Montero, F. Pisano, V. Pleitez, Phys. Rev. D 47 (1993) 2918, arXiv:hep-ph/9212271.

[44] R. Foot, H.N. Long, T.A. Tran, Phys. Rev. D 50 (1994) R34, arXiv:hep-ph/9402243.
[45] R.D. Peccei, in: Sources and Detection of Dark Matter and Dark Energy in the Universe, Proceedings, 4th International Symposium, DM 2000, Marina del Rey, USA, February 23–25, 2000, 2000, pp. 98–104, arXiv:hep-ph/0009030.
[46] C. Patrignani, et al., Particle Data Group, Chin. Phys. C 40 (2016) 100001.
[47] S.R. Coleman, E.J. Weinberg, Phys. Rev. D 7 (1973) 1888.
[48] A.R. Liddle, P. Parsons, J.D. Barrow, Phys. Rev. D 50 (1994) 7222, arXiv:astro-ph/9408015.
[49] D.H. Lyth, A.R. Liddle, The Primordial Density Perturbation: Cosmology, Inflation and the Origin of Structure, 2009.
[50] L.F. Abbott, E. Farhi, M.B. Wise, Phys. Lett. B 117 (1982) 29.
[51] A.D. Linde, Contemp. Concepts Phys. 5 (1990) 1, arXiv:hep-th/0503203.
[52] P. Minkowski, Phys. Lett. B 67 (1977) 421.
[53] M. Gell-Mann, P. Ramond, R. Slansky, in: Supergravity Workshop Stony Brook, New York, September 27–28, 1979, in: Conf. Proc., vol. C790927, 1979, p. 315, arXiv:1306.4669 [hep-th].
[54] R.N. Mohapatra, G. Senjanovic, Phys. Rev. Lett. 44 (1980) 912.
[55] G. Lazarides, Q. Shafi, C. Wetterich, Nucl. Phys. B 181 (1981) 287.
[56] J. Schechter, J.W.F. Valle, Phys. Rev. D 22 (1980) 2227.
[57] M.S. Turner, F. Wilczek, Phys. Rev. Lett. 66 (1991) 5.
[58] A.D. Linde, Phys. Lett. B 259 (1991) 38.
[59] D.T. Huong, P.V. Dong, C.S. Kim, N.T. Thuy, Phys. Rev. D 91 (2015) 055023, arXiv:1501.00543 [hep-ph].
[60] M. Bastero-Gil, A. Berera, R.O. Ramos, J.G. Rosa, Phys. Rev. Lett. 117 (2016) 151301, arXiv:1604.08838 [hep-ph].

9

Evolution of spherical over-densities in tachyon scalar field model

M.R. Setare [a], F. Felegary [b], F. Darabi [b,*]

[a] Department of Science, Campus of Bijar, University of Kurdistan, Bijar, Iran
[b] Department of Physics, Azarbaijan Shahid Madani University, Tabriz, 53714-161, Iran

ARTICLE INFO

Editor: J. Hisano

ABSTRACT

We study the tachyon scalar field model in flat FRW cosmology with the particular potential ϕ^{-2} and the scale factor behavior $a(t)=t^n$. We consider the spherical collapse model and investigate the effects of the tachyon scalar field on the structure formation in flat FRW universe. We calculate $\delta_c(z_c)$, $\lambda(z_c)$, $\xi(z_c)$, $\Delta_V(z_c)$, $\log[\nu f(\nu)]$ and $\log[n(k)]$ for the tachyon scalar field model and compare the results with the results of EdS model and ΛCDM model. It is shown that in the tachyon scalar field model the structure formation may occur earlier, in comparison to the other models.

1. Introduction

The last cosmological and astrophysical data of Large Scale structure, the observations of type Ia and Cosmic Microwave Background radiation have demonstrated that currently there is an acceleration expansion phase in the universe [4,18]. The cosmic expansion can be well described by a negative pressure so-called dark energy (DE). The simplest candidate for DE is the cosmological constant. However, the cosmological constant suffers from the fine-tuning and the cosmic coincidence problems [6,24]. Therefore, to avoid these problems, different models for dark energy have been proposed such as quintessence, K-essence, tachyon [20], ghost [27], phantom, quintom [5], and the quantum gravity models, as well as holographic [28] and new agegraphic models [6,14]. The tachyon model as a scalar field model arises in particle physics and string theory. Thus, it can be considered as one of the potential candidates to describe the nature of the DE.

On the other hand, the problem of structure formation in the universe is a very important issue in theoretical cosmology. A simple model of structure formation is the spherical collapse model. The spherical collapse model was presented by Gunn and Gott [8]. This model studies the evolution of growth of overdense structures with respect to the dynamics of scale factor or cosmic redshift. The dynamics of overdense structures depends on the dynamics of the background Hubble flow and expansion of the universe. In the frame of general relativity, the spherical collapse model has been studied [7,9,19]. In this paper, we study the spherical collapse and the evolution of spherical overdensities in the framework of tachyon scalar field model and compare the results with the results of Einstein-de Sitter (EdS) and Λ-Cold Dark Matter (ΛCDM) models.

2. Cosmology with Tachyon scalar field

The Lagrangian of tachyon scalar field over a cosmological background is given by [26]

$$\mathcal{L}=-V(\phi)\sqrt{1-\partial_a\phi\,\partial^a\phi}, \tag{1}$$

where ϕ and $V(\phi)$ are the tachyon scalar field and tachyon potential, respectively, and we consider the Friedmann–Robertson–Walker (FRW) metric having the cosmic time t dependent scale factor $a(t)$. For a homogeneous field, the equation of motion is obtained as

$$\frac{\ddot{\phi}}{1-\dot{\phi}^2}+3H\dot{\phi}+\frac{\acute{V}(\phi)}{V(\phi)}=0, \tag{2}$$

where the symbols $\cdot$ and $'$ denote the derivatives with respect to t and ϕ, respectively, and $H=\dot{a}/a$ is called the Hubble parameter. In the flat FRW universe, the energy density ρ_Λ and the pressure p_Λ of the tachyon field read as

$$\rho_\Lambda=\frac{V(\phi)}{\sqrt{1-\dot{\phi}^2}}, \tag{3}$$

$$p_\Lambda=-V(\phi)\sqrt{1-\dot{\phi}^2}. \tag{4}$$

For the pressureless matter and tachyon scalar field matter, the Friedmann equation is given by

* Corresponding author.
E-mail addresses: rezakord@ipm.ir (M.R. Setare), falegari@azaruniv.ac.ir (F. Felegary), f.darabi@azaruniv.ac.ir (F. Darabi).

$$H^2 = \frac{1}{3M_{pl}^2}(\rho_m + \rho_\Lambda), \tag{5}$$

where ρ_Λ is the energy density of tachyon scalar field, and ρ_m is the density of pressureless matter. We suppose that there is no interaction between ρ_Λ and ρ_m, so the continuity equations are given separately by

$$\dot{\rho}_\Lambda + 3H\rho_\Lambda(1+\omega_\Lambda) = 0, \tag{6}$$

$$\dot{\rho}_m + 3H\rho_m = 0. \tag{7}$$

Using Eqs. (3) and (4) and also $p_\Lambda = \omega_\Lambda \rho_\Lambda$, the equation of state parameter (EoS) for tachyon scalar field is obtained as

$$\omega_\Lambda = \dot{\phi}^2 - 1. \tag{8}$$

The requirement for a real ρ_Λ results in $0 < \dot{\phi}^2 < 1$ according to which ω_Λ should vary as $-1 < \omega_\Lambda < 0$. The fractional energy densities are defined by

$$\Omega_\Lambda = \frac{\rho_\Lambda}{3M_{pl}^2 H^2}, \tag{9}$$

$$\Omega_m = \frac{\rho_m}{3M_{pl}^2 H^2}. \tag{10}$$

Taking time derivative of Eq. (9) and using Eq. (6) yields

$$\dot{\Omega}_\Lambda = -\Omega_\Lambda H\Big[3(1+\omega_\Lambda) + 2\frac{\dot{H}}{H^2}\Big]. \tag{11}$$

Also, taking time derivative of Eq. (5) and using Eqs. (6) and (7) yields

$$2\frac{\dot{H}}{H^2} = -3(1+\omega_\Lambda\Omega_\Lambda). \tag{12}$$

Using Eq. (12) and inserting Eq. (11), we obtain

$$\Omega'_\Lambda = 3\omega_\Lambda\Omega_\Lambda(\Omega_\Lambda - 1). \tag{13}$$

Here, the prime is the derivative with respect to $x = \ln a$ where $a = (1+z)^{-1}$ and z is the cosmic redshift. Using $\frac{d}{dx} = -(1+z)\frac{d}{dz}$ and Eq. (8), one finds

$$\frac{d\Omega_\Lambda}{dz} = -3\Omega_\Lambda(\Omega_\Lambda - 1)(\dot{\phi}^2 - 1)(1+z)^{-1}. \tag{14}$$

The differential equation for the evolution of dimensionless Hubble parameter, $E(z) = \frac{H}{H_0}$, in tachyon scalar field model, is obtained by using Eqs. (6), (7), (8) and (12) as follows

$$\frac{dE}{dz} = \frac{3}{2}\frac{E}{(1+z)}\Big[1 + \Omega_\Lambda(\dot{\phi}^2 - 1)\Big]. \tag{15}$$

Now, we consider the following particular potential which results in the scalar field with linear time dependence and the scale factor with suitable power law behavior, as follows [26]

$$V(\phi) = \frac{2n}{M_{pl}^2}(1 - \frac{2}{3n})^{\frac{1}{2}}\frac{1}{\phi^2}, \tag{16}$$

$$\phi = \sqrt{\frac{2}{3n}}t, \tag{17}$$

$$a(t) = t^n. \tag{18}$$

Taking time derivative of Eq. (17), inserting in Eq. (8) and using Eq. (18), we can obtain the equation of state parameter for tachyon scalar field model

$$\omega_\Lambda = \frac{2}{3n} - 1. \tag{19}$$

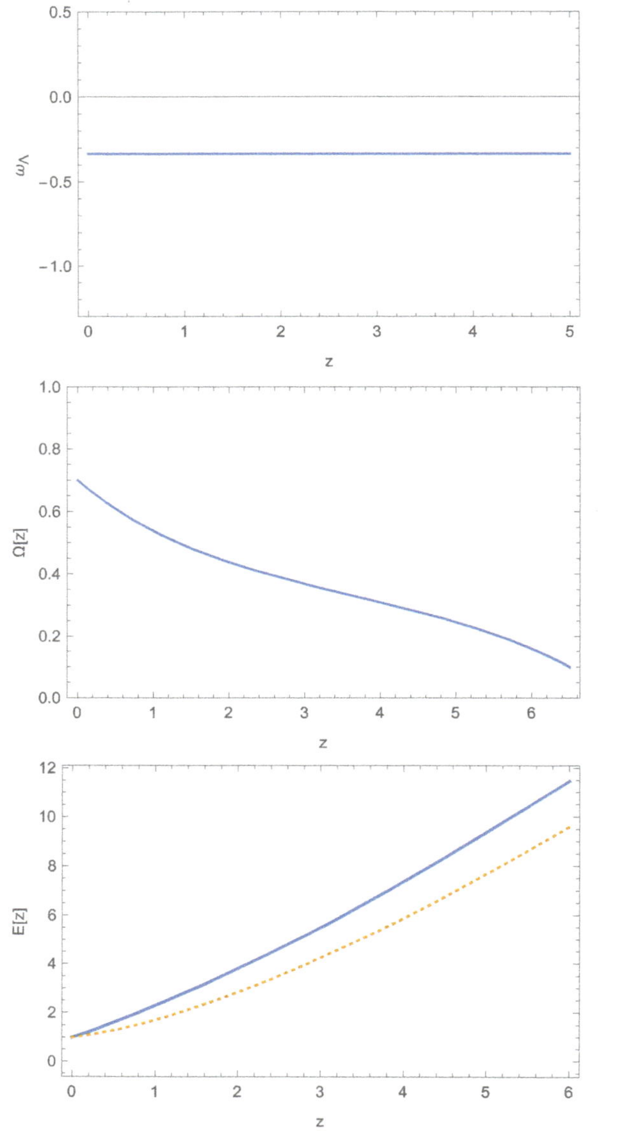

Fig. 1. The evolution of EoS parameter (top), dark energy density parameter (middle), and dimensionless Hubble parameter (down) of tachyon scalar field model with respect to the redshift parameter z. The thick line represents the tachyon scalar field model for $n = 1$ and the dotted line shows the ΛCDM model.

In Eq. (19), we can see that if $n \geq \frac{2}{3}$, then we will have $-1 < \omega_\Lambda < 0$. Using Eq. (19) and inserting it in Eqs. (14), (15) we can get the evolution of EoS parameter (ω_Λ), the density parameter of dark energy (Ω_Λ), and the dimensionless Hubble parameter ($E(z)$) in tachyon scalar field model as a function of cosmic redshift. In Fig. 1, assuming the present values $\Omega_{\Lambda_0} \approx 0.7$, $\Omega_{m_0} \approx 0.3$ and $H_0 \approx 67.8\frac{\text{km}}{\text{s Mpc}}$, we have shown the evolution of EoS parameter, the evolution of density parameter and the evolution of dimensionless Hubble parameter of tachyon scalar field model with respect to the redshift parameter z for the typical value $n = 1$.

3. Linear perturbation theory

In this section, we study the linear growth of perturbation of nonrelativistic dust matter by computing the evolution of growth factor $g(a)$ in tachyon scalar field model, and then compare it with the evolution of growth factor in EdS and ΛCDM models. The dif-

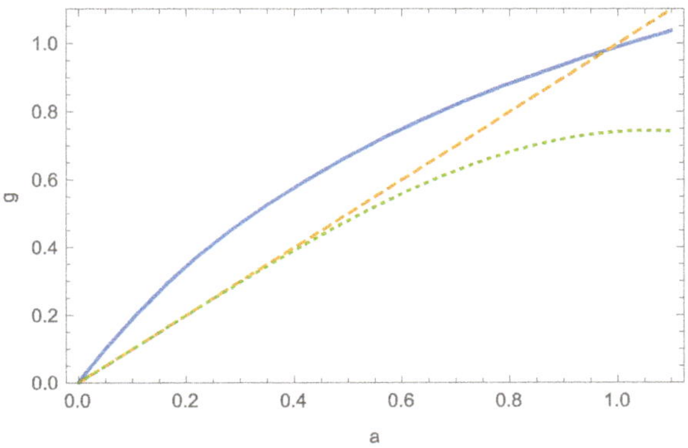

Fig. 2. Time evolution of the growth factor as a function of the scale factor. The thick line represents the tachyon scalar field model for $n = 1$. The dotted line indicates the ΛCDM model and the dashed line represent the *EdS* model.

ferential equation for the evolution of the growth factor $g(a)$ is given by [11,12,16] (see Appendix I)

$$g''(a) + (\frac{3}{a} + \frac{E'(a)}{E(a)})g'(a) - \frac{3}{2}\frac{\Omega_{m0}}{a^5 E^2(a)}g(a) = 0. \tag{20}$$

In order to study the linear growth in tachyon scalar field model, using Eqs. (14), (15) and (19) for $n = 1$, we solve numerically Eq. (20). To obtain the linear growth of structures in the EdS model and the ΛCDM model, we use the procedure used in Ref. [28]. In Fig. 2, we have plotted the evolution of growth factor $g(a)$ with respect to the scale factor. At first, namely for small scale factors, the growth factor in the tachyon scalar field model is larger than those of EdS and ΛCDM models. However, for rather larger scale factors, the growth factor in the tachyon scalar field model becomes smaller than the EdS model while it is still larger enough than that of ΛCDM model. This means that, at the beginning, the tachyon scalar field model predicts structure formation more efficient than EdS and ΛCDM models. For later times, however, the structure formation in the tachyon scalar field model is dropped behind that of EdS model, whereas it precedes the structure formation in the ΛCDM model.

4. Spherical collapse in the tachyon scalar field model

The structure formation is described by a non-linear differential equation for the evolution of the matter perturbation δ in a matter dominated universe [3,13]. In [1] this differential equation was generalized to the case of evolution of δ in a universe including a dark energy component. Now, we consider the non-linear differential equation which is given by [11] (see Appendix I)

$$\delta'' + \left(\frac{3}{a} + \frac{E'(a)}{E(a)}\right)\delta' - \frac{4}{3}\frac{\delta'^2}{1+\delta} - \frac{3}{2}\frac{\Omega_{m0}}{a^5 E^2(a)}\delta(1+\delta) = 0, \tag{21}$$

where $'$ denotes the derivative with respect to a. In the linear regime, we have

$$\delta'' + \left(\frac{3}{a} + \frac{E'(a)}{E(a)}\right)\delta' - \frac{3}{2}\frac{\Omega_{m0}}{a^5 E^2(a)}\delta = 0. \tag{22}$$

In EdS model, we consider the initial conditions $\delta_i = 20.9 \times 10^{-4}$, $\acute{\delta}_i = 0$ and $a_i = 10^{-4}$ [11]. In order to study the linear growth of density perturbation and the non-linear growth of density perturbation in tachyon scalar field model, using Eqs. (14), (15) and (19) for $n = 1$, we solve numerically Eqs. (21) and (22) (see Appendix II). To obtain the linear growth of density perturbation δ in the EdS model and the ΛCDM model, we use the procedure used in Ref. [28]. The Fig. 3(a) shows that the linear growth factor in the tachyon scalar field model is larger than those of EdS and the ΛCDM models, and the Fig. 3(b) shows that the non-linear growth factor in the tachyon scalar field model is larger than the EdS model.

5. Determination of δ_c and Δ_V

As time passes, the perturbation is growing and one can no longer use the linear regime. At this stage, the radius of perturbation region becomes maximal $R = R_{max}$ and the perturbation stops growing. This condition is called turn-around which points to the epoch when the grows of perturbation decouples from the Hubble flow of the homogenous background. After the turn-around the perturbation starts contracting. For a perfect spherical symmetry and perfect pressureless matter, the perturbation would collapse to a single point becoming infinitely dense. Since there is hardly any perfect spherical symmetric overdensity in the universe, the perturbation does not collapse to a single point and finally a virialized object of a certain finite size in equilibrium state is formed that is called halo.

We call (z_c, R_c) and (z_{ta}, R_{ta}) as the redshift and radius corresponding to virialization and the turn-around epochs, respectively. Now, we peruse two characterizing quantities of the spherical collapse model for the tachyon scalar field model: the virial overdensity Δ_V and the linear overdensity parameter δ_c. We consider a spherical overdense region with matter density ρ in a surrounding universe described by its background dynamics and density ρ_b. The virial overdensity Δ_V is defined by [10]

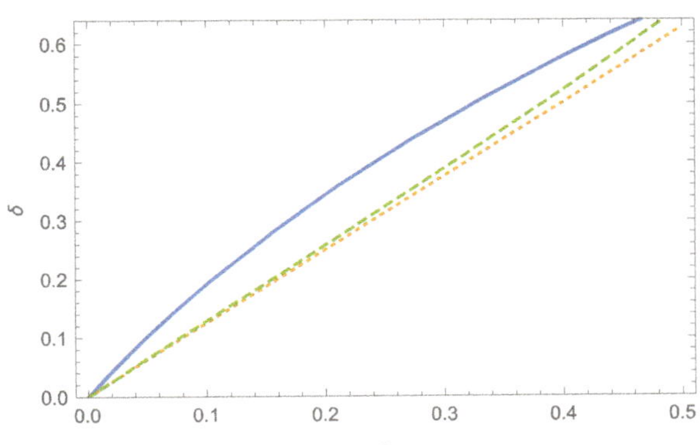

(a) The linear growth of density perturbation δ in terms of a.

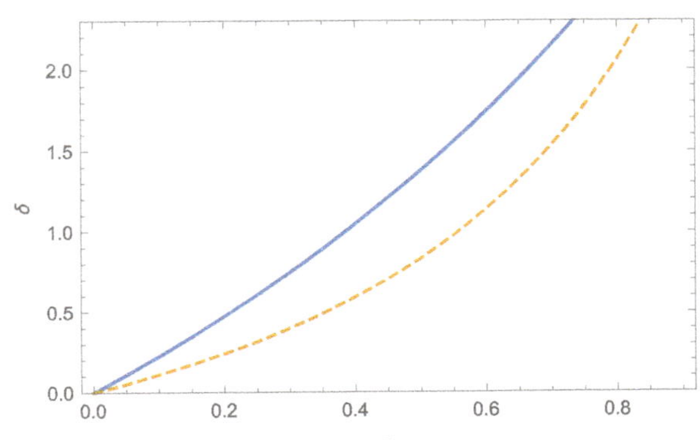

(b) The non-linear growth of density perturbation δ in terms of a.

Fig. 3. The thick line represents the tachyon scalar field model for $n = 1$. The dotted line indicates the ΛCDM model and the dashed line indicates the EdS model.

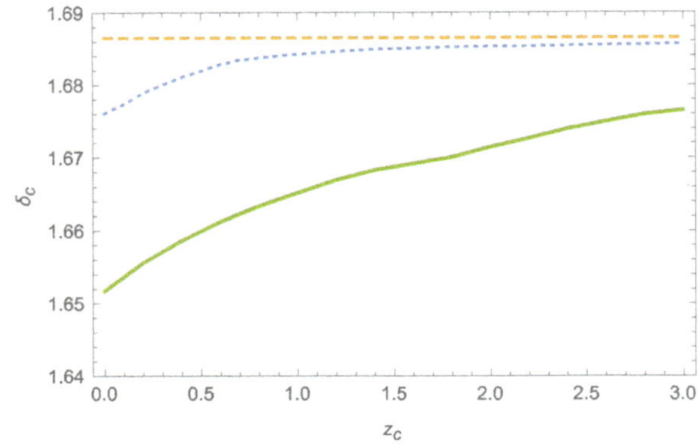

Fig. 4. The time evolution of the linear overdensity, $\delta_c(z)$, in terms of a function of the collapse redshift for the ΛCDM model, the EdS model and the tachyon scalar field model. The thick line represents the tachyon scalar field model for $n=1$. The dotted line indicates the ΛCDM model and the dashed line indicates the EdS model.

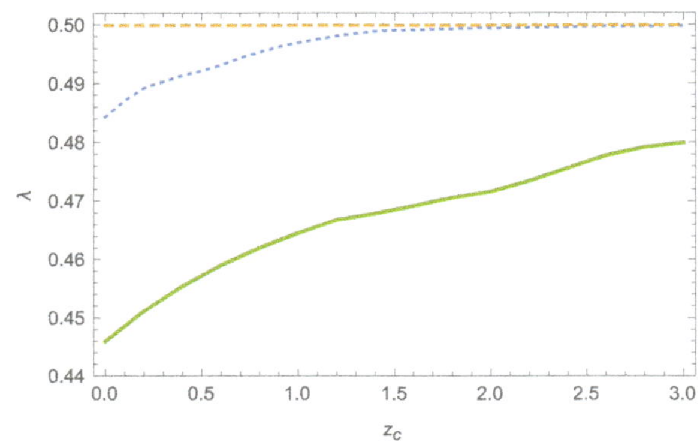

Fig. 5. The virial radius $\lambda(z_c)$ in terms of the collapse redshift z_c for the ΛCDM model, the EdS model and the tachyon scalar field model. The thick line represents the tachyon scalar field model for $n=1$. The dotted line indicates the ΛCDM model and the dashed line indicates the EdS model.

$$\Delta_V = \frac{\rho}{\rho_b}\frac{R_c}{a_c}, \tag{23}$$

which is a function of scale factor and redshift. We can rewrite the virial overdensity Δ_V as follows [10]

$$\Delta_V = 1+\delta(a_c) = \xi\left(\frac{x_c}{\lambda}\right)^3, \tag{24}$$

where

$$x_c = \frac{a_c}{a_{ta}}, \tag{25}$$

$$\xi = \frac{\rho(R_{ta})}{\rho_b(a_{ta})} = 1+\delta(a_{ta}). \tag{26}$$

Here, λ is the virial radius which is given by [23]

$$\lambda = \frac{1-\frac{\eta_v}{2}}{2+\eta_t-\frac{3}{2}\eta_v}, \tag{27}$$

where η_t and η_v are the (Wang–Steinhardt) WS parameters

$$\eta_t = \frac{2}{\xi}\frac{\Omega_\Lambda(a_{ta})}{\Omega_m(a_{ta})}, \tag{28}$$

$$\eta_v = \frac{2}{\xi}\frac{\Omega_\Lambda(a_c)}{\Omega_m(a_c)}\left(\frac{a_{ta}}{a_c}\right). \tag{29}$$

Now, we discuss the results obtained for the linear overdensity parameter and the virial overdensity for the models introduced in this work. The Fig. 4 shows the time evolution of linear overdensity, $\delta_c(z)$ in terms of a function of the collapse redshift for the ΛCDM model, the EdS model and the tachyon scalar field model. In the EdS model, δ_c is independent of the redshift, hence it has a constant value i.e. $\delta_c = 1.686$. In the ΛCDM model, δ_c is smaller than 1.686 but the time evolution of the linear overdensity approaches the value of the EdS model at high redshifts.

In fact, at high redshifts we have a matter dominated universe (dust matter), but at lower redshifts we have a dark energy dominated universe, thus the structure formation must occur earlier. In the tachyon scalar field model, δ_c drives more slowly than the ΛCDM and the EdS models because in Fig. 1, the Hubble parameter in the tachyon scalar field model is larger than that of ΛCDM model.

In the Fig. 5, we represent $\lambda(z_c)$ in terms of z_c for the ΛCDM model, the EdS model and the tachyon scalar field model. In the EdS model, $\lambda(z_c)$ is independent of the redshift, thus it has a constant value i.e. $\lambda = 0.5$. In the ΛCDM model, $\lambda(z_c)$ is smaller than 0.5 but it approaches the value of the EdS model at high redshifts. In the tachyon scalar field model, $\lambda(z_c)$ drives more slowly than the ΛCDM and the EdS models but its value approaches the value of the EdS model at high redshifts. Therefore, we can conclude that the size of structures in the ΛCDM model is larger than the tachyon scalar field model.

In the Fig. 6(a), we represent $\xi(z_c)$ in terms of z_c for the ΛCDM model, the EdS model and the tachyon scalar field model. In the EdS model, $\xi(z_c)$ is independent of redshift thus it has a constant

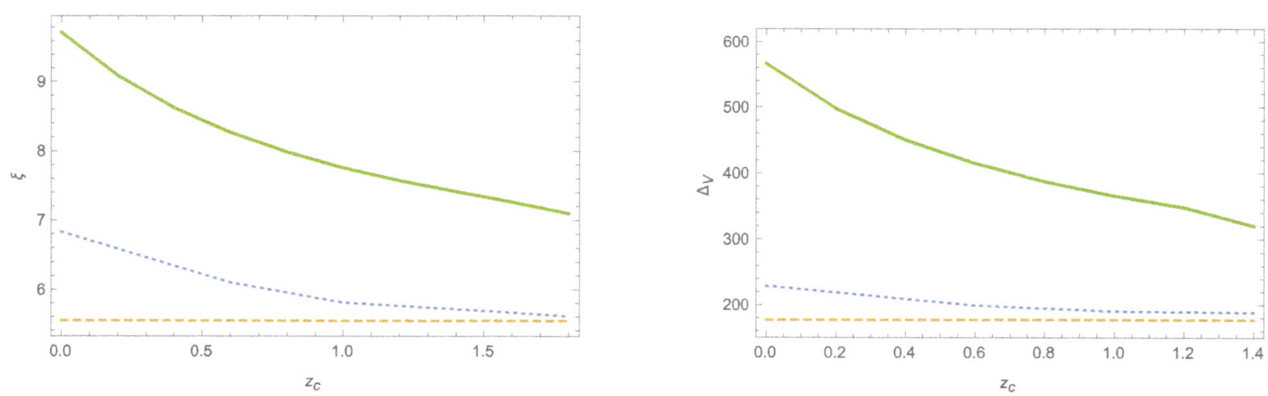

(a) The variation of $\xi(z_c)$ - z_c for the ΛCDM model, the EdS model and the tachyon scalar field model.

(b) The variation of $\Delta_V(z_c)$ - z_c for the ΛCDM model, the EdS model and the tachyon scalar field model.

Fig. 6. The blue, green and red lines represent the tachyon scalar field model for $n = 0.9, 1, 1.1$, respectively. The dotted line indicates the ΛCDM model and the dashed line indicates the EdS model. (For interpretation of the references to color in this figure legend, the reader is referred to the web version of this article.)

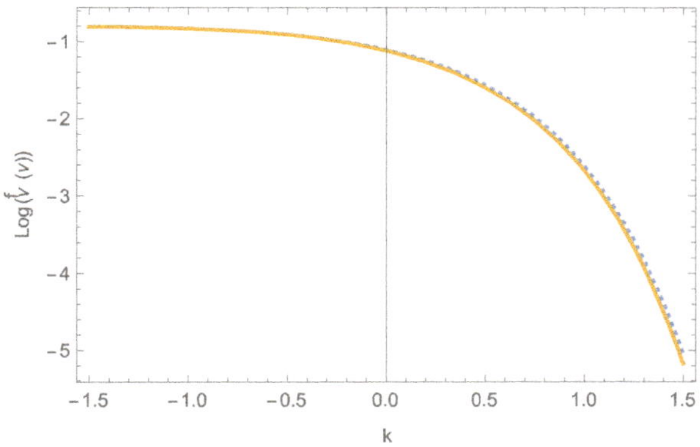

(a) The evolution of the mass function with respect to k for tachyon scalar field model and ΛCDM model in the case $z = 0$.

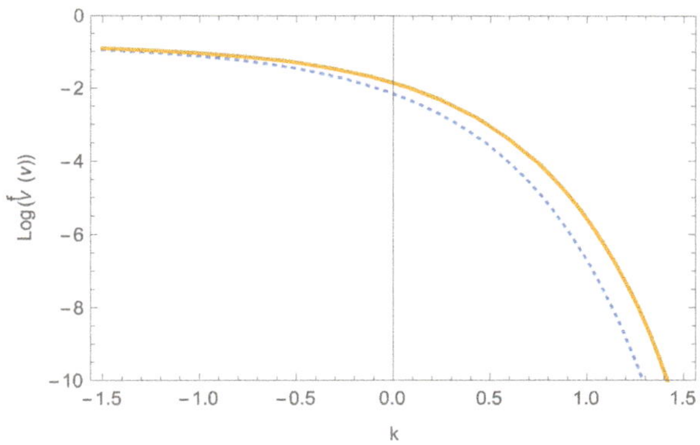

(b) The evolution of the mass function with respect to k for tachyon scalar field model and ΛCDM model in the case $z = 1$.

Fig. 7. The thick line represents the tachyon scalar field model for $n=1$ and The dotted line indicates the ΛCDM model.

value i.e. $\xi = 5.6$. In the ΛCDM model, $\xi(z_c)$ is larger than 5.6 but its value approaches the value of the EdS model in terms of high redshifts. In the tachyon scalar field model, $\xi(z_c)$ drives faster than the ΛCDM model and the EdS model but its value approaches the value of the EdS model in terms of high redshifts. Therefore, we can conclude that in the tachyon scalar field model the overdense spherical regions in terms of z_c are denser than the ΛCDM model and the EdS model.

In the Fig. 6(b), we show the virial overdensity $\Delta_V(z_c)$ in terms of z_c for the ΛCDM model, the EdS model and the tachyon scalar field model. In the EdS model, $\Delta_V(z_c)$ is independent of redshift, thus it has a constant value, $\Delta_V = 178$. In the ΛCDM model, $\Delta_V(z_c)$ drives more faster than 178, but its value approaches the value of the EdS model in terms of high redshifts. In the tachyon scalar field model, $\Delta_V(z_c)$ drives faster than the ΛCDM model and its value approaches the value of the EdS model in terms of high redshifts. The evolution of virial overdensity parameter $\Delta_V(z_c)$ is the main quantity for the halo size. Therefore, we can conclude that in the tachyon scalar field model the halo size is larger than the EdS model and the ΛCDM model.

6. Number density and mass function

The average comoving number density of halos of mass M is given by [2,15]

$$n(M,z) = (\frac{\rho}{M^2})\frac{d\log \upsilon}{d\log M}\nu f(\nu), \tag{30}$$

where $f(\upsilon)$ and ρ are the multiplicity function and the background density, respectively and ν is given by

$$\nu = \frac{\delta_c^2}{\sigma^2(M)}. \tag{31}$$

Here $\sigma(M)$ is the r.m.s of the mass fluctuation in sphere of mass M. We can use the formula given by [22]

$$\sigma(M,z) = \sigma_8(z)\Big(\frac{M}{M_8}\Big)^{-\frac{\gamma(M)}{3}}, \tag{32}$$

where σ_8 is the mass variance of the overdensity on the scale of R_8, $M_8 = 6\times 10^{14}\Omega_{m_0}$ $h^{-1} M_\odot$ and $R_8 = 8$ h^{-1} Mpc are the mass and the radius inside a sphere. Also, $\sigma_8(z)$ is given by

$$\sigma_8(z) = g(z)\sigma_8(M, z=0), \tag{33}$$

where $g(z)$ is the linear growth factor, $\sigma_{8,DE}(M, z=0) = 0.8\times\Big(\frac{\delta_{c,DE}(z=0)}{\delta_{c,\Lambda CDM}(z=0)}\Big)$ and

$$\gamma(M) = (0.3\Gamma+0.2)\Big[2.92+\frac{1}{3}\log(\frac{M}{M_8})\Big], \tag{34}$$

where $\Gamma = \Omega_{m_0} h\exp(-\Omega_b - \frac{\Omega_b}{\Omega_{m_0}})$. Eqs. (32) and (34) have a validation range [22]. They express that the fitting formula predicts higher values of the variance for $M < M_8$ and the fitting formula predicts lower values of the variance for $M > M_8$. Following [21], we apply ST mass function

$$\nu f_{ST}(\nu) = 0.3222\sqrt{\frac{0.707\nu}{2\pi}}\Big[1+(0.707\nu)^{-0.3}\Big]\exp(-\frac{0.707\nu}{2}). \tag{35}$$

We use the mass function introduced by del Popolo (PO mass function) [17]

$$\nu f(\nu) = 1.75\sqrt{\frac{0.707\nu}{2\pi}}\Big[1+\frac{0.1218}{(0.707\nu)^{0.585}}+\frac{0.0079}{(0.707\nu)^{0.4}}\Big] \times\exp\Big[-0.4019\times 0.707\nu\Big(1+\frac{0.5526}{(0.707\nu)^{0.585}} +\frac{0.02}{(0.707\nu)^{0.4}}\Big)^2\Big]. \tag{36}$$

Also, we use the mass function (YNY mass function) presented in [25]

$$\nu f(\nu) = 0.298\Big[1+(0.893\sqrt{\frac{\nu}{2}})^{1.39}\Big]\nu^{(\frac{0.408}{2})}\exp\Big[-(0.893\sqrt{\frac{\nu}{2}})^2\Big]. \tag{37}$$

Now, we represent the evolution of the ST mass function with respect to k ($k = \log(\frac{M}{M_8})$) in Fig. 7 for the tachyon scalar field model and the ΛCDM model. We can see that the evolution of ST mass function with respect to k is the same for tachyon scalar field and the ΛCDM models in the $z=0$ case, but it is different for tachyon scalar field and ΛCDM models in the $z=1$ case.

Using Eqs. (30) and (35), for the tachyon scalar field model and the ΛCDM model, we obtain the average comoving number density of halos of mass M in the cases $z=0,1$. In Fig. 8, we can see explicitly the differences for the cases $z=0$ and $z=1$. We can see that difference of the number densities of halo objects is negligible

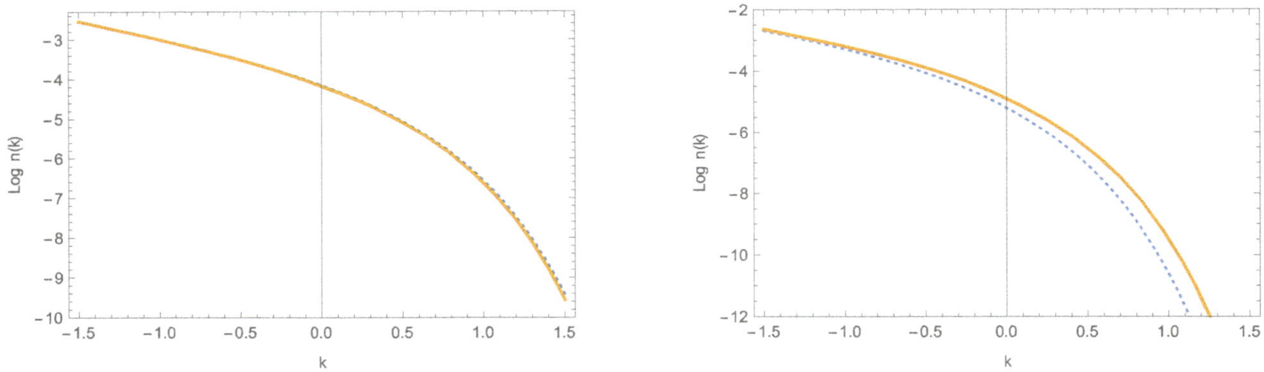

(a) The evolution the number densities of halo objects for the tachyon scalar field model and the ΛCDM model in the case $z = 0$.

(b) The evolution the number densities of halo objects for the tachyon scalar field model and the ΛCDM model in the case $z = 1$.

Fig. 8. The thick line represents the tachyon scalar field model for $n = 1$ and the dotted line indicates the ΛCDM model.

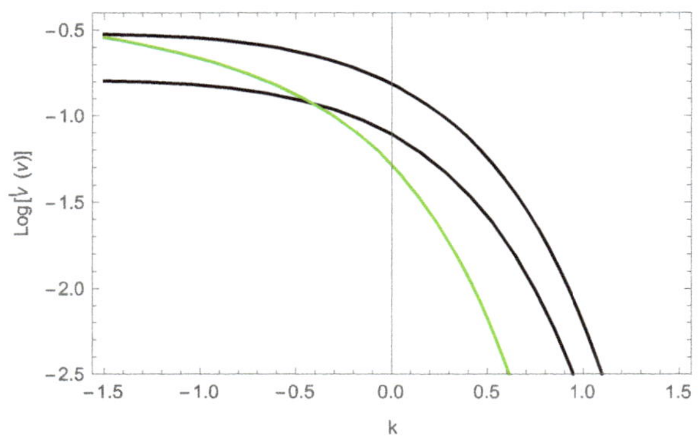

Fig. 9. The evolution of the various mass functions with respect to k for the tachyon scalar field model, $n = 1$, in the case $z = 0$. The blue thick line represent ST mass function, the red thick line represent PO mass function and the green thick line represent YNY mass function. (For interpretation of the references to color in this figure legend, the reader is referred to the web version of this article.)

for small objects in the case $z = 1$. Therefore, we can obtain the number density of halo objects for high mass, and we find that the number of objects per unit mass is increasing for high mass in the tachyon scalar field model. Also, using Eqs. (35), (36) and (37), we can compare the various mass functions at $k = 0$ in Fig. 9. We can see that the PO mass function is larger than ST mass function and YNY mass function for all mass scales.

7. Conclusion

In this paper, we have studied the evolution of spherical overdensities in tachyon scalar field model by assuming a particular potential and the scale factor with power law behavior. We have shown the evolution of the EoS parameter, the evolution of the density parameter and the evolution of the dimensionless Hubble parameter of tachyon scalar field model with respect to a function of z, for a typical value $n = 1$. We have also shown that at early times of the scale factor evolution, the growth factor in the tachyon scalar field model drives faster than the EdS and ΛCDM models. So, it is concluded that in the tachyon scalar field model the structure formation may occur sooner than in the other models. At later times, however, we have shown that the growth factor in the tachyon scalar field model drives more slower than that of EdS model. Also, in the EdS model, δ_c is independent of the redshift and thus it has a constant value $\delta_c = 1.686$. In the ΛCDM model, δ_c is smaller than 1.686, but the time evolution of the linear overdensity approaches the value of EdS model at high redshifts. In fact, at high redshifts, we have a matter dominated universe (dust matter), but at lower redshifts we have a dark energy dominated universe, thus the structure formation occurs earlier. In the tachyon scalar field model, δ_c is driven more slower than that of ΛCDM and EdS models, because in Fig. 1, the Hubble parameter in the tachyon scalar field model is larger than that of ΛCDM model.

Also, we have shown that in the EdS model, $\lambda(z_c)$ is independent of the redshift, thus it has a constant value i.e. $\lambda = 0.5$. Moreover, the size of structures in the ΛCDM model was larger than that of tachyon scalar field model. In the EdS model, $\xi(z_c)$ is independent of the redshift, hence it has a constant value i.e. $\xi = 5.6$. We have shown that in tachyon scalar field model, the overdense spherical regions with respect to z_c are denser than those of ΛCDM and EdS models. In the EdS model, $\Delta_V(z_c)$ is independent of the redshift, so it has a constant value, $\Delta_V = 178$. The evolution of virial overdensity parameter $\Delta_V(z_c)$ is the main quantity for the halo size. Therefore, we have found that in the tachyon scalar field model the halo size is larger than the EdS and ΛCDM models.

Finally, we have shown that the evolution of the ST mass function with respect to k is the same for tachyon scalar field and ΛCDM models in the $z = 0$ case, but it is not the same for tachyon scalar field model and the ΛCDM model in the $z = 1$ case. Also, the evolution of the number density with respect to k is the same for the tachyon scalar field and ΛCDM models in the $z = 0$ case, but its evolution is not the same for the tachyon scalar field and ΛCDM models in the $z = 1$ case. The difference of number densities of halo objects is negligible for small objects in the $z = 1$ case. Therefore, in obtaining the number density of halo objects for high mass, we find that the number of objects per unit mass is increasing for high mass in the tachyon scalar field model.

Acknowledgement

We would like to thank M. Malekjani for giving us useful comments that helped us to improve the scientific content of the manuscript.

Appendix I

The Lagrangian of tachyon scalar field over a cosmological background is given by [26]

$$\mathcal{L} = -V(\phi)\sqrt{1 - \partial_a\phi\, \partial^a\phi}\,, \tag{38}$$

where ϕ and $V(\phi)$ are the tachyon scalar field and tachyon potential, respectively. One can obtain the energy-momentum tensor of the tachyon scalar field as follows [29]

$$T_{\mu\nu} = \frac{V(\phi)\partial_\mu\phi\partial_\nu\phi}{\sqrt{1+g^{\alpha\beta}\partial_\alpha\phi\partial_\beta\phi}} - g_{\mu\nu}V(\phi)\sqrt{1+g^{\alpha\beta}\partial_\alpha\phi\partial_\beta\phi}\,. \tag{39}$$

Now, considering (39) as describing a perfect fluid, the energy density ρ and the pressure p for the tachyon scalar field are given by

$$\rho = -T_0^0 = \frac{V(\phi)}{\sqrt{1-\dot{\phi}^2}},$$

$$p = T_i^i = -V(\phi)\sqrt{1-\dot{\phi}^2}.$$

The fundamental equations for cosmic fluid in Newtonian gravity are defined as follows [11]

$$\frac{\partial\rho}{\partial t} + \nabla_{\vec{r}}.(\rho\vec{v}) + \frac{p}{c^2}\nabla_{\vec{r}}.\vec{v} = 0, \tag{40}$$

$$\frac{\partial\vec{v}}{\partial t} + (\vec{v}.\nabla_{\vec{r}})\vec{v} + \nabla_{\vec{r}}\Phi + \frac{c^2\nabla_{\vec{r}}p + \vec{v}\dot{p}}{\rho c^2 + p} = 0, \tag{41}$$

$$\nabla^2\Phi = 4\pi G\Big(\rho + \frac{3p}{c^2}\Big), \tag{42}$$

$$\dot{\bar{\rho}} + 3H\Big(\bar{\rho} + \frac{p}{c^2}\Big) = 0, \tag{43}$$

where $\vec{v}$ is the velocity in three-space, Φ is the Newtonian gravitational potential, $\vec{r}$ is the physical coordinate and $\bar{\rho}$ is the density of cosmic background. Now, we use the comoving coordinates as follows [11]:

$$\vec{r} = a\vec{x}. \tag{44}$$

Here $\vec{r}$, a and $\vec{x}$ are the physical coordinates, the scale factor and the comoving coordinates, respectively. Taking time derivative of Eq. (44), one can obtain

$$\vec{v}(\vec{x},t) = a\Big[H(a)\vec{x} + \vec{u}(\vec{x},t)\Big], \tag{45}$$

where

$$\vec{v}(\vec{x},t) = \frac{dr(\vec{x},t)}{dt}, \tag{46}$$

$$\vec{u}(\vec{x},t) = \frac{dx(\vec{x},t)}{dt}. \tag{47}$$

Here, $H(a)$ is the Hubble function and $\vec{u}(\vec{x},t)$ is the comoving peculiar velocity. Next, one can introduce the following definitions

$$\nabla_{\vec{r}} = \frac{1}{a}\nabla_{\vec{x}}\,, \tag{48}$$

$$\frac{\partial}{\partial t}\Big|_r = \frac{\partial}{\partial t}\Big|_x - \frac{1}{a}\vec{v}.\nabla_{\vec{x}}\,, \tag{49}$$

$$\rho(\vec{x},t) = \bar{\rho}(1+\delta(\vec{x},t))\,, \tag{50}$$

$$p = \omega\rho(\vec{x},t)c^2\,, \tag{51}$$

$$\Phi(\vec{x},t) = \Phi_0(\vec{x},t) + \phi(\vec{x},t)\,, \tag{52}$$

where ω is the equation of state parameter. Now, using Eqs. (45), (48), (49), (50), (51), (52) and inserting Eqs. (40), (41), (42) and (43), one can obtain [11]

$$\dot{\delta} + (1+\omega)(1+\delta)\nabla_{\vec{x}}.\vec{u} = 0, \tag{53}$$

$$\frac{\partial\vec{u}}{\partial t} + 2H\vec{u} + (\vec{u}.\nabla_{\vec{x}})\vec{u} + \frac{1}{a^2}\nabla_{\vec{x}}\phi = 0, \tag{54}$$

$$\nabla^2_{\vec{x}}\phi - 4\pi G\Big(1+3\omega\Big)a^2\bar{\rho}\delta = 0. \tag{55}$$

We take the divergence of the Eq. (54) and represent the analysis [11]

$$\nabla.\Big[(\vec{u}.\nabla)\vec{u}\Big] = \frac{1}{3}\theta^2 + \sigma^2 - w^2, \tag{56}$$

where

$$\theta = \nabla_{\vec{x}}.\vec{u}\,, \tag{57}$$

$$\sigma_{ij} = \frac{1}{2}\Big(\frac{\partial u^j}{\partial x^i} + \frac{\partial u^i}{\partial x^j}\Big) - \frac{1}{3}\theta\delta_{ij}\,, \tag{58}$$

$$w_{ij} = \frac{1}{2}\Big(\frac{\partial u^j}{\partial x^i} - \frac{\partial u^i}{\partial x^j}\Big). \tag{59}$$

Here, $\sigma^2 = \sigma_{ij}\sigma^{ij}$ is the shear tensor and $w^2 = w_{ij}w^{ij}$ is the rotation tensor. Taking time derivative of Eq. (53) and using Eqs. (53), (54), (55) and (56), one can obtain [11]

$$\ddot{\delta} + \Big(2H - \frac{\dot{\omega}}{1+\omega}\Big)\dot{\delta} - \frac{4+3\omega}{3(1+\omega)}\frac{\dot{\delta}^2}{1+\delta} - 4\pi G\bar{\rho}(1+\omega)(1+3\omega)\delta(1+\delta) - (1+\omega)(1+\delta)(\sigma^2 - w^2) = 0. \tag{60}$$

Now, we can introduce the following definition

$$\frac{\partial}{\partial t} = aH(a)\frac{\partial}{\partial a}. \tag{61}$$

Using Eq. (61) and inserting Eq. (60), one can rewrite Eq. (60) as follows [11]

$$\delta'' + \Big(\frac{3}{a} + \frac{E'}{E} - \frac{\omega'}{1+\omega}\Big)\delta' - \frac{4+3\omega}{3(1+\omega)}\frac{\delta'^2}{1+\delta} - \frac{3}{2}\frac{\Omega_{fluid,0}}{a^2E^2(a)}h(a)(1+\omega)(1+3\omega)\delta(1+\delta) - \frac{1}{aH^2(a)}(1+\omega)(1+\delta)(\sigma^2 - w^2) = 0, \tag{62}$$

where $E = H/H_0$ is the dimensionless Hubble parameter, $\Omega_{fluid,0} = 8\pi G\bar{\rho}/3H^2$ is the density parameter of the fluid at $a_0 = 1$, $h(a)$ is a function that describes the time evolution of dark energy with scale factor, H_0 is the Hubble parameter at the present time and the prime sign denotes the derivative with respect to the scale factor.

For the collapse of a homogeneous sphere, one can ignore the shear and rotation tensors. Also, we limit ourselves to the spherical perturbation filled with dust $\omega = 0$, for which

$$h(a) = a^{-3}.$$

Therefore, one can obtain the non-linear and linear perturbation equations as follows [11,30]

$$\delta'' + \Big(\frac{3}{a} + \frac{E'}{E}\Big)\delta' - \frac{4}{3}\frac{\delta'^2}{1+\delta} - \frac{3}{2}\frac{\Omega_{m0}}{a^5E^2(a)}\delta(1+\delta) = 0, \tag{63}$$

$$\delta'' + \Big(\frac{3}{a} + \frac{E'}{E}\Big)\delta' - \frac{3}{2}\frac{\Omega_{m0}}{a^5E^2(a)}\delta = 0. \tag{64}$$

Appendix II

In the numerical study of solving Eqs. (21) and (22), we have used the procedure used in Ref. [11] to obtain the initial conditions for drawing the curves of Fig. 3, in tachyon scalar field model, as follows

$$g(a_i) = 3.615\times10^{-4},\ g'(a_i) = 0,\ a_i = 10^{-4}. \tag{65}$$

The scale factor a at the redshift z is defined as follows

$$a = (1+z)^{-1}. \tag{66}$$

Taking derivative of Eq. (66), one can obtain

$$\frac{d}{dz} = -a^2 \frac{d}{da}. \tag{67}$$

Using Eqs. (8), (19) and (67) and inserting Eqs. (14), (15), we can obtain

$$\frac{d\Omega_\Lambda}{da} + \frac{3\Omega_\Lambda}{a}(1-\Omega_\Lambda)(\frac{2}{3n}-1) = 0, \tag{68}$$

$$\frac{dE}{da} + \frac{3}{2}\frac{E}{a}\Big[1+\Omega_\Lambda(\frac{2}{3n}-1)\Big] = 0. \tag{69}$$

Using Eqs. (68), (69), (65) and inserting Eq. (22), we can plot the curve of Fig. 3(a) for tachyon scalar field model using the "Mathematica" in the linear case. Also, using Eqs. (5), (9), (10) and $\rho_m = \rho_{m_0} a^{-3}$ and $\rho_\Lambda = \rho_{\Lambda_0} a^{-3(1+\omega_\Lambda)}$, one can obtain

$$E(a) = \sqrt{\Omega_{m_0} a^{-3} + \Omega_{\Lambda_0} a^{-3(1+\omega_\Lambda)}}. \tag{70}$$

Also, using Eqs. (19), (65), (70) and inserting Eq. (21), we can plot the curve of Fig. 3(b) for tachyon scalar field model using the "Mathematica" in the non-linear case.

References

[1] L.R. Abramo, R.C. Batista, L. Liberato, R. Rosenfeld, J. Cosmol. Astropart. Phys. 11 (2007) 12.
[2] R.J. Bond, S. Cole, G. Efstathiou, N. Kaiser, Astrophys. J. 379 (1991) 440.
[3] F. Bernardeau, Astrophys. J. 433 (1994) 1.
[4] P. De Bernardis, et al., Nature 404 (2000) 955;
S. Perlmutter, et al., Astrophys. J. 598 (2003) 102;
U. Seljak, et al., Phys. Rev. D 71 (2005) 103515.
[5] Y.F. Cai, E.N. Saridakis, M.R. Setare, J.Q. Xia, Phys. Rep. 493 (2010) 1.
[6] E.J. Copeland, M. Sami, T.S. Tsujikawa, Int. J. Mod. Phys. D 15 (2006) 1753.
[7] A.J. Fillmore, P. Goldreich, Astrophys. J. 281 (1984) 1.
[8] J.E. Gunn, R.J. Gott, Astrophys. J. 176 (1972) 1.
[9] Y. Hoffman, J. Shaham, Astrophys. J. 297 (1985) 16.
[10] S. Meyer, F. Pace, M. Bartelmann, Phys. Rev. D 86 (2012) 103002.
[11] F. Pace, J.C. Waizmann, M. Bartelmann, Mon. Not. R. Astron. Soc. 406 (2010) 1865.
[12] F. Pace, L. Moscardini, R. Crittenden, M. Bartelmann, V. Pettorino, Mon. Not. R. Astron. Soc. 437 (2014) 547.
[13] T. Padmanabhan, Cosmology and Astrophysics Through Problems, Cambridge University Press, 1996.
[14] T. Padmanabhan, Phys. Rep. 380 (2003) 235.
[15] H.W. Press, P. Schechter, Astrophys. J. 187 (1974) 425.
[16] W.J. Percival, Annu. Rev. Astron. Astrophys. 443 (2005) 819.
[17] A. Del Popolo, Astrophys. J. 637 (2006) 12;
A. Del Popolo, Annu. Rev. Astron. Astrophys. 448 (2006) 439.
[18] A.G. Riess, et al., Astron. J. 116 (1998) 1009;
S Perlmutter, et al., Astrophys. J. 517 (1999) 565.
[19] S.B. Ryden, E.J. Gunn, Astrophys. J. 318 (1987) 15.
[20] M.R. Setare, Phys. Lett. B 653 (2007) 116;
M.R. Setare, J. Sadeghi, A.R. Amani, Phys. Lett. B 673 (2009) 241.
[21] R.K. Sheth, G. Tormen, Mon. Not. R. Astron. Soc. 308 (1999) 119;
R.K. Sheth, G. Tormen, Mon. Not. R. Astron. Soc. 329 (2002) 61.
[22] P.T.P. Viana, A.R. Liddle, Mon. Not. R. Astron. Soc. 281 (1996) 323.
[23] L. Wang, P.J. Steinhardt, Astro. Phys. J. 508 (1998) 483.
[24] S. Weinberg, Rev. Mod. Phys. 61 (1989) 1.
[25] H. Yahagi, M. Nagashima, Y. Yoshii, Astrophys. J. 605 (2004) 709.
[26] T. Padmanabhan, Phys. Rev. D 66 (2002) 021301.
[27] M. Malekjani, T. Naderi, F. Pace, Mon. Not. R. Astron. Soc. 453 (2015) 4148.
[28] T. Naderi, M. Malekjani, F. Pace, Mon. Not. R. Astron. Soc. 447 (2015) 1873.
[29] J. Zhang, X. Zhang, H. Liu, Phys. Lett. B 651 (2007) 84.
[30] M. Malekjani, J. Lu, N. Nazari-Pooya, L. Xu, D. Mohammad-Zadeh Jassur, M. Honari-Jafarpour, Astrophys. Space Sci. 360 (2015) 24.

10

Friedmann–Lemaître–Robertson–Walker braneworlds

P. Michel L.T. da Silva, A. de Souza Dutra*, J.M. Hoff da Silva

Departamento de Física e Química, Universidade Estadual Paulista, Av. Dr. Ariberto Pereira da Cunha, 333, Guaratinguetá, SP, Brazil

ARTICLE INFO

Editor: M. Trodden

ABSTRACT

We study the cosmological evolution with nonsingular branes generated by a bulk scalar field coupled to gravity. The specific setup investigated leads to branes with a time-dependent warp factor. We calculate the effective Hubble parameter and the effective scale factor for the FLRW branes obtained solutions. The spatially dependent branes solutions also were found.

1. Introduction

Braneworld models were introduced as a genuine branch of research in high energy physics since the outstanding works presented in Refs. [1–3]. After those works, several authors tried to construct models compatible with the Friedmann–Lemaître–Robertson–Walker (FLRW) cosmology. In fact, the modeling of realistic braneworld scenarios from the large scale physics point of view must include, at least in some level, the description encoded in the cosmological standard model. The attempts put forward to this program can be roughly separated into two categories: one dealing with infinitely thin branes, in which the extra dimensional effects enters as corrections to the Einstein equations via the presence of the Weyl tensor (the so-called 'dark fluid') and quadratic contributions to the stress tensor, and another one, whose modeling arrive at thick branes described by one or more bulk scalar fields. We refer the reader to the references [4] and [5] for quite complete reviews about thin and thick braneworld models and cosmological consequences, respectively. In the former approach, the corrections to the gravitational equations comes from the Gauss–Codazzi procedure and, potentially, all the relevant aspects of the four-dimensional cosmology are revisited [6]. It turns out, however, that an infinitely thin brane seems to be only an approximation of the more realistic case, at best. In fact, by keeping in mind the simple fact that at very short scales a classic gravitational theory must be replaced by its quantum counterpart (whatever it is), it is mandatory some thickness to the brane itself.

Unfortunately, in the thick brane context it is not possible, at the best of our knowledge, to apply the Gauss–Codazzi formalism. The reason is that it is not clear what (if any) are the Israel–Darmois junction conditions in a thick brane context, for early attempts see [7]. The junction conditions are at the heart of the projection procedure, and its lack makes the whole program fall apart. In this vein, the investigation of a five-dimensional thick braneworld setup whose four-dimensional part describes a FLRW universe is, indeed, in order. There are, nevertheless, only a few works addressing this crucial point. The cumbersome algebraic task inevitably present in this endeavor can be attributed as a cause for such. In Refs. [8,9] a more or less recent attempt to find out thick braneworld scenarios whose four dimensional part describes relevant cosmology, was taken into account. The general approach used in both cases take advantage of a well known functional form to the spacial part of the warp factor, as the one presented in Ref. [10]. While this is a clever starting point, it would be desirable to have a more powerful method arriving at relevant cosmological setups. In this paper we address ourselves to this task, starting from a five dimensional scalar field whose solution describes a FLRW brane, i.e., a braneworld whose four-dimensional part is given by a FLRW universe. The main braneworld characteristic, the warp factor, is also taken into account in the solution. Part of the solutions obtained via a given ansatz in Refs. [8,9] are recovered here as particular cases of a broad approach.

As a general procedure, we shall use a separable warp factor function. This might be seen as an oversimplification. In fact, there is an interesting discussion about non separable warp factors generating traveling-like defects in five dimensional space-times [11]. We stress, however, that in order to reproduce four-dimensional cosmological solutions (which is not the aim of Ref. [11]) separable warp factors are indeed in order. Besides, the adoption of a separable warp factor (in addiction to being mathematically consistent, although not the most general possibility) is in line with the hierarchy problem approach via thick branes [12].

* Corresponding author.
E-mail addresses: pmichel@fc.unesp.br (P.M.L.T. da Silva), dutra@feg.unesp.br (A. de Souza Dutra), hoff@feg.unesp.br (J.M. Hoff da Silva).

We organize this paper as follows. In Section 2, we present the mathematical preliminaries supporting the metric ansatz with the four-dimensional background metric given by the Friedmann–Lemaître–Robertson–Walker. Then we derive gravitational equations and the expressions for the scalar field and its potential. In Section 3, we find the time-dependent FLRW brane solutions. We solve the field equations for two cases with respect to spatial curvature, i.e., $k = 0$ and $k \neq 0$. Going further we determine the spatial part of the set of equations in Section 4. In Section 5 we present the effective Hubble parameter as well as the effective scale factor for all the possibilities found previously. In the final section we conclude.

2. Field equations

The assumption of isotropy and homogeneity implies the large scale geometry described by a metric of the form

$$ds^2 = -dt^2 + a^2(t)\left[\frac{dr^2}{1-kr^2} + r^2(d\theta^2 + sen^2\theta d\phi^2)\right], \tag{1}$$

in a synchronized coordinate system (a suitable set of coordinates called comoving coordinates). The comoving observers, also called Hubble observers, are the ones located at spacelike hypersurfaces accompanying the cosmic fluid, which is at rest with respect to such hypersurfaces. Here $a(t)$ is an arbitrary function of the cosmic time called scale factor and $k = 0, \pm 1$, denotes the spatial curvature of the universe for Minkowski, Riemann and Lobachevsky geometry, respectively.

Let us consider 5D spacetimes for which the metric takes the following form

$$ds^2 = a^2(t,y)\left\{-dt^2 + u^2(t)\left[\frac{dr^2}{1-kr^2} + r^2(d\theta^2 + sen^2\theta d\phi^2)\right]\right\} + b^2(t,y)dy^2, \tag{2}$$

where the background metric in $4D$ is given by the Friedmann–Lemaître–Robertson–Walker line element (1). The metric signature is given by $(-++++)$. The function $a(t,y)$ is an warp factor with time and extra dimension dependence, while $u(t)$ performs the usual scale factor for an homogeneous and isotropic universe. The function $u(t)$ can, indeed, be absorbed in a rescaling of the time coordinate. For the sake of exposition, however, we shall keep it explicitly along the text in order to be in touch with the FLRW cosmological model. The function $b(t,y)$ shows the dynamics of the extra dimension at different times and positions in the bulk.

Let us consider the 5D action in the presence of a bulk scalar field with the potential $V(\phi)$ minimally coupled to the gravitational sector

$$S = \int d^5x\sqrt{-g}\left\{2M^3R - \frac{1}{2}g^{MN}\nabla_M\phi\nabla_N\phi - V(\phi)\right\}, \tag{3}$$

where M is the Planck mass and R is the five-dimensional Ricci scalar. In general we suppose that the scalar field ϕ depends only on time and the extra dimension y.

The Einstein equations read

$$R_{MN} - \frac{1}{2}g_{MN}R = \frac{1}{4M^3}T_{MN}, \tag{4}$$

and the energy momentum-tensor T_{MN} for the scalar field $\phi(t,y)$ is

$$T_{MN} = \nabla_M\phi\nabla_N\phi - g_{MN}\left(\frac{1}{2}g^{AB}\nabla_A\phi\nabla_B\phi + V(\phi)\right). \tag{5}$$

The time–time component of the field equations for the spacetime under consideration is given by

$$3\left[\frac{1}{a^2}\frac{\dot{a}^2}{a^2} - \frac{1}{b^2}\left(\frac{a''}{a} + \frac{a'^2}{a^2}\right) + \frac{1}{b^2}\frac{a'}{a}\frac{b'}{b} + \frac{1}{a^2}\left(2\frac{\dot{a}}{a}\frac{\dot{u}}{u} + \frac{\dot{a}}{a}\frac{\dot{b}}{b} + \frac{\dot{u}^2}{u^2} + \frac{\dot{b}}{b}\frac{\dot{u}}{u} + \frac{k}{u^2}\right)\right] = \frac{1}{2b^2}\phi'^2 + \frac{1}{2a^2}\dot{\phi}^2 + V(\phi), \tag{6}$$

whilst the space components give

$$\frac{1}{a^2}\left[2\frac{\ddot{a}}{a} - \frac{\dot{a}^2}{a^2} + \frac{\dot{a}}{a}\frac{\dot{b}}{b} + 4\frac{\dot{a}}{a}\frac{\dot{u}}{u} + 2\frac{\dot{b}}{b}\frac{\dot{u}}{u} + \frac{\ddot{b}}{b} + 2\frac{\ddot{u}}{u} + \frac{\dot{u}^2}{u^2} + \frac{k}{u^2}\right] - \frac{3}{b^2}\left[\frac{a''}{a} + \frac{a'^2}{a^2} - \frac{a'}{a}\frac{b'}{b}\right] = \frac{1}{2b^2}\phi'^2 - \frac{1}{2a^2}\dot{\phi}^2 + V(\phi). \tag{7}$$

The extra dimensional part contributes with

$$\left[\frac{a'}{a}\frac{\dot{a}}{a} + \frac{a'}{a}\frac{\dot{b}}{b} - \frac{\dot{a}'}{a}\right] = \frac{1}{3}\dot{\phi}\phi', \tag{8}$$

$$3\left[2\frac{a'^2}{a^2}\frac{1}{b^2} - \frac{\ddot{a}}{a}\frac{1}{a^2} - 3\frac{\dot{a}}{a}\frac{\dot{u}}{u}\frac{1}{a^2} - \frac{\dot{u}^2}{u^2}\frac{1}{a^2} - \frac{\ddot{u}}{u}\frac{1}{a^2} - \frac{k}{a^2u^2}\right] = \frac{1}{2a^2}\dot{\phi}^2 + \frac{1}{2b^2}\phi'^2 - V(\phi). \tag{9}$$

Finally, the scalar equation of motion reads

$$\frac{1}{b^2}\phi'' - \frac{1}{a^2}\ddot{\phi} + \frac{4}{b^2}\frac{a'}{a}\phi' - \frac{2}{a^2}\frac{\dot{a}}{a}\dot{\phi} - \frac{1}{a^2}\frac{\dot{b}}{b}\dot{\phi} - \frac{3}{a^2}\frac{\dot{u}}{u}\dot{\phi} - \frac{1}{b^2}\frac{b'}{b}\phi' - \frac{dV}{d\phi} = 0, \tag{10}$$

where a dot denotes a derivative with respect to t, and a prime represents a derivative with respect to the extra dimension y.

By combining Eqs. (6)–(7) and (9), we arrive at convenient expressions for $\dot{\phi}$, ϕ' and $V(\phi)$:

$$\dot{\phi}^2 = 2\left[2\frac{\dot{a}^2}{a^2} - \frac{\ddot{a}}{a} + \frac{\dot{a}}{a}\frac{\dot{b}}{b} + \frac{\dot{a}}{a}\frac{\dot{u}}{u} + \frac{1}{2}\frac{\dot{b}}{b}\frac{\dot{u}}{u} - \frac{1}{2}\frac{\ddot{b}}{b} - \frac{\ddot{u}}{u} + \frac{\dot{u}^2}{u^2} + \frac{k}{u^2}\right], \tag{11}$$

$$\phi'^2 = b^2\left[\frac{1}{a^2}\left(-\frac{\ddot{a}}{a} - \frac{\dot{a}^2}{a^2} - 5\frac{\dot{a}}{a}\frac{\dot{u}}{u} + \frac{\dot{a}}{a}\frac{\dot{b}}{b} + 2\frac{\dot{b}}{b}\frac{\dot{u}}{u} - 2\frac{\dot{u}^2}{u^2} - \frac{\ddot{u}}{u} + \frac{\ddot{b}}{b} - \frac{2k}{u^2}\right)\right] - 3\left(-\frac{a'^2}{a^2} + \frac{a''}{a} + \frac{a'}{a}\frac{b'}{b}\right), \tag{12}$$

$$V(\phi) = \frac{3}{2}\left[\frac{1}{a^2}\left(\frac{\ddot{a}}{a} + \frac{\dot{a}^2}{a^2} + 5\frac{\dot{a}}{a}\frac{\dot{u}}{u} + \frac{\dot{a}}{a}\frac{\dot{b}}{b} + 2\frac{\dot{u}^2}{u^2} + \frac{\ddot{u}}{u} + \frac{\dot{b}}{b}\frac{\dot{u}}{u} + \frac{2k}{u^2}\right) - \frac{3}{b^2}\frac{a'^2}{a^2} - \frac{1}{b^2}\frac{a''}{a} + \frac{1}{b^2}\frac{a'}{a}\frac{b'}{b}\right]. \tag{13}$$

In what follows we shall depict several important cases resulting from some relevant particularizations. Let us assume that $\phi(t,y) \equiv \phi(y)$ and $b(t,y) \equiv 1$. Thus the ${G^0}_5$ component of the field equation becomes

$$\frac{a'}{a}\frac{\dot{a}}{a} - \frac{\dot{a}'}{a} = 0, \tag{14}$$

implying the possibility of spatial and temporal separation, i.e., $a(y,t) \equiv \alpha(y)\beta(t)$. Therefore, the equations (10)–(13) take the form

$$\phi'' + 4\left(\frac{\alpha'}{\alpha}\right)\phi' - \frac{dV}{d\phi} = 0, \tag{15}$$

$$\frac{\dot{\phi}^2}{2} = \left[2\left(\frac{\dot\beta}{\beta}\right)^2 - \left(\frac{\ddot\beta}{\beta}\right) + \left(\frac{\dot\beta}{\beta}\right)\left(\frac{\dot u}{u}\right) - \left(\frac{\ddot u}{u}\right) + \left(\frac{\dot u}{u}\right)^2 + \left(\frac{k}{u^2}\right)\right] \equiv 0, \tag{16}$$

$$\begin{aligned}\phi'^2 &= \frac{1}{\alpha^2\beta^2}\left[-\left(\frac{\ddot\beta}{\beta}\right) - \left(\frac{\dot\beta}{\beta}\right)^2 - 5\left(\frac{\dot\beta}{\beta}\right)\left(\frac{\dot u}{u}\right) - 2\left(\frac{\dot u}{u}\right)^2 - \left(\frac{\ddot u}{u}\right) - \left(\frac{2k}{u^2}\right)\right] - 3\left[\left(\frac{\alpha''}{\alpha}\right) - \left(\frac{\alpha'}{\alpha}\right)^2\right] \\ &= -\frac{\Delta}{\alpha^2} - 3\left[\left(\frac{\alpha''}{\alpha}\right) - \left(\frac{\alpha'}{\alpha}\right)^2\right],\end{aligned} \tag{17}$$

$$\begin{aligned}\frac{2}{3}V(\phi) &= \frac{1}{\alpha^2\beta^2}\left[\left(\frac{\ddot\beta}{\beta}\right) + \left(\frac{\dot\beta}{\beta}\right)^2 + 5\left(\frac{\dot\beta}{\beta}\right)\left(\frac{\dot u}{u}\right) + 2\left(\frac{\dot u}{u}\right)^2 + \left(\frac{\ddot u}{u}\right) + \left(\frac{2k}{u^2}\right)\right] - 3\left(\frac{\alpha'}{\alpha}\right)^2 - \left(\frac{\alpha''}{\alpha}\right) \\ &= \frac{\Sigma}{\alpha^2} - 3\left(\frac{\alpha'}{\alpha}\right)^2 - \left(\frac{\alpha''}{\alpha}\right).\end{aligned} \tag{18}$$

At this point we need to consider separately some different regimes of the above equations. Afterwards, we solve the time-dependent part of the solutions and, then, the spatial part is analyzed.

3. Time-dependent part of the solutions

From now on we shall evince all the relevant solutions, solving the time dependent part of the solutions, after what we concentrate in the spacial part of the solutions.

3.1. The case $\Delta \equiv \Sigma \equiv 0$ with $k = 0$ and $k \neq 0$

Within the specifications outlined in this subsection epigraph, the time dependence of the equations (16)–(18) must vanish.[1] Thus, the relevant equations read

$$\left(\frac{\ddot\beta}{\beta}\right) - 2\left(\frac{\dot\beta}{\beta}\right)^2 - \left(\frac{\dot\beta}{\beta}\right)\left(\frac{\dot u}{u}\right) + \left(\frac{\ddot u}{u}\right) - \left(\frac{\dot u}{u}\right)^2 - \left(\frac{k}{u^2}\right) = 0, \tag{19}$$

$$\left(\frac{\ddot\beta}{\beta}\right) + \left(\frac{\dot\beta}{\beta}\right)^2 + 5\left(\frac{\dot\beta}{\beta}\right)\left(\frac{\dot u}{u}\right) + 2\left(\frac{\dot u}{u}\right)^2 + \left(\frac{\ddot u}{u}\right) + \left(\frac{2k}{u^2}\right) = -\Delta\beta^2 = 0, \tag{20}$$

$$\left(\frac{\ddot\beta}{\beta}\right) + \left(\frac{\dot\beta}{\beta}\right)^2 + 5\left(\frac{\dot\beta}{\beta}\right)\left(\frac{\dot u}{u}\right) + 2\left(\frac{\dot u}{u}\right)^2 + \left(\frac{\ddot u}{u}\right) + \left(\frac{2k}{u^2}\right) = \Sigma\beta^2 = 0. \tag{21}$$

When $k = 0$, Eqs. (19)–(21) constraint $\beta(t)$ and $u(t)$ as

$$\beta(t) = \frac{C}{u(t)}, \tag{22}$$

where C is an arbitrary constant of integration.

It is interesting to notice that the set of equations (19)–(21) with $k \neq 0$ and $u = 1$, recover the result obtained by [8], whose solution is

$$\beta(t) = \beta_0 e^{\pm\sqrt{-k}t}. \tag{23}$$

On the other hand, for the case in which $k \neq 0$ e $u \neq 1$ and using the following redefinition

$$u(t) = f(t)\beta(t) \tag{24}$$

where $f(t)$ is an arbitrary function, we have

$$\beta(t) = \frac{1}{\sqrt{f(t)}}\left[C \pm \sqrt{-k}\int_0^t \frac{dt'}{\sqrt{f(t')}}\right]. \tag{25}$$

3.2. The case $\Delta = -\Sigma$ with $k = 0$ and $k \neq 0$

Now, for the case specified here, the set of equations to be solved is

$$\left(\frac{\ddot\beta}{\beta}\right) - 2\left(\frac{\dot\beta}{\beta}\right)^2 - \left(\frac{\dot\beta}{\beta}\right)\left(\frac{\dot u}{u}\right) + \left(\frac{\ddot u}{u}\right) - \left(\frac{\dot u}{u}\right)^2 - \left(\frac{k}{u^2}\right) = 0, \tag{26}$$

$$\left(\frac{\ddot\beta}{\beta}\right) + \left(\frac{\dot\beta}{\beta}\right)^2 + 5\left(\frac{\dot\beta}{\beta}\right)\left(\frac{\dot u}{u}\right) + 2\left(\frac{\dot u}{u}\right)^2 + \left(\frac{\ddot u}{u}\right) + \left(\frac{2k}{u^2}\right) = -\Delta\beta^2, \tag{27}$$

$$\left(\frac{\ddot\beta}{\beta}\right) + \left(\frac{\dot\beta}{\beta}\right)^2 + 5\left(\frac{\dot\beta}{\beta}\right)\left(\frac{\dot u}{u}\right) + 2\left(\frac{\dot u}{u}\right)^2 + \left(\frac{\ddot u}{u}\right) + \left(\frac{2k}{u^2}\right) = \Sigma\beta^2. \tag{28}$$

All together, these equations implies $\Delta = -\Sigma$ and therefore the solution is given by

$$\beta(t) = \left[u(t)\left(C_1 \pm \sqrt{\frac{\Sigma}{3}}\int_0^t \frac{dt'}{u(t')}\right)\right]^{-1}. \tag{29}$$

Note that as far as $k = 0$ and $u(t) = 1$, we recover result obtained in [8], i.e.,

$$\beta(t) \propto \frac{1}{t},$$

as expected.

For $k \neq 0$, the solution is given by

$$\beta(t) = -\frac{1}{u(t)}\sqrt{\frac{3}{\Sigma}}\cot\left[C_1\sqrt{3k} - \sqrt{k}\int_0^t \frac{dt'}{u(t')}\right] \times \sqrt{k + k\tan^2\left[C_1\sqrt{3k} - \sqrt{k}\int_0^t \frac{dt'}{u(t')}\right]}. \tag{30}$$

Making use of the usual trigonometric relation $\sec^2 x = 1 + \tan^2 x$, we have

[1] Notice that as an equation on the spatial variable, Eq. (15) shall be faced in the next Section.

$$\beta(t) = -\frac{1}{u(t)}\sqrt{\frac{3}{\Sigma}}\cot\left[C_1\sqrt{3k} - \sqrt{k}\int\limits_0^t \frac{dt'}{u(t')}\right]$$
$$\times\sqrt{k\sec^2\left[C_1\sqrt{3k} - \sqrt{k}\int\limits_0^t \frac{dt'}{u(t')}\right]}, \tag{31}$$

or simply

$$\beta(t) = -\frac{1}{u(t)}\sqrt{\frac{3k}{\Sigma}}\frac{1}{\sin\left(C_1\sqrt{3k} - \sqrt{k}\int_0^t \frac{dt'}{u(t')}\right)}. \tag{32}$$

Thus Eq. (30) can be rewritten as

$$\beta(t) = -\frac{1}{u(t)}\sqrt{\frac{3k}{\Sigma}}\sec\left(\sqrt{k}\int\limits_0^t \frac{dt'}{u(t')}\right), \tag{33}$$

where use was made of

$$\pm\sin x = \pm\cos(\pi/2 - x) \text{ with}$$
$$x = -\sqrt{k}\int\limits_0^t \frac{dt'}{u(t')} \text{ and } C_1\sqrt{3k} = \pi/2. \tag{34}$$

Note that, similarly to the case of Eq. (23), one can reproduce the results obtained in [8], in which the solutions given in (33), for $k = 1$, $k = -1$ (both for $u = 1$) are respectively

$$\beta(t) \propto \sec(t), \tag{35}$$

and

$$\beta(t) \propto \mathrm{sech}(t). \tag{36}$$

We shall investigate the cosmological outputs of the obtained solutions in Section 5.

4. Spatial-dependent part of the solutions

Now we shall consider the spatial part of the equations (15), (17) and[2] (18):

$$\phi'' + 4\left(\frac{\alpha'}{\alpha}\right)\phi' - \frac{dV}{d\phi} = 0, \tag{37}$$

$$\phi'^2 = -\frac{\Delta}{\alpha^2} - 3\left[\left(\frac{\alpha''}{\alpha}\right) - \left(\frac{\alpha'}{\alpha}\right)^2\right], \tag{38}$$

$$\frac{2}{3}V(\phi) = \frac{\Sigma}{\alpha^2} - 3\left(\frac{\alpha'}{\alpha}\right)^2 - \left(\frac{\alpha''}{\alpha}\right). \tag{39}$$

There are two interesting cases, concerning the separation constants which we are going to investigate in detail. The first one is given by the vanishing of both.

4.1. The case $\Delta = \Sigma = 0$

For better dealing with the system of equations we use the redefinition,

$$\alpha(y) = e^{A(y)}. \tag{40}$$

In the light of (40), the Equations (38)–(37) become

$$\frac{1}{3}\phi'^2 = -A'', \ \frac{2}{3}V(\phi) = -A'' - 4\left(A'\right)^2, \tag{41}$$

and

$$\phi'' + 4A'\phi' = \frac{dV}{d\phi}. \tag{42}$$

Assuming, for ulterior convenience, that

$$\phi' = rW_\phi(\phi), \ A' = sW(\phi), \tag{43}$$

and substituting the equations (43) in (42) we find $s = -1/3$. Therefore the potential acquires the form

$$V(\phi) = r^2\left[\frac{1}{2}\left(W_\phi\right)^2 - \frac{2}{3}W(\phi)^2\right]. \tag{44}$$

The derivative of Eq. (44) with respect to ϕ, V_ϕ, reads

$$\begin{aligned} V_\phi &= r^2\left[W_{\phi\phi}W_\phi - \frac{4}{3}W_\phi W(\phi)\right] \\ &= r^2\left[\frac{d}{d\phi}\left(\frac{W_\phi^2}{2}\right) - \frac{4}{3}\frac{d}{d\phi}\left(\frac{W(\phi)^2}{2}\right)\right] \\ &= \frac{d}{d\phi}\left[r^2\left(\frac{W_\phi^2}{2} - \frac{2}{3}W(\phi)^2\right)\right]. \end{aligned} \tag{45}$$

Therefore the potential itself can be expressed as

$$V(\phi) = \frac{W_\phi^2}{2} - \frac{2}{3}W(\phi)^2, \tag{46}$$

and the Eqs. (43) become

$$\phi' = W_\phi(\phi), \ A' = -\frac{1}{3}W(\phi), \tag{47}$$

with

$$A(y) = -\frac{1}{3}\int W\left[\phi(y)\right]dy. \tag{48}$$

Now we turn ourselves to a different arrangement of the separation constants.

4.2. The case $\Delta = -\Sigma \neq 0$

Making use of the expression (40) in (38)–(37), one gets

$$\phi'^2 = \Sigma e^{-2A} - 3A'', \ \frac{2}{3}V(\phi) = \Sigma e^{-2A} - 4\left(A'\right)^2 - A'', \tag{49}$$

and

$$\phi'' + 4A'\phi' = \frac{dV}{d\phi}. \tag{50}$$

Combining the equations (49), we obtain

$$V(\phi) = 3[A'' - 2\left(A'\right)^2] + \frac{3}{2}\phi'^2. \tag{51}$$

By assuming [13] that

$$\phi' = aW_{1\phi}(\phi), \ A' = bW_2(\phi), \tag{52}$$

$$\phi'' = a^2W_{1\phi\phi}W_{1\phi}, \ A'' = abW_{2\phi}W_{1\phi}, \tag{53}$$

and substituting Eqs. (52) and (53) into (51), we have

$$V(\phi) = 3[abW_{2\phi}W_{1\phi} - 2b^2W_2^2] + \frac{3}{2}a^2W_{1\phi}^2. \tag{54}$$

Taking the derivative of (54) with respect to ϕ one gets

[2] Notice that the equation (16) has not spatial-dependence.

$$\frac{dV}{d\phi} = 3[abW_{2\phi\phi}W_{1\phi} + abW_{1\phi\phi}W_{2\phi} - 4b^2 W_2 W_{2\phi}] + 3a^2 W_{1\phi}W_{1\phi\phi}, \tag{55}$$

and inserting (55) in the scalar field equation (50), one arrives at the following consistence equation for the superpotential

$$a[4bW_2 - 3bW_{2\phi\phi} - 2aW_{1\phi\phi}]W_{1\phi} = 3b[aW_{1\phi\phi} - 4bW_2]W_{2\phi}, \tag{56}$$

where $a = 1$ and $b = 1/3$.

Now we define the following quantities

$$W_1 = W + \lambda Z \quad \text{and} \quad W_2 = W + \sigma Z, \tag{57}$$

from which one can see that some terms can be written as a total derivative, and the equation (56) takes the form

$$\frac{d}{d\phi}\left[\frac{2}{3}W^2 + \frac{1}{2}W_\phi^2 - \frac{2}{3}\lambda\sigma Z^2 + \frac{\alpha}{2}(2\sigma - \lambda)Z_\phi^2 + \frac{4}{3}\sigma^2 Z^2 + \sigma W_\phi Z_\phi + \frac{4}{3}\sigma W Z\right] + \frac{4}{3}(\sigma - \lambda)WZ_\phi = 0. \tag{58}$$

In order to deal with a concrete and exact case, we assume that

$$W = C_1 + C_2 Z + C_3 Z_{\phi\phi}. \tag{59}$$

For simplicity, some authors consider $Z(\phi) = W(\phi)$ [13,14]. Here we shall consider a more general case, substituting (59) into Eq. (58). This procedure leads to

$$\begin{aligned}&\frac{2}{3}C_1^2 - C_4 + \frac{4}{3}C_1C_3Z_{\phi\phi} + \frac{2}{3}C_3Z_{\phi\phi}^2 + \frac{1}{2}C_3^2Z_{\phi\phi\phi}^2 \\ &+ Z_\phi Z_{\phi\phi\phi}(C_2C_3 + C_3\beta) \\ &+ Z_\phi^2\left(\frac{C_2^2}{2} - \frac{2C_3\lambda}{3} - \frac{\lambda^2}{2} + C_2\sigma + \frac{2}{3}C_3\sigma + \lambda\sigma\right) \\ &+ Z^2\left(\frac{2C_2^2}{3} - \frac{2C_2\lambda}{3} + 2C_2\sigma - \frac{2\lambda\sigma}{3} + \frac{4\sigma^2}{3}\right) \\ &+ Z\left[\frac{4C_1C_2}{3} - \frac{4C_1\lambda}{3} + \frac{8C_1\sigma}{3} + Z_{\phi\phi}\left(\frac{4C_2C_3}{3} + \frac{4C_3\sigma}{3}\right)\right] \\ &= 0.\end{aligned} \tag{60}$$

In order to keep some similarity with the well known literature, we look for a solution as the one presented in Ref. [15], for instance. We choose thereof

$$Z(\phi) = Z_0 \cos(v\phi + s). \tag{61}$$

In the light of Eq. (60), gathering Z, Z_ϕ^2 and Z^2 terms together, we have

$$\begin{aligned}&\frac{2}{3}C_1^2 - C_4 + \frac{4Z}{3}\left(C_2 - C_3v^2 - \lambda + 2\sigma\right)C_1 + Z_\phi^2\Bigg[\frac{C_2^2}{3} + \frac{C_3v^4}{2} \\ &- \frac{2C_3\lambda}{3} - \frac{\lambda^2}{2} + C_2\sigma + \frac{2C_3\sigma}{3} + \lambda\sigma - v^2(C_2C_3 + C_3\sigma)\Bigg] \\ &+ Z^2\Bigg[\frac{2C_2^2}{3} + \frac{2C_3v^4}{3} - \frac{2C_2\lambda}{3} + 2C_2\sigma - \frac{2\lambda\sigma}{3} + \frac{4\sigma^2}{3} \\ &- v^2\left(\frac{4C_2C_3}{3} + \frac{4C_3\sigma}{3}\right)\Bigg] = 0.\end{aligned} \tag{62}$$

By taking the Z coefficient equal to zero

$$C_2 - C_3v^2 - \lambda + 2\sigma = 0, \tag{63}$$

we get as the solution for v, the following expression

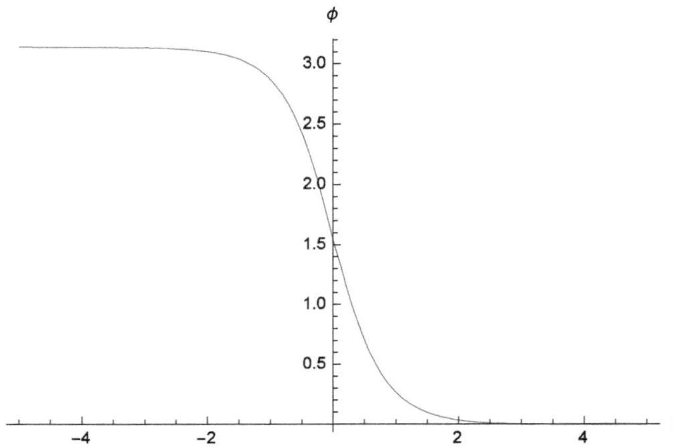

Fig. 1. Behavior of $\phi(y)$ as a function of y for $\Delta \neq \Sigma$, $\lambda = 2$ and $\sigma = 1$.

$$v = \sqrt{\frac{C_2 - \lambda + 2\sigma}{C_3}}, \tag{64}$$

where $C_1 = \pm\sqrt{3/2C_4}$ and $C_2 = C_3 + \lambda - 2\sigma$ if $v = 1$, in order to full accomplish the consistence constraint. In this vein, the expression (59) for W, becomes

$$W = C_1 + (\lambda - 2\sigma)\cos\phi, \tag{65}$$

with the choice $Z_0 = 1$ and $s = 0$ in equation (61).

Using Eq. (65) into (52) and (53), we obtain

$$\phi' = 2(\sigma - \lambda)\sin\phi, \tag{66}$$

and

$$A' = \frac{1}{3}[C_1 + (\lambda - \sigma)\cos\phi]. \tag{67}$$

Therefore, the solutions for ϕ and A, are respectively given by

$$\phi(y) = 2\ \mathrm{arccot}[e^{2y(\lambda-\sigma)}], \tag{68}$$

and

$$A(y) = \frac{1}{3}y(C_1 + \lambda - \sigma) + \frac{1}{6}\ln(1 + e^{4y(\lambda-\sigma)}). \tag{69}$$

By means of Eqs. (68) and (69) for the expression (51), we obtain the following shape for the potential

$$V(\phi) = -\frac{1}{3}(\lambda - \sigma)^2\Big\{-11 + 13\cos\big[4\mathrm{arccot}[\cot(\phi/2)]\big]\Big\}. \tag{70}$$

In Figs. 1 and 2 we depict the profiles of $\phi(y)$ and $Exp[A(y)]$ in the relevant range where the scalar field is also varying. Before to delve into the effective quantities study, we remark by passing that, despite the rather non-trivial functional form of the obtained solutions, the resulting spacetime is after all well behaved. In fact, all the Kretschmann scalars associated to the solutions are finite.

5. Effective Hubble parameter and scale factor

In this section, from the found solutions for $\beta(t)$, we obtain the effective Hubble parameter as well as the effective scale factor. The Hubble parameter, $H = \dot{a}/a$, is used to measure the expansion rate of the universe. The time elapsed in this scenario is the so called proper time or cosmic time.

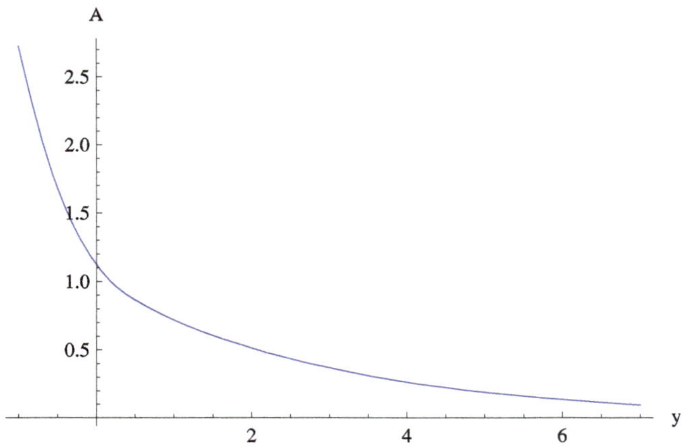

Fig. 2. The exponential of the warp factor function $A(y)$ as a function of y for $\Delta \neq \Sigma$, $\lambda = 2$, $\sigma = 1$ and $C_1 = -1$.

In view of the transformation

$$d\tau = \beta(t)dt, \tag{71}$$

it is possible to construct the effective Hubble parameter

$$H_{eff} = \frac{1}{\beta(t)}\frac{d}{dt}[\ln(\beta(t)u(t)] = \frac{1}{a_{eff}}\frac{da_{eff}}{d\tau} = \frac{\dot{a}_{eff}}{a_{eff}}, \tag{72}$$

where

$$a_{eff}(t) = \beta(t)u(t). \tag{73}$$

5.1. The case $\Delta = \Sigma = 0$

Consider for a while the case that $k = 0$, so that

$$\beta(t) = \frac{C}{u(t)} \tag{74}$$

and therefore

$$H_{eff}(\tau) = 0 \quad \text{and} \quad a_{efe} = constant. \tag{75}$$

This case leads, then, to the typical Hubble parameter describing a static universe, sometimes called Einstein's universe.

Within this case ($\Delta = \Sigma = 0$) but now with $k \neq 0$, the solution is given by

$$\beta(t) = \frac{1}{\sqrt{f(t)}}\left[C \pm \frac{\sqrt{-k}}{\sqrt{f(t)}}\int_0^t \frac{dt'}{\sqrt{f(t')}}\right]. \tag{76}$$

Hence we have

$$d\tau = \beta(t)dt = \frac{dt}{\sqrt{f(t)}}\left[C \pm \frac{\sqrt{-k}}{\sqrt{f(t)}}\int_0^t \frac{dt'}{\sqrt{f(t')}}\right] = dz[C \pm \sqrt{-k}z], \tag{77}$$

where we defined that $dz = dt/\sqrt{f(t)}$. Thus, we find the expression for the cosmic time τ as given by

$$\tau = Cz \pm z^2\sqrt{-k}/2. \tag{78}$$

Choosing, for simplicity, $C = 0$ and using $f(t) \equiv e^{-2at}$, we have $z = (1/a)\,e^{at}$ and, consequently

$$\tau = \pm\frac{\sqrt{-k}}{2}\frac{1}{a^2}e^{2at}. \tag{79}$$

Therefore

$$t = \frac{1}{2a}\ln\left(\pm\frac{2a^2\tau}{\sqrt{-k}}\right), \tag{80}$$

and it is possible see that returning to Eq. (76) one arrives at

$$\beta(t) = \pm\frac{\sqrt{-k}}{a}e^{2at}. \tag{81}$$

Finally, substituting the expressions (81) and (80) into (72), one can verify that the effective Hubble parameter decays with the inverse of τ, while that scale factor is linearly growing and real as far as $k < 0$.

$$H_{eff}(\tau) = \frac{1}{\tau} \text{ and } a_{eff}(\tau) = 2\sqrt{-k}\,\tau. \tag{82}$$

It is reasonable that the behavior of $H_{eff}(\tau)$ is the one expected both for the matter and radiation dominated phase of the universe. Unfortunately, however, the corresponding scale factor grows much faster than it should in a realistic scenario.

5.2. The case $\Delta = -\Sigma \neq 0$

Now let us consider the case in which $\Delta = -\Sigma \neq 0$ both for $k = 0$ and $k \neq 0$. Firstly, we analyze the simplest case with $k = 0$, leading to

$$\beta(t) = \frac{1}{u(t)}\frac{1}{C_1 \pm \sqrt{\frac{\Sigma}{3}}\int_0^t \frac{dt'}{u(t')}}, \tag{83}$$

and from Eq. (67),

$$d\tau = \beta(t)dt = \frac{dt}{u(t)(C_1 \pm bz)} \rightarrow \tau = \mp\frac{1}{b}\ln(C_1 \pm bz) \tag{84}$$

where $b = \sqrt{\Sigma/3}$ and $z = \int_0^t dt'/u(t')$. Inserting Eqs. (83) and (84) into the expression (72), one gets

$$\begin{aligned} H_{eff}(t) &= u(t)(C_1 \pm bz)\frac{d}{dt}\left[\ln\left(\frac{1}{(C_1 \pm bz)}\right)\right] = \\ &= u(t)(C_1 \pm bz)\frac{d}{dt}(\pm b\tau) \\ &= u(t)(C_1 \pm bz) \pm b\frac{1}{u(t)(C_1 \pm bz)} \\ &= \pm b, \end{aligned} \tag{85}$$

and therefore

$$H_{eff}(\tau) = \pm\sqrt{\frac{\Sigma}{3}}, \tag{86}$$

whose integration leads to

$$a_{eff}(\tau) = a_0 \exp\left(\pm\sqrt{\frac{\Sigma}{3}}\tau\right). \tag{87}$$

We can note that the above solution to $a_{efe}(\tau)$ with $k = 0$ is similar to the solution found for an usual universe (without any brane), and dominated by the vacuum energy, i.e.,

$$a(t) \propto \exp\left(\pm\sqrt{\frac{\Lambda}{3}}t\right). \tag{88}$$

where $\Lambda > 0$ is the cosmological constant. This fact is more important that it may sound. In fact, in the context of thick braneworlds, as discussed in the Introduction, there is no how to directly relate the four-dimensional cosmological constant with some counterpart quantity in five dimensions, or even some property of the brane modeling. In this approach, however, we see that the separation constant (necessary to solve the gravitational system endowed with a brane) takes the place of the four-dimension cosmological constant. The solution given in (87) could, indeed, represent the current phase of accelerated expansion of our universe in a ΛCDM (Lambda Cold Dark Matter) model.

Finally, for the case in which $k \neq 0$, and remembering that

$$\beta(t) = -\frac{1}{u(t)}\sqrt{\frac{3k}{\Sigma}}\sec\left(\sqrt{k}z\right), \tag{89}$$

where $z \equiv \int_0^t \frac{dt'}{u(t')}$, it can be seen that

$$d\tau = \beta(t)dt = -\sqrt{\frac{3k}{\Sigma}}\sec\left(\sqrt{k}z\right)dz. \tag{90}$$

By taking $dz = dt/u(t)$ and integrating the above expression, we obtain

$$\tau = -\sqrt{\frac{3k}{\Sigma}}\ln\left|\sec\left(\sqrt{k}z\right) + \tan\left(\sqrt{k}z\right)\right|. \tag{91}$$

Now, by inverting the Eq. (91) for z as a function of τ, one gets

$$z = \frac{2}{\sqrt{k}}\arccos\left\{\frac{\left[1+\exp(\tau\sqrt{\Sigma/3})\right]^{1/2}}{\sqrt{2}}\right\}. \tag{92}$$

The Hubble parameter, then, reads

$$H_{eff} = \frac{1}{\beta(t)}\frac{d}{dt}[\ln(\beta(t)u(t)] = \frac{d}{d\tau}\left[\ln\left(-\sqrt{\frac{3k}{\Sigma}}\sec\left(\sqrt{k}z\right)\right)\right]. \tag{93}$$

In order to reach Eq. (93) we have used $d\tau = \beta(t)dt$ in the first equality. Substituting Eq. (91) in (92) one arrives at

$$H_{eff}(\tau) = \frac{d}{d\tau}\left[\ln\left(-\sqrt{\frac{3k}{\Sigma}}\sec\left(2\arccos\left\{\frac{\left[1+\exp(\tau\sqrt{\Sigma/3})\right]^{1/2}}{\sqrt{2}}\right\}\right)\right)\right], \tag{94}$$

and therefore

$$H_{eff}(\tau) = \sqrt{\frac{\Sigma}{3}}\tanh\left[\sqrt{\frac{\Sigma}{3}}\tau\right]. \tag{95}$$

By means of (95) the effective scale factor is given by

$$a_{eff}(\tau) = \ln\left[\cosh\sqrt{\frac{\Sigma}{3}}\tau\right]. \tag{96}$$

The above Hubble parameter, despite the fact that it presents an expected behavior at large values of τ, is again too fast at lower values of τ.

6. Concluding remarks

We have investigated exact solutions for a FLRW braneworld, whose brane is performed by a bulk scalar field. The general idea was to find out explicit solutions which could, at least in some regime, to perform the large scale dynamics of the universe. In the course of our analysis a plenty of possibilities had appeared. Among them, we believe we pay attention to the most physically appealing cases.

In some aspects, the physical outputs can model a specific era of the known universe, as in the case represented by Eq. (82) in which the matter and radiation phases can be reached. By the same token, in the specific $k = 0$, $\Delta \neq -\Sigma$ case, the separation constant Σ can mimic a four-dimensional cosmological constant for a de-Sitter-like universe. Therefore, the late-time acceleration can be modeled without regarding to any type of dark energy.

Currently we are delving into the possibility to describe more aspects of the cosmic evolution. To accomplish that, more bulk scalar fields, as well as different potentials may be in order. We shall postpone these generalizations for a future work.

Acknowledgements

ASD and JMHS thanks to CNPq 301073/2016-3, 304629/2015-4, 445385/2014-6 for financial support. PMLTS acknowledge CAPES for financial support.

References

[1] L. Randall, R. Sundrum, Phys. Rev. Lett. 83 (1999) 46900.
[2] L. Randall, R. Sundrum, Phys. Rev. Lett. 83 (1999) 3370.
[3] M. Gogberashvili, Europhys. Lett. 49 (2000) 396.
[4] R. Maartens, K. Koyama, Living Rev. Relativ. 13 (2010) 5.
[5] V. Dzhunushaliev, V. Folomeev, M. Minamitsuji, Rep. Prog. Phys. 73 (2010) 066901.
[6] T. Shiromizu, K. Maeda, M. Sasaki, Phys. Rev. D 62 (2000) 043523.
[7] C.T. Hill, D.N. Schramm, J.N. Fray, Comm. Nucl. Part. Sci. 19 (1989) 25;
V. Silveira, Phys. Rev. D 38 (1988) 3832;
A.K. Raychaudhuri, G. Mukherjee, Phys. Rev. Lett. 59 (1987) 1504;
R. Geroch, J. Traschen, Phys. Rev. D 36 (1987) 1017;
L. Widrow, Phys. Rev. D 39 (1989) 3571.
[8] A. Ahmed, B. Grzadkowski, J. Wudka, J. High Energy Phys. 061 (2014) 1404.
[9] A.E. Bernardini, R.T. Cavalcanti, R. da Rocha, Gen. Relativ. Gravit. 47 (2015) 1840.
[10] M. Gremm, Phys. Lett. B 478 (2000) 434.
[11] M. Giovannini, Phys. Rev. D 76 (2007) 124017.
[12] A. de Souza Dutra, G.P. de Brito, J.M. Hoff da Silva, Europhys. Lett. 108 (2014) 11001.
[13] D. Bazeia, F.A. Brito, L. Losano, J. High Energy Phys. 64 (2006) 11.
[14] J. Sadeghi, A. Mohammadi, Eur. Phys. J. C 49 (2007) 859.
[15] M. Gremm, Phys. Rev. D 62 (2000) 044017.

11

Geometric horizons

Alan A. Coley [a], David D. McNutt [b,*], Andrey A. Shoom [c]

[a] *Department of Mathematics and Statistics, Dalhousie University, Halifax, Nova Scotia, B3H 3J5, Canada*
[b] *Faculty of Science and Technology, University of Stavanger, N-4036 Stavanger, Norway*
[c] *Department of Mathematics and Statistics, Memorial University, St. John's, Newfoundland and Labrador, A1C 5S7, Canada*

ARTICLE INFO

Editor: M. Trodden

ABSTRACT

We discuss black hole spacetimes with a geometrically defined quasi-local horizon on which the curvature tensor is algebraically special relative to the alignment classification. Based on many examples and analytical results, we conjecture that a spacetime horizon is always more algebraically special (in all of the orders of specialization) than other regions of spacetime. Using recent results in invariant theory, such geometric black hole horizons can be identified by the alignment type **II** or **D** discriminant conditions in terms of scalar curvature invariants, which are not dependent on spacetime foliations. The above conjecture is, in fact, a suite of conjectures (isolated vs dynamical horizon; four vs higher dimensions; zeroth order invariants vs higher order differential invariants). However, we are particularly interested in applications in four dimensions and especially the location of a black hole in numerical computations.

1. Introduction

Black holes, which are exact solutions in general relativity (GR) (representing, for example, physical objects formed out of the gravitational collapse of fuel-exhausted stars), are characterized by the boundary of the region from where light can still travel to asymptotic null infinity, called the event horizon, which is usually identified as the surface of the black hole and relates its area to the entropy. The event horizon is essentially a global (*teleological*) object, since it depends on the entire future history of the spacetime [1].

There has been much effort to give a general quasi-local description of a dynamical black hole [1,2]. Of particular interest are quasi-local objects called *marginally trapped tubes* (MTTs) or *trapping horizons*, and the special cases of *dynamical horizons* or *future outer trapping horizons* (FOTHs); in numerical work, these are also called *apparent horizons*. MTTs are hypersurfaces foliated by (closed compact space-like two-dimensional (2D) submanifolds without boundary) *marginally trapped surfaces* (MTSs) in which the expansion of one of the null normals vanishes and the other is non-positive. A dynamical horizon is a smooth 3D submanifold of spacetime foliated by MTSs such that the expansion of one future-directed null normal to the foliation vanishes, while the expansion of the other future directed null normal is negative (a FOTH has the additional condition that the directional derivative along the second null direction is negative).

A dynamical horizon is particularly well-suited to analyze dynamical processes involving black holes, such as black hole growth and coalescence. The area of a dynamical horizon necessarily increases with time [1]. An explicit example of a dynamical horizon is given by the Vaidya spacetime which admits spherically symmetric MTSs [3,4]. For a given mass function, the Vaidya spacetime also provides explicit examples of the transition from the dynamical to isolated horizons. If a hypersurface admits a dynamical horizon structure, it is unique. However, because a spacetime may have several distinct black holes, it may admit several distinct dynamical horizons. For dynamical horizons which are also FOTHs, two non-intersecting horizons generally either coincide or one is contained in the other [1].

It is believed that closed MTSs constitute an important ingredient in the formation of black holes, which motivates the idea of using MTTs as viable replacements for the event horizon of black holes [2]. Unfortunately, since the 2D apparent horizons depend on the choice of a reference foliation of spacelike hypersurfaces, MTTs and consequently trapping horizons and dynamical horizons are highly non-unique [5]. There have been some attempts to provide a physically sound criterion for selecting a preferred MTT such as, for example, in which the shear scalars along ingoing/outgoing null directions foliated by 2D spacelike surfaces vanish [6].

* Corresponding author.
E-mail addresses: aac@mathstat.dal.ca (A.A. Coley), david.d.mcnutt@uis.no (D.D. McNutt), ashoom@ualberta.ca (A.A. Shoom).

Realistic black holes interact with their environment and are consequently dynamical. The gravitational collapse leading to black hole formation is also a highly dynamical process. It is crucial to locate a black hole locally, which may not rely on the existence of an event horizon alone. A significant fraction of research in numerical relativity aims at predicting with high precision the waveforms of gravitational waves generated in the merger of compact-object binary systems or in stellar collapse to form black holes. Comparison with templates played a crucial role in the recent observations of gravitational waves from black hole mergers by the LIGO Collaboration [7].

In numerical studies of time-dependent collapse, it is often more practical to track apparent horizons or trapping horizons [8]. In contrast with the event horizon, which is a global concept defined using the global structure of spacetime, the apparent horizon is a quasi-local concept and is intrinsically foliation-dependent. In this paper we propose a foliation invariant and more geometrical approach, which is possible due to recent results in invariant theory.

2. Scalar polynomial curvature invariants

The algebraic classification of the Weyl tensor and the Ricci tensor in arbitrary dimensions using the boost weight decomposition [9] can be refined utilizing the restricted eigenvector and eigenvalue structure of their associated curvature operators [10], allowing for necessary conditions to be defined for a particular algebraic type in terms of a set of discriminants. A *scalar polynomial curvature invariant of order* k (or, in short, a scalar polynomial invariant or SPI) is a scalar obtained by contraction from a polynomial in the Riemann tensor and its covariant derivatives up to the order k. Black hole spacetimes are completely characterized by their $SPIs$ [11]. In particular, we can use discriminants to study the necessary conditions in arbitrary dimensions, in terms of simple $SPIs$, allowing for the algebraic classification of the higher dimensional Weyl and Ricci tensor when treated as curvature operators, for the spacetime to be of algebraic type **II** or **D** [12].

For example, in 5D the necessary condition for the trace-free Ricci tensor, $S_{ab} = R_{ab} - \frac{1}{5}Rg_{ab}$, to be of algebraic type **II/D** is that the discriminant (SPI) ${}^5_S D_5$ is zero, and the necessary conditions for the Weyl tensor to be of type **II/D** is that the $SPIs$ ${}^{10}_W D_i$ ($i = 8, 9, 10$) vanish [12]. As an illustration, the 5D rotating black ring [13] is generally of type $\mathbf{I_i}$, but on the horizon the discriminant ${}^5_T D_5$ of the trace-free part of the operator $T^a{}_b = C^{acde}C_{bcde}$ vanishes (and ${}^5_T D_4 > 0$), which signals that the spacetime is of Weyl type **II** on the horizon.

We are primarily interested in the 4D case here. The necessary type **II/D** discriminant condition ${}^4D_4 = 0$ for the trace-free ($s_1 = 0$) symmetric Ricci tensor S in 4D is [12]:

$$\mathcal{D} \equiv {}^4D_4 = -s_3^2(4s_2^3 - 144s_2s_4 + 27s_3^2) + s_4(16s_2^4 - 128s_4s_2^2 + 256s_4^2) = 0, \tag{1}$$

where

$$s_2 \equiv -\frac{1}{2}S^a{}_bS^b{}_a,\quad s_3 \equiv -\frac{1}{3}S^a{}_bS^b{}_cS^c{}_a,$$
$$s_4 \equiv \frac{1}{4}(2s_2^2 - S^a{}_bS^b{}_cS^c{}_dS^d{}_a). \tag{2}$$

Similar conditions hold for any trace-free symmetric tensor T_{ab}.

The necessary real conditions for the Weyl tensor to be of type **II/D** are [12]:

$$\mathcal{W}_1 \equiv -11W_2^3 + 33W_2W_4 - 18W_6 = 0, \tag{3}$$

$$\mathcal{W}_2 \equiv (W_2^2 - 2W_4)(W_2^2 + W_4)^2 + 18W_3^2(6W_6 - 2W_3^2 - 9W_2W_4 + 3W_2^3) = 0, \tag{4}$$

where

$$W_2 = \frac{1}{8}C_{abcd}C^{abcd},\quad W_3 = \frac{1}{16}C_{abcd}C^{cd}{}_{pq}C^{pqab},$$
$$W_4 = \frac{1}{32}C_{abcd}C^{cd}{}_{pq}C^{pq}{}_{rs}C^{rsab},$$
$$W_6 = \frac{1}{128}C_{abcd}C^{cd}{}_{pq}C^{pq}{}_{rs}C^{rs}{}_{tu}C^{tu}{}_{vw}C^{vwab}. \tag{5}$$

These 2 real conditions are equivalent to the real and imaginary parts of the complex syzygy $I^3 - 27J^2 = 0$ in terms of the complex Weyl tensor in the Newman–Penrose (NP) formalism [14].

Alternatively we can use the discriminant analysis to provide the type **II/D** syzgies expressed in terms of the $SPIs$ W_i ($i \in [1,6]$, defined above) by treating the Weyl tensor as a trace-free operator acting on the 6-dimensional vector space of bivectors [12] (however, these conditions are very large). More practical *necessary* conditions can be obtained by considering the trace-free symmetric operator $C_{abcd}C^{ebcd} - 2W_2\delta_a{}^e$; applying the discriminant analysis we find the coefficients of the characteristic equation are $w_2 = 8(W_2{}^2 - 2W_4)$ (and similarly for w_3, w_4) and so the necessary condition for this operator to be type **II/D** is given by equation (1) (with the s_i replaced by w_i).

The alignment classification can be applied to any rank tensor. To consider whether the covariant derivatives of the Ricci tensor, $R_{ab;cd...}$, are of type **II** or **D**, we can use the eigenvalue structure of the operators associated with the derivatives of the Ricci curvature and impose the type **II/D** necessary conditions. This can be repeated for the Weyl tensor and in arbitrary dimensions [15]. For example, for the covariant derivative of the Weyl tensor, $C_{abcd;e}$, in 4D we can consider the second order symmetric and trace-free operator ${}^1T^a{}_b$ defined by:

$${}^1T^a_{\ b} \equiv C^{cdef;a}C_{cdef;b} - \frac{1}{4}\delta^a{}_b{}^1I_2 \tag{6}$$

where ${}^1I_2 \equiv C^{abcd;e}C_{abcd;e}$, and we have the corresponding 4th, 6th, 8th order invariants ${}^1I_4$, ${}^1I_6$, ${}^1I_8$. Computing the coefficients of the characteristic equation we obtain ${}^1s_2 = -\frac{1}{2}{}^1I_4 + \frac{1}{8}{}^1I_2^2$ (and similarly for ${}^1s_3$, ${}^1s_4$). The necessary condition for this operator to be of type **II/D** (of the form $\mathcal{D} \equiv {}^4D_4 = 0$) is equivalent in form to the condition given in equation (1) with s_i replaced by ${}^1s_i$, and can be expanded out explicitly. For example, for the operator ${}^1T^a{}_b$ defined above for the type **D** Kerr metric, the vanishing of ${}^4_T D_4$ implies that $C_{abcd;e}$ is of type **D/II** on the horizon.

It is known that differential invariants, constructed from $SPIs$ in terms of the Weyl tensor and its covariant derivatives, detect horizons for several type **D** stationary solutions [16]. The construction of the class of Page–Shoom $SPIs$ that detect the horizons of stationary black holes exploits the fact that on the horizon the timelike Killing vector becomes null and is, in fact, a generator of the horizon [17]. Noting that stationary horizons are a special case of weakly isolated horizons, the type **II/D** $SPIs$ arising from the discriminant analysis of $C_{abcd;e}$ vanish on the horizon, and it can be explicitly shown that the type **II/D** discriminants share common zeros with the Page–Shoom invariant **W** for the Kerr spacetime [18]. Similar results using Cartan invariants are possible [19].

3. Examples and motivation

There are many examples (some briefly discussed in this paper, but see also [19]) that support the geometric conjectures to follow. Indeed, all of the known exact black hole solutions are algebraically special of Weyl (curvature) type **II/D** on the horizon [20]. This led to an earlier conjecture that asserted that stationary black holes in higher dimensions, possibly with the additional conditions of vacuum or asymptotic flatness, must be of Weyl type **D** [20].

There is also motivation for the conjectures from analytical results. Quasi-local isolated horizons, which account for equilibrium states of black holes and cover all essential local features of event horizons [1], are essentially defined as a 3D null surface (submanifold) with topology $S_2 \times R$ with an outgoing expansion rate which vanishes on the horizon. It follows that the null normal vector is a local time-translational Killing vector field on the horizon, and requires neither asymptotic structures nor foliations of spacetime. Every such Killing horizon is an isolated horizon [1]. In particular, this implies that the event horizon of the Kerr geometry is an isolated horizon. However, in general, spacetimes with isolated horizons need not admit any Killing vector fields even in a neighborhood. In [21] it was proven that if a stationary, real analytic, asymptotically flat vacuum black hole spacetime of dimension $D \geq 4$ contains a non-degenerate horizon with compact cross sections that are transverse to the stationarity generating Killing vector field then, for each connected component of the black hole's horizon, there is a Killing vector field which is tangent to the generators of the horizon.

In the 4D case, and assuming the "mild energy condition" implied by the dominant energy condition, the existence of an induced degenerate metric tensor which locally acts as a metric tensor on the 2D tangent space, and the induced covariant derivative, which constitute the geometry of a nonexpanding null surface, was demonstrated; this then leads to the conditions that on the nonexpanding weakly isolated horizon the Ricci and Weyl tensors are of type **II/D** [22]. This local result has been generalized to nonexpanding null surfaces in arbitrary dimensions (and the result is applicable to surfaces of any topology); indeed, it was shown that if the expansion of a null surface vanishes, then the shear must also vanish and a covariant derivative can be induced on each non-expanding null surface [23]. It can also be shown [18] that for any weakly-isolated horizon the Riemann tensor and the covariant derivatives of the Riemann tensor are of type **II** on the horizon.

We note that when a star collapses to form a black hole, the exterior of the black hole eventually settles down to a stationary state, most likely described by the Kerr metric. Despite what the interior of the black hole settles down to, this leads by continuity to the expectation that there will be a region of the interior near the horizon that should be close to the interior Kerr metric. Inside the black hole event horizon the Kerr metric has an inner horizon which is also a null surface. However, the inner horizon is unstable, so for a spacetime that begins close to the Kerr metric, the inner horizon should be replaced by something else, perhaps even a singularity [24–27]. There are a variety of analytic arguments, mathematical results, and numerical simulations that indicate that this singularity maintains the inner horizon's character as a null surface [1,22].

This supports the notion that the horizon is smooth and unique at later times and, in principle, can be identified by algebraic/geometrical conditions. It is possible that as we follow this unique, smooth surface back in time (during the physics of collapse or merger), this surface suffers a bifurcation and this surface is no longer unique or smooth (or even differentiable). But it is plausible that there exists a unique, smooth geometric horizon that shields all other horizons (or at least identifies the region of interest).

4. The geometric horizon conjecture

This consequently motivates us to conjecture that there is a geometrically defined unique, locally determinable, smooth (dynamical) horizon on which the curvature tensor is algebraically special. In particular, this implies that a spacetime horizon is always more algebraically special (in all of the orders of specialization) than (all) other regions of spacetime. Such geometric black hole horizons can be identified and located by the type **II/D** conditions in terms of SPIs, which are not dependent on spacetime foliations. To state the conjectures, we will say a tensor **T** is *nth-order algebraically special* if **T** and all covariant derivatives of **T** up to order n are of algebraic type **II** or more special.

Conjecture Part I: *If a black hole spacetime is zeroth-order algebraically general, then on the geometric horizon the spacetime is algebraically special. We can identify this geometric horizon using scalar curvature invariants.*

This is the more practical part of the Conjecture and will hopefully be useful to numerical relativists who study the collapse or merger of real black holes, which are generically of general algebraic type away from the horizon. The conjecture might be qualitatively different for isolated and dynamical horizons. The issue then becomes one of finding effective ways to do computations.

Conjecture Part II: *If a black hole spacetime is zeroth-order algebraically special (and on the horizon the spacetime is thus also algebraically special), then if the black hole spacetime is first-order algebraically general, then on the horizon the spacetime is first-order algebraically special. We can identify this geometric horizon using differential scalar curvature invariants.*

If necessary, this can be repeated for higher order covariant derivatives. This is the more theoretical and analytical part of the Conjecture, and can be applied to exact solutions. Note that in general we may not wish that the covariant derivatives be algebraically special (i.e., of type **II**/**D**) to each order (i.e., of type D^k) on the black hole horizon, as this might be too restrictive.

Comments: The algebraic conditions expressed in terms of SPIs essentially define a *geometric horizon*. In order to make the definition more precise, we need to focus on physical black hole solutions (both exact black hole solutions and generic physical collapse and black hole coalescences) and in order to prove definitive results we need to append some physical conditions to the definition such as, for example, energy conditions, a particular theory of gravity (e.g., we assume GR; in principle some conditions may be different for different theories), and perhaps other asymptotic conditions. We also note that SPIs may not specify the geometric horizon completely in the sense that they may also vanish at fixed points of any isometries and along any axes of symmetry. However, we expect that the identification of a smooth surface for physical situtations is always possible. Unlike apparent horizons, a geometric horizon does not depend on a chosen foliation in the spacetime.

Although this conjecture is also intended to apply in higher dimensions [19], we are primarily interested in applications in 4D, and particularly in numerical computations. Indeed, the above conjectures are, in fact, a suite of conjectures (isolated vs dynamical horizon; 4D vs higher dimensions; zeroth order invariants vs higher order differential invariants). In physical problems with dynamical evolution the horizon might not be unique, or may not exist at all, and amendments to the conjecture may be necessary (e.g., it may be appropriate to replace the vanishing of invariants in the definition of a geometric horizon as an algebraically special hypersurface, with the conditions that the magnitudes of certain SPIs take their smallest values).

In a sense the conjecture refers to "peeling properties" (of the geometrical curvature) close to the horizon. The curvature is of algebraically special type **II** close to the horizon but it is plausible that, as gravitational wave modes (of algebraic types **III** and **N**) fall off more quickly to infinity, the horizon eventually settles down to be type **D** under some reasonable asymptotic conditions.

5. Discussion

The question of whether these definitions and conjectures are useful will have to be further evaluated. Although we have attempted to support the conjectures with analytical results and practical examples (also see [18]), further work is required and perhaps additional refinement of the conjectures will be necessary. In particular, it is of primary importance to study geometric horizons numerically in 4D in physically relevant asymmetric collapse and black hole coalescences.

Dynamical horizons: The conjecture is intended to apply to dynamical horizons. It is much more difficult to study dynamical horizons, but let us discuss some preliminary encouraging results [18].

Let us first consider the imploding spherically symmetric metric in advanced coordinates [14]:

$$ds^2 = -e^{2\beta(v,r)}\left(1 - \frac{2m(v,r)}{r}\right)dv^2 + 2e^{\beta(v,r)}dvdr + r^2 d\Omega^2$$

where $m(v,r)$ is the mass function and $\beta(v,r)$ is an arbitrary function. It is known that the unique spherically symmetric FOTH is given by the surface $r - 2m(v,r) = 0$, which is expansion-free.

Using the NP formalism, the Riemann tensor for any spherically symmetric black hole solution is found to be of type **II/D** on the horizon; i.e., at the algebraic level the Ricci tensor always detects the horizon [18]. However, the covariant derivatives of Ricci and Weyl will generally be of type **I**. Note that for this class of metrics the Ricci tensor cannot be globally type **D** unless $m_{,v} = 0$, in which case the horizon is isolated.

Let us assume that $m_{,v} \neq 0$ and consider dynamical horizons. Since no field equations have been imposed, we need additional conditions. For the class of spherically symmetric metrics admitting a geodesic-lined horizon which is a shear-free MTT [2], with the additional condition that the frame vectors normal to the surface are geodesics, requires that the metric function β satisfies $\beta_{,v} = 0$ [18]. It can then be shown that the Riemann tensor and its covariant derivatives for any spherically symmetric black hole solution with $\beta_{,v} = 0$ are of type **II/D** on the horizon [18]. Hence the imploding exact Vaidya solution and the class of exact Tolman–Bondi solutions admitting MTTs [8] satisfy the type **II/D** conditions on the horizon.

We next consider the Kastor–Traschen dynamical two-black-hole solution [28], which describes two charge-equal-to-mass black holes in a spacetime with a positive cosmological constant. The spatial part of the metric is written in coordinates centered at each of the black hole positions ($r_i = 0$, $i = 1,2$) and represents a 3D infinite cylinder with 2D cross-sectional area of $4\pi m_i^2$, where m_1 and m_2 are the black hole masses. When the sum of the black hole masses does not exceed a critical mass, the black holes coalesce and form a larger black hole.

At earliest times $\mathcal{W}_1 \to 0$ as $\tau \to -\infty$ (in these spacetimes $\tau \in (-\infty, 0)$), and there are two 3D geometric horizons enclosing the 2 black holes. It is found [29] that $\mathcal{W}_2$ is identically zero, and the type **II/D** discriminant $\mathcal{W}_1$ vanishes on segments of the symmetry-axis, at the black hole coordinate locations $r_i = 0$ and on an additional "dynamical" 2D (cylindrical) surface around the symmetry-axis (this 2D surface appears concurrently with the two 3D black hole horizons in the center of mass plane). At the earlier stages of the coalescence the 2D surface has a finite cross-sectional radius (from the symmetry-axis), but at later stages this surface expands as the two black holes move toward one another. There is a measure of the black hole separation that can be introduced such that as $\tau \to 0^-$ this measure approaches zero as the two black holes merge [29] and the 2D surface forms around the two black holes, so that it is possible to identify the location of a geometric horizon in the dynamical regime. As $\tau \to 0^-$, $\mathcal{W}_1 \to 0$ and in the quasi-stationary regime there will be a single 3D horizon; after merger the spacetime eventually settles down to a type **D** Reissner–Nordstrom–de Sitter black hole of mass $m_1 + m_2$ (which is known to have a 3D geometric horizon [19]).

In addition, it is found that there are 3D surfaces located at a finite distance from the axis of symmetry for which the type **II/D** discriminant for the trace-less Ricci tensor vanishes, and where the Ricci tensor is consequently of type **II/D**. There is also numerical evidence for a minimal 3D geometric surface evolving in time where the invariant $\mathcal{W}_1$ locally takes on a constant non-zero minimum value. These results are suggestive and lend support for the existence of a geometric horizon in the dynamical regime in these exact spacetimes, but further analysis is necessary which will be presented in the future.

Computability: The calculation of gravitational wave signals in the theoretical modelling of 4D sources in the framework of GR is well understood [30]. In higher dimensions, numerical simulations of rapidly spinning objects have been studied [31]. Of course, if our ultimate aim is to provide potentially useful results for numerical relativists, computability is an important issue. In this regard using Cartan invariants in the NP approach is certainly an advantage. Indeed, we have already used the NP approach to address certain problems, proving its utility, and we aim to develop this further in future research [18]. As an illustration, in the 4D Kerr-NUT-AdS metric [14] the cohomogeneity is 2D, and the Page–Shoom invariant **W** produces a degree eight, first order SPI that detects the horizon [17]. On the other hand, the two NP spin coefficients (Cartan scalars) ρ and μ, vanish on the event horizon, which also implies that the first covariant derivative of the Weyl spinor is of type **D** on the event horizon. These Cartan invariants are easier to compute than the related SPIs [19].

Acknowledgements

This work was supported through NSERC (A.A.C.), NSERC Discovery Grant 261429-2013 (A.A.S.), the AARMS collaborative research group (A.A.S.), and the Research Council of Norway, Toppforsk grant no. 250367: Pseudo-Riemannian Geometry and Polynomial Curvature Invariants: Classification, Characterisation and Applications (D.D.M.). GRTensorII was used in some of our calculations. We also would like to thank Ivan Booth, Kayll Lake and Malcolm MacCallum for discussions at the early stages of this work.

References

[1] A. Ashtekar, B. Krishnan, Phys. Rev. Lett. 89 (2002) 261101;
A. Ashtekar, B. Krishnan, Phys. Rev. D 68 (2003) 104030; see also, Living Rev. Relativ. 7 (2004) 10, http://www.livingreviews.org/lrr-2004-10.
[2] J.M.M. Senovilla, Int. J. Mod. Phys. D 20 (2011) 21392168, arXiv:1107.1344.
[3] P.C. Vaidya, Proc. Indian Acad. Sci. A 33 (1951) 264; reprinted Gen. Relativ. Gravit. 31 (1999) 119.
[4] W.B. Bonnor, P.C. Vaidya, Gen. Relativ. Gravit. 1 (1970) 127.
[5] A. Ashtekar, G.J. Galloway, Adv. Theor. Math. Phys. 9 (2005) 1.

[6] N. Cipriani, J.M.M. Senovilla, J. Van der Veken, Umbilical properties of spacelike co-dimension two submanifolds, arXiv:1604.06375, 2016.
[7] B.P. Abbott, et al., LIGO Scientific Collaboration Virgo Collaboration, Observation of gravitational waves from a binary black hole merger, Phys. Rev. Lett. 116 (2016) 061102, arXiv:1606.04856.
[8] I. Booth, Can. J. Phys. 83 (2005) 1073, arXiv:gr-qc/0508107.
[9] A. Coley, R. Milson, V. Pravda, A. Pravdova, Class. Quantum Gravity 21 (2004) L35, arXiv:gr-qc/0401008;
A. Coley, R. Milson, V. Pravda, A. Pravdova, Class. Quantum Gravity 21 (2004) 5519, arXiv:gr-qc/0410070.
[10] A. Coley, S. Hervik, Class. Quantum Gravity 27 (2010) 015002, arXiv:0909.1160.
[11] A. Coley, S. Hervik, N. Pelavas, Class. Quantum Gravity 26 (2009) 025013, arXiv:0901.0791.
[12] A. Coley, S. Hervik, Gen. Relativ. Gravit. 43 (2011) 2199; see also arXiv: 1011.2175.
[13] R. Emparan, H.S. Reall, Phys. Rev. Lett. 88 (2002) 101101;
R. Emparan, H.S. Reall, Living Rev. Relativ. 11 (2008) 6.
[14] H. Stephani, D. Kramer, M.A.H. MacCallum, C.A. Hoenselaers, E. Herlt, Exact Solutions of Einstein's Field Equations, second edition, Cambridge University Press, Cambridge, 2003.
[15] A. Coley, S. Hervik, N. Pelavas, Class. Quantum Gravity 27 (2010) 102001, arXiv: 1003.2373.
[16] M. Abdelqader, K. Lake, Phys. Rev. D 91 (2015) 084017, arXiv:1412.8757.
[17] D.N. Page, A.A. Shoom, Phys. Rev. Lett. 114 (2015) 141102, arXiv:1501.03510.
[18] A.A. Coley, D.D. McNutt, Identification of black hole horizons using discriminating scalar curvature invariants, preprint, 2017.
[19] D. Brooks, P. Chavy-Waddy, A. Coley, A. Forget, D. Gregoris, M.A.H. MacCallum, D. McNutt, Cartan Invariants as Event Horizon Detectors, preprint, 2016;
A.A. Coley, D.D. McNutt, Class. Quantum Gravity 34 (2016) 035008, arXiv: 1704.03055.
[20] A. Coley, N. Pelavas, Gen. Relativ. Gravit. 38 (2006) 445.
[21] J. Isenberg, V. Moncrief, Class. Quantum Gravity 9 (1992) 1683.
[22] A. Ashtekar, C. Beetle, J. Lewandowski, Class. Quantum Gravity 19 (2002) 1195, arXiv:gr-qc/0111067.
[23] J. Lewandowski, T. Pawlowski, Class. Quantum Gravity 22 (2005) 1573, arXiv:gr-qc/0410146.
[24] P.P. Avelino, A.J.S. Hamilton, C.A.R. Herdeiro, M. Zilhao, Phys. Rev. D 84 (2011) 024019, arXiv:1105.4434 [gr-qc].
[25] E. Poisson, W. Israel, Phys. Rev. D 41 (1990) 1796.
[26] A. Ori, Phys. Rev. Lett. 67 (1991) 789;
A. Ori, Phys. Rev. Lett. 68 (1992) 2117.
[27] D. Marolf, A. Ori, Phys. Rev. D 86 (2012) 124026.
[28] D. Kastor, J. Traschen, Phys. Rev. D 47 (1993) 5370.
[29] A.A. Coley, D.D. McNutt, A.A. Shoom, Geometric horizons in the Kastor–Traschen solutions, 2017, in preparation.
[30] J. Thornburg, Living Rev. Relativ. 10 (2007) 3, http://www.livingreviews.org/lrr-2007-3.
[31] W.G. Cook, U. Sperhake, Gravitational wave extraction in higher dimensional numerical relativity using the Weyl tensor, arXiv:1609.01292.

Gravitational instabilities of the cosmic neutrino background with non-zero lepton number

Neil D. Barrie *, Archil Kobakhidze

ARC Centre of Excellence for Particle Physics at the Terascale, School of Physics, The University of Sydney, NSW 2006, Australia

ARTICLE INFO

Editor: J. Hisano

ABSTRACT

We argue that a cosmic neutrino background that carries non-zero lepton charge develops gravitational instabilities. Fundamentally, these instabilities are related to the mixed gravity-lepton number anomaly. We have explicitly computed the gravitational Chern–Simons term which is generated quantum-mechanically in the effective action in the presence of a lepton number asymmetric neutrino background. The induced Chern–Simons term has a twofold effect: (i) gravitational waves propagating in such a neutrino background exhibit birefringent behaviour leading to an enhancement/suppression of the gravitational wave amplitudes depending on the polarisation, where the magnitude of this effect is related to the size of the lepton asymmetry; (ii) Negative energy graviton modes are induced in the high frequency regime, which leads to very fast vacuum decay producing, e.g., positive energy photons and negative energy gravitons. From the constraint on the present radiation energy density, we obtain an interesting bound on the lepton asymmetry of the universe.

1. Introduction

Along with the Cosmic Microwave Background radiation (CMB), the existence of the Cosmic Neutrino Background (CνB) is an inescapable prediction of the standard hot big bang cosmology (see e.g. [1] for a review). It is assumed to be a highly homogeneous and isotropic distribution of relic neutrinos with the temperature:

$$T_\nu = \left(\frac{4}{11}\right)^{\frac{1}{3}} T_\gamma \approx 1.945\ \mathrm{K}\,, \tag{1}$$

where $T_\gamma = 2.725$ K is the temperature of the CMB today. Unlike the CMB though, the CνB is extremely hard to detect and its properties are largely unknown. Namely, the CνB may exhibit a neutrino–antineutrino asymmetry

$$\eta_{\nu_\alpha} = \frac{n_{\nu_\alpha} - \bar{n}_{\nu_\alpha}}{n_\gamma} = \frac{\pi^2}{12\zeta(3)}\left(\xi_\alpha + \frac{\xi_\alpha^3}{\pi^2}\right), \tag{2}$$

for each neutrino flavour $\alpha = e,\ \mu,\ \tau$. Here $\xi_\alpha = \mu_\alpha/T$ is the degeneracy parameter, μ_α being the chemical potential for α-neutrinos. In fact, such an asymmetry is generically expected to be of the order of the observed baryon–antibaryon asymmetry, $\eta_B = (n_B - \bar{n}_B)/n_\gamma \sim 10^{-10}$, due to the equilibration by sphalerons of lepton and baryon asymmetries in the very early universe. However, there are also models [2,3] which predict an asymmetry in the neutrino sector that are many orders of magnitude larger than η_B. If so, this would have interesting cosmological implications for the QCD phase transition [4] and/or large-scale magnetic fields [5].

The most stringent bound on the neutrino asymmetry comes from the successful theory of big bang nucleosynthesis (BBN). BBN primarily constrains the electron neutrino asymmetry. However, this bound applies to all flavours, since neutrino oscillations below $\sim$10 MeV are sizeable enough to lead to an approximate flavour equilibrium before BBN, $\mu_e \approx \mu_\mu \approx \mu_\tau (\equiv \mu_\nu)$ [6–8].[1] The updated analysis presented in [9] leads to the following bound on the common degeneracy parameter:

$$|\xi_\nu| \lesssim 0.049 \tag{3}$$

In this paper, we argue that the lepton asymmetry in the active neutrino sector leads to gravitational instabilities. These instabilities originate from the gravity-lepton number chiral quantum

* Corresponding author.
E-mail addresses: neil.barrie@sydney.edu.au (N.D. Barrie), archilk@physics.usyd.edu.au (A. Kobakhidze).

[1] See, however, a recent analysis in [10], where a larger η_{ν_μ,ν_τ} asymmetry is found to be allowed.

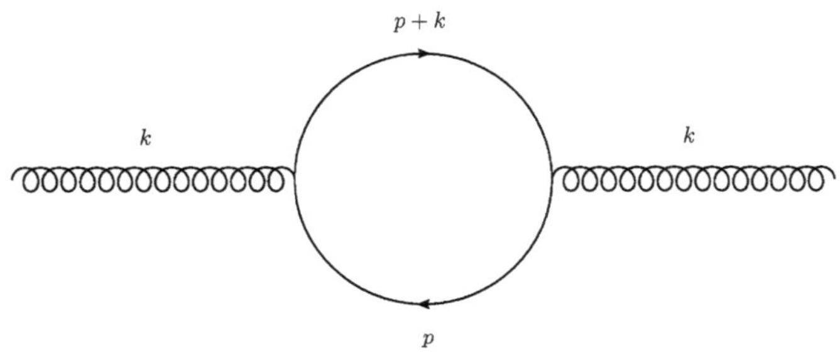

Fig. 1. Parity violating contribution to the fermion propagator.

anomaly, which is present in the Standard Model when considering Majorana neutrinos. Indeed, in the case of Majorana neutrinos, a non-zero lepton asymmetry for active neutrinos implies an imbalance between neutrinos of left-handed chirality and antineutrinos of right-handed chirality which, as we demonstrate explicitly below, leads to the inducement of the gravitational Chern–Simons term in the effective action. This is analogous to the inducement of Chern–Simons terms in gauge theories [11].

The induced Chern–Simons term causes birefringence of gravitational waves propagating in the lepton asymmetric neutrino background, which can be sizeable for gravitational waves generated at very early times. More importantly, short-scale gravitational fluctuations exhibit negative energy modes, which lead to a rapid decay of the vacuum state, e.g., into negative energy graviton and photons. Since the graviton energy is not bounded from below, the phase space for this process is formally infinite, that is, the instability is expected to develop very rapidly. Conservatively, we introduce a comoving cut-off Λ and compute the spectrum of produced photons as a function of neutrino chemical potential. From the constraint on the radiation energy density today, we then obtain an interesting bound on the neutrino degeneracy parameter:

$$\xi_\nu \lesssim 2 \cdot 10^{-41} \left(\frac{T_a}{10^{15}\,\text{GeV}} \right)^{4/3} \left(\frac{M_p}{\Lambda} \right)^{17/3} , \tag{4}$$

provided that the lepton asymmetry has been generated above $T_* \gtrsim \frac{440}{\sqrt{\xi}} \sqrt{M_p/\Lambda}$ GeV (here $M_p \approx 2.4 \cdot 10^{18}$ GeV is the reduced Planck mass), where T_a is the temperature at which the asymmetry is generated.

This paper is organised as follows; in Section 2 we describe the calculation of the graviton polarisation tensor in the presence of a lepton asymmetric CνB, and consider the associated effective action. Section 3 illustrates the birefringent behaviour of gravitational waves in such a background, while in Section 4 we derive constraints on the CνB lepton asymmetry through the induced gravitational instabilities, before concluding in Section 5.

2. Graviton polarisation tensor in the lepton asymmetric CνB

We calculate the inducement of the Chern–Simons like terms in the effective graviton Lagrangian through the 1-loop graviton polarization diagram depicted in Fig. 1, influenced by a lepton asymmetric neutrino background. The lepton asymmetry is enforced in the Lagrangian through a chiral chemical potential $\mathcal{L}_{\mu_\nu} = \bar{\nu} \not{b} \gamma^5 \nu = \mu_\nu \bar{\nu} \gamma_0 \gamma^5 \nu$, for which we have considered the frame in which the CνB is at rest ($\not{b} = \mu_\nu \gamma_0$). The neutrino propagator is altered as follows:

$$S(p) = \frac{i}{\not{p} - m - \not{b}\gamma^5} = \frac{i}{\not{p} - m} \sum_{n=0}^{\infty} \left(-i\not{b}\gamma^5 \frac{i}{\not{p} - m} \right)^n$$
$$\equiv S_0(p) + \sum_{n=1}^{\infty} S_n(p) , \tag{5}$$

where $S_0(p)$ is the usual fermion propagator in vacuum. The above modified neutrino propagator to first-order in μ_ν is given by $S(p) \approx S_0(p) - i\mu_\nu \frac{i}{\not{p}-m} \gamma_0 \gamma^5 \frac{i}{\not{p}-m}$. The higher order terms in b_μ, or μ, are neglected because we are only interested in the linear terms in b_μ, which will result in a Chern–Simons like term. Taking this, and using the standard Feynman rules, we find that the parity odd part of the full graviton polarization tensor is:

$$\Pi_{\mu\nu\rho\sigma} = - \int \frac{d^4 p}{(2\pi)^4} (2p+k)_\nu (2p+k)_\sigma$$
$$\times \Big[Tr(\gamma_\mu S_0(p+k) \gamma_\rho S_1(p))$$
$$+ Tr(\gamma_\rho S_0(p) \gamma_\mu S_1(p+k)) \Big]$$
$$+ (\mu \leftrightarrow \nu) + (\rho \leftrightarrow \sigma) + (\mu \leftrightarrow \nu, \rho \leftrightarrow \sigma) . \tag{6}$$

To evaluate the divergent loop integral in (6) we employ the dimensional regularization method ($d = 4 - \epsilon$, $\epsilon \to 0$) and utilise the relations given in Appendix A. We hence obtain:

$$\Pi_{\mu\nu\rho\sigma} = \frac{\mu_\nu}{8\pi^2} k^\alpha \varepsilon_{\mu\rho\alpha 0} \int_0^1 dx \left[\frac{4\pi^2 \lambda^2}{M^2} \right]^\epsilon$$
$$\times \Big[8x^2(1-x)^2(1-2x)^2 \frac{k^2}{M^2} \Gamma(1+\epsilon) k_\nu k_\sigma$$
$$+ (24x^2 - 44x + 18)\Gamma(\epsilon - 1) M^2 \eta_{\nu\sigma}$$
$$- 16x^2(1-x)^2 \Gamma(\epsilon) k^2 \eta_{\nu\sigma}$$
$$- (80x^4 - 192x^3 + 156x^2 - 50x + 5)\Gamma(\epsilon) k_\nu k_\sigma \Big] \tag{7}$$
$$+ (\mu \leftrightarrow \nu) + (\rho \leftrightarrow \sigma) + (\mu \leftrightarrow \nu, \rho \leftrightarrow \sigma) , \tag{8}$$

where $M^2 = m^2 - x(1-x)k^2$ and the limit $\epsilon \to 0$ has been assumed. In simplifying this result we find a divergent quantity of the following form:

$$\Pi^{(div)}_{\mu\nu\rho\sigma} = -\frac{1}{\epsilon} \frac{\mu_\nu}{2\pi^2} k^\alpha \varepsilon_{\mu\rho\alpha 0} m^2 \eta_{\nu\sigma}$$
$$+ (\mu \leftrightarrow \nu) + (\rho \leftrightarrow \sigma) + (\mu \leftrightarrow \nu, \rho \leftrightarrow \sigma) , \tag{9}$$

where γ is Euler's constant. A straightforward inspection reveals that this divergent term does not satisfy the gravitational Ward identity, $k^\nu \Pi^{(div)}_{\mu\nu\rho\sigma} \neq 0$, and hence violates the gauge invariance of the effective gravitational action. This has also been observed previously in a somewhat related calculation in Ref. [12]. The origin of this violation is rooted in the method of dimensional regularization, which violates Local Lorentz invariance explicitly through the extrapolation to non-integer spacetime dimensions $d = 4 - \epsilon$. Therefore, following the standard lore, we introduce non-invariant counter-terms to renormalise away this unphysical divergent term. The polarisation tensor then takes the following simple form:

$$\Pi_{\mu\nu\rho\sigma} = \mu_\nu \varepsilon_{\mu\rho\alpha 0} k^\alpha [k_\nu k_\sigma - k^2 \eta_{\nu\sigma}] C(k^2)$$
$$+ (\mu \leftrightarrow \nu) + (\rho \leftrightarrow \sigma) + (\mu \leftrightarrow \nu, \rho \leftrightarrow \sigma) , \tag{10}$$

where

$$C(k^2) = \frac{1}{192\pi^2} - \frac{m^2}{16\pi^2 (k^2)^{3/2}}$$
$$\times \left[\sqrt{k^2} - \sqrt{4m^2 - k^2} \tan^{-1} \left(\frac{\sqrt{k^2}}{\sqrt{4m^2 - k^2}} \right) \right] . \tag{11}$$

This further reduces to:

$$C(k^2) = \begin{cases} -\frac{1}{1920\pi^2}\frac{k^2}{m^2}, & \text{if } k^2/m^2 \ll 1 \\ \frac{1}{192\pi^2}, & \text{if } k^2/m^2 \gg 1 \end{cases} \tag{12}$$

We wish to investigate the second of these two possible limiting cases, $k^2/m_\nu^2 \gg 1$. In this limit we obtain the following contribution to the graviton action,

$$\begin{aligned} S_{eff} &= -\frac{\mu_\nu}{192\pi^2}\int d^4x \varepsilon_{\mu\rho\alpha 0} h^{\mu\nu}\partial^\alpha(\Box h^{\rho\sigma}\eta_{\nu\sigma} - \partial_\nu\partial_\sigma h^{\rho\sigma}) \\ &= \frac{\mu_\nu}{48\pi^2}\int d^4x\, K^0 , \end{aligned} \tag{13}$$

which contains the same number of derivatives as the standard graviton kinetic term in the weak field approximation. In fact, K^0 is the linearised 0th component of the four dimensional Chern–Simons topological current:

$$K^\beta = \varepsilon^{\beta\alpha\mu\nu}(\Gamma^\sigma_{\alpha\rho}\partial_\mu\Gamma^\rho_{\nu\sigma} - \frac{2}{3}\Gamma^\sigma_{\alpha\rho}\Gamma^\rho_{\mu\lambda}\Gamma^\lambda_{\nu\sigma}) . \tag{14}$$

Therefore, the presence of an asymmetry in the CνB replicates Chern–Simons modified gravity:

$$S_{CS} = \int d^4x\,(\partial_\mu\theta)K^\mu = \int d^4x\,\theta({}^*RR) , \tag{15}$$

where $\partial_\mu\theta = \frac{\mu_\nu}{48\pi^2}$.

3. Gravitational wave propagation in asymmetric CνB

The Chern–Simons term in Eq. (13) is found to induce a birefringence effect on the propagation of gravitational waves. The planned gravitational wave detectors, such as eLISA, DECIGO and BBO, can potentially measure the polarization of gravitational waves and hence this birefringence effect. With this in mind, we consider the propagation of gravitational waves in a lepton asymmetric background over cosmological distances. To this end we parametrise the gravitational waves as: $h_{ij} = \frac{\mathcal{A}_{ij}}{a(\eta)}\exp[-i(\phi(\eta) - \mathbf{k}\cdot\mathbf{x})]$, which can be decomposed into the two circularly polarised states: e^R_{ij} and e^L_{ij}. These two possible circularly polarised states are defined as: $e^R_{ij} = \frac{1}{\sqrt{2}}(e^+_{ij} + ie^\times_{ij})$ and $e^L_{ij} = \frac{1}{\sqrt{2}}(e^+_{ij} - ie^\times_{ij})$, which satisfy $n_i\varepsilon^{ijk}e^{R,L}_{kl} = i\lambda_{R,L}(e^j_l)^{R,L}$, where $\lambda_{R,L} = \pm 1$. The phase factor $\lambda_{R,L}$ leads to exponential suppression or enhancement of the left and right circular polarisations of the propagating gravitational waves, the magnitude of which we shall now calculate.

From the equations of motion for the action $S = S_{EH} + S_{eff}$ [13] we obtain:

$$\begin{aligned} &(i\phi^{R,L}_{,\eta\eta} + (\phi^{R,L}_{,\eta})^2 + \mathcal{H}_{,\eta} + \mathcal{H}^2 - |\mathbf{k}|^2)\left(1 - \frac{\lambda_{R,L}\kappa\theta_{,\eta}}{a^2}\right) \\ &= \frac{i\lambda_{R,L}|\mathbf{k}|}{a^2}(\theta_{,\eta\eta} - 2\mathcal{H}\theta_{,\eta})(\phi^{R,L}_{,\eta} - i\mathcal{H}) . \end{aligned} \tag{16}$$

We will first solve the above equation, assuming propagation in the matter dominated epoch $a(\eta) = a_0\eta^2 = \frac{a_0}{1+z}$. The accumulated phase over the length of propagation, to first order in θ, is given by,

$$\Delta\phi^{R,L}_{mat} = i\lambda_{R,L}|\mathbf{k}|H_0\int_\eta^1\left[\frac{1}{4}\theta_{,\eta\eta} - \frac{1}{\eta}\theta_{,\eta}\right]\frac{d\eta}{\eta^4} . \tag{17}$$

In the case considered in this manuscript, we make the following identification $\theta_{,\eta} = \left(\frac{a(\eta_0)}{a(\eta)}\right)^2\frac{\mu_0}{48\pi^2 M_p^2}$, where $\mu_0 = a(\eta)\mu_\nu$ is the present neutrino chemical potential. For this lepton asymmetric CνB,

$$\Delta\phi^{R,L}_{mat} = -i\frac{1}{288\pi^2}\frac{\mu_\nu H_0}{M_p^2}\left(\frac{|\mathbf{k}|}{1\text{ GeV}}\right)(1+z)^4 . \tag{18}$$

Hence the ratio of the amplitudes of each polarisation is given by:

$$\frac{h_R}{h_L} \propto e^{-2|\Delta\phi^{R,L}_{mat}|} . \tag{19}$$

Taking into account the current bounds on the CνB asymmetry parameter, ξ, we find $|i\Delta\phi^{R,L}| \lesssim 10^{-87}\left(\frac{|\mathbf{k}|}{1\text{ GeV}}\right)$, for $z \sim 30$. Therefore, the accumulated phase difference for $z \sim 30$ sources is too small to be observable by any conceivable gravitational wave detector.

The more interesting scenario to consider is the propagation of gravitational waves from sources in the very early in the universe. At early times the chemical potential would have been larger and the longer accumulated propagation time. Conceivably, any early universe sources could provide constraints, if the different polarisations are measurable. Therefore, we now consider gravitational waves produced at very early times, during the radiation dominated epoch after reheating. The accumulated phase now reads:

$$\Delta\phi^{R,L}_{rad} = i\lambda_{R,L}\frac{|\mathbf{k}|}{\Omega_{r,0}H_0^2}\int_\eta^1\left[\frac{1}{2}\theta_{,\eta\eta} - \frac{1}{\eta}\theta_{,\eta}\right]\frac{d\eta}{\eta^2} , \tag{20}$$

where $\Omega_{r,0} \sim 9.2\cdot 10^{-5}$ is the radiation density parameter today. After solving the integral we find:

$$\Delta\phi^{R,L}_{rad} \simeq -i\lambda_{R,L}\xi_\nu\left(\frac{|\mathbf{k}|}{1\text{ GeV}}\right)\left(\frac{T_s}{1\text{ TeV}}\right)^4 , \tag{21}$$

where we have redefined the redshift in terms of the temperature at which the gravitational waves are produced, T_s, or when the asymmetry is generated, whichever is lowest.

From Eq. (21), it can be seen that if an asymmetry in the CνB is present at early times, which equilibrium sphalerons transitions may assure, then it is possible to get significant birefringent behaviour in the propagation of gravitational waves from primordial sources. This is dependent on the momenta $|\mathbf{k}|$ of the gravitational waves and size of the asymmetry.

4. Induced ghost-like modes and vacuum decay

Another interesting consequence of the induced Chern–Simons term in Eq. (13) is that short-scale gravitational fluctuations exhibit negative energy modes, which lead to a rapid decay of a vacuum state e.g., into negative energy graviton and photons [14]. Since the graviton energy is not bounded from below, the phase space for this process is formally infinite [15,16], and as such will develop very rapidly. We investigate the production of two photons and a negative energy graviton via this process, to obtain constraints on the neutrino asymmetry at early times. The relevant effective interaction is of the form:

$$\begin{aligned} S_{\text{int}} &\sim \frac{1}{m_*}\int d^4x h^{can}_{\mu\nu}T^{\mu\nu} \\ &= \frac{1}{m_*}\int d^4x\frac{1}{2}h^{can}F_{\mu\nu}F^{\mu\nu} - h^{can}_{\mu\nu}F^{\mu\alpha}F^\nu_\alpha , \end{aligned} \tag{22}$$

where the canonically normalised graviton field is: $h^{can}_{\mu\nu} = m_{can}h^{can}_{\mu\nu}$, with

$$m_{can} = M_p\sqrt{1 + \lambda_{R,L}\frac{|\mathbf{k}|}{am_{CS}}} , \tag{23}$$

where m_{CS} is the analogous Chern–Simons mass scale:

$$m_{CS}(t) = \frac{M_p^2}{\mu_\nu} = \frac{M_p^2}{\xi T} = \frac{a(t)M_p^2}{\mu_0} \,. \tag{24}$$

4.1. Photon energy spectrum from induced vacuum decay

To obtain a finite result for the decay rate we need to constrain the phase space. In the absence of a fundamental physical reason for such a truncation, we, following [15,16], simply cut-off the three momenta at $|\mathbf{k}|_{max} = \Lambda$. In the analysis that follows, we consider decays into this mode as it will have the largest contribution to the energy density of the generated photons. In addition, we take the reasonable approximation:

$$m_{can} \simeq \sqrt{\frac{|\mathbf{k}|\mu_\nu}{a}} \,, \tag{25}$$

and consider the dynamics of our scenario after reheating and prior to BBN, when the universe is radiation dominated and evolves as follows,

$$a(t) = a_0\sqrt{t} = \sqrt{2\Omega_{\mathrm{r},0}^{1/2} H_0 t} \,, \tag{26}$$

where $\Omega_{\mathrm{r},0} \sim 9.2 \cdot 10^{-5}$ is the radiation density parameter today.

The time at which this ghost term is no longer present will be defined as t_* and can be found in terms of the scale factor:

$$1 \simeq \frac{\Lambda}{a(t_*) m_{CS}(t_*)} \Rightarrow a(t_*) \simeq \sqrt{\frac{\mu_0 \Lambda}{M_p^2}} \text{ or } a(t_*) \simeq \frac{\xi_\nu T_* \Lambda}{M_p^2} \,, \tag{27}$$

where T_* is the temperature at which the ghost terms stop contributing.

This fixes the time at which the ghost modes no longer exist, and decay of the vacuum ceases. We can reinterpret this as a temperature, so that it is possible to associate this with the maximal reheating temperature, and also ensure it does not have adverse implications on BBN. If we assume that the asymmetry is present/produced during the reheating epoch, and prior to BBN, the scale factor has a $\frac{1}{T}$ dependence, if ignoring the decoupling of radiation degrees of freedom. The scale factor takes the following form:

$$a(t) \simeq \left(\frac{90\Omega_{\mathrm{r},0}}{g^* \pi^2}\right)^{\frac{1}{4}} \frac{\sqrt{H_0 M_p}}{T} \,, \tag{28}$$

where $g^* = 106.75$. Equating Eq. (27) and (28) to find the temperature at which this effect ends we find:

$$T_* = \left(\sqrt{\frac{90\Omega_{\mathrm{r},0}}{g^*\pi^2}} \frac{H_0 M_p^3}{\xi_\nu^2}\right)^{\frac{1}{4}} \sqrt{\frac{M_p}{\Lambda}} \simeq \frac{440}{\sqrt{\xi_\nu}} \mathrm{GeV} \sqrt{\frac{M_p}{\Lambda}} \,. \tag{29}$$

Given that the maximum reheating temperature is $\sim 10^{15}$ GeV, Eq. (29) implies we can constrain the production temperature of neutrino asymmetries to be in the range $\xi_\nu \gtrsim 2 \cdot 10^{-25} \frac{M_p}{\Lambda}$, with smaller ξ's not generating ghost like modes after reheating. We also assume here that ξ is approximately constant, and hence is the same parameter currently constrained by BBN measurements in the calculation of the lepton asymmetry stored in the CνB.

Next we compute the spectrum of photons generated by the induced vacuum decay, and subsequently the energy density, which can be constrained by experiment. It is given by:

$$\frac{1}{a^3}\frac{d}{dt}(a^3 n(k,t)) = \Gamma \delta\left(\frac{|\mathbf{k}|}{\Lambda} - 1\right) \,, \tag{30}$$

where $n(k,t)$ is the number of photons per unit logarithmic wave number $|\mathbf{k}|$ and Γ is the total decay width, which we take to approximately be:

$$\Gamma \sim \frac{\Lambda^6}{m_{can}^2} = \frac{a(t)\Lambda^6}{|\mathbf{k}|\mu_\nu} = \frac{a(t)^2\Lambda^5}{\mu_0} \,. \tag{31}$$

Since the above decay rate is much faster than the expansion rate of the universe, we may safely assume that the decay happens instantaneously. Therefore, we fix the scale factor in Eq. (31) at time t_a, when the asymmetric background is first produced. We then integrate Eq. (30) between the time t_a and when the ghost terms are no longer present t_*:

$$|\mathbf{k}| n_*(|\mathbf{k}|) \sim \frac{a(t_*)^2 \Lambda \Gamma_a}{5\Omega_{\mathrm{r},0}^{1/2} H_0} \,. \tag{32}$$

Taking into account the dilution factor due to the expansion of the universe since the end of photon production to today, $\left(\frac{a(t_*)}{a_0}\right)^3 = a(t_*)^3$, we obtain:

$$|\mathbf{k}| n_0(|\mathbf{k}|) \sim \frac{a(t_*)^5 \Lambda \Gamma_a}{5\Omega_{\mathrm{r},0}^{1/2} H_0} \,. \tag{33}$$

Therefore, the energy density for a given momenta k is:

$$\frac{dE}{d^3x d\ln|\mathbf{k}|} \sim |\mathbf{k}| n_0(|\mathbf{k}|) \sim \frac{\xi^4 T_*^5}{10 T_a^2}\sqrt{\frac{M_p^3}{H_0}}\left(\frac{\Lambda}{M_p}\right)^{11} \,. \tag{34}$$

We can obtain a bound on the energy density of the produced photons, through the observation that the universe is not radiation dominated today:

$$\frac{dE}{d^3x d\ln|\mathbf{k}|} \lesssim M_p^2 H_0^2 \,. \tag{35}$$

This means we get the following constraint on ξ_ν, assuming the asymmetry is generated above the characteristic temperature T_*, when requiring consistency with observation:

$$\xi_\nu \lesssim 2 \cdot 10^{-41}\left(\frac{T_a}{10^{15}\,\mathrm{GeV}}\right)^{4/3}\left(\frac{M_p}{\Lambda}\right)^{17/3} \,, \tag{36}$$

for which it is assumed $T_a \gtrsim \frac{440}{\sqrt{\xi_\nu}}\mathrm{GeV}\sqrt{\frac{M_p}{\Lambda}}$. Equivalently,

$$T_* \gtrsim 10^{23}\,\mathrm{GeV}\left(\frac{T_a}{10^{15}\,\mathrm{GeV}}\right)^{-2/3}\left(\frac{\Lambda}{M_p}\right)^{17/6} \,. \tag{37}$$

Thus we arrive at the conclusion that, unless $\Lambda \ll M_p$, the resulting photon energy density from the induced vacuum decay can hardly be accommodated with observation. Substituting the constraint in Eq. (36) into that for the asymmetry stored in the CνB as a function of ξ_ν, in Eq. (2), we find the following bound:

$$\eta_\nu \lesssim 10^{-41}\left(\frac{T_a}{10^{15}\,\mathrm{GeV}}\right)^{4/3}\left(\frac{M_p}{\Lambda}\right)^{17/3} \,. \tag{38}$$

If we instead assume that $T_a \lesssim \frac{440}{\sqrt{\xi_\nu}}\mathrm{GeV}\sqrt{\frac{M_p}{\Lambda}}$, and hence vacuum decay does not occur, then we get the following constraint on η_ν,

$$\eta_\nu \lesssim 0.033\left(\frac{2000\,\mathrm{GeV}}{T_a}\right)^2 \frac{M_p}{\Lambda} \tag{39}$$

where $\eta_\nu \lesssim 0.033$ is the current upper limit from BBN constraints.

5. Conclusions

In this paper we have argued that the relic neutrino background with non-zero lepton number exhibits gravitational instabilities. Fundamentally, these instabilities are related to the gravity-lepton number mixed quantum anomaly. Indeed, in the relevant limit of vanishing neutrino masses, we have explicitly calculated the parity odd part of the graviton polarization tensor in the presence of a lepton asymmetric CνB, which replicates the gravitational Chern–Simons term in the effective action.

The induced Chern–Simons term leads to birefringent behaviour leading to an enhancement/suppression of the gravitational wave amplitudes depending on the polarisation. While this effect is negligibly small for local sources, we demonstrate that it could be sizeable for gravitational waves produced in very early universe, e.g. during a first-order phase transition.

In addition to the above, we have also argued that short-scale gravitational fluctuations in the presence of an asymmetric CνB exhibit negative energy modes, which lead to a rapid decay of a vacuum state e.g., into negative energy graviton and photons. Since the graviton energy is not bounded from below, the phase space for this process is formally infinite, that is the instability is expected to develop very rapidly. From the constraint on the radiation energy density today, we have obtained an interesting bound on the neutrino degeneracy parameter in Eq. (4).

We believe that the findings reported in this paper will prove to be useful for further understanding the properties of the CνB and putting constraints on particle physics models with a lepton asymmetry.

Acknowledgements

We would like to thank Rohana Wijewardhana for useful discussions. This work was supported by the Australian Research Council.

Appendix A. Further details of calculations

Useful equations:

$$Tr(\gamma_\mu\gamma_\alpha\gamma_\rho\gamma_\beta\gamma^5) = -4i\varepsilon_{\mu\alpha\rho\beta} \tag{40}$$

Taking $\epsilon \to 0$:

$$\Gamma(1+\epsilon)|_{\epsilon\to 0} \simeq 1, \quad \Gamma(\epsilon)|_{\epsilon\to 0} \simeq \frac{1}{\epsilon} - \gamma,$$

$$\Gamma(\epsilon-1)|_{\epsilon\to 0} \simeq -\frac{1}{\epsilon} + \gamma - 1 \tag{41}$$

$$\eta_{\mu\nu}\eta^{\mu\nu} \simeq 4 - 2\epsilon \tag{42}$$

$$\left[\frac{4\pi\lambda^2}{M^2}\right]^\epsilon |_{\epsilon\to 0} \simeq 1 + \epsilon \ln\left(\frac{4\pi\lambda^2}{M^2}\right) \tag{43}$$

Dimensional regularisation of loop integrals:

$$i\int \frac{d^N p}{(2\pi)^N}\frac{1}{(p^2-m^2)^2} = -\frac{1}{16\pi^2}\left[\frac{4\pi^2\lambda^2}{M^2}\right]^\epsilon \Gamma(\epsilon) \tag{44}$$

$$i\int \frac{d^N p}{(2\pi)^N}\frac{1}{(p^2-m^2)^3} = \frac{1}{32\pi^2}\left[\frac{4\pi^2\lambda^2}{M^2}\right]^\epsilon \frac{\Gamma(1+\epsilon)}{M^2} \tag{45}$$

$$i\int \frac{d^N p}{(2\pi)^N}\frac{p_\mu p_\nu}{(p^2-m^2)^2} = \frac{1}{32\pi^2}\left[\frac{4\pi^2\lambda^2}{M^2}\right]^\epsilon M^2\Gamma(\epsilon-1)g_{\mu\nu} \tag{46}$$

$$i\int \frac{d^N p}{(2\pi)^N}\frac{p_\mu p_\nu}{(p^2-m^2)^3} = -\frac{1}{64\pi^2}\left[\frac{4\pi^2\lambda^2}{M^2}\right]^\epsilon \Gamma(\epsilon)g_{\mu\nu} \tag{47}$$

$$i\int \frac{d^N p}{(2\pi)^N}\frac{p_\mu p_\nu p_\rho p_\sigma}{(p^2-m^2)^3} = \frac{1}{128\pi^2}\left[\frac{4\pi^2\lambda^2}{M^2}\right]^\epsilon M^2\Gamma(\epsilon-1)(g_{\mu\nu}g_{\rho\sigma} + g_{\mu\sigma}g_{\nu\rho} + g_{\mu\rho}g_{\nu\sigma}) \tag{48}$$

References

[1] J. Lesgourgues, S. Pastor, Phys. Rev. D 60 (1999) 103521, http://dx.doi.org/10.1103/PhysRevD.60.103521, arXiv:hep-ph/9904411.
[2] J. March-Russell, H. Murayama, A. Riotto, J. High Energy Phys. 9911 (1999) 015, http://dx.doi.org/10.1088/1126-6708/1999/11/015, arXiv:hep-ph/9908396.
[3] J. McDonald, Phys. Rev. Lett. 84 (2000) 4798, http://dx.doi.org/10.1103/PhysRevLett.84.4798, arXiv:hep-ph/9908300.
[4] D.J. Schwarz, M. Stuke, J. Cosmol. Astropart. Phys. 0911 (2009) 025, http://dx.doi.org/10.1088/1475-7516/2009/11/025, arXiv:0906.3434 [hep-ph];
D.J. Schwarz, M. Stuke, J. Cosmol. Astropart. Phys. 1010 (2010) E01, http://dx.doi.org/10.1088/1475-7516/2010/10/E01 (Erratum).
[5] V.B. Semikoz, D.D. Sokoloff, J.W.F. Valle, Phys. Rev. D 80 (2009) 083510, http://dx.doi.org/10.1103/PhysRevD.80.083510, arXiv:0905.3365 [hep-ph].
[6] A.D. Dolgov, S.H. Hansen, S. Pastor, S.T. Petcov, G.G. Raffelt, D.V. Semikoz, Nucl. Phys. B 632 (2002) 363, http://dx.doi.org/10.1016/S0550-3213(02)00274-2, arXiv:hep-ph/0201287.
[7] Y.Y.Y. Wong, Phys. Rev. D 66 (2002) 025015, http://dx.doi.org/10.1103/PhysRevD.66.025015, arXiv:hep-ph/0203180.
[8] K.N. Abazajian, J.F. Beacom, N.F. Bell, Phys. Rev. D 66 (2002) 013008, http://dx.doi.org/10.1103/PhysRevD.66.013008, arXiv:astro-ph/0203442.
[9] G. Mangano, G. Miele, S. Pastor, O. Pisanti, S. Sarikas, Phys. Lett. B 708 (2012) 1, http://dx.doi.org/10.1016/j.physletb.2012.01.015, arXiv:1110.4335 [hep-ph].
[10] G. Barenboim, W.H. Kinney, W.I. Park, arXiv:1609.01584 [hep-ph].
[11] A.N. Redlich, L.C.R. Wijewardhana, Phys. Rev. Lett. 54 (1985) 970, http://dx.doi.org/10.1103/PhysRevLett.54.970.
[12] M. Gomes, T. Mariz, J.R. Nascimento, E. Passos, A.Y. Petrov, A.J. da Silva, Phys. Rev. D 78 (2008) 025029, http://dx.doi.org/10.1103/PhysRevD.78.025029, arXiv:0805.4409 [hep-th].
[13] S. Alexander, L.S. Finn, N. Yunes, Phys. Rev. D 78 (2008) 066005, http://dx.doi.org/10.1103/PhysRevD.78.066005, arXiv:0712.2542 [gr-qc].
[14] S. Dyda, E.E. Flanagan, M. Kamionkowski, Phys. Rev. D 86 (2012) 124031, http://dx.doi.org/10.1103/PhysRevD.86.124031, arXiv:1208.4871 [gr-qc].
[15] S.M. Carroll, M. Hoffman, M. Trodden, Phys. Rev. D 68 (2003) 023509, http://dx.doi.org/10.1103/PhysRevD.68.023509, arXiv:astro-ph/0301273.
[16] J.M. Cline, S. Jeon, G.D. Moore, Phys. Rev. D 70 (2004) 043543, http://dx.doi.org/10.1103/PhysRevD.70.043543, arXiv:hep-ph/0311312.

13

Gravitational time advancement under gravity's rainbow

Xue-Mei Deng [a], Yi Xie [b,c,*]

[a] Purple Mountain Observatory, Chinese Academy of Sciences, Nanjing 210008, China
[b] School of Astronomy and Space Science, Nanjing University, Nanjing 210093, China
[c] Key Laboratory of Modern Astronomy and Astrophysics, Nanjing University, Ministry of Education, Nanjing 210093, China

ARTICLE INFO

Editor: N. Lambert

Keywords:
Gravity's rainbow
Gravitational time advancement

ABSTRACT

Under gravity's rainbow, we investigate its effects on the gravitational time advancement, which is a natural consequence of measuring proper time span for a photon's round trip. This time advancement can be complementary to the time delay for testing the gravity's rainbow, because they are sensitive to different modified dispersion relations (MDRs). Its observability on ranging a spacecraft far from the Earth by two radio and a laser links is estimated at superior conjunction (SC) and inferior conjunction (IC). We find that (1) the IC is more favorable than the SC for measurement on the advancement caused by the rainbow; (2) a specific type of MDR has a significantly larger effect on the advancement than others in both SC and IC cases; and (3) a combination of available optical clocks and the realization of planetary laser ranging in the future will benefit distinguishing the gravity's rainbow from GR by measuring the gravitational time advancement.

1. Introduction

Einstein's general relativity (GR) has been the most successful theory of gravitation when it faces all of astronomical observations and physical experiments [1,2]. However, it seems that GR might be incomplete since it still cannot be rigorously unified with quantum mechanics. If GR is indeed the classical limit of a theory of quantum gravity, there should also have a semiclassical limit or an effective field theory [3,4]. The leading order of such an effective theory can go back to GR and the next-to-leading-order might phenomenologically have corrections depending on the Planck energy E_p or the Planck length l_p [5,6]. In the present work, we focus on those corrections associated with E_p, which can yield modified dispersion relations (MDRs).

In order to incorporate MDRs into curved spacetime, an approach called *gravity's rainbow* was proposed [7]. It is assumed that the geometry of spacetime is also determined by the ratio of the energy of a test particle to E_p, which leads to a rainbow metric. Cosmology in the gravity's rainbow scenario has been intensively studied [8–19]. In the rainbow spacetime, black holes and neutron stars [20–36], thermodynamics and Hawking radiation [37–49], its quantum properties [50–54] and its application in modified gravity [55–59] are also widely discussed. Dynamics of massive and massless particles in the gravity's rainbow is as well investigated [60–63]. In [64], a proposal for testing gravity's rainbow in the Solar System is raised and upper bounds on the parameters of the rainbow functions are obtained based on the experiments on light deflection, photon time delay, gravitational redshift and the weak equivalence principle.

Recently, a new type observable of the Solar System experiments, which is called gravitational time advancement, has been proposed and studied [65,66]. The gravitational time advancement is a natural consequence of a curved spacetime if an observer, who is located at a stronger gravitational field, measures the proper time span for the round trip of a photon passing through a weaker field [65]. It was found [66] that dark energy and dark matter can affect the gravitational time advancement, whose magnitude is small but in principle observable.

In this work, as an extension of the previous works [65,66], we will investigate the gravitational time advancement under the gravity's rainbow and examine its possible observables. In Sect. 2, the rainbow metric we adopt is briefly reviewed for completeness. We detailedly investigate the gravitational time advancement under this rainbow spacetime in Sect. 3, in which two generic configurations for the observer and the turning point of the round trip of a photon are considered. Observability of the time advancement within the rainbow scenario is discussed in Sect. 4. Finally, in Sect. 5, we summarize our results.

* Corresponding author at: School of Astronomy and Space Science, Nanjing University, Nanjing 210093, China.

E-mail addresses: xmd@pmo.ac.cn (X.-M. Deng), yixie@nju.edu.cn (Y. Xie).

2. Rainbow spacetime

For the observations and experiments conducted in the Solar System, the dominance of the Sun ensures that a spherically symmetric spacetime is a sufficiently good approximation. In the framework of the gravity's rainbow, the Sun's Schwarzschild metric is extended as [7]

$$ds^2 = -B(r)dt^2 + A(r)dr^2 + C(r)d\Omega^2, \tag{1}$$

where r is the radial distance from the origin and $d\Omega^2 = d\theta^2 + \sin^2\theta d\phi^2$ and the metric coefficients are

$$B(r) = f(E)^{-2}\left(1-\frac{2GM}{r}\right), \tag{2}$$

$$A(r) = g(E)^{-2}\left(1-\frac{2GM}{r}\right)^{-1}, \tag{3}$$

$$C(r) = r^2 g(E)^{-2}. \tag{4}$$

Here, $f(E)$ and $g(E)$ are the rainbow functions determined by MDRs and their forms are based on phenomenological motivations:

1. Originated from loop quantum gravity and noncommutative spacetime, the rainbow functions are [5,67]

$$f(E/E_p) = 1, \quad g(E/E_p) = \sqrt{1-\eta E/E_p}, \tag{5}$$

where η is a model parameter. Following the notation of [64], we denote it as MDR1 for short.
2. In order to explain the hard spectra of gamma-ray bursts at cosmological distances, the rainbow functions are proposed to be [68]

$$f(E/E_p) = \frac{e^{\alpha E/E_p}-1}{\alpha E/E_p}, \quad g(E/E_p) = 1, \tag{6}$$

where α is a model parameter. Following the notation of [64], we denote it as MDR2.
3. Providing a constant speed of light and a solution to the horizon problem [7], the rainbow functions are proposed to be [69]

$$f(E/E_p) = g(E/E_p) = \frac{1}{1-\lambda E/E_p}, \tag{7}$$

where λ is a model parameter. Following the notation of [64], we denote also it as MDR3.

Based on the rainbow metric (1) and MDRs (5)–(7), we can calculate the gravitational time advancement under the gravity's rainbow.

3. Gravitational time advancement under rainbow

The gravitational time delay is the fourth test of GR by measuring the time delay between transmission of radar pulses towards either Venus or Mercury and detection of the echoes [70]. This delay is caused by the dependence of the (average) speed of a light ray on the strength of the gravitational potential along its path. However, if we consider that an observer is located at a place closer to the Sun and the turning point of the round trip of the light ray is farther to the Sun, then measurement of the proper time span of the light's round trip can give the gravitational time advancement in GR [65] and in the presence of dark energy and dark matter [66].

For a photon moving in the gravity's rainbow spacetime (1), its null worldline leads to [71]

$$0 = -B(r)\dot{t}^2 + A(r)\dot{r}^2 + C(r)\dot{\phi}^2, \tag{8}$$

where the dot mean derivative against an affine parameter. Because of the spherical symmetry of the gravitational field, the orbit of the photon is confined to the equatorial plane $\theta = \pi/2$. Along the light trajectory, we have two conserved quantities [71]:

$$\mathcal{E} = B(r)\dot{t} \quad \text{and} \quad \mathcal{L} = C(r)\dot{\phi}. \tag{9}$$

We can have [71]

$$\frac{d\phi}{dr} = \pm\frac{1}{C(r)}\left[\frac{1}{A(r)B(r)}\left(\frac{1}{b^2}-\frac{B(r)}{C(r)}\right)\right]^{-1/2}, \tag{10}$$

where $b \equiv \mathcal{L}/\mathcal{E}$. At the closest approach d, $dr/d\phi = 0$ gives

$$b = \sqrt{\frac{C(d)}{B(d)}}. \tag{11}$$

Then, the relationship between t and r for light can be obtained as [71]

$$\frac{dt}{dr} = \pm\frac{1}{b}\left[\frac{B(r)}{A(r)}\left(\frac{1}{b^2}-\frac{B(r)}{C(r)}\right)\right]^{-1/2}, \tag{12}$$

which leads to a generic expression for the time span of a photon from d to r under the gravity's rainbow as [71,64]

$$\begin{aligned} t(r,d) &\equiv \int_d^r \frac{1}{b}\left[\frac{B(r)}{A(r)}\left(\frac{1}{b^2}-\frac{B(r)}{C(r)}\right)\right]^{-1/2} dr \\ &= \frac{f(E)}{g(E)}\left[\sqrt{r^2-d^2} + GM\sqrt{\frac{r-d}{r+d}} \right. \\ &\quad \left. + 2GM\ln\left(\frac{r+\sqrt{r^2-d^2}}{d}\right)\right] + \mathcal{O}(G^2). \end{aligned} \tag{13}$$

Since there is a plus sign in the front of the logarithmic correction in Eq. (13), the photon is always delayed with respect to the one in absence of the Sun, i.e., $f(E)[g(E)]^{-1}\sqrt{r^2-d^2}$. When $f(E) = g(E)$, Eq. (13) can effectively return to the one in GR [71]. It also means that the gravity's rainbow with MDR3 does not affect the gravitational time delay [64].

Now, we consider two points A and B in the spacetime (1). Either A or B can be set as the location of the observer and the other will be the location of the turning point of the round trip of a photon. Without loss of generality, we assume r_A is always larger than r_B, i.e. $r_A > r_B$, where r_A and r_B are respectively the radial coordinates of the points A and B. There are two cases: (i) as seen from the point B, the point A is on the opposite side of the Sun, which is denoted as "A-⊙-B", and (ii) the points A and B are on the same side of the Sun, denoted as "A-B-⊙". See Fig. 1 for details.

3.1. A-⊙-B: opposite sides case

According to Eq. (13), the total coordinate time required for the time duration of a photon travelling from the point A to the point B and back to A is given by

$$\begin{aligned} \Delta t_{A\odot B} &= 2t(r_A,d) + 2t(r_B,d) \\ &= \frac{f(E)}{g(E)}\left[2\sqrt{r_A^2-d^2} + 2\sqrt{r_B^2-d^2}\right. \end{aligned}$$

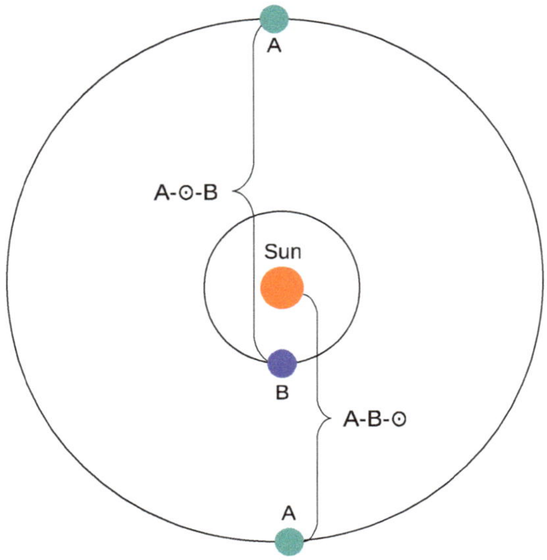

Fig. 1. Schematic diagram (not to scale) of configurations "A-⊙-B" and "A-B-⊙".

$$+2GM\sqrt{\frac{r_A-d}{r_A+d}}+2GM\sqrt{\frac{r_B-d}{r_B+d}}$$
$$+4GM\ln\left(\frac{r_A+\sqrt{r_A^2-d^2}}{d}\right)$$
$$+4GM\ln\left(\frac{r_B+\sqrt{r_B^2-d^2}}{d}\right)\Bigg]+\mathcal{O}(G^2), \tag{14}$$

and its proper time span measured by an observer at the point A is

$$\begin{aligned}\Delta\tau_{A\odot B} &= \frac{1}{f(E)}\left(1-\frac{GM}{r_A}\right)\Delta t_{A\odot B}\\ &= \frac{1}{g(E)}\Bigg[2\sqrt{r_A^2-d^2}+2\sqrt{r_B^2-d^2}\\ &\quad+2GM\sqrt{\frac{r_A-d}{r_A+d}}+2GM\sqrt{\frac{r_B-d}{r_B+d}}\\ &\quad+4GM\ln\left(\frac{r_A+\sqrt{r_A^2-d^2}}{d}\right)\\ &\quad+4GM\ln\left(\frac{r_B+\sqrt{r_B^2-d^2}}{d}\right)\\ &\quad-2GM\frac{\sqrt{r_A^2-d^2}}{r_A}-2GM\frac{\sqrt{r_B^2-d^2}}{r_A}\Bigg]\\ &\quad+\mathcal{O}(G^2).\end{aligned} \tag{15}$$

The signal takes more time for the round trip than the one in absence of the Sun, i.e. $\Delta\tau_{A\odot B}(M=0)$, and its delay is positive for $r_A > r_B \geq d$. Unlike the case of coordinate time span (14) which is immune to MDR3, the proper time span under the gravity's rainbow (15) depends only on the rainbow function $g(E)$, which means that it cannot be influenced by MDR2.

Nevertheless, if the observer is located at the point B which is closer to the Sun than the point A, the coordinate time delay for the round trip from B to A and back to B will remain the same as

Table 1
Detectability on MDRs for gravitational time delay and advancement.

MDR	Delay	Advancement
1	Yes	Yes
2	Yes	No
3	No	Yes

Eq. (14) but the proper time span measured by the observer at B will become

$$\begin{aligned}\Delta\tau_{A\odot B} &= \frac{1}{f(E)}\left(1-\frac{GM}{r_B}\right)\Delta t_{A\odot B}\\ &= \frac{1}{g(E)}\Bigg[2\sqrt{r_A^2-d^2}+2\sqrt{r_B^2-d^2}\\ &\quad+2GM\sqrt{\frac{r_A-d}{r_A+d}}+2GM\sqrt{\frac{r_B-d}{r_B+d}}\\ &\quad+4GM\ln\left(\frac{r_A+\sqrt{r_A^2-d^2}}{d}\right)\\ &\quad+4GM\ln\left(\frac{r_B+\sqrt{r_B^2-d^2}}{d}\right)\\ &\quad-2GM\frac{\sqrt{r_A^2-d^2}}{r_B}-2GM\frac{\sqrt{r_B^2-d^2}}{r_B}\Bigg]\\ &\quad+\mathcal{O}(G^2).\end{aligned} \tag{16}$$

Since the last two terms in Eq. (16) dominate those terms proportional to GM due to their dependence of r_B^{-1}, this proper time span can be effectively decreased and even be less than the proper time span in the absence of the Sun, i.e. $\Delta\tau_{A\odot B}(M=0)$, if r_A is sufficiently larger than r_B by a specific value depending on r_B and d. This effect is the gravitational time advancement (negative time delay) under the gravity's rainbow, which is caused by the fact that clocks run differently at different positions in the gravitational field [65]. When gravity's rainbow vanishes, i.e., $g(E)=1$, our result (16) can return to the one in GR given by [65].

If we consider the configuration of superior conjunction (SC) that the points A and B are on the opposite sides of the Sun and $r_A > r_B \gg d$ which might happen in radio tracking a spacecraft, the gravitational time advancement (16) can be reduced to a simpler form as

$$\begin{aligned}\Delta\tau_{A\odot B}^{\mathrm{SC}} &= \frac{1}{g(E)}\Bigg[2(r_A+r_B)+2GM+4GM\ln\frac{4r_Ar_B}{d^2}\\ &\quad-2GM\frac{r_A}{r_B}\Bigg]+\mathcal{O}\left(G^2,\frac{d}{r_A},\frac{d}{r_B}\right).\end{aligned} \tag{17}$$

We can see that, from Eqs. (16) and (17), MDR1 and MDR3 can affect the gravitational time advancement but MDR2 cannot. If $\eta > 0$ and $g(E) < 1$, MDR1 can amplify the advancement and make it larger than the one in GR. On the contrary to the behavior of MDR1, MDR3 can make the advancement smaller than the one in GR when $\lambda > 0$ and $g(E) > 1$. However, the time delay (14) cannot distinguish MDR3 from the others. It suggests that measurements on the gravitational time delay and advancement can be complementary to each other for constraining MDRs (see Table 1 for a summary).

3.2. A-B-⊙: the same side case

If the points A and B are on the same side of the Sun and $r_A > r_B$, the coordinate time span for a photon travelling from A to B and back to A can be worked out based on Eq. (13) as

$$\begin{aligned}\Delta t_{AB\odot} &= 2t(r_A,d) - 2t(r_B,d)\\ &= \frac{f(E)}{g(E)}\Bigg[2\sqrt{r_A^2-d^2} - 2\sqrt{r_B^2-d^2}\\ &\quad + 2GM\sqrt{\frac{r_A-d}{r_A+d}} - 2GM\sqrt{\frac{r_B-d}{r_B+d}}\\ &\quad + 4GM\ln\left(\frac{r_A+\sqrt{r_A^2-d^2}}{d}\right)\\ &\quad - 4GM\ln\left(\frac{r_B+\sqrt{r_B^2-d^2}}{d}\right)\Bigg] + \mathcal{O}(G^2),\end{aligned} \tag{18}$$

where the minus sign on the right hand side at the first line is physically caused by such a configuration. When an observer is at the point A, the measured proper time span is

$$\begin{aligned}\Delta\tau_{AB\odot} &= \frac{1}{f(E)}\left(1-\frac{GM}{r_A}\right)\Delta t_{AB\odot}\\ &= \frac{1}{g(E)}\Bigg[2\sqrt{r_A^2-d^2} - 2\sqrt{r_B^2-d^2}\\ &\quad + 2GM\sqrt{\frac{r_A-d}{r_A+d}} - 2GM\sqrt{\frac{r_B-d}{r_B+d}}\\ &\quad + 4GM\ln\left(\frac{r_A+\sqrt{r_A^2-d^2}}{d}\right)\\ &\quad - 4GM\ln\left(\frac{r_B+\sqrt{r_B^2-d^2}}{d}\right)\\ &\quad - 2GM\frac{\sqrt{r_A^2-d^2}}{r_A} + 2GM\frac{\sqrt{r_B^2-d^2}}{r_A}\Bigg]\\ &\quad + \mathcal{O}(G^2).\end{aligned} \tag{19}$$

Like one of the A-⊙-B cases that the observer is at the point A [see Eq. (15)], the signal takes more time for the round trip and the delay is positive for $r_A > r_B \geq d$.

If an observer in the A-B-⊙ is at the point B instead of A, the coordinate time delay for the round trip from B to A and back to B is the same as Eq. (18) and the proper time span measured by the observer at B reads as

$$\begin{aligned}\Delta\tau_{AB\odot} &= \frac{1}{f(E)}\left(1-\frac{GM}{r_B}\right)\Delta t_{AB\odot}\\ &= \frac{1}{g(E)}\Bigg[2\sqrt{r_A^2-d^2} - 2\sqrt{r_B^2-d^2}\\ &\quad + 2GM\sqrt{\frac{r_A-d}{r_A+d}} - 2GM\sqrt{\frac{r_B-d}{r_B+d}}\\ &\quad + 4GM\ln\left(\frac{r_A+\sqrt{r_A^2-d^2}}{d}\right)\\ &\quad - 4GM\ln\left(\frac{r_B+\sqrt{r_B^2-d^2}}{d}\right)\\ &\quad - 2GM\frac{\sqrt{r_A^2-d^2}}{r_B} + 2GM\frac{\sqrt{r_B^2-d^2}}{r_B}\Bigg]\\ &\quad + \mathcal{O}(G^2).\end{aligned} \tag{20}$$

Like the situation of Eq. (16) for A-⊙-B, those terms depending on r_B^{-1} dominate others proportional to GM in the above equation so that this proper time span can be smaller than the one in the absence of the Sun if r_A is sufficiently larger than r_B. It can be easily checked that, when we consider a special case that $r_A = r_B + \Delta R$, $\Delta R \ll r_B$, $d = 0$ and $g(E) = 1$, Eq. (20) can give the equation for the "small distance travel" in GR discussed in [65].

Another interesting case, which was not discussed in [65,66], is the configuration of inferior conjunction (IC) of A-B-⊙ where $r_A > r_B \gg d$ so that the gravitational time advancement (20) becomes to

$$\begin{aligned}\Delta\tau_{AB\odot}^{\mathrm{IC}} &= \frac{1}{g(E)}\Bigg[2(r_A-r_B) + 2GM + 4GM\ln\frac{r_A}{r_B}\\ &\quad - 2GM\frac{r_A}{r_B}\Bigg] + \mathcal{O}\left(G^2, \frac{d}{r_A}, \frac{d}{r_B}\right).\end{aligned} \tag{21}$$

Unlike the case of small distance travel that r_A is comparable with r_B, r_A in the IC condition of Eq. (21) can be much larger than r_B, which can be used to describe ranging measurement on a spacecraft in deep space far beyond the Earth orbit.

4. Observability of time advancement under gravity's rainbow

After working out the equations for the gravitational time advancement under gravity's rainbow, we discuss its observability in this section.

4.1. A-⊙-B

In the A-⊙-B configuration, a SC condition is favorable for measurement on the time advancement due to the smallness of d. According to Eq. (17), we can find that the time advancement caused by the gravity's rainbow is given by

$$\delta\tau_{A\odot B}^{\mathrm{SC}} \equiv \Delta\tau_{A\odot B}^{\mathrm{SC}} - \Delta\tau_{A\odot B}^{\mathrm{SC}}\Big|_{M=0}, \tag{22}$$

the time advancement in GR is

$$\bar{\delta}\tau_{A\odot B}^{\mathrm{SC}} \equiv \Delta\tau_{A\odot B}^{\mathrm{SC}}\Big|_{g=1} - \Delta\tau_{A\odot B}^{\mathrm{SC}}\Big|_{\substack{g=1\\M=0}}, \tag{23}$$

and their relative deviation is defined as

$$r_{A\odot B}^{\mathrm{SC}} \equiv \frac{\delta\tau_{A\odot B}^{\mathrm{SC}} - \bar{\delta}\tau_{A\odot B}^{\mathrm{SC}}}{\Delta\tau_{A\odot B}^{\mathrm{SC}}}. \tag{24}$$

Since the time advancement is defined as negative time delay, $\delta\tau_{A\odot B}^{\mathrm{SC}} - \bar{\delta}\tau_{A\odot B}^{\mathrm{SC}} > 0$ means that the advancement caused by the gravity's rainbow is smaller than the one in GR, and vice versa. $r_{A\odot B}^{\mathrm{SC}}$ represents the theoretical resolution for time measurement required to distinguish the gravity's rainbow from GR.

We consider a SC condition that an observer on the Earth with $r_B = 1$ au[1] conducts two radio-tracking measurements on X-band (7.2 GHz) and Ka-band (34.3 GHz) to range a spacecraft at a distance of 40 au from the Sun, $r_A = 40$ au, which is close to the

[1] We use lower-case 'au' to represent the astronomical unit, according to International Astronomical Union 2012 Resolution B2: http://www.iau.org/static/resolutions/IAU2012_English.pdf.

Table 2
Estimation of observability on the gravitational time advancement in SC condition with links of X-band, Ka-band and visible laser where $r_A = 40$ au, $r_B = 1$ au and $d = 1.5$ $R_\odot$. The parameters in the rainbow functions are taken as $\eta = 1.3 \times 10^{20}$ and $\lambda = 8.5 \times 10^{21}$ [64] and their uncertainties are set as 10%.

MDR	Band	Frequency (GHz)	$\delta\tau^{SC}_{A\odot B}$ (μs)	$\bar{\delta}\tau^{SC}_{A\odot B}$ (μs)	$\delta\tau^{SC}_{A\odot B} - \bar{\delta}\tau^{SC}_{A\odot B}$ (s)	$r^{SC}_{A\odot B}$
1	X	7.2	$-88.5348375693597 \pm (1.4 \times 10^{-12})$	-88.5348375693464	$-(1.4 \pm 0.1) \times 10^{-17}$	$-(3.4 \pm 0.3) \times 10^{-22}$
1	Ka	34.3	$-88.5348375694125 \pm (6.7 \times 10^{-12})$	ibid.	$-(6.7 \pm 0.7) \times 10^{-17}$	$-(1.6 \pm 0.2) \times 10^{-21}$
1	Visible	6×10^5	$-88.5348387388309 \pm (1.16949 \times 10^{-7})$	ibid.	$-(1.2 \pm 0.1) \times 10^{-12}$	$-(2.9 \pm 0.3) \times 10^{-17}$
3	X	7.2	$-88.534837568 \pm (1.8 \times 10^{-8})$	ibid.	$(1.8 \pm 0.2) \times 10^{-15}$	$(4.5 \pm 0.5) \times 10^{-20}$
3	Ka	34.3	$-88.534837561 \pm (8.7 \times 10^{-8})$	ibid.	$(8.7 \pm 0.9) \times 10^{-15}$	$(2.1 \pm 0.2) \times 10^{-19}$
3	Visible	6×10^5	$-88.534684637 \pm (1.529 \times 10^{-5})$	ibid.	$(1.5 \pm 0.2) \times 10^{-10}$	$(3.7 \pm 0.4) \times 10^{-15}$

Table 3
Estimation of observability on the gravitational time advancement in IC condition with links of X-band, Ka-band and visible laser where $r_A = 40$ au and $r_B = 1$ au. The parameters in the rainbow functions are taken as $\eta = 1.3 \times 10^{20}$ and $\lambda = 8.5 \times 10^{21}$ [64] and their uncertainties are set as 10%.

MDR	Band	Frequency (GHz)	$\delta\tau^{IC}_{AB\odot}$ (μs)	$\bar{\delta}\tau^{IC}_{AB\odot}$ (μs)	$\delta\tau^{IC}_{AB\odot} - \bar{\delta}\tau^{IC}_{AB\odot}$ (s)	$r^{IC}_{AB\odot}$
1	X	7.2	$-311.4174782731487 \pm (4.9 \times 10^{-12})$	-311.4174782730994	$-(4.9 \pm 0.5) \times 10^{-17}$	$-(1.3 \pm 0.1) \times 10^{-21}$
1	Ka	34.3	$-311.4174782733345 \pm (2.35 \times 10^{-11})$	ibid.	$-(2.4 \pm 0.2) \times 10^{-16}$	$-(6.0 \pm 0.6) \times 10^{-21}$
1	Visible	6×10^5	$-311.4174823867135 \pm (4.11361 \times 10^{-7})$	ibid.	$-(4.1 \pm 0.4) \times 10^{-12}$	$-(1.0 \pm 0.1) \times 10^{-16}$
3	X	7.2	$-311.41747826664 \pm (6.4 \times 10^{-10})$	ibid.	$(6.5 \pm 0.7) \times 10^{-15}$	$(1.7 \pm 0.2) \times 10^{-19}$
3	Ka	34.3	$-311.41747824234 \pm (3.08 \times 10^{-9})$	ibid.	$(3.1 \pm 0.3) \times 10^{-14}$	$(7.9 \pm 0.8) \times 10^{-19}$
3	Visible	6×10^5	$-311.41694033895 \pm (5.37934 \times 10^{-5})$	ibid.	$(5.4 \pm 0.5) \times 10^{-10}$	$(1.4 \pm 0.1) \times 10^{-14}$

semi-major axis of Pluto, and the closest approach of the radio signals is $d = 1.5\ R_\odot$ where $R_\odot$ is radius of the Sun. Following a trend of developing interplanetary laser ranging [72–74], we take a laser link with 600 THz into account. The values of parameters η for MDR1 (5) and λ for MDR3 (7) are respectively taken as 1.3×10^{20} and 8.5×10^{21} based on the results from the time delay and redshift experiments according to [64], because these measurements are of the same kind we discuss here.

Our results of observability for this case are listed in Table 2. It is found that, in this SC condition, the gravitational time advancements under the gravity's rainbow and the one in GR can reach about −88 microsecond (μs). With MDR1, the contribution caused by the gravity's rainbow in the time advancement ranges from -1.4×10^{-17} s to -1.2×10^{-12} s, where the minus signs mean MDR1 enlarge the time advancement in GR and the absolute values depend on the frequency of ranging signal; and the time resolution for distinguishing the gravity's rainbow from GR needs to be from -3.4×10^{-22} to -2.9×10^{-17}. With MDR3, the contribution caused by the gravity's rainbow in the time advancement has values from 1.8×10^{-15} s to 1.5×10^{-10} s, where the positive values mean MDR3 lessen the time advancement in GR; and the time resolution is required to be from 4.5×10^{-20} to 3.7×10^{-15}.

It shows that, the time advancement caused by the gravity's rainbow with MDR3 has a bigger deviation from the one in GR than the advancement due to MDR1 in such a SC case by nearly 2 orders of magnitude. It also suggests that if the planetary laser ranging become available in the future, the measurement on the gravitational advancement might be able to detect the gravity's rainbow and obtain its new constraints, given the fact that optical clocks on the ground have achieved the accuracy and stability at the 10^{-18} level [75–77].

4.2. A-B-⊙

According to Eq. (21), the time advancement at IC condition caused by the gravity's rainbow in the A-B-⊙ configuration is given by

$$\delta\tau^{IC}_{AB\odot} \equiv \Delta\tau^{IC}_{AB\odot} - \Delta\tau^{IC}_{AB\odot}\Big|_{M=0}, \tag{25}$$

the time advancement in GR is

$$\bar{\delta}\tau^{IC}_{AB\odot} \equiv \Delta\tau^{IC}_{AB\odot}\Big|_{g=1} - \Delta\tau^{IC}_{AB\odot}\Big|_{\substack{g=1\\M=0}}, \tag{26}$$

and their relative deviation is defined as

$$r^{IC}_{AB\odot} \equiv \frac{\delta\tau^{IC}_{AB\odot} - \bar{\delta}\tau^{IC}_{AB\odot}}{\Delta\tau^{IC}_{AB\odot}}, \tag{27}$$

which represents the time resolution required for distinguishing the gravity's rainbow from GR in such an IC condition.

We consider an IC condition that an observer is at $r_B = 1$ au who conducts two radio ranging measurements (X-band and Ka-band) and a laser ranging (600 THz) on a spacecraft at $r_A = 40$ au. Our results of observability for this case are listed in Table 3. We find that, in this IC condition, the gravitational time advancements under the gravity's rainbow and the one in GR can reach about −311 μs, which is nearly 3.5 times larger than those in the SC condition we discuss before. It demonstrates that the IC condition is more favorable than the SC condition for measurement on the gravitational advancement. With MDR1, the contribution caused by the gravity's rainbow in the time advancement ranges from -4.9×10^{-17} s to -4.1×10^{-12} s and the time resolution for distinguishing the gravity's rainbow from GR needs to be from -1.3×10^{-21} to -1.1×10^{-16}, which also depend on the frequency. With MDR3, the contribution caused by the gravity's rainbow in the time advancement has values from 6.5×10^{-15} s to 5.4×10^{-10} s and the time resolution is required to be from 1.7×10^{-19} to 1.4×10^{-14}.

Like the case of SC condition, it shows the time advancement caused by the gravity's rainbow with MDR3 has a larger deviation from the one in GR than the advancement due to MDR1 in this IC case by nearly 2 orders of magnitude. It also suggests that the planetary laser ranging will benefit the detection on the gravity's rainbow in the future.

5. Conclusions

Under the gravity's rainbow with three various MDRs, we investigate its effects on the gravitational time advancement. If an observer measures the proper time span for the round trip of a photon passing through a weaker gravitational field, then such a time advancement will be a natural consequence. We find that this time advancement can be complementary to the classical test of Shapiro time delay because they are sensitive to different MDRs (see Table 1).

Considering ranging a spacecraft at a distance of Pluto from the Earth, we estimate its observability on the time advancement under SC and IC configurations (see Fig. 1). We also assume that two radio links (at X-band and Ka-band) and a laser link (600 THz) are used in the ranging. It is found that (1) the IC configuration is more favorable for measuring the time advancement; and (2) the time advancement caused by MDR3 is significantly larger than others (see Tables 2 and 3 for details). We expect that, with a combination of optical clocks and planetary laser ranging, measurements on the gravitational time advancement will benefit detecting the gravity's rainbow in the future.

Acknowledgements

The work of XMD is supported by the National Natural Science Foundation of China under Grant Nos. 11473072 and 11533004. The work of YX is funded by the Natural Science Foundation of China under Grant No. 11573015.

References

[1] C.M. Will, Theory and Experiment in Gravitational Physics, Cambridge University Press, Cambridge, England, 1993.
[2] C.M. Will, The confrontation between general relativity and experiment, Living Rev. Relativ. 17 (2014) 4, http://dx.doi.org/10.12942/lrr-2014-4, arXiv:1403.7377.
[3] I.L. Buchbinder, S.D. Odintsov, I.L. Shapiro, Effective Action in Quantum Gravity, Institute of Physics, Bristol, 1992.
[4] C.P. Burgess, Quantum gravity in everyday life: general relativity as an effective field theory, Living Rev. Relativ. 7 (2004) 5, http://dx.doi.org/10.12942/lrr-2004-5, arXiv:gr-qc/0311082.
[5] G. Amelino-Camelia, Quantum-spacetime phenomenology, Living Rev. Relativ. 16 (2013) 5, http://dx.doi.org/10.12942/lrr-2013-5, arXiv:0806.0339.
[6] S. Hossenfelder, Minimal length scale scenarios for quantum gravity, Living Rev. Relativ. 16 (2013) 2, http://dx.doi.org/10.12942/lrr-2013-2, arXiv:1203.6191.
[7] J. Magueijo, L. Smolin, Gravity's rainbow, Class. Quantum Gravity 21 (2004) 1725–1736, http://dx.doi.org/10.1088/0264-9381/21/7/001, arXiv:gr-qc/0305055.
[8] B. Majumder, Quantum rainbow cosmological model with perfect fluid, Int. J. Mod. Phys. D 22 (2013) 1350079, http://dx.doi.org/10.1142/S021827181350079X, arXiv:1307.5273.
[9] C. Corda, Primordial inflation from gravity's rainbow, in: T.E. Simos, G. Psihoyios, C. Tsitouras (Eds.), AIP Conf. Ser., vol. 1281, 2010, pp. 847–851, arXiv:1007.4087.
[10] G. Amelino-Camelia, M. Arzano, G. Gubitosi, J. Magueijo, Rainbow gravity and scale-invariant fluctuations, Phys. Rev. D 88 (4) (2013) 041303, http://dx.doi.org/10.1103/PhysRevD.88.041303, arXiv:1307.0745.
[11] A. Awad, A.F. Ali, B. Majumder, Nonsingular rainbow universes, J. Cosmol. Astropart. Phys. 10 (2013) 052, http://dx.doi.org/10.1088/1475-7516/2013/10/052, arXiv:1308.4343.
[12] J.D. Barrow, J. Magueijo, Intermediate inflation from rainbow gravity, Phys. Rev. D 88 (10) (2013) 103525, http://dx.doi.org/10.1103/PhysRevD.88.103525, arXiv:1310.2072.
[13] R. Garattini, M. Sakellariadou, Does gravity's rainbow induce inflation without an inflaton?, Phys. Rev. D 90 (4) (2014) 043521, http://dx.doi.org/10.1103/PhysRevD.90.043521.
[14] S. Wang, Z. Chang, Nearly scale-invariant power spectrum and quantum cosmological perturbations in the gravity's rainbow scenario, Eur. Phys. J. C 75 (2015) 259, http://dx.doi.org/10.1140/epjc/s10052-015-3457-y, arXiv:1412.3600.
[15] G. Santos, G. Gubitosi, G. Amelino-Camelia, On the initial singularity problem in rainbow cosmology, J. Cosmol. Astropart. Phys. 8 (2015) 005, http://dx.doi.org/10.1088/1475-7516/2015/08/005, arXiv:1502.02833.
[16] A. Ashour, M. Faizal, A.F. Ali, F. Hammad, Branes in gravity's rainbow, Eur. Phys. J. C 76 (2016) 264, http://dx.doi.org/10.1140/epjc/s10052-016-4124-7, arXiv:1602.04926.
[17] A. Chatrabhuti, V. Yingcharoenrat, P. Channuie, Starobinsky model in rainbow gravity, Phys. Rev. D 93 (4) (2016) 043515, http://dx.doi.org/10.1103/PhysRevD.93.043515, arXiv:1510.09113.
[18] S.H. Hendi, M. Momennia, B. Eslam Panah, M. Faizal, Nonsingular universes in Gauss–Bonnet gravity's rainbow, Astrophys. J. 827 (2016) 153, http://dx.doi.org/10.3847/0004-637X/827/2/153, arXiv:1703.00480.
[19] S.H. Hendi, M. Momennia, B. Eslam Panah, S. Panahiyan, Nonsingular universe in massive gravity's rainbow, Phys. Dark Universe 16 (2017) 26–33, http://dx.doi.org/10.1016/j.dark.2017.04.001, arXiv:1705.01099.
[20] H. Li, Y. Ling, X. Han, Modified (A)dS Schwarzschild black holes in rainbow spacetime, Class. Quantum Gravity 26 (6) (2009) 065004, http://dx.doi.org/10.1088/0264-9381/26/6/065004, arXiv:0809.4819.
[21] A.F. Ali, M. Faizal, M.M. Khalil, Remnant for all black objects due to gravity's rainbow, Nucl. Phys. B 894 (2015) 341–360, http://dx.doi.org/10.1016/j.nuclphysb.2015.03.014.
[22] A.F. Ali, M. Faizal, M.M. Khalil, Absence of black holes at LHC due to gravity's rainbow, Phys. Lett. B 743 (2015) 295–300, http://dx.doi.org/10.1016/j.physletb.2015.02.065.
[23] A.F. Ali, M. Faizal, B. Majumder, Absence of an effective horizon for black holes in gravity's rainbow, Europhys. Lett. 109 (2015) 20001, http://dx.doi.org/10.1209/0295-5075/109/20001, arXiv:1406.1980.
[24] A.F. Ali, M. Faizal, B. Majumder, R. Mistry, Gravitational collapse in gravity's rainbow, Int. J. Geom. Methods Mod. Phys. 12 (2015) 1550085, http://dx.doi.org/10.1142/S0219887815500851.
[25] S.H. Hendi, M. Faizal, Black holes in Gauss–Bonnet gravity's rainbow, Phys. Rev. D 92 (4) (2015) 044027, http://dx.doi.org/10.1103/PhysRevD.92.044027, arXiv:1506.08062.
[26] C. Liu, H. Liu, Spectroscopy of a charged black hole in gravity's rainbow via an action invariance, Astrophys. Space Sci. 357 (2015) 114, http://dx.doi.org/10.1007/s10509-015-2342-3.
[27] Y. Gim, W. Kim, Black hole complementarity in gravity's rainbow, J. Cosmol. Astropart. Phys. 05 (2015) 002, http://dx.doi.org/10.1088/1475-7516/2015/05/002, arXiv:1501.04702.
[28] S.H. Hendi, Asymptotically charged BTZ black holes in gravity's rainbow, Gen. Relativ. Gravit. 48 (2016) 50, http://dx.doi.org/10.1007/s10714-016-2044-3, arXiv:1507.04733.
[29] S. Hendi, G. Bordbar, B.E. Panah, S. Panahiyan, Modified TOV in gravity's rainbow: properties of neutron stars and dynamical stability conditions, J. Cosmol. Astropart. Phys. 2016 (09) (2016) 013, http://dx.doi.org/10.1088/1475-7516/2016/09/013.
[30] S.H. Hendi, M. Faizal, B.E. Panah, S. Panahiyan, Charged dilatonic black holes in gravity's rainbow, Eur. Phys. J. C 76 (2016) 296, http://dx.doi.org/10.1140/epjc/s10052-016-4119-4, arXiv:1508.00234.
[31] S.H. Hendi, B. Eslam Panah, S. Panahiyan, A. Sheykhi, Dilatonic BTZ black holes with power-law field, Phys. Lett. B 767 (2017) 214–225, http://dx.doi.org/10.1016/j.physletb.2017.01.066, arXiv:1703.03403.
[32] S.H. Hendi, A. Dehghani, M. Faizal, Black hole thermodynamics in Lovelock gravity's rainbow with (A)dS asymptote, Nucl. Phys. B 914 (2017) 117–137, http://dx.doi.org/10.1016/j.nuclphysb.2016.10.022, arXiv:1702.02431.
[33] S.H. Hendi, B. Eslam Panah, S. Panahiyan, M. Momennia, F(R) gravity's rainbow and its Einstein counterpart, Adv. High Energy Phys. 2016 (2016) 9813582, http://dx.doi.org/10.1155/2016/9813582, arXiv:1607.03383.
[34] S.H. Hendi, B. Eslam Panah, S. Panahiyan, Topological charged black holes in massive gravity's rainbow and their thermodynamical analysis through various approaches, Phys. Lett. B 769 (2017) 191–201, http://dx.doi.org/10.1016/j.physletb.2017.03.051, arXiv:1602.01832.
[35] S.H. Hendi, S. Panahiyan, S. Upadhyay, B. Eslam Panah, Charged BTZ black holes in the context of massive gravity's rainbow, Phys. Rev. D 95 (2017) 084036, http://dx.doi.org/10.1103/PhysRevD.95.084036.
[36] S.H. Hendi, S. Panahiyan, B. Eslam Panah, M. Faizal, M. Momennia, Critical behavior of charged black holes in Gauss–Bonnet gravity's rainbow, Phys. Rev. D 94 (2) (2016) 024028, http://dx.doi.org/10.1103/PhysRevD.94.024028, arXiv:1607.06663.
[37] P. Galán, G.A. Mena Marugán, Entropy and temperature of black holes in a gravity's rainbow, Phys. Rev. D 74 (4) (2006) 044035, http://dx.doi.org/10.1103/PhysRevD.74.044035, arXiv:gr-qc/0608061.
[38] Y. Ling, X. Li, H. Zhang, Thermodynamics of modified black holes from gravity's rainbow, Mod. Phys. Lett. A 22 (2007) 2749–2756, http://dx.doi.org/10.1142/S0217732307022931, arXiv:gr-qc/0512084.
[39] C.-Z. Liu, J.-Y. Zhu, Hawking radiation and black hole entropy in a gravity's rainbow, Gen. Relativ. Gravit. 40 (2008) 1899–1911, http://dx.doi.org/10.1007/s10714-008-0607-7, arXiv:gr-qc/0703055.
[40] J.-J. Peng, S.-Q. Wu, Covariant anomaly and Hawking radiation from the modified black hole in the rainbow gravity theory, Gen. Relativ. Gravit. 40 (2008) 2619–2626, http://dx.doi.org/10.1007/s10714-008-0642-4, arXiv:0709.0167.
[41] Y.-Q. Yuan, X.-X. Zeng, Fermions tunneling and entropy correction of black hole in gravity's rainbow space time, Int. J. Theor. Phys. 48 (2009) 1937–1944, http://dx.doi.org/10.1007/s10773-009-9967-3.
[42] C.-Z. Liu, Hawking radiation via tunneling of massive particles from a gravity's rainbow, Mod. Phys. Lett. A 25 (2010) 3229–3240, http://dx.doi.org/10.1142/S0217732310034341.
[43] A.F. Ali, Black hole remnant from gravity's rainbow, Phys. Rev. D 89 (10) (2014) 104040, http://dx.doi.org/10.1103/PhysRevD.89.104040, arXiv:1402.5320.
[44] Y. Gim, W. Kim, Thermodynamic phase transition in the rainbow Schwarzschild black hole, J. Cosmol. Astropart. Phys. 10 (2014) 003, http://dx.doi.org/10.1088/1475-7516/2014/10/003, arXiv:1406.6475.
[45] E.J. Son, W. Kim, Note on uncertainty relations in doubly special relativity and gravity's rainbow, Mod. Phys. Lett. A 30 (2015) 1550178, http://dx.doi.org/10.1142/S0217732315501783, arXiv:1504.06941.

[46] S.H. Hendi, S. Panahiyan, B.E. Panah, M. Momennia, Thermodynamic instability of nonlinearly charged black holes in gravity's rainbow, Eur. Phys. J. C 76 (2016) 150, http://dx.doi.org/10.1140/epjc/s10052-016-3994-z, arXiv:1512.05192.

[47] B. Mu, P. Wang, H. Yang, Thermodynamics and luminosities of rainbow black holes, J. Cosmol. Astropart. Phys. 11 (2015) 045, http://dx.doi.org/10.1088/1475-7516/2015/11/045, arXiv:1507.03768.

[48] Y. Gim, W. Kim, Hawking, fiducial, and free-fall temperature of black hole on gravity's rainbow, Eur. Phys. J. C 76 (2016) 166, http://dx.doi.org/10.1140/epjc/s10052-016-4025-9, arXiv:1509.06846.

[49] Y.-W. Kim, Y.-J. Park, Local free-fall temperature of modified Schwarzschild black hole in rainbow spacetime, Mod. Phys. Lett. A 31 (2016) 1650106, http://dx.doi.org/10.1142/S0217732316501066, arXiv:1508.07439.

[50] P. Galán, G.A. Marugán, Quantum time uncertainty in a gravity's rainbow formalism, Phys. Rev. D 70 (12) (2004) 124003, http://dx.doi.org/10.1103/PhysRevD.70.124003, arXiv:gr-qc/0411089.

[51] P. Galán, G.A. Marugán, Length uncertainty in a gravity's rainbow formalism, Phys. Rev. D 72 (4) (2005) 044019, http://dx.doi.org/10.1103/PhysRevD.72.044019, arXiv:gr-qc/0507098.

[52] R. Garattini, G. Mandanici, Modified dispersion relations lead to a finite zero point gravitational energy, Phys. Rev. D 83 (8) (2011) 084021, http://dx.doi.org/10.1103/PhysRevD.83.084021, arXiv:1102.3803.

[53] R. Garattini, F.S.N. Lobo, Gravity's rainbow induces topology change, Eur. Phys. J. C 74 (2014) 2884, http://dx.doi.org/10.1140/epjc/s10052-014-2884-5, arXiv:1303.5566.

[54] R. Garattini, B. Majumder, Electric charges and magnetic monopoles in gravity's rainbow, Nucl. Phys. B 883 (2014) 598–614, http://dx.doi.org/10.1016/j.nuclphysb.2014.04.005, arXiv:1305.3390.

[55] R. Garattini, Distorting general relativity: gravity's rainbow and f(R) theories at work, J. Cosmol. Astropart. Phys. 6 (2013) 017, http://dx.doi.org/10.1088/1475-7516/2013/06/017, arXiv:1210.7760.

[56] A.F. Ali, M. Faizal, M.M. Khalil, Remnants of black rings from gravity's rainbow, J. High Energy Phys. 12 (2014) 159, http://dx.doi.org/10.1007/JHEP12(2014)159.

[57] R. Garattini, E.N. Saridakis, Gravity's rainbow: a bridge towards Hořava–Lifshitz gravity, Eur. Phys. J. C 75 (2015) 343, http://dx.doi.org/10.1140/epjc/s10052-015-3562-y, arXiv:1411.7257.

[58] G.G. Carvalho, I.P. Lobo, E. Bittencourt, Extended disformal approach in the scenario of rainbow gravity, Phys. Rev. D 93 (4) (2016) 044005, http://dx.doi.org/10.1103/PhysRevD.93.044005, arXiv:1511.00495.

[59] P. Rudra, M. Faizal, A.F. Ali, Vaidya spacetime for Galileon gravity's rainbow, Nucl. Phys. B 909 (2016) 725–736, http://dx.doi.org/10.1016/j.nuclphysb.2016.06.002, arXiv:1606.04529.

[60] J. Hackett, Asymptotic flatness in rainbow gravity, Class. Quantum Gravity 23 (2006) 3833–3841, http://dx.doi.org/10.1088/0264-9381/23/11/010, arXiv:gr-qc/0509103.

[61] C. Leiva, J. Saavedra, J. Villanueva, Geodesic structure of the Schwarzschild black hole in rainbow gravity, Mod. Phys. Lett. A 24 (2009) 1443–1451, http://dx.doi.org/10.1142/S0217732309029983, arXiv:0808.2601.

[62] R. Garattini, G. Mandanici, Particle propagation and effective space–time in gravity's rainbow, Phys. Rev. D 85 (2) (2012) 023507, http://dx.doi.org/10.1103/PhysRevD.85.023507, arXiv:1109.6563.

[63] A.F. Grillo, E. Luzio, F. Méndez, F. Torres, Gravitational lensing in AN energy-dependent spacetime metric, Int. J. Mod. Phys. D 21 (2012) 1250007, http://dx.doi.org/10.1142/S0218271812500071.

[64] A.F. Ali, M.M. Khalil, A proposal for testing gravity's rainbow, Europhys. Lett. 110 (2015) 20009, http://dx.doi.org/10.1209/0295-5075/110/20009, arXiv:1408.5843.

[65] A. Bhadra, K.K. Nandi, Gravitational time advancement and its possible detection, Gen. Relativ. Gravit. 42 (2010) 293–302, http://dx.doi.org/10.1007/s10714-009-0842-6, arXiv:0808.3729.

[66] S. Ghosh, A. Bhadra, Influences of dark energy and dark matter on gravitational time advancement, Eur. Phys. J. C 75 (2015) 494, http://dx.doi.org/10.1140/epjc/s10052-015-3719-8, arXiv:1508.05745.

[67] G. Amelino-Camelia, J. Ellis, N.E. Mavromatos, D.V. Nanopoulos, Distance measurement and wave dispersion in a Liouville–String approach to quantum gravity, Int. J. Mod. Phys. A 12 (1997) 607–623, http://dx.doi.org/10.1142/S0217751X97000566, arXiv:hep-th/9605211.

[68] G. Amelino-Camelia, J. Ellis, N.E. Mavromatos, D.V. Nanopoulos, S. Sarkar, Tests of quantum gravity from observations of γ-ray bursts, Nature 393 (1998) 763–765, http://dx.doi.org/10.1038/31647, arXiv:astro-ph/9712103.

[69] J. Magueijo, L. Smolin, Lorentz invariance with an invariant energy scale, Phys. Rev. Lett. 88 (19) (2002) 190403, http://dx.doi.org/10.1103/PhysRevLett.88.190403, arXiv:hep-th/0112090.

[70] I.I. Shapiro, Fourth test of general relativity, Phys. Rev. Lett. 13 (1964) 789–791, http://dx.doi.org/10.1103/PhysRevLett.13.789.

[71] S. Weinberg, Gravitation and Cosmology: Principles and Applications of the General Theory of Relativity, Wiley, New York, 1972.

[72] D. Smith, M. Zuber, X. Sun, G. Neumann, J. Cavanaugh, J. McGarry, T. Zagwodzki, Two-way laser link over interplanetary distance, Science 311 (2006) 53, http://dx.doi.org/10.1126/science.1120091.

[73] D. Dirkx, L.L.A. Vermeersen, R. Noomen, P.N.A.M. Visser, Phobos laser ranging: numerical geodesy experiments for Martian system science, Planet. Space Sci. 99 (2014) 84–102, http://dx.doi.org/10.1016/j.pss.2014.03.022.

[74] D. Dirkx, R. Noomen, P.N.A.M. Visser, S. Bauer, L.L.A. Vermeersen, Comparative analysis of one- and two-way planetary laser ranging concepts, Planet. Space Sci. 117 (2015) 159–176, http://dx.doi.org/10.1016/j.pss.2015.06.005.

[75] C.W. Chou, D.B. Hume, J.C.J. Koelemeij, D.J. Wineland, T. Rosenband, Frequency comparison of two high-accuracy al^+ optical clocks, Phys. Rev. Lett. 104 (2010) 070802, http://dx.doi.org/10.1103/PhysRevLett.104.070802.

[76] C.W. Chou, D.B. Hume, T. Rosenband, D.J. Wineland, Optical clocks and relativity, Science 329 (2010) 1630, http://dx.doi.org/10.1126/science.1192720.

[77] B.J. Bloom, T.L. Nicholson, J.R. Williams, S.L. Campbell, M. Bishof, X. Zhang, W. Zhang, S.L. Bromley, J. Ye, An optical lattice clock with accuracy and stability at the 10^{-18} level, Nature 506 (2014) 71–75, http://dx.doi.org/10.1038/nature12941, arXiv:1309.1137.

14

Gravitino production in a thermal Universe revisited

Richa Arya [a,b,*], Namit Mahajan [a], Raghavan Rangarajan [a]

[a] *Theoretical Physics Division, Physical Research Laboratory, Navrangpura, Ahmedabad 380 009, India*
[b] *IIT Gandhinagar, Palaj, Gandhinagar 382 355, India*

ARTICLE INFO

Editor: M. Trodden

ABSTRACT

We study the production of spin 1/2 gravitinos in a thermal Universe. Taking into account supersymmetry breaking due to the finite thermal energy density of the Universe, there is a large enhancement in the cross section of production of these gravitino states. We consider gravitinos with zero temperature masses of 0.1 eV, 1 keV, 100 GeV and 30 TeV as representative of gauge mediated, gravity mediated and anomaly mediated supersymmetry breaking scenarios. We find that the abundance of gravitinos produced in the early Universe is very high for gravitinos of mass 1 keV and 100 GeV. The gravitino abundances can be sufficiently suppressed if the reheat temperature is less than 100 GeV and 4×10^4 GeV respectively. However such low reheat temperatures will rule out many models of baryogenesis including those via leptogenesis.

1. Introduction

Local supersymmetry, or supergravity, gives us a massless spin 2 particle that one can identify with the graviton, the intermediate boson for gravitational interactions. The superpartner of the graviton is the massless gravitino with spin states $\pm 3/2$. When supersymmetry breaks, the gravitino gains mass and spin $\pm 1/2$ states via a super-Higgs mechanism. The spin $\pm 1/2$ states of the gravitino are often referred to as goldstino modes.

Gravitinos are produced in the early Universe either in the radiation dominated Universe after reheating [1–17], or during standard (perturbative) reheating [15–20], by the scattering of thermalised inflaton decay products. Gravitinos can also be produced during preheating [15,21–29] or via direct inflaton decay [30–33], or during and after inflation in warm inflation scenarios [34,35]. As argued in Ref. [12], the gravitino production rate in supersymmetric QCD via scattering at high temperature is proportional to

$$\frac{1}{M_P^2}\left(1+\frac{m_{\tilde{g}}^2}{3m_{\tilde{G}}^2}\right) \tag{1}$$

where $M_P \simeq 2.4 \times 10^{18}$ GeV is the reduced Planck mass, $m_{\tilde{g}}$ is the explicit supersymmetry breaking gluino mass and $m_{\tilde{G}}$ is the gravitino mass. The first term within the parentheses is associated with spin 3/2 gravitino production while the second term is associated with spin 1/2 gravitino production.

Excessive abundance of gravitinos creates cosmological problems. A very light ($m_{\tilde{G}} \ll 1$ MeV) and stable gravitino acts as an additional relativistic degree of freedom during primordial nucleosynthesis and can affect the expansion rate and thereby the light nuclear abundances (depending on its contribution to the effective relativistic degrees of freedom). For a stable gravitino of mass greater than 1 keV, its energy density today turns out to be higher than the critical density and it can overclose the Universe. A gravitino of mass between 100 GeV $\lesssim m_{\tilde{G}} \lesssim 10$ TeV decays into energetic particles after nucleosynthesis which dissociate light nuclei created during primordial nucleosynthesis. The extent of impact of the gravitinos on the cosmology of our Universe depends directly on its abundance.

The initial calculation of the gravitino abundance done in Refs. [1–5] considered gravitino production in the radiation dominated Universe after reheating for spin 3/2 states. It was found that the abundance $Y_{\tilde{G}} \equiv n_{\tilde{G}}/s$, where $n_{\tilde{G}}$ is the gravitino number density and s is the entropy density, is proportional to the reheat temperature T_{reh}. This then gave an upper bound on the reheat temperature. Many subsequent estimates of the gravitino abundance created in the radiation dominated Universe after reheating considered different channels for gravitino decay as a function of the gravitino mass and obtained associated upper bounds on the abundance or reheat temperature [6,16]. Again, these works also considered only spin 3/2 states. As mentioned in Ref. [16] consid-

* Corresponding author.
E-mail addresses: richaarya@prl.res.in, richa.arya@iitgn.ac.in (R. Arya), nmahajan@prl.res.in (N. Mahajan), raghavan@prl.res.in (R. Rangarajan).

ering only the spin 3/2 states gives a conservative estimate of the gravitino abundance. Furthermore, for gravity mediated supersymmetry breaking the second term in the parentheses in Eq. (1) is of $O(1)$ and so the abundance obtained is of the right order in this case.

In the present work we study the production of spin 1/2 gravitinos in the radiation dominated Universe after reheating. Spin 1/2 gravitinos are associated with goldstino modes and, as we argue below, their production cross section should not be Planck mass suppressed but instead suppressed by the supersymmetry breaking scale in the hidden sector. We then argue that the finite energy density of a thermal Universe also breaks supersymmetry, and in scenarios where interactions in the thermal bath are mediated by light particles the finite energy density affects both the fermion-boson mass squared splitting and the gravitino mass, and thus the goldstino production cross section.

In the standard picture of hidden sector supersymmetry breaking we have a hidden sector with fields $[H]$, a visible sector with fields $[V]$ and a messenger sector that mediates the supersymmetry breaking with fields $[X]$ with mass M_X. Supersymmetry breaks in the hidden sector, say, by F-term breaking with $\langle F_H\rangle = f_H$. The soft supersymmetry breaking mass in the visible sector that gets generated due to the interaction between the visible and hidden sectors mediated by the messenger sector is

$$m_{\rm soft} \sim \frac{1}{M_X}\langle F_H\rangle . \tag{2}$$

For phenomenological reasons, we require $m_{\rm soft} \sim 100\,{\rm GeV}$. This then, depending upon the mediation mechanism, sets the scale for $\langle F_H\rangle$.

The goldstino coupling to matter will be proportional to the mass squared splitting between particles and their superpartners. From Eq. (1) the production rate for spin 1/2 gravitino states is

$$\Gamma_s \sim \frac{1}{M_P^2}\frac{m_{\tilde g}^2}{m_{\tilde G}^2} \tag{3}$$

$$\sim \frac{1}{M_P^2}\frac{m_{\rm soft}^2}{m_{\tilde G}^2} \tag{4}$$

$$\sim \frac{1}{M_P^2}\frac{m_{\rm soft}^2}{(M_S^2/M_P)^2}$$

$$\sim \frac{1}{M_S^2}\frac{m_{\rm soft}^2}{M_S^2} \tag{5}$$

where $m_{\rm soft}^2$ is the mass squared splitting between superpartners while $M_S = \sqrt{\langle F_H\rangle}$ is the scale of supersymmetry breaking. It can be further noted that the goldstino production rate above is not Planck mass suppressed but suppressed by the supersymmetry breaking scale M_S. The production rate goes to zero in the supersymmetric case.

It is known that supersymmetry is broken by non-zero temperature T. It has also been shown that the effect of the non-zero temperature is to split the boson and fermion masses, with the splitting $m_{{\rm soft},T}^2 \sim g^2T^2$, where g is a generic coupling constant, which we refer to as soft mass generation due to finite temperature effects. That supersymmetry is broken by finite temperature effects can also be seen by the following argument: in the high temperature limit we know that the theory gets dimensionally reduced to a lower dimension. All the fermion Matsubara modes (recall that there is no $n = 0$ mode for fermions) become heavy while for bosons all modes become heavy except the $n = 0$ mode. Thus, the low energy effective theory will only contain a bosonic field and no fermionic field, and therefore supersymmetry is broken. Moreover in Refs. [36,37] it has been shown, by invoking a thermal superspace approach and applying it to systems of thermal fields, that supersymmetry is explicitly broken at finite temperature and that the thermal action is not invariant under thermal supersymmetry. Unlike in other works, we consider supersymmetry breaking due to the finite energy density of the radiation dominated Universe, $\rho = (\pi^2/30)g_*T^4$, where $g_* \sim 228.75$ is the effective number of relativistic degrees of freedom.

Let us now consider the effect of having a finite energy density, $\rho \sim T^4$. Consider three chiral superfields, S, Φ and Y having a coupling $\lambda S\Phi Y$, where λ is the coupling, Φ belongs to $[V]$ and S could belong to $[V]$, $[H]$ or $[X]$. We assume that S contributes to the radiation energy density of the Universe. In the superfield language, the four point amplitude $\Phi^\dagger\Phi S^\dagger S$ reads (we have employed the off-diagonal component of the GRS propagator for Y and retained the external fields)

$$\mathcal{A}^{(4)} = |\lambda|^2\Phi^\dagger S^\dagger\left[\int d^4x d^4x' d^4\theta d^4\theta' \frac{1}{\Box^2 - m_Y^2}\delta(z-z')\right]\Phi S \tag{6}$$

where $z = (x, \theta, \bar\theta)$ and $\delta(z-z') = \delta^4(x-x')\delta^2(\theta-\theta')\delta^2(\bar\theta-\bar\theta')$. In a thermal bath, the typical Y momentum $q \sim Q$, i.e. there is a distribution peaked at $Q \sim T$ (or $Q \sim \sqrt{Tm_Y}$), for $T \gg m_Y$ (for $T \ll m_Y$), which for simplicity we take to be $\delta(q-Q)$. For $T \ll m_Y$ we then get $\mathcal{A}^{(4)} \sim |\lambda|^2/m_Y^2\,(\Phi^\dagger S^\dagger\Phi S)$, while for $T \gg m_Y$, we obtain $\mathcal{A}^{(4)} \sim |\lambda|^2/T^2(\Phi^\dagger S^\dagger\Phi S)$. This is equivalent to having a term in the effective Lagrangian as

$$\mathcal{L}_{\rm eff} = \frac{|\lambda|^2}{m_Y^2, T^2}\,\Phi^\dagger S^\dagger\Phi S . \tag{7}$$

Now we expand $\mathcal{L}_{\rm eff}$ in all powers of θ and $\bar\theta$. The relevant $\theta\theta\bar\theta\bar\theta$ term for us will be, in a thermal environment,

$$\frac{|\lambda|^2}{m_Y^2, T^2}\,\langle S^\dagger S|_{\theta\theta\bar\theta\bar\theta}\rangle_{\rm thermal}\phi^\dagger\phi , \tag{8}$$

where ϕ is the scalar component of Φ. The term $\langle S^\dagger S|_{\theta\theta\bar\theta\bar\theta}\rangle_{\rm thermal}$ above is $\langle i\partial_m\bar\psi_S\bar\sigma^m\psi_S + s^*\Box s + F_S^*F_S\rangle_{\rm thermal}$. The first two terms can be identified with the kinetic energy terms in the Lagrangian for the fermionic and scalar components of the superfield S. $|F_S|^2 = |\partial W/\partial s|^2 = V(s)$ and one may consider, say, a quartic potential for s. Then $\langle S^\dagger S|_{\theta\theta\bar\theta\bar\theta}\rangle_{\rm thermal} \sim T^4$. Therefore we obtain

$$m_{\rm soft}^2 \sim |\lambda|^2\frac{T^4}{m_Y^2} \quad \text{for } T \ll m_Y \tag{9}$$

and

$$m_{\rm soft}^2 \sim |\lambda|^2 T^2 \quad \text{for } T \gg m_Y . \tag{10}$$

(The above example can be suitably extended to vector superfields. Also, there will be, in general, more than one such contribution to the soft masses. We have chosen the simplest one to bring out the essence of the argument.)

Comparing Eq. (9) and Eq. (2), the two forms are quite similar, and when S belongs to, say, $[H]$, and the amplitude is mediated by a heavy field it is quite natural to assume $m_Y \sim M_X$. In such a case, it appears that the finite temperature effects essentially look like an additional contribution to the F-term breaking. For $f_H \gg T^2$ one does not have any large temperature dependent contribution to the soft breaking masses, and therefore no enhanced contribution to goldstino couplings. A similar conclusion is reached if S is one of the visible sector fields and the interaction between S and Φ is mediated by a third visible sector field which is very massive. Refs. [20,38,39] also argue that finite temperature effects will not lead to enhanced goldstino production as originally argued in Ref. [40].

Now consider the case when the four point superfield amplitude is mediated by a light Y superfield. This is quite reasonable to expect since in thermal equilibrium different fields in the visible sector, for example, Φ and S, can interact via, say, gauge/Yukawa interactions such that the mediator is a massless/light field. In this case, i.e. when the amplitude is mediated by the light field, the soft supersymmetry breaking mass in Eq. (10) contributes to the scale of the mass splitting between the bosonic and fermionic partners.

From the above discussion we observe that depending up on the mass scale of the field that mediates the four point amplitude, the finite temperature contribution to the soft breaking mass takes the form

$$m_{\rm soft}^2 \sim \frac{1}{M_X^2}(f_H^2 + \delta^2 T^4), \;\; m_Y \sim M_X \gg T \tag{11}$$

where δ is some parameter, or

$$m_{\rm soft}^2 \sim \frac{1}{M_X^2} f_H^2 + \delta T^2, \;\; m_Y \ll T\,. \tag{12}$$

Below we shall assume that S is one of the visible sector fields and there is naturally a massless/light field that mediates the four point amplitude, and therefore what is relevant is Eq. (12). We would like to emphasize that this is exactly where we differ from the usual treatment of finite temperature effects in the context of gravitinos. Then in Eq. (3)

$$m_{\tilde g}^2 \to m_{\tilde g}^2 - m_g^2 \sim \delta_3 T^2 + m_0^2\,, \tag{13}$$

where m_g is the gluon mass, δ_3 is some parameter and $m_0 \sim$ 100 GeV represents the zero temperature mass splitting, while

$$\begin{aligned} m_{\tilde G} &\sim \sqrt{\rho}/(\sqrt{3}M_P) + m_{\tilde G 0} \\ &= \delta' T^2/(\sqrt{3}M_P) + m_{\tilde G 0} \end{aligned} \tag{14}$$

where $m_{\tilde G 0}$ is the zero temperature gravitino mass (which depends on the supersymmetry breaking mechanism relevant at low temperatures), and δ' is another parameter. Then the factor in the scattering rate in Eq. (4)

$$\gamma_3 \equiv \frac{m_{\rm soft}^2}{3m_{\tilde G}^2} = \frac{\delta_3 T^2 + m_0^2}{3[\delta' T^2/(\sqrt{3}M_P) + m_{\tilde G 0}]^2}\,. \tag{15}$$

When $\delta_3 T^2 \gg m_0^2$ and $\delta' T^2/(\sqrt{3}M_P) \gg m_{\tilde G 0}$

$$\gamma_3 \approx \frac{\delta_3 T^2}{(\delta' T^2/M_P)^2} = \frac{\delta_3}{\delta'^2}\frac{M_P^2}{T^2} \tag{16}$$

which can be much larger than 1. This can be much larger than the zero temperature limit $m_0^2/(3m_{\tilde G 0}^2)$ in Γ_s, and therefore necessitates a fresh look at the calculation of the gravitino abundance. (It may be noted, however, that our final results will depend only on the zero temperature form of the gravitino mass.)

Unlike in the standard calculations of the gravitino abundance, in our scenario gravitinos will be in thermal equilibrium in the early Universe because of the enhanced scattering rates. The gravitinos decouple when they are relativistic, and hence they have a large abundance as a hot relic. Below we shall consider gravitinos with zero temperature masses of 0.1 eV, 1 keV, 100 GeV, 30 TeV. Typically one can obtain such masses in gauge mediated (0.1 eV, 1 keV), gravity mediated and anomaly mediated supersymmetry breaking scenarios respectively. We find that the very light gravitinos ($m_{\tilde G 0} = 0.1$ eV) have $\Omega_{\tilde G} \ll 1$, and hence will not overclose the Universe. The very heavy gravitinos ($m_{\tilde G 0} = 30$ TeV) have a very large abundance but have a short lifetime and decay before nucleosynthesis and do not greatly alter the cosmology of the Universe. However, the gravitinos with mass ~ 1 keV have $\Omega_{\tilde G} \sim 1$, so can affect the cosmology of our Universe. We further find that the abundance of $m_{\tilde G 0} = 100$ GeV gravitinos is orders of magnitude higher than the cosmological upper bound. The abundances can be suppressed by considering a low reheat temperature less than 100 GeV and 4×10^4 GeV for the 1 keV and 100 GeV gravitinos respectively. Such low reheat temperatures will rule out models of high scale baryogenesis including those via leptogenesis. Very low scale baryogenesis scenarios, and electroweak baryogenesis and low scale leptogenesis models respectively will then be the preferred mechanisms for generating the matter–antimatter asymmetry of the Universe.

In an earlier work we had studied the production of spin 1/2 gravitinos in the presence of supersymmetric flat directions which give mass to some gauge bosons, gauginos, and sfermions and had found that there is resonant production of (spin 1/2) gravitinos leading to an extremely large abundance which is orders of magnitude larger than the cosmological bound [41]. Below we do not consider the presence of supersymmetric flat directions.[1]

2. Gravitino production

As mentioned earlier, gravitinos are produced by the scattering of the thermalised decay products of the inflaton. The different processes that produce gravitinos are listed in Table 1 in Refs. [5,11] and Table 4.3 of Ref. [45]. These processes include, for example, $g\tilde g \to g\tilde G$, $\tilde q g \to q\tilde G$, $q\bar q \to \tilde g\tilde G$, etc. Besides these scattering processes there are also annihilation processes, $\tilde G\tilde G \to \gamma\gamma$ and $\tilde G\tilde G \to f\bar f$, as we shall see below.

The thermally averaged cross section $\langle\Sigma_{\rm tot}|v|\rangle$ for the scattering processes in Refs. [5,11] is given by [14]

$$\begin{aligned} \langle\Sigma_{\rm tot}|v|\rangle &\equiv \frac{\alpha}{M_P^2} \\ &= \frac{1}{M_P^2}\frac{3\pi}{16\zeta(3)}\sum_{i=1}^{3}\left[1+\frac{m_i^2}{3m_{\tilde G}^2}\right]c_i g_i^2 \ln\left(\frac{k_i}{g_i}\right) \end{aligned} \tag{17}$$

where $i = 1,2,3$ refers to the three gauge groups $U(1)_Y$, $SU(2)_L$ and $SU(3)_c$ respectively, $g_i(T)$ are the gauge coupling constants (evaluated at the most relevant temperature), and $c_{1,2,3} =$ 11, 27, 72 and $k_{1,2,3} = 1.266, 1.312, 1.271$ are constants associated with the gauge groups (see Table 1 of Ref. [17]). The above expression includes corrections to the cross section for gravitino production obtained earlier in Refs. [12] and [16]. We have also replaced the gaugino mass squared in the original expression with $m_i^2 = \delta_i T^2 + m_0^2 = m_{\rm soft}^2$ representing the difference in gaugino and gauge boson masses squared. (In the current analysis we have ignored the possibility of a Breit–Wigner resonance associated with incoming particles of energy $\sim T$ and the intermediate supersymmetric particle having a mass of $O(T)$. We hope to return to this issue in a future publication.)

The rate of production of gravitinos for the processes listed in Table 1 of [5,11] is given by

$$\Gamma_s = n\langle\Sigma_{\rm tot}|v|\rangle$$

[1] If the supersymmetric flat direction gives mass to *all* gauge bosons it delays thermalization of the inflaton decay products leading to suppressed gravitino production, as discussed in Refs. [42–44]. In these works, the thermal energy density contribution to supersymmetry breaking was not included.

where $n = (\zeta(3)/\pi^2)T^3$ is the number density of the scatterers, and the Riemann zeta function $\zeta(3) = 1.2020....$ Then, taking all $m_i = m_3$, we get

$$\Gamma_s = \frac{3T^3}{16\pi M_P^2}\left(1 + \frac{m_3^2}{3m_{\tilde{G}^2}}\right)\sum_{i=1}^{3} c_i g_i^2 \ln\left(\frac{k_i}{g_i}\right). \quad (18)$$

We now consider different cases for γ_3 defined in Eq. (15).

- **Region I:** $\delta_3 T^2 > m_0^2$ and $\delta' T^2/(\sqrt{3}M_P) > m_{\tilde{G}0}$. Then

$$\gamma_3 \approx \frac{\delta_3 M_P^2}{\delta'^2 T^2}.$$

- **Region II:** $\delta_3 T^2 > m_0^2$ and $\delta' T^2/(\sqrt{3}M_P) < m_{\tilde{G}0}$. Then

$$\gamma_3 \approx \frac{\delta_3 T^2}{3m_{\tilde{G}0}^2}.$$

- **Region III:** $\delta_3 T^2 < m_0^2$ and $\delta' T^2/(\sqrt{3}M_P) < m_{\tilde{G}0}$. Then

$$\gamma_3 \approx \frac{m_0^2}{3m_{\tilde{G}0}^2}.$$

We shall take $\delta_3, \delta' \sim 0.1$. As we shall see below, scattering processes will maintain the gravitinos in thermal equilibrium in our scenario till they freeze out. Thereafter annihilation processes such as $\tilde{G}\tilde{G} \to \gamma\gamma$ and $\tilde{G}\tilde{G} \to f\bar{f}$ become relevant.

For the process $\tilde{G}\tilde{G} \to \gamma\gamma$, if $\sqrt{s} \gg m_{\tilde{\gamma}}$, where $\sqrt{s} \sim T$ is the centre of mass energy, then the annihilation cross section is given by [46]

$$\sigma_{\tilde{G}\tilde{G}\to\gamma\gamma} = \frac{1}{1728\pi}\frac{\kappa^4}{m_{\tilde{G}}^4} m_{\tilde{\gamma}}^4 s \quad (19)$$

where $\kappa = \frac{1}{M_{Pl}}$ and $M_{Pl} = 1.2 \times 10^{19}$ GeV is the Planck mass. If $\sqrt{s} \ll m_{\tilde{\gamma}}$ then the annihilation cross section is

$$\sigma_{\tilde{G}\tilde{G}\to\gamma\gamma} = \frac{1}{576\pi}\frac{\kappa^4}{m_{\tilde{G}}^4} m_{\tilde{\gamma}}^2 s^2 = \sigma_{\gamma\gamma} s^2, \quad (20)$$

where $\sigma_{\gamma\gamma} \equiv \frac{1}{576\pi}\frac{\kappa^4}{m_{\tilde{G}}^4} m_{\tilde{\gamma}}^2$. For the process $\tilde{G}\tilde{G} \to f\bar{f}$, if $\sqrt{s} \gg m_{\tilde{f}}$ then from Ref. [46]

$$\sigma_{\tilde{G}\tilde{G}\to f\bar{f}} = \frac{1}{180\pi}\frac{\kappa^4}{m_{\tilde{G}}^4} m_{\tilde{f}}^4 s, \quad (21)$$

and if $\sqrt{s} \ll m_{\tilde{f}}$ then

$$\sigma_{\tilde{G}\tilde{G}\to f\bar{f}} = \frac{1}{180\pi}\frac{\kappa^4}{m_{\tilde{G}}^4} s^3 = \sigma_{f\bar{f}} s^3, \quad (22)$$

where $\sigma_{f\bar{f}} \equiv \frac{1}{180\pi}\frac{\kappa^4}{m_{\tilde{G}}^4}$. The total annihilation cross section is given by

$$\sigma_A = \sigma_{\gamma\gamma} s^2 + \sum_f \sigma_{f\bar{f}} s^3$$

where all possible fermion pairs in the final state have been summed over. The dominant contribution to the total cross section for $\sqrt{s} \ll m_{\tilde{\gamma}}$, as can be seen from Eqs. (20) and (22), is from $\sigma_{\tilde{G}\tilde{G}\to\gamma\gamma}$, and the thermally averaged cross section times velocity is approximately given by [46]

$$\langle \sigma v_{\text{Moller}} \rangle = 1800 \frac{\zeta(5)^2}{\zeta(3)^2} \sigma_{\gamma\gamma} T^4$$

where $\zeta(5) = 1.0369....$ The annihilation rate for gravitinos for $\sqrt{s} \ll m_{\tilde{\gamma}}$ is given by

$$\Gamma_A = n_{\tilde{G}} \langle \sigma_A v_{\text{Moller}} \rangle$$

where $n_{\tilde{G}} = \frac{3\zeta(3)}{2\pi^2} T^3$. Then

$$\Gamma_A = \frac{75}{16\pi^3}\frac{\zeta(5)^2}{\zeta(3)}\frac{\kappa^4}{m_{\tilde{G}}^4} m_{\tilde{\gamma}}^2 T^7 \simeq 0.135 \frac{\kappa^4}{m_{\tilde{G}}^4} m_{\tilde{\gamma}}^2 T^7. \quad (23)$$

For $\sqrt{s} \ll m_{\tilde{\gamma}} \sim \delta_1 T^2 + m_0^2$, $T \ll m_0$ and $\delta' T^2/(\sqrt{3}M_P) \ll m_{\tilde{G}0}$ and so

$$\Gamma_A = 0.135 \frac{\kappa^4}{m_{\tilde{G}0}^4} m_0^2 T^7. \quad (24)$$

For higher temperatures we use the appropriate expressions for the annihilation rate.

3. Calculation of the gravitino freeze out temperature

In the standard scenario of gravitino production the rate for production is small compared to the Hubble parameter H. Therefore one does not produce very many gravitinos. The small number density of gravitinos then implies that the inverse scattering process is also suppressed. However, in our scenario, because of the enhanced gravitino production rate, $\Gamma_s > H$. Moreover, the inverse process will be unsuppressed because of the large gravitino abundance. Therefore the gravitinos maintain a thermal distribution, till their interactions freeze out. (We can use the expression for Γ_s till $T \sim \max\,(m_{\tilde{G}}, m_{\tilde{g}})$ since the cross section in Eq. (17) presumes the particles are relativistic.)

The freeze out condition is $\Gamma_s(T_f) = H(T_f) = 5\frac{T_f^2}{M_P}$. We also consider the gravitino annihilation processes discussed above. For each (zero temperature) gravitino mass we consider each of the three regions discussed in the previous section.

3.1. Very light gravitino

$m_{\tilde{G}0} = 0.1$ eV.

Region I: $T > 300$ GeV and $T > 6.5 \times 10^4$ GeV. We find that $\Gamma_s > \Gamma_A$ and also $\Gamma_s > H$. Hence, gravitinos are in thermal equilibrium in this domain and maintain a thermal abundance with $n_{\tilde{G}} = \frac{3\zeta(3)}{2\pi^2} T^3$.

Region II: $T > 300$ GeV and $T < 6.5 \times 10^4$ GeV. We find that $\Gamma_s > \Gamma_A$ and also $\Gamma_s > H$. Hence, gravitinos are in thermal equilibrium in this domain and maintain a thermal abundance.

Region III: $T < 300$ GeV. Till $T \sim m_0 = 100$ GeV, $\Gamma_s > \Gamma_A$ and $\Gamma_s > H$. For $T < 100$ GeV, scattering processes given in Table 1 in Refs. [5,11] are kinematically forbidden and $\Gamma_{\tilde{G}\tilde{G}\to\gamma\gamma}$ is the relevant process. However we find that $\Gamma_{\tilde{G}\tilde{G}\to\gamma\gamma} < H$.

Therefore the freeze out temperature $T_f = 100$ GeV. The abundance of gravitinos at freeze out is given by

$$Y_{\tilde{G}f} = Y_{\tilde{G}}(T_f) = \frac{n(T_f)}{s(T_f)} = \frac{3\zeta(3)}{2\pi^2}\frac{45}{2\pi^2 g_{*s}(T_f)}.$$

For the MSSM particle content, $g_{*s} \sim 228.75$. Then

$$Y_{\tilde{G}f} = 1.8 \times 10^{-3}.$$

The lifetime of the gravitinos is given by [47]

$$t=\frac{M_P^2}{m_{\tilde{G}}^3}=1.2\times 10^{35}\ \text{yr}$$

which is much larger than the age of Universe. The density parameter of thermally produced gravitinos is given by

$$\Omega_{\tilde{G}}=\rho_{\tilde{G}}/\rho_c=m_{\tilde{G}}Y_{\tilde{G}f}s(T_0)/\rho_c. \quad (25)$$

Taking $\rho_c/s(T_0)=1.95\times 10^{-9}$ GeV, we get $\Omega_{\tilde{G}}\approx 0.92\times 10^{-4}$, which implies that the gravitinos will not overclose the Universe. Furthermore, there is no constraint from primordial nucleosynthesis because the effective number of (nearly) massless neutrino flavors over and above the Standard Model value, $\Delta N_\nu=[g_*(1\,\text{MeV})/g_*(T_f)]^{4/3}$ will be 0.02, which is less than the current upper bound of 0.4 from Planck 2015 [48]. However the Cosmic Microwave Background Stage 4 experiments hope to probe ΔN_ν down to an accuracy of 0.027 [49].

3.2. Light gravitino

$$m_{\tilde{G}0}=1\ \text{keV}.$$

Region I: $T>300$ GeV and $T>6.5\times 10^6$ GeV. We find that $\Gamma_s>\Gamma_A$ and also $\Gamma_s>H$. Hence, gravitinos are in thermal equilibrium in this domain and maintain a thermal abundance.

Region II: $T>300$ GeV and $T<6.5\times 10^6$ GeV. We find that $\Gamma_s>\Gamma_A$ in this temperature range. But $\Gamma_s<H$ for $T<600$ GeV. At this temperature, the gravitinos freeze out.

Region III: $T<300$ GeV. Till $T\sim 100$ GeV, $\Gamma_s>\Gamma_A$ but $\Gamma_s<H$. For $T<100$ GeV, $\Gamma_{\tilde{G}\tilde{G}\to\gamma\gamma}$ is the relevant process. But $\Gamma_{\tilde{G}\tilde{G}\to\gamma\gamma}<H$. Hence, gravitinos are out of equilibrium in this domain.

Thus the freeze out temperature $T_f=600$ GeV and the abundance of gravitinos is given by

$$Y_{\tilde{G}f}=1.8\times 10^{-3}.$$

The lifetime of the gravitinos

$$t=\frac{M_P^2}{m_{\tilde{G}}^3}=1.2\times 10^{23}\ \text{yr}$$

which is much larger than the age of Universe. The density parameter in Eq. (25) $\Omega_{\tilde{G}}\approx 0.92$, which is in conflict with observations. In order to avoid this, the gravitino mass was bounded to be less than a keV in Refs. [47,50]. Again, $\Delta N_\nu=0.02$ which is less than the current upper bound.

3.3. Heavy gravitino

$$m_{\tilde{G}0}=100\ \text{GeV}.$$

Region I: $T>300$ GeV and $T>6.5\times 10^{10}$ GeV. We find that $\Gamma_s>\Gamma_A$ and also $\Gamma_s>H$. Hence, gravitinos are in thermal equilibrium in this domain and maintain a thermal abundance.

Region II: $T>300$ GeV and $T<6.5\times 10^{10}$ GeV. We find that $\Gamma_s>\Gamma_A$ in this temperature range. But $\Gamma_s<H$ for $T<1.2\times 10^8$ GeV. At this temperature, the gravitinos freeze.

Region III: $T<300$ GeV. Till $T\sim 100$ GeV, $\Gamma_s>\Gamma_A$ but $\Gamma_s<H$. Thereafter $\Gamma_{\tilde{G}\tilde{G}\to\gamma\gamma}$ is the relevant process. But $\Gamma_{\tilde{G}\tilde{G}\to\gamma\gamma}<H$ which implies that the gravitinos are out of equilibrium in this domain.

Thus the freeze out temperature $T_f=1.2\times 10^8$ GeV and the abundance of gravitinos at freeze out is given by

$$Y_{\tilde{G}f}=1.8\times 10^{-3}.$$

This is much larger than the cosmological upper bound on the gravitino abundance of 10^{-14} [51].

3.4. Very heavy gravitino

$$m_{\tilde{G}0}=30\ \text{TeV}.$$

Region I: $T>300$ GeV and $T>1\times 10^{12}$ GeV. We find that $\Gamma_s>\Gamma_A$ and also $\Gamma_s>H$. Hence, gravitinos are in thermal equilibrium in this domain and maintain a thermal abundance.

Region II: $T>300$ GeV and $T<1\times 10^{12}$ GeV. We find that $\Gamma_s>\Gamma_A$ in this temperature range. But $\Gamma_s<H$ for $T<5.5\times 10^9$ GeV. At this temperature, the gravitinos freeze out.

Region III: $T<300$ GeV. In this domain, the scattering processes are kinematically forbidden as $T<m_{\tilde{G}0}$. Hence, $\Gamma_{\tilde{G}\tilde{G}\to\gamma\gamma}$ is the relevant process. But $\Gamma_{\tilde{G}\tilde{G}\to\gamma\gamma}<H$ which implies that the gravitinos are out of equilibrium in this domain.

Thus the freeze out temperature $T_f=5.5\times 10^9$ GeV and the abundance of gravitinos at freeze out is given by

$$Y_{\tilde{G}f}=1.8\times 10^{-3}.$$

The lifetime of the gravitinos is

$$t=\frac{M_P^2}{m_{\tilde{G}}^3}=0.1\ \text{s}$$

which implies that the gravitinos would have decayed before nucleosynthesis and not lead to any cosmological problem.

In all the cases considered above, $T_f\gtrsim m_{\tilde{G}},\ m_{\tilde{g}}$. Then the use of the expression for $\langle\Sigma_{\text{tot}}|v|\rangle$ in Eq. (17), which presumes relativistic incoming and outgoing particles, is justified. Note that in all cases above freeze out occurs in *Region II* or *III* for which the zero temperature gravitino mass is the relevant mass.

4. Discussion

For the (zero temperature) gravitino masses of 0.1 eV, 1 keV, 100 GeV and 30 TeV that we have considered, the freeze out temperature is higher than the gravitino mass (including thermal effects). Then the gravitinos are hot relics and their abundance at freeze out is $Y_{\tilde{G}f}\sim 1.8\times 10^{-3}$. For the gravitino with zero temperature masses of 0.1 eV and 1 keV there is no constraint from primordial nucleosynthesis because $\Delta N_\nu=0.02$ which is less than the current upper bound. The abundance of the 0.1 eV gravitinos today will not overclose the Universe. However, for the 1 keV gravitino, $\Omega_{\tilde{G}}\sim 1$.

The decay products of gravitinos with zero temperature mass of 100 GeV will modify the light nuclei abundances adversely – the corresponding upper bound on the gravitino abundance is 10^{-14} [51] which is 11 orders of magnitude lower than the abundance obtained above. The gravitinos with zero temperature mass of 30 TeV will decay before nucleosynthesis and will not modify the cosmology substantially.

Thus one needs to consider the cases of the 1 keV and 100 GeV mass gravitinos carefully. Now, in the above analysis it was presumed that $T_{\rm reh} > T_f$ which allowed the gravitinos to be in thermal equilibrium. So to suppress the high abundance of the gravitinos as a hot relic one may consider scenarios with $T_{\rm reh} < T_f$ for the 1 keV and 100 GeV gravitinos. Then one has to consider out of equilibrium gravitino production till $T \sim m_0$ (when the scattering processes will be Boltzmann or kinematically suppressed) using the Boltzmann equation, as in the standard calculation of the gravitino abundance.

We now consider the out of equilibrium production of gravitinos with a zero temperature mass of 100 GeV and $T_{\rm reh} < T_f = 1.2 \times 10^8$ GeV, and with a zero temperature mass of 1 keV and $T_{\rm reh} < T_f = 600$ GeV.

5. Out of equilibrium production of gravitinos

The gravitino production rate is given by the integrated Boltzmann equation

$$\frac{dn_{\tilde{G}}}{dt} + 3Hn_{\tilde{G}} = \langle \Sigma_{\rm tot}|v|\rangle n^2. \tag{26}$$

It is presumed that $n_{\tilde{G}} = 0$ at the beginning of the radiation dominated era after reheating and the gravitinos are then produced through thermal scattering of the inflaton decay products. We can rewrite Eq. (26) as

$$\dot{T}\frac{dY_{\tilde{G}}}{dT} = n\langle \Sigma_{\rm tot}|v|\rangle Y = \Gamma_s Y \tag{27}$$

where $Y = n/s$ is the abundance of the scatterers. $T \propto \frac{1}{a}$, where a is the scale factor of Universe. So

$$\frac{\dot{T}}{T} = -\frac{\dot{a}}{a} = -H = -\sqrt{\frac{8\pi G_N \rho}{3}} \tag{28}$$

$$= -\sqrt{\frac{8\pi G_N}{3}\frac{\pi^2}{30} g_* T^4}. \tag{29}$$

This gives

$$\dot{T} = -\sqrt{\frac{g_*\pi^2}{90}}\frac{T^3}{M_P}. \tag{30}$$

Then, on substituting Eqs. (18) and (30) in Eq. (27), we obtain for spin 1/2 gravitinos

$$\frac{dY_{\tilde{G}}}{dT} = -\frac{\beta\gamma_3}{M_P}, \tag{31}$$

where γ_3 is defined as in Eq. (15) and

$$\beta = \left(\frac{90}{g_*\pi^2}\right)^{1/2}\left(\frac{45}{2\pi^2 g_{*s}}\right)\left(\frac{\zeta(3)}{\pi^2}\right)^2 \tag{32}$$

$$\times \sum_{i=1}^{3}\frac{3\pi}{16\zeta(3)} c_i g_i^2 \ln\left(\frac{k_i}{g_i}\right). \tag{33}$$

Hereafter we shall assume β to be independent of temperature and evaluate it at the dominant temperature limit in the integrals invoked below.

We are analyzing the case when $T_f > T_{\rm reh} \gg T$, which for $m_{\tilde{G}0} = 100$ GeV can correspond to *Regions II* or *III*. Consider *Region II* where $\delta_3 T^2 > m_0^2$ and $\delta' T^2/(\sqrt{3}M_P) < m_{\tilde{G}0}$. Then $\gamma_3 = \delta_3 T^2/(3m_{\tilde{G}0}^2)$ which gives

$$\frac{dY_{\tilde{G}}}{dT} \approx -\beta\frac{\delta_3 T^2}{3M_P m_{\tilde{G}0}^2}. \tag{34}$$

On integrating from $T_{\rm reh}$ to $T \ll T_{\rm reh}$, we get

$$Y_{\tilde{G}}^{(1)}(T) = \beta\frac{\delta_3}{3M_P m_{\tilde{G}0}^2}\frac{1}{3}\left(T_{\rm reh}^3 - T^3\right) \tag{35}$$

$$\approx \beta\frac{\delta_3}{9M_P m_{\tilde{G}0}^2}T_{\rm reh}^3. \tag{36}$$

Now consider *Region III* when $\delta_3 T^2 < m_0^2$ and $\delta' T^2/(\sqrt{3}M_P) < m_{\tilde{G}0}$, i.e. $T < 3m_0 = 300$ GeV. Then $\gamma_3 = \frac{m_0^2}{3m_{\tilde{G}0}^2} \approx \frac{1}{3}$, as in the standard calculation. On integrating Eq. (31) from $T \sim 3m_0$ to $T \sim m_0$, we get the gravitino abundance

$$Y_{\tilde{G}}^{(2)} = \frac{1}{3}\frac{\beta}{M_P}2m_0. \tag{37}$$

Then the total gravitino abundance will be

$$Y_{\tilde{G}} = Y_{\tilde{G}}^{(1)} + Y_{\tilde{G}}^{(2)}$$

$$= \beta\frac{\delta_3}{9M_P m_{\tilde{G}0}^2}T_{\rm reh}^3 + \frac{2}{3}\frac{\beta}{M_P}m_0 \tag{38}$$

$$\approx \beta\frac{\delta_3}{9M_P m_{\tilde{G}0}^2}T_{\rm reh}^3, \tag{39}$$

as $Y_{\tilde{G}}^{(2)}$ is much less than $Y_{\tilde{G}}^{(1)}$.

We find that the abundance of gravitinos is proportional to $T_{\rm reh}^3$. In order that the abundance lie within the cosmological bound of order 10^{-14} as given in Ref. [51], $T_{\rm reh}$ should be less than 4×10^4 GeV.

For the 1 keV zero temperature mass gravitinos with $T_f = 600$ GeV, let $T_{\rm reh}$ be 300 GeV. This will correspond to *Region III* with $\gamma_3 = \frac{m_0^2}{3m_{\tilde{G}0}^2} = 3 \times 10^{15}$. The gravitino abundance generated from $T_{\rm reh} = 300$ GeV to $T \sim m_0 = 100$ GeV will be

$$Y_{\tilde{G}} = \gamma_3\frac{\beta}{M_P}2m_0 = 1 \times 10^{-4}. \tag{40}$$

Then from Eq. (25)

$$\Omega_{\tilde{G}} = 0.05. \tag{41}$$

This is large and inconsistent with current observations. This then implies that $T_{\rm reh}$ must be less than $m_0 = 100$ GeV to shut off this mode of gravitino production.

6. Results and conclusion

By considering supersymmetry breaking due to the finite energy density of the Universe we find that there is enhanced production of the spin 1/2 states of gravitinos (goldstino modes). We have considered gravitinos with zero temperature masses of 0.1 eV, 1 keV, 100 GeV and 30 TeV as representative of gauge mediated (0.1 eV, 1 keV), gravity mediated and anomaly mediated supersymmetry breaking scenarios respectively and find that the production processes are in thermal equilibrium in the early Universe. By studying the freeze out temperature for the gravitinos we have shown that the gravitinos decouple as hot relics with large abundances. In particular, the 1 keV and 100 GeV mass gravitinos have a very high abundance that can respectively close the Universe or affect light nuclear abundances through their decay products.

For both these cases one can suppress the abundance by lowering the reheat temperature $T_{\rm reh}$ below the freeze out temperature for gravitino production. Therefore we have further considered gravitino production from $T_{\rm reh}$ below T_f till $T \sim m_0 = 100\,\text{GeV}$ (when the production shuts off) using the Boltzmann equation. For the 100 GeV gravitino with a freeze out temperature T_f of $10^8\,\text{GeV}$, we find that the abundance is proportional to $T_{\rm reh}^3$ and that the reheat temperature must be less than $4 \times 10^4\,\text{GeV}$ to satisfy cosmological constraints. Such a low reheat temperature will be inconsistent with models of high scale baryogenesis including those via leptogenesis. Models of electroweak baryogenesis and low scale leptogenesis [52–55] will then be preferred mechanisms for generating the matter–antimatter asymmetry of the Universe.

For the 1 keV gravitino with a freeze out temperature T_f of $600\,\text{GeV}$, we first chose a reheat temperature of 300 GeV and calculated the abundance generated till $T \sim m_0$. We found that the gravitinos will contribute 5% of the total energy density of the Universe today which is inconsistent with observations. This implies that the reheat temperature should be less than m_0. Such a low reheat temperature may be obtained in models of electroweak scale inflation [56] but will rule out electroweak baryogenesis and leptogenesis scenarios. Then the preferred models of baryogenesis will be very low scale scenarios such as in Ref. [57] or those involving neutron–antineutron oscillations [58] or black hole evaporation [59–62].

The above analysis clearly provides a new manifestation of the gravitino problem.

Acknowledgement

R. Rangarajan would like to thank Chandan Hati for useful discussions.

References

[1] D.V. Nanopoulos, K.A. Olive, M. Srednicki, Phys. Lett. B 127 (1983) 30.
[2] L.M. Krauss, Nucl. Phys. B 227 (1983) 556.
[3] I.V. Falomkin, G.B. Pontecorvo, M.G. Sapozhnikov, M.Y. Khlopov, F. Balestra, G. Piragino, Nuovo Cimento A 79 (1984) 193, Yad. Fiz. 39 (1984) 990.
[4] M.Y. Khlopov, A.D. Linde, Phys. Lett. B 138 (1984) 265.
[5] J.R. Ellis, J.E. Kim, D.V. Nanopoulos, Phys. Lett. B 145 (1984) 181.
[6] K. Kohri, T. Moroi, A. Yotsuyanagi, Phys. Rev. D 73 (2006) 123511, http://dx.doi.org/10.1103/PhysRevD.73.123511, arXiv:hep-ph/0507245.
[7] R. Juszkiewicz, J. Silk, A. Stebbins, Phys. Lett. B 158 (1985) 463.
[8] J.R. Ellis, D.V. Nanopoulos, S. Sarkar, Nucl. Phys. B 259 (1985) 175.
[9] M.Y. Khlopov, Y.L. Levitan, E.V. Sedelnikov, I.M. Sobol, Phys. At. Nucl. 57 (1994) 1393, Yad. Fiz. 57 (1994) 1466.
[10] T. Moroi, H. Murayama, M. Yamaguchi, Phys. Lett. B 303 (1993) 289.
[11] M. Kawasaki, T. Moroi, Prog. Theor. Phys. 93 (1995) 879, arXiv:hep-ph/9403364, arXiv:hep-ph/9403061.
[12] M. Bolz, A. Brandenburg, W. Buchmuller, Nucl. Phys. B 606 (2001) 518;
M. Bolz, A. Brandenburg, W. Buchmuller, Nucl. Phys. B 790 (2008) 336 (Erratum), arXiv:hep-ph/0012052.
[13] R.H. Cyburt, J.R. Ellis, B.D. Fields, K.A. Olive, Phys. Rev. D 67 (2003) 103521, arXiv:astro-ph/0211258.
[14] J. Pradler, F.D. Steffen, Phys. Rev. D 75 (2007) 023509, arXiv:hep-ph/0608344.
[15] G.F. Giudice, A. Riotto, I. Tkachev, J. High Energy Phys. 9911 (1999) 036, arXiv:hep-ph/9911302.
[16] M. Kawasaki, K. Kohri, T. Moroi, Phys. Rev. D 71 (2005) 083502, arXiv:astro-ph/0408426.
[17] J. Pradler, F.D. Steffen, Phys. Lett. B 648 (2007) 224, arXiv:hep-ph/0612291.
[18] R. Rangarajan, N. Sahu, Mod. Phys. Lett. A 23 (2008) 427, arXiv:hep-ph/0606228.
[19] R. Rangarajan, N. Sahu, Phys. Rev. D 79 (2009) 103534, arXiv:0811.1866 [hep-ph].
[20] V.S. Rychkov, A. Strumia, Phys. Rev. D 75 (2007) 075011, arXiv:hep-ph/0701104.
[21] A.L. Maroto, A. Mazumdar, Phys. Rev. Lett. 84 (2000) 1655, arXiv:hep-ph/9904206.
[22] R. Kallosh, L. Kofman, A.D. Linde, A. Van Proeyen, Phys. Rev. D 61 (2000) 103503, arXiv:hep-th/9907124.
[23] S. Tsujikawa, B.A. Bassett, F. Viniegra, J. High Energy Phys. 0008 (2000) 019, arXiv:hep-ph/0006354.
[24] H.P. Nilles, M. Peloso, L. Sorbo, J. High Energy Phys. 0104 (2001) 004, arXiv:hep-th/0103202.
[25] H.P. Nilles, K.A. Olive, M. Peloso, Phys. Lett. B 522 (2001) 304, arXiv:hep-ph/0107212.
[26] P.B. Greene, K. Kadota, H. Murayama, Phys. Rev. D 68 (2003) 043502, arXiv:hep-ph/0208276.
[27] D.I. Podolsky, G.N. Felder, L. Kofman, M. Peloso, Phys. Rev. D 73 (2006) 023501, arXiv:hep-ph/0507096.
[28] H.P. Nilles, M. Peloso, L. Sorbo, Phys. Rev. Lett. 87 (2001) 051302, arXiv:hep-ph/0102264.
[29] H. Fujisaki, K. Kumekawa, M. Yoshimura, M. Yamaguchi, Phys. Rev. D 54 (1996) 2494, arXiv:hep-ph/9511381.
[30] M. Kawasaki, F. Takahashi, T.T. Yanagida, Phys. Lett. B 638 (2006) 8, arXiv:hep-ph/0603265.
[31] M. Kawasaki, F. Takahashi, T.T. Yanagida, AIP Conf. Proc. 903 (2007) 677, arXiv:hep-ph/0611166.
[32] M. Endo, F. Takahashi, T.T. Yanagida, Phys. Rev. D 76 (2007) 083509, arXiv:0706.0986 [hep-ph].
[33] K. Nakayama, F. Takahashi, T.T. Yanagida, Phys. Lett. B 718 (2012) 526.
[34] A.N. Taylor, A.R. Liddle, Phys. Rev. D 64 (2001) 023513, arXiv:astro-ph/0011365.
[35] S. Bartrum, A. Berera, J.G. Rosa, Phys. Rev. D 86 (2012) 123525, arXiv:1208.4276 [hep-ph].
[36] J.P. Derendinger, C. Lucchesi, Nucl. Phys. B 536 (1998) 483, arXiv:hep-ph/9807403.
[37] C. Lucchesi, arXiv:hep-ph/9808435.
[38] R.G. Leigh, R. Rattazzi, Phys. Lett. B 352 (1995) 20, arXiv:hep-ph/9503402.
[39] J.R. Ellis, D.V. Nanopoulos, K.A. Olive, S.J. Rey, Astropart. Phys. 4 (1996) 371, arXiv:hep-ph/9505438.
[40] W. Fischler, Phys. Lett. B 332 (1994) 277, arXiv:hep-th/9404044.
[41] N. Mahajan, R. Rangarajan, A. Sarkar, Phys. Rev. D 90 (2014) 023522, arXiv:1310.5872 [astro-ph.CO].
[42] R. Allahverdi, A. Mazumdar, arXiv:hep-ph/0505050.
[43] R. Allahverdi, A. Mazumdar, J. Cosmol. Astropart. Phys. 0610 (2006) 008, arXiv:hep-ph/0512227.
[44] R. Rangarajan, A. Sarkar, Astropart. Phys. 48 (2013) 37, arXiv:1205.5408 [astro-ph.CO].
[45] T. Moroi, Ph.D. Thesis, Tohoku University, 1995, arXiv:hep-ph/9503210.
[46] T. Gherghetta, Nucl. Phys. B 485 (1997) 25.
[47] S. Weinberg, Phys. Rev. Lett. 48 (1982) 1303.
[48] K.A. Olive, et al., Particle Data Group Collaboration, Chin. Phys. C 38 (2014) 090001, see Ch. 23.
[49] http://cmb-s4.org/CMB-S4workshops/index.php/File:CMB-S4_ScienceBook_1stEdition-160801.pdf (see p. 7).
[50] H. Pagels, J.R. Primack, Phys. Rev. Lett. 48 (1982) 223.
[51] R.H. Cyburt, J. Ellis, B.D. Fields, F. Luo, K.A. Olive, V.C. Spanos, J. Cosmol. Astropart. Phys. 0910 (2009) 021, arXiv:0907.5003 [astro-ph.CO].
[52] E.J. Chun, Phys. Rev. D 69 (2004) 117303, arXiv:hep-ph/0404029.
[53] L. Boubekeur, T. Hambye, G. Senjanovic, Phys. Rev. Lett. 93 (2004) 111601, arXiv:hep-ph/0404038.
[54] P.S.B. Dev, R.N. Mohapatra, Phys. Rev. D 92 (2015) 016007, arXiv:1504.07196 [hep-ph].
[55] P.S.B. Dev, Springer Proc. Phys. 174 (2016) 245, arXiv:1506.00837 [hep-ph].
[56] L. Knox, M.S. Turner, Phys. Rev. Lett. 70 (1993) 371.
[57] K. Kohri, A. Mazumdar, N. Sahu, Phys. Rev. D 80 (2009) 103504, arXiv:0905.1625 [hep-ph].
[58] D.G. Phillips II, et al., Phys. Rep. 612 (2016) 1, arXiv:1410.1100 [hep-ex].
[59] A.S. Majumdar, P. Das Gupta, R.P. Saxena, Int. J. Mod. Phys. D 4 (1995) 517.
[60] Y. Nagatani, Phys. Rev. D 59 (1999) 041301, arXiv:hep-ph/9811485.
[61] N. Upadhyay, P. Das Gupta, R.P. Saxena, Phys. Rev. D 60 (1999) 063513, arXiv:astro-ph/9903253.
[62] R. Rangarajan, S. Sengupta, A.M. Srivastava, Astropart. Phys. 17 (2002) 167, arXiv:hep-ph/9911488.

15

Improving the single scalar consistency relation

D.J. Brooker [a], N.C. Tsamis [b], R.P. Woodard [a,*]

[a] Department of Physics, University of Florida, Gainesville, FL 32611, United States
[b] Institute of Theoretical Physics & Computational Physics, Department of Physics, University of Crete, GR-710 03 Heraklion, Greece

ARTICLE INFO

Editor: M. Trodden

ABSTRACT

We propose a test of single-scalar inflation based on using the well-measured scalar power spectrum to reconstruct the tensor power spectrum, up to a single integration constant. Our test is a sort of integrated version of the single-scalar consistency relation. This sort of test can be used effectively, even when the tensor power spectrum is measured too poorly to resolve the tensor spectral index. We give an example using simulated data based on a hypothetical detection with tensor-to-scalar ratio $r = 0.01$. Our test can also be employed for correlating scalar and tensor features in the far future when the data is good.

1. Introduction

The theory of primordial inflation [1–8] has had a profound effect on cosmology and fundamental theory. Particularly striking is the prediction that primordial tensor [9] and scalar [10] perturbations derive from quantum gravitational fluctuations which fossilized near the end of inflation. This not only affords us access to quantum gravity at an intoxicating energy scale [11–13], it also provides information about the mechanism that powered inflation. This information can be accessed by comparing observations of the two power spectra, $\Delta^2_{\mathcal{R}}(k)$ and $\Delta^2_h(k)$, to predictions from the many models [14–16]. For example, the simplest models of inflation are driven by the potential of a single, minimally coupled scalar. These models all obey the single-scalar consistency relation [17–19],

$$r \approx -8n_t \; , \tag{1}$$

where r is the tensor-to-scalar ratio and n_t is the tensor spectral index,

$$r(k) \equiv \frac{\Delta^2_h(k)}{\Delta^2_{\mathcal{R}}(k)} \qquad , \qquad n_t(k) \equiv \frac{\partial \ln(\Delta^2_h(k))}{\partial \ln(k)} \; . \tag{2}$$

A statistically significant violation of (1) would falsify the entire class of single-scalar models, as well as all models which are related to them by conformal transformation, such as $f(R)$ inflation [20].

Although the single-scalar consistency relation was a brilliant theoretical insight, the progress of observation has rendered it somewhat inconvenient. The scalar power spectrum was first resolved in 1992 [21], and is now quite well measured [22–25]. The tensor power spectrum has not yet been resolved [26,27], but polarization measurements are now providing the strongest limits on it [28]. It is not known if the current generation of polarization experiments [29–33] can resolve the tensor power spectrum at all, and it is very unlikely that they will measure it well enough to constrain the tensor spectral index with much accuracy.

In view of the observational situation, it makes sense to develop a test of single-scalar inflation that is based primarily on the abundant data for $\Delta^2_{\mathcal{R}}(k)$, and does not require taking derivatives of the sparse data for $\Delta^2_h(k)$ likely to result from the first positive detections. There is no reason not to do this because the close relation between the tensor and scalar mode functions of single-scalar inflation implies that either power spectrum determines the other, up to some integration constants. That is the purpose of this paper. In the next section we fix notation, recall the relation between the two power spectra, and infer the tensor power spectrum from the scalar one. Section 3 gives a comparison between the single scalar consistency relation and the scatter test we propose, using simulated data based on a hypothetical detection of $r = 0.01$ with the same number of data points and the same fractional error as was in fact reported by the recent spurious BICEP2 detection [34]. The final section mentions applications.

* Corresponding author.
E-mail addresses: djbrooker@ufl.edu (D.J. Brooker), tsamis@physics.uoc.gr (N.C. Tsamis), woodard@phys.ufl.edu (R.P. Woodard).

2. Constructing $\Delta^2_h(k)$ from $\Delta^2_{\mathcal{R}}(k)$

We work in spatially flat, co-moving coordinates with scale factor $a(t)$, Hubble parameter $H(t)$ and first slow roll parameter $\epsilon(t)$,

$$ds^2 = -dt^2 + a^2(t) d\vec{x}\cdot d\vec{x} \quad \Longrightarrow \quad H(t) \equiv \frac{\dot{a}}{a} \quad , \quad \epsilon(t) \equiv -\frac{\dot{H}}{H^2} \, . \tag{3}$$

We assume $a(t)$ is known, with the scalar background and potential determined to enforce the background Einstein equations [35–39],

$$\varphi_0(t) = \varphi_0(t_i) \pm \int_{t_i}^{t} dt' H(t') \sqrt{\frac{\epsilon(t')}{4\pi G}} \quad \Longleftrightarrow \quad t(\varphi) \, , \tag{4}$$

$$V(\varphi) = \frac{[3-\epsilon(t)]H^2(t)}{8\pi G}\Bigg|_{t=t(\varphi)} \, . \tag{5}$$

We fix the gauge so that the full scalar agrees with its background value and the graviton field h_{ij} is transverse, with g_{00} and g_{0i} regarded as constraints. The two dynamical fields are h_{ij} and ζ, which reside in the 3-metric $g_{ij} = a^2 e^{2\zeta}[e^h]_{ij}$. At quadratic order their Lagrangian is [40],

$$\mathcal{L}_2 = \frac{a^3}{64\pi G}\Big[\dot{h}_{ij}\dot{h}_{ij} - \frac{h_{ij,k}h_{ij,k}}{a^2}\Big] + \frac{\epsilon a^3}{8\pi G}\Big[\dot{\zeta}^2 - \frac{\zeta_{,k}\zeta_{,k}}{a^2}\Big] \, . \tag{6}$$

The spatial plane wave mode functions of the graviton are $u(t,k)$, with exactly the same polarization tensors as in flat space. From (6) we see that the evolution equation, Wronskian and asymptotically early form of the tensor mode functions $u(t,k)$ are,

$$\ddot{u} + 3H\dot{u} + \frac{k^2}{a^2}u = 0 \, , \; u\dot{u}^* - \dot{u}u^* = \frac{i}{a^3} \, ,$$

$$u(t,k) \longrightarrow \frac{\exp[-ik\int_{t_i}^{t}\frac{dt'}{a(t')}]}{\sqrt{2ka^2(t)}} \, . \tag{7}$$

The scalar perturbation ζ has spatial plane wave mode functions $v(t,k)$. From (6) we see that their evolution equation, Wronskian and asymptotically early form are,

$$\ddot{v} + \Big(3H + \frac{\dot{\epsilon}}{\epsilon}\Big)\dot{v} + \frac{k^2}{a^2}v = 0 \, , \; v\dot{v}^* - \dot{v}v^* = \frac{i}{\epsilon a^3} \, ,$$

$$v(t,k) \longrightarrow \frac{\exp[-ik\int_{t_i}^{t}\frac{dt'}{a(t')}]}{\sqrt{2k\epsilon(t)a^2(t)}} \, . \tag{8}$$

The two power spectra are determined (at tree order) by evolving their respective mode functions from their early forms through the time t_k of first horizon crossing ($k \equiv H(t_k)a(t_k)$), after which they approach constants,

$$\Delta^2_{\mathcal{R}}(k) = \frac{k^3}{2\pi^2} \times 4\pi G \times \Big|v(t,k)\Big|^2_{t\gg t_k} \approx \frac{GH^2(t_k)}{\pi\epsilon(t_k)} \, , \tag{9}$$

$$\Delta^2_h(k) = \frac{k^3}{2\pi^2} \times 32\pi G \times 2 \times \Big|u(t,k)\Big|^2_{t\gg t_k} \approx \frac{16GH^2(t_k)}{\pi} \, . \tag{10}$$

The relations (7) which define $u(t,k)$ are carried into the relations (8) which define $v(t,k)$ by making simultaneous changes in the scale factor and the co-moving time [41,42],

$$a(t) \longrightarrow \sqrt{\epsilon(t)}\, a(t) \quad , \quad \frac{\partial}{\partial t} \longrightarrow \frac{1}{\sqrt{\epsilon(t)}}\frac{\partial}{\partial t} \, . \tag{11}$$

To understand what this means for the power spectra we must consider them as nonlocal functionals of the expansion history $a(t)$, which will involve integrals and derivatives with respect to time. We denote this functional dependence with square brackets, so relation (11) implies,

$$\Delta^2_{\mathcal{R}}\Big[a, dt\Big](k) = \frac{1}{16}\Delta^2_h\Big[\sqrt{\epsilon}a, \sqrt{\epsilon}dt\Big](k) \, . \tag{12}$$

Relation (12) is easy to check at leading slow roll order by comparing the slow roll approximation for $\Delta^2_{\mathcal{R}}(k)$ on the right hand side of (9) with the effect of making transformation (11) on the Hubble parameter in the right hand side of expression (10),

$$H(t) \equiv \frac{\partial}{\partial t}\ln[a(t)] \longrightarrow \frac{1}{\sqrt{\epsilon}}\frac{\partial}{\partial t}\ln\Big[\sqrt{\epsilon}\, a\Big] = \frac{H + \frac{\dot{\epsilon}}{2\epsilon}}{\sqrt{\epsilon}} \, . \tag{13}$$

However, we stress that relation (12) is exact, not just valid at leading slow roll order, provided one employs the exact expressions for $\Delta^2_h(k)$ and $\Delta^2_{\mathcal{R}}(k)$.

We should also point out that very accurate functional expressions are now available for the power spectra of single scalar inflation, valid to all orders in the slow roll parameter $\epsilon(t_k)$, and even including nonlocal effects from times before and after first horizon crossing [43,44]. These expressions take the form [45],

$$\Delta^2_{\mathcal{R}}(k) \simeq \frac{GH^2(t_k)}{\pi\epsilon(t_k)} \times C\Big(\epsilon(t_k)\Big) \times \exp\Big[\sigma[\epsilon](k)\Big] \, , \tag{14}$$

$$\Delta^2_h(k) \simeq \frac{16GH^2(t_k)}{\pi} \times C\Big(\epsilon(t_k)\Big) \times \exp\Big[\tau[\epsilon](k)\Big] \, , \tag{15}$$

where the local slow roll correction factor is,

$$C(\epsilon) \equiv \frac{1}{\pi}\Gamma^2\Big(\frac{1}{2} + \frac{1}{1-\epsilon}\Big)\Big[2(1-\epsilon)\Big]^{\frac{2}{1-\epsilon}} \approx 1 - \epsilon \, . \tag{16}$$

For the nonlocal corrections $\sigma[\epsilon](k)$ and $\tau[\epsilon](k)$ it is best to abuse the notation by writing the first slow parameter $\epsilon(n) \equiv \epsilon(t(n))$ as a function of $n \equiv \ln[a(t)/a_i]$, the number of e-foldings since the start of inflation,

$$\begin{aligned} \sigma[\epsilon](k) &= \int_0^{n_k} dn \Big[\partial_n^2 \ln[\epsilon(n)] + \frac{1}{2}\Big(\partial_n \ln[\epsilon(n)]\Big)^2 + 3\partial_n \ln[\epsilon(n)]\Big] G\Big(e^{\Delta n}\Big) \\ &\quad - \partial_{n_k}\ln[\epsilon(n_k)]\, G(1) + \int_{n_k}^{\infty} dn\, \partial_n \ln[\epsilon(n)] \frac{2G(e^{\Delta n})}{1+e^{2\Delta n}} \, , \end{aligned} \tag{17}$$

$$\begin{aligned} \tau[\epsilon](k) &= \int_0^{n_k} dn \Big[\mathcal{E}_1\Big(e^{\Delta n}\Big)\epsilon''(n) + \mathcal{E}_2\Big(e^{\Delta n}\Big)\Big(\epsilon'(n)\Big)^2 \\ &\quad + \mathcal{E}_3\Big(e^{\Delta n}\Big)\epsilon'(n)\Big] G\Big(e^{\Delta n}\Big) - \epsilon'(n_k)\mathcal{E}_1(1)G(1) \\ &\quad - \int_{n_k}^{\infty} dn \Big\{\Delta\epsilon(n) + \Big(\frac{4+2e^{2\Delta n}}{1+e^{2\Delta n}}\Big)\int_{n_k}^{n} dm\, \Delta\epsilon(m)\Big\} \frac{2G(e^{\Delta n})}{1+e^{2\Delta n}} \, . \end{aligned} \tag{18}$$

Here $\Delta n \equiv n - n_k$, $\Delta\epsilon(m) \equiv \epsilon(m) - \epsilon_k$, and the functions of $x \equiv e^{\Delta n}$ are,

$$G(x) = \frac{1}{2}\Big(x + x^3\Big)\sin\Big[\frac{2}{x} - 2\arctan\Big(\frac{1}{x}\Big)\Big] \, , \tag{19}$$

$$\mathcal{E}_1(x) \simeq= \frac{\frac{1}{2}x^2 - 1.8x^4 + 1.5x^6 - 0.63x^8}{1+x^2} \, , \tag{20}$$

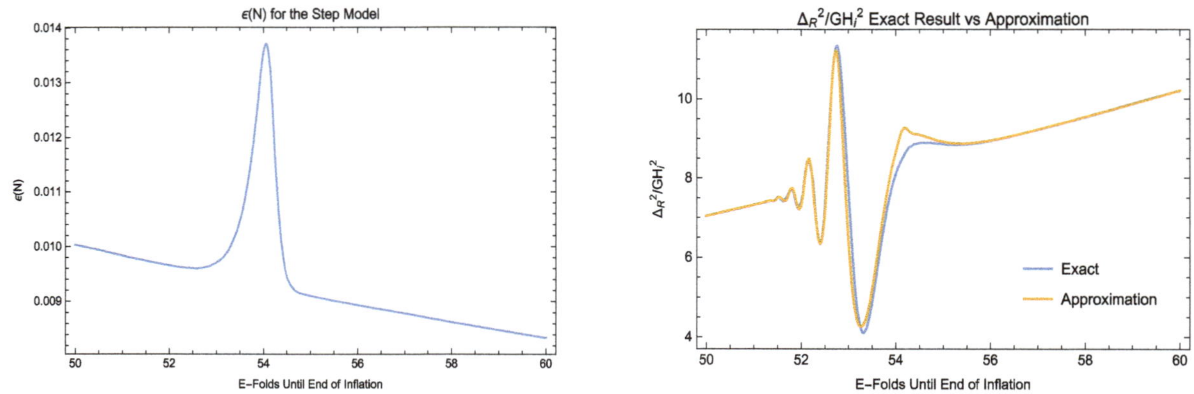

Fig. 1. The left hand figure shows the first slow roll parameter for a model which was proposed [46,47] to explain the observed features in the scalar power spectrum at $\ell \approx 22$ and $\ell \approx 40$ which are visible in the data reported from both WMAP [48,49] and PLANCK [50,51]. The right hand figure shows the resulting scalar power spectrum (in blue), with the result of our analytic approximation (14) (in yellow). The slow roll approximation (9) does not give a very accurate fit even to the main feature in the range $54.5 < N < 53$ e-foldings before the end of inflation, and it completely misses the secondary oscillations visible in the range $53.5 < N < 51.5$. The nonlocal contributions (17) are essential for correctly reproducing these features. (For interpretation of the references to color in this figure legend, the reader is referred to the web version of this article.)

$$\mathcal{E}_2(x) \simeq \frac{2.8x^4 - 7x^6 + 3x^8 + 1.8x^{10} - 2.3x^{12} + 0.95x^{14} - 0.20x^{16}}{(1+x^2)^2}, \tag{21}$$

$$\mathcal{E}_3(x) \simeq \frac{\frac{9}{2}x^2 - 11.9x^4 + 7.1x^6 - 1.3x^8 - 1.9x^{10}}{(1+x^2)^2}. \tag{22}$$

The 95% confidence bound on the tensor-to-scalar ratio of $r < 0.12$ [27,28] implies $\epsilon < 0.0075$, so $\tau[\epsilon](k)$ is about a hundred times smaller than $\sigma[\epsilon](k)$. Models with smooth potentials typically have $\epsilon' \sim \epsilon^2$ and $\epsilon'' \sim \epsilon^3$, so the leading contributions in $\sigma[\epsilon](k)$ come from the 3rd and 5th terms of expression (17). In particular the 5th (final) term is needed to correct for a systematic under-prediction of the local slow roll approximation [45]. For models with features the leading contributions to $\sigma[\epsilon](k)$ come from the 1st, 3rd and 4th terms of expression (17) [45]. These corrections can be very important for realistic models such as the one depicted in Fig. 1.

To keep the analysis simple, we illustrate the procedure for predicting $\Delta_h^2(k)$ from $\Delta_{\mathcal{R}}^2(k)$ using only the leading slow roll terms in expressions (14)–(15), without either of the nonlocal corrections or even the slow roll factor $C(\epsilon)$. The conversion from wave number to time is,

$$k = H(t_k)a(t_k) \qquad \Longrightarrow \qquad \frac{dk}{k} = (1-\epsilon)Hdt \approx Hdt\,. \tag{23}$$

The leading slow roll approximation (14) for the scalar power spectrum can be recognized as a differential equation for the Hubble parameter,

$$\Delta_{\mathcal{R}}^2(k) \simeq \frac{GH^2(t_k)}{\pi\epsilon(t_k)} = -\frac{GH^4(t_k)}{\pi\dot{H}(t_k)}\,. \tag{24}$$

We can integrate this equation from some arbitrary time t_* to t_k,

$$d\Big(\frac{1}{H^2}\Big) \simeq \frac{2G\,d\ln(k)}{\pi\Delta_{\mathcal{R}}^2(k)}$$

$$\Longrightarrow \quad \frac{1}{H^2(t_k)} - \frac{1}{H^2(t_*)} \simeq \frac{2G}{\pi}\int_0^{\ln(k/k_*)} \frac{d\ln(k')}{\Delta_{\mathcal{R}}^2(k')}\,. \tag{25}$$

Substituting the reconstructed Hubble parameter (25) into the leading slow roll approximation (10) for the tensor power spectrum gives,

$$\Delta_h^2(k) \simeq \frac{16GH^2(t_k)}{\pi}$$

$$\simeq \Delta_h^2(k_*)\Bigg[1 + \frac{r(k_*)}{8}\int_0^{\ln(k/k_*)} d\ln(k')\frac{\Delta_{\mathcal{R}}^2(k_*)}{\Delta_{\mathcal{R}}^2(k')}\Bigg]^{-1}. \tag{26}$$

Equation (26) is in some sense an integrated form of the single-scalar consistency relation (1) which can be applied more reliably. Both relations are valid to leading slow roll order, but whereas (1) compares a single value of the high quality data in $\Delta_{\mathcal{R}}^2(k)$ with a derivative of the poor data on $\Delta_h^2(k)$, our relation (26) combines a single measurement of the tensor power spectrum at $k = k_*$ with the high quality scalar data to predict what $\Delta_h^2(k)$ should be for other wave numbers. This seems to be a better way of exploiting the sparse data on $\Delta_h^2(k)$ which is likely to persist for some years after a first positive detection.

3. Comparison using simulated data

It is illuminating to compare the single scalar consistency relation with the method we propose using simulated data. Let us suppose that the actual tensor power spectrum corresponds to single scalar inflation with $r = \frac{1}{100}$, and which implies $n_t = -\frac{1}{800}$. We further suppose the simplest possible k dependence,

$$\Delta_h^2(k) = rA_S\Big(\frac{k}{k_0}\Big)^{n_t}$$

$$\Longrightarrow \quad \ln\Big[\Delta_h^2(k)\Big] = \ln\Big[rA_S\Big] + n_t \times \ln\Big[\frac{k}{k_0}\Big], \tag{27}$$

where the scalar amplitude (at the tensor pivot k_0) is $A_S = 2.5 \times 10^{-9}$. Let us assume that the first positive detection of this tensor power spectrum consists of results for five binned wave numbers, the same as was in fact reported for the spurious BICEP2 detection [34]. To simplify matters we assume a linear relation for logarithms of the observed wave numbers, $\ln[k_{i+1}/k_i] = \frac{1}{3}$, and that each measurement of $\ln[\Delta_h^2]$ has the same 1-sigma uncertainty of $\sigma = \frac{1}{4}$. These numbers are roughly consistent with what BICEP2 actually reported [34]. Hence the detection consists of five observations y_i obeying the relation,

Table 1
Simulated data from relation (28), representing a hypothetical first detection of the tensor power spectrum with $r = \frac{1}{100}$ and $n_t = -\frac{1}{800}$. The random errors e_i follow a normal distribution with mean zero and standard deviation $\sigma = \frac{1}{4}$.

i	$\ln(2.5 \times 10^{-11})$	$-\frac{i}{2400}$	e_i	y_i
1	−24.412145	−0.000417	+0.226742	−24.185820
2	−24.412145	−0.000833	−0.176041	−24.589020
3	−24.412145	−0.001250	−0.091555	−24.504950
4	−24.412145	−0.001667	−0.164330	−24.578142
5	−24.412145	−0.002083	+0.331640	−24.082589

$$y_i = \ln\Big[2.5 \times 10^{-11}\Big] - \frac{i}{2400} + e_i \quad , \quad i \in \Big\{1, 2, 3, 4, 5\Big\} , \tag{28}$$

where the e_i are independent Gaussian random variables with mean zero and standard deviation $\sigma = \frac{1}{4}$. Table 1 simulates the five data points using a random number generator to find the e_i.

Because the relation (27) is linear we can use least squares to determine the parameters. The least squares fit for N data points obeying the relation $y_i = \alpha + \beta x_i$ (with $x_i = i/3$) is,

$$\alpha = \frac{\sum_{i=1}^N x_i^2 \sum_{j=1}^N y_j - \sum_{i=1}^N x_i \sum_{j=1}^N x_j y_j}{N \sum_{i=1}^N x_i^2 - (\sum_{i=1}^N x_i)^2} = \frac{\sum_{i=1}^N \sum_{j=1}^N x_i (x_i - x_j) y_j}{\sum_{i=2}^N \sum_{j=1}^{i-1} (x_i - x_j)^2} , \tag{29}$$

$$\beta = \frac{N \sum_{i=1}^N x_i y_i - \sum_{i=1}^N x_i \sum_{j=1}^N y_j}{N \sum_{i=1}^N x_i^2 - (\sum_{i=1}^N x_i)^2} = \frac{\sum_{i=2}^N \sum_{j=1}^{i-1} (x_i - x_j)(y_i - y_j)}{\sum_{i=2}^N \sum_{j=1}^{i-1} (x_i - x_j)^2} . \tag{30}$$

Even in this general form it is obvious that expression (29) for α represents a sort of average whereas expression (30) is a kind of numerical derivative. So we expect the fractional error on β to be larger than that on α. That becomes even more apparent when specializing to $N = 5$ and $x_i = i/3$,

$$\alpha \longrightarrow \frac{(8y_1 + 5y_2 + 2y_3 - y_4 - 4y_5)}{10} \simeq -24.453306 \pm 0.262202 , \tag{31}$$

$$\beta \longrightarrow \frac{(-6y_1 - 3y_2 + 3y_4 + 6y_5)}{10} \simeq +0.065202 \pm 0.237171 . \tag{32}$$

Hence the simulated data of Table 1 implies a reasonably accurate reconstruction of the tensor-to-scalar ratio,

$$r = \exp\Big[\alpha - \ln\Big(2.5 \times 10^{-9}\Big)\Big] = 0.0096 \pm 0.0027 , \tag{33}$$

but a miserably inaccurate value for the tensor spectral index,

$$n_t = \beta = 0.065 \pm 0.237 . \tag{34}$$

The resulting check of the single scalar consistency relation is not very sensitive,

$$0.010 \pm 0.003 = -0.522 \pm 1.897 . \tag{35}$$

Because of the large (but statistically allowed) positive fluctuation e_5 the measured tensor spectral index (34) does not even have the right sign!

We propose to instead use the much better measured scalar spectral index to predict the tensor spectral index, up to an integration constant, and then to compare the fluctuation of the observed data around this prediction. For the model in question this might amount to assuming predictions of the form,

$$z_i = \alpha - \frac{r}{24} \times i , \tag{36}$$

where α is (31) and r is (33). Table 2 reports these predictions, along with the difference between each simulated observation y_i and the associated prediction z_i. Of course the parameter r comes from the parameter α through relation (33), so the final column of Table 2 represents four statistically independent measurements. The resulting estimate for the scatter between measurement and prediction is,

Table 2
Predicted results according to relation (36), with the parameters α and r taken from expressions (31) and (33), respectively.

i	α	$-\frac{r}{24} \times i$	z_i	$y_i - z_i$
1	−24.453306	−0.000400	−24.453706	+0.267886
2	−24.453306	−0.000800	−24.454106	−0.134914
3	−24.453306	−0.001200	−24.454506	−0.050445
4	−24.453306	−0.001599	−24.454906	−0.123236
5	−24.453306	−0.001999	−24.455306	+0.372717

$$\sqrt{\frac{1}{4} \sum_{i=1}^5 (y_i - z_i)^2} \simeq 0.246614 . \tag{37}$$

This is quite consistent with our assumed 1-sigma fluctuation of $\sigma = \frac{1}{4}$ for each observation.

4. Discussion

Resolving the tensor power spectrum $\Delta_h^2(k)$ is crucial for the progress of inflation because it constrains the causative mechanism. This is already evident from the angst [52–55] elicited by the increasingly tight bounds on the tensor-to-scalar ratio r [56]. A positive detection at several different wave lengths has the potential to falsify entire classes of models. For example, any model in which inflation is driven by the potential of a minimally coupled scalar must obey relation (1) between r and the tensor spectral index n_t [17–19]. Unfortunately, relation (1) requires taking a derivative of $\Delta_h^2(k)$, and the first generation of detections will probably be too sparse to provide a good bound because numerical differentiation makes bad data worse.

It makes more sense to integrate the high quality data we already possess for $\Delta_{\mathcal{R}}^2(k)$. If the leading slow roll expressions (9)–(10) are assumed then the prediction (26) from $\Delta_{\mathcal{R}}^2(k)$ requires only a single integration constant from $\Delta_h^2(k)$. (The same thing would be true even if the more accurate approximations (14)–(15) were employed [45].) Fixing this constant uses up one combination of whatever data we have for $\Delta_h^2(k)$, leaving the scatter of the remaining data about the prediction as a legitimate test of single scalar inflation. Hence relation (26) is a sort of integrated form of the single-scalar consistency relation (1) which can be applied more reliably. Section 3 compares this sort of scatter test with checking $r = -8n_t$ for simulated data based on a hypothetical detection of $r = 0.01$ at five wave lengths with fractional errors similar to those reported in the spurious BICEP2 detection [34]. Of course no massaging of poorly resolved data is going to extract a precision bound, but the scatter test seems clearly better.

Note that it is simple to adapt the scatter test to data fits. For example, the usual parameterization of the scalar data [22–25] implies,

$$\Delta^2_{\mathcal{R}}(k) \simeq A_s\Big(\frac{k}{k_0}\Big)^{n_s-1}$$
$$\Longrightarrow \quad \Delta^2_h(k) \simeq \Delta^2_h(k_*)\left[1+\frac{r(k_*)}{8(1-n_s)}\Big[\Big(\frac{k}{k_*}\Big)^{1-n_s}-1\Big]\right]^{-1} . \tag{38}$$

Here A_s is the scalar amplitude, n_s is the scalar spectral index, and k_0 is a fiducial wave number.

Finally, we can look forward to the day, in the far future, when the tensor power spectrum is well resolved. Then the sort of scatter test we propose could be employed to search for correlations between features in the two power spectra. For example, Fig. 1 depicts the bump in the first slow roll parameter from a model [46,47] introduced to explain the scalar power spectrum's dip at $\ell \approx 22$ and peak at $\ell \approx 40$ [48–51]. These features are caused by the way the scalar nonlocal corrections (17) depend upon derivatives of $\epsilon(n)$. The tensor nonlocal corrections (18) involve the same derivatives — although lacking the large factors of $1/\epsilon$ — so it is obvious there will be corresponding features [45]. Resolving this sort of correlation probes the functional relation between the two power spectra far more deeply than the single scalar consistency relation.

Acknowledgements

This work was partially supported by the European Union's Seventh Framework Programme (FP7-REGPOT-2012-2013-1) under grant agreement number 316165; by the European Union's Horizon 2020 Programme under grant agreement 669288-SM-GRAV-ERC-2014-ADG; by a travel grant from the University of Florida International Center, College of Liberal Arts and Sciences, Graduate School and Office of the Provost; by NSF grant PHY-1506513; and by the Institute for Fundamental Theory at the University of Florida. One of us (NCT) would like to thank AEI at the University of Bern for its hospitality while this work was partially completed.

References

[1] R. Brout, F. Englert, E. Gunzig, Ann. Phys. 115 (1978) 78, http://dx.doi.org/10.1016/0003-4916(78)90176-8.
[2] A.A. Starobinsky, Phys. Lett. B 91 (1980) 99, http://dx.doi.org/10.1016/0370-2693(80)90670-X.
[3] D. Kazanas, Astrophys. J. 241 (1980) L59, http://dx.doi.org/10.1086/183361.
[4] K. Sato, Mon. Not. R. Astron. Soc. 195 (1981) 467.
[5] A.H. Guth, Phys. Rev. D 23 (1981) 347, http://dx.doi.org/10.1103/PhysRevD.23.347.
[6] A.D. Linde, Phys. Lett. B 108 (1982) 389, http://dx.doi.org/10.1016/0370-2693(82)91219-9.
[7] A. Albrecht, P.J. Steinhardt, Phys. Rev. Lett. 48 (1982) 1220, http://dx.doi.org/10.1103/PhysRevLett.48.1220.
[8] A.D. Linde, Phys. Lett. B 129 (1983) 177, http://dx.doi.org/10.1016/0370-2693(83)90837-7.
[9] A.A. Starobinsky, JETP Lett. 30 (1979) 682, Pis'ma Zh. Eksp. Teor. Fiz. 30 (1979) 719.
[10] V.F. Mukhanov, G.V. Chibisov, JETP Lett. 33 (1981) 532, Pis'ma Zh. Eksp. Teor. Fiz. 33 (1981) 549.
[11] R.P. Woodard, Rep. Prog. Phys. 72 (2009) 126002, http://dx.doi.org/10.1088/0034-4885/72/12/126002, arXiv:0907.4238 [gr-qc].
[12] A. Ashoorioon, P.S. Bhupal Dev, A. Mazumdar, Mod. Phys. Lett. A 29 (30) (2014) 1450163, http://dx.doi.org/10.1142/S0217732314501636, arXiv:1211.4678 [hep-th].
[13] L.M. Krauss, F. Wilczek, Phys. Rev. D 89 (4) (2014) 047501, http://dx.doi.org/10.1103/PhysRevD.89.047501, arXiv:1309.5343 [hep-th].
[14] V.F. Mukhanov, H.A. Feldman, R.H. Brandenberger, Phys. Rep. 215 (1992) 203, http://dx.doi.org/10.1016/0370-1573(92)90044-Z.
[15] A.R. Liddle, D.H. Lyth, Phys. Rep. 231 (1993) 1, http://dx.doi.org/10.1016/0370-1573(93)90114-S, arXiv:astro-ph/9303019.
[16] J.E. Lidsey, A.R. Liddle, E.W. Kolb, E.J. Copeland, T. Barreiro, M. Abney, Rev. Mod. Phys. 69 (1997) 373, http://dx.doi.org/10.1103/RevModPhys.69.373, arXiv:astro-ph/9508078.
[17] D. Polarski, A.A. Starobinsky, Phys. Lett. B 356 (1995) 196, http://dx.doi.org/10.1016/0370-2693(95)00842-9, arXiv:astro-ph/9505125.
[18] J. Garcia-Bellido, D. Wands, Phys. Rev. D 52 (1995) 6739, http://dx.doi.org/10.1103/PhysRevD.52.6739, arXiv:gr-qc/9506050.
[19] M. Sasaki, E.D. Stewart, Prog. Theor. Phys. 95 (1996) 71, http://dx.doi.org/10.1143/PTP.95.71, arXiv:astro-ph/9507001.
[20] D.J. Brooker, S.D. Odintsov, R.P. Woodard, Nucl. Phys. B 911 (2016) 318, http://dx.doi.org/10.1016/j.nuclphysb.2016.08.010, arXiv:1606.05879 [gr-qc].
[21] G.F. Smoot, et al., Astrophys. J. 396 (1992) L1, http://dx.doi.org/10.1086/186504.
[22] G. Hinshaw, et al., WMAP Collaboration, Astrophys. J. Suppl. 208 (2013) 19, http://dx.doi.org/10.1088/0067-0049/208/2/19, arXiv:1212.5226 [astro-ph.CO].
[23] Z. Hou, et al., Astrophys. J. 782 (2014) 74, http://dx.doi.org/10.1088/0004-637X/782/2/74, arXiv:1212.6267 [astro-ph.CO].
[24] J.L. Sievers, et al., Atacama Cosmology Telescope Collaboration, J. Cosmol. Astropart. Phys. 1310 (2013) 060, http://dx.doi.org/10.1088/1475-7516/2013/10/060, arXiv:1301.0824 [astro-ph.CO].
[25] P.A.R. Ade, et al., Planck Collaboration, Astron. Astrophys. 571 (2014) A16, http://dx.doi.org/10.1051/0004-6361/201321591, arXiv:1303.5076 [astro-ph.CO].
[26] R. Adam, et al., Planck Collaboration, Astron. Astrophys. 586 (2016) A133, http://dx.doi.org/10.1051/0004-6361/201425034, arXiv:1409.5738 [astro-ph.CO].
[27] P.A.R. Ade, et al., BICEP2 Collaboration, Planck Collaboration, Phys. Rev. Lett. 114 (2015) 101301, http://dx.doi.org/10.1103/PhysRevLett.114.101301, arXiv:1502.00612 [astro-ph.CO].
[28] P.A.R. Ade, et al., Planck Collaboration, arXiv:1502.01589 [astro-ph.CO].
[29] K. Hattori, et al., Nucl. Instrum. Methods A 732 (2013) 299, http://dx.doi.org/10.1016/j.nima.2013.07.052, arXiv:1306.1869 [astro-ph.IM].
[30] J. Lazear, et al., Proc. SPIE Int. Soc. Opt. Eng. 9153 (2014) 91531L, http://dx.doi.org/10.1117/12.2056806, arXiv:1407.2584 [astro-ph.IM].
[31] A.S. Rahlin, et al., Proc. SPIE Int. Soc. Opt. Eng. 9153 (2014) 915313, http://dx.doi.org/10.1117/12.2055683, arXiv:1407.2906 [astro-ph.IM].
[32] Z. Ahmed, et al., BICEP3 Collaboration, Proc. SPIE Int. Soc. Opt. Eng. 9153 (2014) 91531N, http://dx.doi.org/10.1117/12.2057224, arXiv:1407.5928 [astro-ph.IM].
[33] K. MacDermid, et al., Proc. SPIE Int. Soc. Opt. Eng. 9153 (2014) 915311, http://dx.doi.org/10.1117/12.2056267, arXiv:1407.6894 [astro-ph.IM].
[34] P.A.R. Ade, et al., BICEP2 Collaboration, Phys. Rev. Lett. 112 (24) (2014) 241101, http://dx.doi.org/10.1103/PhysRevLett.112.241101, arXiv:1403.3985 [astro-ph.CO].
[35] N.C. Tsamis, R.P. Woodard, Ann. Phys. 267 (1998) 145, http://dx.doi.org/10.1006/aphy.1998.5816, arXiv:hep-ph/9712331.
[36] T.D. Saini, S. Raychaudhury, V. Sahni, A.A. Starobinsky, Phys. Rev. Lett. 85 (2000) 1162, http://dx.doi.org/10.1103/PhysRevLett.85.1162, arXiv:astro-ph/9910231.
[37] S. Capozziello, S. Nojiri, S.D. Odintsov, Phys. Lett. B 634 (2006) 93, http://dx.doi.org/10.1016/j.physletb.2006.01.065, arXiv:hep-th/0512118.
[38] R.P. Woodard, Lect. Notes Phys. 720 (2007) 403, http://dx.doi.org/10.1007/978-3-540-71013-4_14, arXiv:astro-ph/0601672.
[39] Z.K. Guo, N. Ohta, Y.Z. Zhang, Mod. Phys. Lett. A 22 (2007) 883, http://dx.doi.org/10.1142/S0217732307022839, arXiv:astro-ph/0603109.
[40] R.P. Woodard, Int. J. Mod. Phys. D 23 (09) (2014) 1430020, http://dx.doi.org/10.1142/S0218271814300201, arXiv:1407.4748 [gr-qc].
[41] N.C. Tsamis, R.P. Woodard, Class. Quantum Gravity 21 (2003) 93, http://dx.doi.org/10.1088/0264-9381/21/1/007, arXiv:astro-ph/0306602.
[42] M.G. Romania, N.C. Tsamis, R.P. Woodard, J. Cosmol. Astropart. Phys. 1208 (2012) 029, http://dx.doi.org/10.1088/1475-7516/2012/08/029, arXiv:1207.3227 [astro-ph.CO].
[43] D.J. Brooker, N.C. Tsamis, R.P. Woodard, Phys. Rev. D 93 (4) (2016) 043503, http://dx.doi.org/10.1103/PhysRevD.93.043503, arXiv:1507.07452 [astro-ph.CO].
[44] D.J. Brooker, N.C. Tsamis, R.P. Woodard, Phys. Rev. D 94 (4) (2016) 044020, http://dx.doi.org/10.1103/PhysRevD.94.044020, arXiv:1605.02729 [gr-qc].
[45] D.J. Brooker, N.C. Tsamis, R.P. Woodard, arXiv:1708.03253 [gr-qc].
[46] J.A. Adams, B. Cresswell, R. Easther, Phys. Rev. D 64 (2001) 123514, http://dx.doi.org/10.1103/PhysRevD.64.123514, arXiv:astro-ph/0102236.
[47] M.J. Mortonson, C. Dvorkin, H.V. Peiris, W. Hu, Phys. Rev. D 79 (2009) 103519, http://dx.doi.org/10.1103/PhysRevD.79.103519, arXiv:0903.4920 [astro-ph.CO].
[48] L. Covi, J. Hamann, A. Melchiorri, A. Slosar, I. Sorbera, Phys. Rev. D 74 (2006) 083509, http://dx.doi.org/10.1103/PhysRevD.74.083509, arXiv:astro-ph/0606452.
[49] J. Hamann, L. Covi, A. Melchiorri, A. Slosar, Phys. Rev. D 76 (2007) 023503, http://dx.doi.org/10.1103/PhysRevD.76.023503, arXiv:astro-ph/0701380.
[50] D.K. Hazra, A. Shafieloo, G.F. Smoot, A.A. Starobinsky, J. Cosmol. Astropart. Phys. 1408 (2014) 048, http://dx.doi.org/10.1088/1475-7516/2014/08/048, arXiv:1405.2012 [astro-ph.CO].
[51] D.K. Hazra, A. Shafieloo, G.F. Smoot, A.A. Starobinsky, J. Cosmol. Astropart. Phys. 1609 (09) (2016) 009, http://dx.doi.org/10.1088/1475-7516/2016/09/009, arXiv:1605.02106 [astro-ph.CO].

[52] A. Ijjas, P.J. Steinhardt, A. Loeb, Phys. Lett. B 723 (2013) 261, http://dx.doi.org/10.1016/j.physletb.2013.05.023, arXiv:1304.2785 [astro-ph.CO].
[53] A.H. Guth, D.I. Kaiser, Y. Nomura, Phys. Lett. B 733 (2014) 112, http://dx.doi.org/10.1016/j.physletb.2014.03.020, arXiv:1312.7619 [astro-ph.CO].
[54] A. Linde, http://dx.doi.org/10.1093/acprof:oso/9780198728856.003.0006, arXiv:1402.0526 [hep-th].
[55] A. Ijjas, P.J. Steinhardt, A. Loeb, Phys. Lett. B 736 (2014) 142, http://dx.doi.org/10.1016/j.physletb.2014.07.012, arXiv:1402.6980 [astro-ph.CO].
[56] P.A.R. Ade, et al., Planck Collaboration, arXiv:1502.02114 [astro-ph.CO].

16

Inception of self-interacting dark matter with dark charge conjugation symmetry

Ernest Ma

Physics and Astronomy Department, University of California, Riverside, CA 92521, USA

ARTICLE INFO

Editor: A. Ringwald

ABSTRACT

A new understanding of the stability of self-interacting dark matter is pointed out, based on the simplest spontaneously broken Abelian $U(1)$ gauge model with one complex scalar and one Dirac fermion. The key is the imposition of dark charge conjugation symmetry. It allows the possible existence of two stable particles: the Dirac fermion and the vector gauge boson which acts as a light mediator for the former's self-interaction. Since this light mediator does not decay, it avoids the strong cosmological constraints recently obtained for all such models where the light mediator decays into standard-model particles.

1. Introduction

The Lagrangian of the simplest spontaneously broken Abelian $U(1)$ gauge model was written down by Peter Higgs over 50 years ago [1]. Its particle content consists of a vector gauge boson (call it Z_D) and a complex scalar (call it σ). By itself it has automatic charge conjugation invariance, i.e. $Z_D \to -Z_D$, $\sigma \to \sigma^*$, resulting in $g_D \to -g_D$. After spontaneous symmetry breaking, the above still holds, i.e. $Z_D \to -Z_D$, $\sigma_R \to \sigma_R$, and $\sigma_I \to -\sigma_I$ which becomes the longitudinal component of the now massive Z_D. This fact has been used [2–5] to suggest that Z_D may be dark matter.

The existence of two $U(1)$ gauge factors allows for the gauge-invariant kinetic mixing [6] of the two associated gauge bosons, so Z_D may mix with the $U(1)_Y$ gauge boson of the standard model (SM), of which the photon is a component. This has led to many theoretical studies of a possible light dark photon, and the experiments which may be relevant in finding it [7]. However, this kinetic mixing term breaks the dark charge conjugation symmetry, so the former may be absolutely forbidden if the latter is chosen to be exact.

In the Higgs model, Z_D is the sole dark matter. Suppose a Dirac fermion (call it N) is added, transforming also under $U(1)_D$, then the Lagrangian is also invariant under dark charge conjugation, as well as the global $U(1)$ transformation operating on N, i.e. dark fermion number. Hence N is a dark-matter candidate. What about Z_D? If $m_{Z_D} > 2m_N$, then Z_D will decay into $N\bar{N}$ through the vector current $\bar{N}\gamma_\mu N$ which has charge conjugation $\mathcal{C} = -1$, but if $m_{Z_D} < 2m_N$, then Z_D will be stable. Further, if Z_D is much lighter than N, then it may act as a stable light mediator for N self-interactions. Note that if Z_D is unstable and decays to SM particles, then very strong constraints exist [8] which basically rule out [9] this scenario for explaining [10] the core-cusp anomaly observed in dwarf galaxies [11]. As for the dark Higgs boson $h_D = \sqrt{2}Re(\sigma)$, it may also be light, but it has an unavoidable mixing with the SM Higgs boson h, so it will not be stable. In the following, $m_{h_D} < m_{Z_D}$ will be assumed.

With $m_N \sim 100$ GeV and $m_{Z_D} \sim 10$ MeV, the $N\bar{N}$ annihilation to Z_DZ_D is assumed to have the right cross section for N to be the main component of dark matter. The subsequent Z_DZ_D annihilation to h_Dh_D is assumed to have a large enough cross section, so that the relic abundance of Z_D is small compared to that of N. In direct-search experiments, N does not interact with quarks, so there will be no signal. As for the small Z_D component, it interacts through h_D-h mixing, but since Z_D is very light, current experiments are not sensitive to its presence. On the other hand, the h_D-h mixing has to be large enough for it to decay away before big bang nucleosynthesis (BBN). Even so, h_D may be produced at late times through Z_DZ_D annihilation, and affects the cosmic microwave background (CMB) through its decay. However, there is no Sommerfeld enhancement [12] of this cross section, unlike the case of $N\bar{N}$ annihilation through a light mediator which decays. Hence the proposed model is a natural resolution of this conundrum, as detailed below.

2. Dark $U(1)_D$ model

This model assumes $U(1)_D$ gauge symmetry, implying thus a vector gauge boson Z_D. It is spontaneously broken by a com-

E-mail address: ma@phyun8.ucr.edu.

plex scalar σ with charge g_D. A Dirac fermion N also exists with charge g_N. The complete Lagrangian before symmetry breaking is

$$\mathcal{L} = -\frac{1}{4}(\partial^\mu Z_D^\nu - \partial^\nu Z_D^\mu)(\partial_\mu Z_{D\nu} - \partial_\nu Z_{D\mu}) + (\partial^\mu \sigma - i g_D Z_D^\mu \sigma)(\partial_\mu \sigma^* + i g_D Z_{D\mu}\sigma^*) + \mu_D^2 \sigma^*\sigma - \frac{1}{2}\lambda_D(\sigma^*\sigma)^2 + i\bar{N}\gamma_\mu(\partial^\mu - i g_N Z_D^\mu)N - m_N \bar{N}N. \tag{1}$$

In the above, if we replace g_D by $-g_D$, σ by σ^*, g_N by $-g_N$, and N by its dark charge conjugate, we have exactly the same physical theory. The spontaneous breaking of $U(1)_D$ with $\langle\sigma\rangle = v_D/\sqrt{2}$ changes the Lagrangian to

$$\mathcal{L} = -\frac{1}{4}(\partial^\mu Z_D^\nu - \partial^\nu Z_D^\mu)(\partial_\mu Z_{D\nu} - \partial_\nu Z_{D\mu}) + \frac{1}{2}m_{Z_D}^2 Z_D^\mu Z_{D\mu} + \frac{1}{2}(\partial^\mu h_D)(\partial_\mu h_D) - \frac{1}{2}m_{h_D}^2 h_D^2 - \frac{m_{h_D}^2}{2v_D}h_D^3 - \frac{m_{h_D}^2}{8v_D^2}h_D^4 + g_D^2 v_D h_D (Z_D^\mu Z_{D\mu}) + \frac{1}{2}g_D^2 h_D^2 (Z_D^\mu Z_{D\mu}) + i\bar{N}\gamma_\mu \partial^\mu N - m_N\bar{N}N + g_N Z_D^\mu \bar{N}\gamma_\mu N, \tag{2}$$

where $v_D^2 = 2\mu_D^2/\lambda_D$, $m_{Z_D} = g_D v_D$, and $m_{h_D}^2 = \lambda_D v_D^2$. The crucial interaction terms are $g_N Z_D^\mu \bar{N}\gamma_\mu N$, $g_D^2 v_D h_D (Z_D^\mu Z_{D\mu})$, and $(1/2)g_D^2 h_D^2 (Z_D^\mu Z_{D\mu})$. We assume in the following $m_N \sim 100$ GeV, with $Z_D, h_D \sim 10$ MeV, with $m_{h_D} < m_{Z_D}$. Note that g_N is independent of g_D.

3. Three new particles

There are only three new particles beyond those of the standard model. Each serves a purpose and is an essential ingredient of this two-component dark-matter model. The dark fermion N is a Dirac particle with a conserved dark fermion number. It is the dominant component of the observed dark matter of the Universe. It has a dark gauge interaction mediated by Z_D which is light, thus realizing the requirement of a sufficiently large interaction to affect the core-cusp discrepancy of dwarf galaxies. The imposition of dark charge conjugation symmetry means that Z_D has $C = -1$. It couples to the vector current $\bar{N}\gamma_\mu N$ which also has $C = -1$, so it may decay into $N\bar{N}$, but if it is lighter than $2m_N$ as assumed, then it is itself stable. As such, it may be overproduced in the early Universe. However, it is also assumed that the dark Higgs boson h_D, which breaks the $U(1)_D$ gauge symmetry and provides Z_D with a mass through its vacuum expectation value v_D, is lighter than Z_D. Hence the $Z_D Z_D \to h_D h_D$ annihilation should be strong enough to make it a very small fraction of the observed dark matter of the Universe. As for h_D, which has $C = +1$, it must be unstable through its allowed mixing with the SM Higgs boson h, and decays away early without affecting the standard BBN.

Consider the extended scalar potential involving both σ and the SM Higgs doublet $\Phi = (\phi^+, \phi^0)$:

$$V = -\mu_D^2\sigma^*\sigma + \frac{1}{2}\lambda_D(\sigma^*\sigma)^2 - \mu_h^2\Phi^\dagger\Phi + \frac{1}{2}\lambda_h(\Phi^\dagger\Phi)^2 + \lambda_{hD}(\sigma^*\sigma)(\Phi^\dagger\Phi). \tag{3}$$

Using $\phi^0 = (v_h + h)/\sqrt{2}$, the 2×2 mass-squared matrix spanning (h_D, h) is given by

$$\mathcal{M}_{h_D,h}^2 = \begin{pmatrix} \lambda_D v_D^2 & \lambda_{hD} v_D v_h \\ \lambda_{hD} v_D v_h & \lambda_h v_h^2 \end{pmatrix}. \tag{4}$$

Assuming $m_{h_D} << m_h = 125$ GeV, the h_D-h mixing is then $\theta_{hD} = \lambda_{hD} v_D v_h / m_h^2$. For a light h_D of order 10 MeV in mass, its dominant decay is to e^-e^+ with the decay rate

$$\Gamma(h_D \to e^-e^+) = \frac{m_{h_D} m_e^2}{8\pi v_h^2}\theta_{hD}^2, \tag{5}$$

where $v_h = 246$ GeV. Assuming that $\Gamma^{-1} < 0.05$ s, the constraint

$$\left(\frac{m_{h_D}}{10\ \text{MeV}}\right)\theta_{hD}^2 > 7.66 \times 10^{-9} \tag{6}$$

is obtained. The SM Higgs boson h also decays into $h_D h_D$ with coupling $\lambda_{hD} v_h$. Its decay rate is

$$\Gamma(h \to h_D h_D) = \frac{\lambda_{hD}^2 v_h^2}{16\pi m_h} = \lambda_{hD}^2(9.63\ \text{GeV}). \tag{7}$$

Assuming that this is no more than 10% of the Higgs boson width in the SM (4.12 MeV), this gives a bound of

$$\lambda_{hD} < 0.0066. \tag{8}$$

Comparing Eqs. (6) and (7), the constraint

$$\left(\frac{v_D}{\text{GeV}}\right) > 0.85\sqrt{\frac{10\ \text{MeV}}{m_{h_D}}} \tag{9}$$

is obtained.

4. $Z_d Z_d$ annihilation

Consider first the process $Z_d Z_d \to h_D h_D$ at rest. There are four diagrams summing up to the amplitude

$$\mathcal{A} = \left[\frac{2g_D^2(2+r)}{2-r} - \frac{6g_D^2 r}{4-r}\right](\vec{\epsilon}_1\cdot\vec{\epsilon}_2) + \frac{8g_D^2}{m_{Z_D}^2(2-r)}(\vec{\epsilon}_1\cdot\vec{k})(\vec{\epsilon}_2\cdot\vec{k}), \tag{10}$$

where $r = m_{h_D}^2/m_{Z_D}^2$ and the center-of-mass variables $\vec{k}$ (momentum of h_D) and $\vec{\epsilon}_{1,2}$ (polarizations of Z_D) have been used. The resulting cross section × relative velocity is given by

$$\sigma(Z_D Z_D \to h_D h_D) \times v_{rel} = \frac{g_D^4\sqrt{1-r}}{64\pi m_{Z_D}^2}\left[\frac{4[r^2 + 4(2-r)^2]}{(4-r)^2} - \frac{24r(2+r)}{9(2-r)(4-r)} + \frac{8(2+r)^2}{9(2-r)^2}\right]. \tag{11}$$

Let $m_{Z_D} = 10$ MeV and $m_{h_D} = 8$ MeV, then $r = 0.64$. The coupling g_D is adjustable. Let $g_D = 0.005$ for example, then

$$\sigma(Z_D Z_D \to h_D h_D) \times v_{rel} = 1.1 \times 10^{-24}\ \text{cm}^3/\text{s}, \tag{12}$$

which is 37 times the canonical $\sigma_0 \times v_{rel} = 3 \times 10^{-26}$ cm^3/s for obtaining the correct dark-matter relic abundance of the Universe. This means that Z_D will be underproduced and forms only a small component of the observed dark matter, which will be mainly N as discussed in the next section. Note also that $g_D = 0.005$ and $m_{Z_D} = 10$ MeV imply that $v_D = 2$ GeV, which is perfectly consistent with Eq. (9).

5. $N\bar{N}$ annihilation

The annihilation $N\bar{N} \to Z_D Z_D$ is analogous to $e^-e^+ \to \gamma\gamma$. The cross section at rest $\times$ relative velocity is given by

$$\sigma(N\bar{N} \to Z_D Z_D) \times v_{rel} = \frac{g_N^4}{16\pi m_N^2}. \qquad (13)$$

For $m_N = 100$ GeV, this would be equal to $2\sigma_0 \times v_{rel} = 6 \times 10^{-26}$ cm^3/s if $g_N = 0.225$. For the light mediator with $m_{Z_D} = 10$ MeV, Sommerfeld enhancement is expected. However, at the time of thermal freezeout, this effect is only $\mathcal{O}(1)$ [13,14]. The large enhancement will come at late times (because of the decreasing relative velocity of $N\bar{N}$ annihilation) and may be as large as a factor of 10^4. Whereas the fraction of $N\bar{N}$ which would annihilate is still negligible compared to the entire population, the production of an unstable mediator would allow its decay products (photons and electrons) to affect the CMB, thus ruling out (for s-wave annihilation) all models where the self-interactions are large enough to address the small-scale problems of structure formation, as pointed out recently [9].

Here the light mediator Z_D is stable, so it does not affect the CMB. As for h_D, it may also be produced at late times from $Z_D Z_D$ annihilation, but this cross section has no Sommerfeld enhancement, so even though h_D decays to e^-e^+, its effect is small.

6. Thermal history

The dark fermion N is kept in thermal equilibrium with its light mediator Z_D which couples to the dark Higgs boson h_D. The bridge connecting the dark sector with the SM is the quartic scalar interaction term $\lambda_{hD}(\sigma^*\sigma)(\Phi^\dagger\Phi)$ of Eq. (3). Hence h_D is in thermal equilibrium with the SM Higgs boson h, and through the latter, all the SM particles, at temperatures even below m_h. For example, the annihilation amplitude of $h_D h_D \to b\bar{b}$ is proportional to $\lambda_{hD} m_b/m_h^2$. This is strong enough to keep h_D in thermal equilibrium with the SM particles if the corresponding rate is greater than that of the expansion of the Universe, i.e.

$$\frac{\lambda_{hD}^2 m_b^2}{m_h^4} T^3 > \frac{T^2}{M_{\text{Planck}}}. \qquad (14)$$

At $T \sim m_N = 100$ GeV, this means that $\lambda_{hD} > 10^{-7}$ is required, a condition which is easily satisfied by both Eqs. (6) and (8). As the Universe cools below m_N, N freezes out with a relic abundance which accounts for most of the observed dark matter of the Universe. In structure formation, N has a large enough elastic cross section due to the exchange of its light mediator Z_D to explain the flatter density profiles of dwarf galaxies near their centers [10].

The light vector boson Z_D is stable and interacts with h_D to remain in thermal equilibrium until the Universe cools below m_{Z_D}. It then freezes out with a much smaller relic abundance than that of N. The dark Higgs boson h_D decays away quickly at early times through its mixing with the SM Higgs boson h. All these happen before the onset of BBN so that the standard predictions of all relevant cosmological parameters are unchanged. At late times, Z_D re-emerges from $N\bar{N}$ annihilation, but it is stable and will not disturb the CMB. The dark Higgs boson h_D also re-emerges from $Z_D Z_D$ annihilation, but this cross section is not enhanced by the Sommerfeld effect, so even though h_D decays to e^-e^+, its effect on the CMB is harmless.

7. Phenomenological consequences

The model presented has a dark gauge $U(1)_D$ symmetry, with exact dark charge conjugation invariance. It has two stable particles, the dark fermion N with $m_N \sim 100$ GeV and a light vector mediator Z_D with $m_{Z_D} \sim 10$ MeV. As such, it explains the observed relic abundance of dark matter, as well as the cusp-core anomaly of dwarf galaxies. It avoids the strong constraints of decaying particles on the CMB [8,9]. The $U(1)_D$ symmetry is broken with $v_D \sim 2$ GeV as constrained by Eq. (9). The associated dark Higgs boson h_D is lighter than Z_D and mixes with the SM Higgs boson h.

In direct-search experiments, N is essentially invisible because it has only Z_D interactions which do not affect SM particles at tree level. As for Z_D, its relic abundance is suppressed and its mass is only about 10 MeV, so even though it interacts with SM particles through h_D–h mixing, it is insensitive to present underground experiments. This would not be the case if $m_{Z_D} \sim 100$ GeV. In fact, it has been shown [15] that a light mediator would then be ruled out because the direct-detection bound excludes its decay before the onset of BBN. In indirect-search experiments, the $N\bar{N}$ annihilation is Sommerfeld-enhanced, but it only produces Z_D at tree level which cannot be detected. In one loop, SM particles may be produced, but the cross section is very small. Hence neither types of the conventional search for dark matter would have much promise in detecting such dark matter.

Since the light vector boson Z_D has no kinetic mixing with the photon because of the dark gauge conjugation symmetry, there is also no effect on experiments searching for it through this portal.

A possible way to discover h_D is from $h \to h_D h_D$ decay at an accelerator, and the subsequent decay $h_D \to e^-e^+$. The problem is that h_D has a lifetime of about 1 s, so the decay products are far downstream and not easily observed.

8. Remarks

The idea of self-interacting dark matter is faced with a conundrum [9]. If the interaction is strong enough to address the small-scale problems of structure formation, the production of the light mediator at late times would disrupt the cosmic microwave background because of the inherent Sommerfeld enhancement for s-wave annihilation and the apparently inescapable fact that the mediator must decay into electrons or photons. Its resolution in terms of a simple complete renormalizable model is the subject matter of this paper. Unfortunately, this model predicts null or negligible effects in all present attempts to discover the nature of dark matter. On the other hand, it may be the answer to the question of why dark matter has not been seen so far.

Acknowledgement

This work was supported in part by the U.S. Department of Energy Grant No. DE-SC0008541.

References

[1] P.W. Higgs, Phys. Rev. Lett. 13 (1964) 508.
[2] Y. Farzan, A. Rezaei Akbarrieh, J. Cosmol. Astropart. Phys. 1210 (2012) 026.
[3] S. Baek, P. Ko, W.-I. Park, E. Senaha, J. High Energy Phys. 1305 (2013) 036.
[4] A. DiFranzo, P.J. Fox, T.M.P. Tait, J. High Energy Phys. 1604 (2016) 135.
[5] A. DiFranzo, G. Mohlabeng, J. High Energy Phys. 1701 (2017) 080.
[6] B. Holdom, Phys. Lett. B 166 (1986) 196.
[7] See for example J. Alexander, et al., Dark Sectors 2016 Workshop: community report, arXiv:1608.08632 [hep-ph].
[8] S. Galli, F. Iocco, G. Bertone, A. Melchiorri, Phys. Rev. D 80 (2009) 023505.
[9] T. Bringmann, F. Kahlhoefer, K. Schmidt-Hoberg, P. Walia, Phys. Rev. Lett. 118 (2017) 141802.

[10] See for example J.L. Feng, M. Kaplinghat, H. Tu, H.-B. Yu, J. Cosmol. Astropart. Phys. 0907 (2009) 004.
[11] See for example F. Donato, G. Gentile, P. Salucci, C. Frigerio Martins, M.I. Wilkinson, G. Gilmore, E.K. Grebel, A. Koch, R. Wyse, Mon. Not. R. Astron. Soc. 397 (2009) 1169.
[12] A. Sommerfeld, Ann. Phys. (Berlin) 403 (1931) 257.
[13] L.G. van den Aarssen, T. Bringmann, C. Pfrommer, Phys. Rev. Lett. 109 (2012) 231301.
[14] S. Tulin, H.-B. Yu, K.M. Zurek, Phys. Rev. D 87 (2013) 115007.
[15] M. Kaplinghat, S. Tulin, H.-B. Yu, Phys. Rev. D 89 (2014) 035009.

17

Noether symmetry in $f(T)$ teleparallel gravity

Nayem Sk

Dept. of Physics, University of Kalyani, West Bengal, 741235, India

ARTICLE INFO

Editor: M. Trodden

Keywords:
$f(T)$ gravity
Canonical formulation
Noether symmetry

ABSTRACT

Noether symmetry in teleparallel $f(T)$ gravity, where T is the torsion scalar, has been studied in the background of Robertson–Walker space–time. It is found that Noether symmetry admits $f(T) \propto T^{\frac{3}{2}}$ and the associated conserved current is $\Sigma = a\dot{a}T^{\frac{1}{2}}$, in matter dominated era. In the process, the recent claim by Wei et al. [1] that Noether symmetry admits $f(T) \propto T^n$, (where n is an arbitrary constant) is found not to be correct, since the conserved current satisfies the field equations only for a special choice of $n = \frac{3}{2}$. Further, correspondence between $f(R)$ and $f(T)$ theories of gravity has also been established.

1. Introduction

Luminosity-distance versus redshift curve obtained from distant SN1a supernovae data [2,3] unveils its non-linear feature beyond redshift $z = 0.2$. For last two decades attempts have been made to fit such a curve within a viable cosmological model. Almost all the attempts equivocally predict that at present the universe is undergoing an accelerated expansion. Since cosmological constant (Λ) calculated in view of quantum field theory has been found to be nearly 120 order of magnitudes larger than the same required to fit SN1a data, so it is ruled out. Two options therefore are left. The first is to modify the right hand side of Einstein's equation by accommodating one or more scalar fields including tachyonic fields or some more exotic ones having reverse sign in kinetic term with some typical form of potential. Such fields interact with none other than the gravitational field only, and therefore dubbed as dark energy. However, the field mass responsible for late time cosmic acceleration is very small on one hand and the present technology does not support detection of dark energy in any of its form, on the other. Therefore the second option has been advocated in recent years and that is to modify the left hand side of Einstein's equation, viz. the geometry, and in the process bypassing the dark energy issue. Such attempt is dubbed as modified theory of gravity. Several types of modified theory of gravity exists in the literature, such as, $f(R)$ gravity, $f(G)$ (Gauss–Bonnet) gravity, $f(T)$ (Torsion) gravity, combination of all these, Gauss–Bonnet-dilatonic coupled gravity, Lanczos–Lavlock gravity, Horava–Lifschitz gravity and models with extra dimensions including Kaluza–Klein, Randall–Sundrum, DGP and higher co-dimension braneworlds, etc. Out of these, Teleparallel gravity has drawn lot of attention in the recent years.

To consider teleparallelism, one employs the orthonormal tetrad components $e_A(x^\alpha)$, where the index A runs over 0, 1, 2, 3 to the tangent space at each point x^α of the manifold. Their relation to the metric $g_{\alpha\beta}$ is given by

$$g_{\alpha\beta} = \eta_{AB} e^A_\alpha e^B_\beta, \tag{1}$$

where α and β are coordinate indices on the manifold which again run over 0, 1, 2, 3, while e^A_α forms the tangent vector on the tangent space over which the metric η_{AB} is defined. Instead of the torsionless Levi-Civita connection which is used in General Theory of Relativity, in Teleparallelism [4] one considers the curvatureless Weitzenbock connection, whose non-null torsion $T^\rho_{\alpha\beta}$ and contorsion $K^{\alpha\beta}_\rho$ are defined by

$$T^\rho_{\alpha\beta} \equiv e^\rho_A[\partial_\alpha e^A_\beta - \partial_\beta e^A_\alpha], \tag{2}$$

$$K^{\alpha\beta}_\rho \equiv -\frac{1}{2}[T^{\alpha\beta}{}_\rho - T^{\beta\alpha}{}_\rho - T_\rho{}^{\alpha\beta}], \tag{3}$$

respectively. Moreover, instead of the Ricci scalar R, which is used for the Lagrangian density in general relativity, the teleparallel Lagrangian density is represented by the torsion scalar T given by

$$T \equiv S_\rho{}^{\alpha\beta} T^\rho{}_{\alpha\beta}, \tag{4}$$

where,

$$S_\rho{}^{\alpha\beta} \equiv \frac{1}{2}[K^{\alpha\beta}{}_\rho + \delta^\alpha_\rho T^{\theta\beta}{}_\theta - \delta^\beta_\rho T^{\theta\alpha}{}_\theta]. \tag{5}$$

E-mail address: nayemsk1981@gmail.com.

Accordingly, in analogy to the $f(R)$ theory of gravity, recently a new modified theory of gravity, namely the so-called $f(T)$ theory of gravity has been proposed to explain the current accelerated expansion of the cosmos, without invoking dark energy. Such a modified teleparallel action for $f(T)$ gravity is given by

$$\mathbb{A} = \int d^4x \mid e \mid f(T) + S_m, \tag{6}$$

where $|e| = \det e^A_\alpha = \sqrt{-g}$ and the units has been chosen so that $c = 16\pi G = 1$. It may mentioned that this is a generalized version of the teleparallel gravity originally proposed almost a century back by Einstein [5,6].

Now, in order to study the cosmological consequence of the so-called teleparallel gravity, a particular form of $f(T)$ is required. Instead of setting a form of $f(T)$ by hand or reconstruct it from the history of cosmic evolution, it is always desirable to find its form following some physical consideration, viz. in view of the loop quantum gravity or from some symmetry consideration. Since, loop quantum gravity does not provide a term suitable for late time cosmic acceleration, so Noether symmetry is usually preferred.

Noether symmetry was applied for the first time in scalar–tensor theory of gravity by De Ritis and his collaborators [7] to find a form of the potential. Noether symmetry was found to select [7] exponential form of the potential which can trigger inflation in the early universe. This raise immense interest in the scientific community, and thereafter Noether symmetry has been extensively studied in cosmological models with minimally [8,9] and non-minimally coupled [10–14] scalar–tensor theories, higher order theory [15] and $f(R)$ theory [16–19] of gravity. Additionally, the same has also been applied in different anisotropic Bianchi models [20], induced gravity theory [21], Gauss–Bonnet gravity [22] and so on. Quantum cosmological models have also been expatiated in view of Noether symmetry [23]. Here, we are therefore motivated to study Noether symmetry in teleparallel theory of gravity, to find a form of $f(T)$.

Recently, Wei et al. [1] has claimed that Noether symmetry for teleparallel $f(T)$ theory of gravity in the background of spatially flat Robertson–Walker (R–W) metric described by

$$ds^2 = -dt^2 + a^2(t)dX^2, \tag{7}$$

where $a(t)$ is the scale factor, admits $f(T) \propto T^n$, where n is an arbitrary constant in matter domain era. However, in the present study we show that the associated conserved current satisfies the field equations only for a special choice of $n = \frac{3}{2}$. Thus, it is found that Noether Symmetry only admits $f(T) \propto T^{\frac{3}{2}}$ along with a conserved current $\Sigma = a\dot{a}T^{\frac{1}{2}}$.

In the following section, the canonical formulations of $f(T)$ theory of gravity following Lagrange multiplier technique and its scalar–tensor counterpart have been discussed. In section 3, Noether symmetry has been invoked in both the canonical point Lagrangians corresponding to teleparallel $f(T)$ gravity. In section 4, analogy of teleparallel gravity with $f(R)$ theory of gravity has been discussed in some detail. Finally we conclude in section 5.

2. Canonical formulation of $f(T)$ gravity

It is not possible to find solutions to the field equations corresponding to the above action (6) to study cosmological consequence of teleparallel gravity, unless a specific form of $f(T)$ is known a priori. As already mentioned, one can choose a form by hand out of indefinite possibilities, or reconstruct it in view of cosmic evolution history. Nevertheless, it is always desirable to find the form in view of some physical consideration like Noether symmetry. Nevertheless, this requires canonical formulation of the theory under consideration. In fact, there exists two possible techniques towards canonical formulation of $f(T)$ theory of gravity. One is Lagrange multiplier technique, which is applicable with finite degrees of freedom, and the other is scalar-Tensor representation of $f(T)$ gravity.

2.1. Lagrange multiplier technique

Unlike Scalar-Tensor representation of $f(T)$ Theory (as we see next), canonical formulation following Lagrange multiplier technique may be performed with finite degrees of freedom only. Therefore we restricting ourselves to the Robertson–Walker metric (7), we can treat $T + 6\frac{\dot{a}^2}{a^2} = 0$ as a constraint and introduce it in the action (6) through a Lagrange multiplier λ as,

$$\mathbb{A} = 2\pi^2 \int \Big[f(T) - \lambda\Big\{T + 6\Big(\frac{\dot{a}^2}{a^2}\Big)\Big\} - \frac{\rho_{m0}}{a^3}\Big]a^3 dt. \tag{8}$$

Now varying the action with respect to T one gets $\lambda = f'(T)$, where $f'(T)$ is the derivative of $f(T)$ with respect to T. Substituting the form of λ so obtained in the above action (8) the following canonical action is found, viz.

$$\mathbb{A} = 2\pi^2 \int \Big[f(T) - f'(T)\Big\{T + 6\Big(\frac{\dot{a}^2}{a^2}\Big)\Big\} - \frac{\rho_{m0}}{a^3}\Big]a^3 dt. \tag{9}$$

Therefore, the point Lagrangian in the presence of ordinary matter may be expressed in Robertson–Walker metric (7) as

$$L(a, \dot{a}, T, \dot{T}) = \Big[-6a\dot{a}^2 f' + a^3(f - f'T) - \rho_{m0}\Big]. \tag{10}$$

In the above, ρ_{m0} stands for the matter density at the present epoch.

2.2. Scalar–tensor representation of $f(T)$ gravity

As already mentioned, it is also possible to translate the action (6) in its scalar–tensor equivalent form, in analogy to $f(R)$ theory of gravity. The Scalar–Tensor representation [24] of $f(T)$ gravity reads

$$\mathbb{A} = \int d^4x \mid h \mid [\Phi T - U(\Phi)] + S_m. \tag{11}$$

where,

$$\Phi = f'(T); \quad U(\Phi) = Tf'(T) - f(T). \tag{12}$$

The corresponding point Lagrangian in Robertson–Walker (7) space–time reads

$$L(a, \Phi, \dot{a}, \dot{\Phi}) = \Big[6a\dot{a}^2\Phi - a^3 U(\Phi) - \rho_{m0}\Big], \tag{13}$$

where, ρ_{m0} is the matter density at the present epoch, as already mentioned.

3. Noether symmetries

In view of the canonical Lagrangians obtained in the previous subsections, we now move on to explore Noether symmetry. It is well known that Noether symmetry ($\pounds_X L = XL = 0$) in $f(R)$ theory of gravity yields nothing other than $f(R) = f_0 R^{\frac{3}{2}}$ along with a conserved current $\frac{d}{dt}(a\sqrt{R})$ in R–W metric, when coupled to pressure-less dust or in vacuum [16–18]. Despite such unique result, Noether symmetry of $f(R)$ theory of gravity had been reopened by some authors [25,26], who claimed to find new conserved currents in the name of Noether gauge symmetry. Particularly, it was claimed by Hussain et al. [25] that Noether gauge symmetry admits $f(R) \propto R^n$, where n is an arbitrary constant. Jamil et al. [26] on the other hand found $f(R) \propto R^2$ and $V(\phi) \propto \phi^{-4}$, considering Noether gauge symmetry with Tachyonic field. The claim [25] had been reviewed by the present author and his collaborator (Sk and Sanyal) [27] taking both vanishing and non-vanishing gauge into account. It was found that the conserved currents so obtained do not satisfy the field equations, particularly the $\binom{0}{0}$ equation of Einstein, unless $n = \frac{3}{2}$. Thus, the claim that arbitrary power of R generates Noether symmetry is not correct. The claim of Jamil et al. [26] had also been reviewed by the same authors Sk and Sanyal [28] and it was shown that $f(R) \propto R^2$ do not satisfy the Tachyonic field equations. Shamir et al. [29] on the contrary, had claimed that Noether symmetry of $f(R) \propto R^{\frac{3}{2}}$ admits four different generators corresponding to which four different conserved currents exist in the presence of non-zero gauge. In a subsequent communication, the same authors Sk and Sanyal [30] reviewed the work and proved that the claim is not correct, for the same reason that not all the conserved currents satisfy the $\binom{0}{0}$ equation of Einstein. Later, Roshan et al. [31] claimed that Noether symmetry in the context of Palatini $f(\mathfrak{R})$ theory of gravity admits $f(\mathfrak{R}) \propto \mathfrak{R}^n$, (where n is again an arbitrary constant) in matter dominated era. This claim had also been reviewed by the present author [32] and it has been also shown that Noether Symmetry only admits $f(\mathfrak{R}) \propto \mathfrak{R}^{\frac{3}{2}}$ in Palatini gravity. Under such circumstances, it would really be interesting if $f(T)$ theory of gravity yields new forms of $f(T)$ as claimed by Wei et al. [1]. In the following subsections we therefore review the claim [1] in the process of finding Noether symmetries of $f(T)$ theory of gravity, which satisfy the field equations.

3.1. Noether symmetry following Lagrange multiplier technique

The field equations constructed out of the point Lagrangian (13) in the Robertson–Walker metric (7) are,

$$\left(f - f'T + 2f'H^2\right) + 4\left(2f'\frac{\ddot{a}}{a} + Hf''\dot{T}\right) = 0, \tag{14}$$

$$a^3 f''\left(T + 6\frac{\dot{a}^2}{a^2}\right) = 0. \tag{15}$$

In the above $H = \frac{\dot{a}}{a}$ stands for the Hubble parameter. The $\binom{0}{0}$ equation of Einstein is

$$\left[-6a\dot{a}^2 f' + a^3(f - f'T) - \rho_{m0}\right] = 0. \tag{16}$$

Now, Noether theorem state that, if there exists a vector field X, for which the Lie derivative of a given Lagrangian L vanishes i.e. $\pounds_X L = XL = 0$, the Lagrangian admits a Symmetry and thus yields a conserved current. For the Lagrangian (10) under consideration, the configuration space is $M(a, T)$ and the corresponding tangent space is $TM(a, T, \dot{a}, \dot{T})$. Hence the generic infinitesimal generator of the Noether Symmetry is

$$X = \gamma\frac{\partial}{\partial a} + \zeta\frac{\partial}{\partial T} + \dot{\gamma}\frac{\partial}{\partial \dot{a}} + \dot{\zeta}\frac{\partial}{\partial \dot{T}}, \tag{17}$$

where, $\gamma = \gamma(a, T), \zeta = \zeta(a, T)$. The constant of motion is given by

$$\Sigma = \gamma\frac{\partial L}{\partial \dot{a}} + \zeta\frac{\partial L}{\partial \dot{T}}. \tag{18}$$

Finding the Noether equation in view of the existence condition $\pounds_X L = XL = 0$, and equating the coefficients of $\dot{a}^2$, $\dot{T}^2$, $\dot{a}\dot{T}$ along with the term free from derivative respectively to zero as usual, we obtain the following set of partial differential equations,

$$a\gamma' = 0, \quad \gamma f' + \zeta a f'' + 2af'\gamma_{,a} = 0, \qquad 3\gamma\left(f - Tf'\right) - a\zeta T f'' = 0. \tag{19}$$

The above set of partial differential equations admit the following set of solutions, viz.

$$\gamma = \gamma_0 a^{1-s}, \quad \zeta = -2s\gamma_0 a^{-s}T, \quad f(T) = f_0 T^{\frac{3}{2s}}. \tag{20}$$

The corresponding conserved current is

$$\Sigma = a^{2-s}\dot{a}f'(T). \tag{21}$$

It may be trivially checked that the above conserved current satisfies the field equations (14) through (16) only for $s = 1$. The expression of the conserved current (21) for $s = 1$ therefore reads,

$$\Sigma = a\dot{a}f'(T). \tag{22}$$

It is interesting to note that the reduced form of $f(T)$ turns out to be,

$$f(T) = f_0 T^{\frac{3}{2}}. \tag{23}$$

In view of the above form of $f(T)$ and the conserved current (22), $\dot{a}(t)$ turns out to be a constant, and therefore the cosmic scale factor $a(t)$ admits the solution,

$$a(t) = a_1 t + a_0, \tag{24}$$

where a_1, a_0 are constants of integration. In this context we mention that the same solution [32] has also been found in the context of Palatini $f(\mathfrak{R})$ theory of gravity. However, the above coasting solution although fits SnIa data perfectly in the matter dominated era [33], does not fit to other available cosmological data.

3.1.1. Comments on Hao Wei et al. work

It is important to note that the $\binom{0}{0}$ equation of Einstein is essentially the Hamiltonian constraint equation, when expressed in terms of phase-space variables. It is the outcome of diffeomorphic invariance of the theory of gravity. Since Noether equation $\pounds_X L = XL = 0$ does not recognize the constraint, therefore one can not expect that the solutions of Noether equations would satisfy the Hamilton constraint equation automatically. This has been proved by Wald and Zoupas [34]. This means that Noether theorem is not on-shell for constrained system. Conserved current is not an independent equation, but rather it is the first integral of certain combination of the field equations. Thus, it is essential to check if the conserved current satisfies the $\binom{0}{0}$ equation of Einstein. Like earlier authors [25,26,29,31] it has not been checked by the present authors [1]. However, it is not difficult to check that the conserved current satisfies the field equations only under the special choice $n = \frac{3}{2}$. Therefore, the claim of finding $f(T) \propto T^n$ by Wei et al. [1] is not correct.

3.2. Noether symmetry in scalar–tensor representation of $f(T)$ gravity

Let us now turn our attention in this subsection, to explore Noether symmetry in scalar–tensor representation of $f(T)$ theory of gravity. The field equations constructed out of the point Lagrangian (11) in the Robertson–Walker metric (7) are,

$$\left[\frac{\ddot{a}}{a} + \frac{\dot{a}^2}{2a^2} + \frac{\dot{a}\dot{\Phi}}{a\Phi} + \frac{U}{4\Phi}\right] = 0, \tag{25}$$

$$\left[\frac{\dot{a}^2}{a^2} - \frac{U_{,\Phi}}{6}\right] = 0. \tag{26}$$

The $\binom{0}{0}$ equation of Einstein is

$$\left[6a\dot{a}^2\Phi + a^3 U(\Phi) + \rho_{m0}\right] = 0. \tag{27}$$

In order to apply Noether symmetry approach, let us again introduce the lift vector X as an infinitesimal generator of Noether symmetry in the tangent space $[a, \dot{a}, \Phi, \dot{\Phi}]$ as follows

$$X = \gamma\frac{\partial}{\partial a} + \zeta\frac{\partial}{\partial \Phi} + \dot{\gamma}\frac{\partial}{\partial \dot{a}} + \dot{\zeta}\frac{\partial}{\partial \dot{\Phi}}, \tag{28}$$

and the existence condition for symmetry, $XL = 0$, leads to the following system of partial differential equations

$$\gamma_{,\Phi} = 0, \quad \Phi\gamma + a\zeta + 2a\Phi\gamma_{,a} = 0, \quad 3\gamma U + a\zeta U_{,\Phi} = 0. \tag{29}$$

The solution of the above set of equations reads,

$$\gamma = -\gamma_0 d a^{\frac{1-d}{2d}}, \quad \zeta = \gamma_0 a^{\frac{1-3d}{2d}}\Phi, \quad U = U_0\Phi^{3d}, \tag{30}$$

while the expression of conserved current is

$$\Sigma = a^{\frac{1+d}{2d}}\dot{a}\Phi. \tag{31}$$

Again, it has been shown that the above conserved current satisfies the field equations (25) to (27) only for $d = 1$. The expression for the conserved current for $d = 1$ is therefore,

$$\Sigma = a\dot{a}\Phi. \tag{32}$$

Now, using the transformation relations (12), we rewrite $U(\Phi)$ as $U(\Phi) = Tf'(T) - f(T) = U_0\Phi^3 = U_0[f'(T)]^3$. Equation (30) therefore yields the following form of $f(T)$, viz.

$$f(T) = f_0 T^{\frac{3}{2}}, \tag{33}$$

Note that the form of $f(T)$ and the associated conserved current so obtained is identical with those obtained following Lagrange multiplier technique. The solution to the scale factor therefore remains unchanged

$$a(t) = a_1 t + a_0, \tag{34}$$

which as already stated is not a viable solution to fit available cosmological data. Nevertheless, one important issue has been revealed and that is Noether symmetry is independent on the choice of the configuration space variables.

One of the main advantages of Noether conserved current is that one can express the field equations in terms of a cyclic coordinate, so that finding solutions becomes easier, and sometimes the cosmological solution emerges directly from Noether conserved current [35]. Being a first integral, one can even use it to find the solutions without even finding cyclic coordinate. In any case, one has to use the conserved current to find the solutions to the field equations. In a recent article [36], power-law teleparallel $f(T)$ gravity is discussed in details. The authors first applied Noether symmetry to find the form $f(T) \propto f_0 T^n$, and the associated conserved current. Thereafter, they explored the cosmological solution of the above mentioned form of $f(T)$ analytically with the help of the field equations and claimed the solutions to be outcome of Noether symmetry. One can easily check that the solutions do not satisfy Noether conserved current. Therefore such solutions cannot be an outcome of Noether symmetry, rather, it is like setting a form of $f(T) \propto f_0 T^n$ by hand, and solving the field equations.

4. Analogy with $f(R)$ gravity

The teleparallel $f(T)$ gravity is not equivalent to metric $f(R)$ gravity in general, since they differ by an appropriate boundary term ($\mathbb{B}$) [37,38]. The relation between torsion scalar (T) and the Ricci scalar (R) is given by

$$R = -T + \frac{2}{e}\partial_\rho(eT^\rho) = -T + \mathbb{B}, \tag{35}$$

where, $\mathbb{B} = \frac{2}{e}\partial_\rho(eT^\rho)$ is the boundary term. The action generally can be expressed as

$$\mathbb{A}_{\mathbb{B},\mathbb{T}} = \int d^4x \mid e \mid f(T, \mathbb{B}) + S_m. \tag{36}$$

It has been mentioned in a recent article [38] that both the metric $f(R)$ and the teleparallel $f(T)$ gravity can be recovered from $f(T, \mathbb{B})$ theory of gravity under suitable limit. Now, the expressions of torsion scalar (T) and boundary term ($\mathbb{B}$) in a flat R–W metric are, $T = -6(\frac{\dot{a}^2}{a^2})$ and $\mathbb{B} = -6(\frac{\ddot{a}}{a} + 2\frac{\dot{a}^2}{a^2})$. Therefore, the Ricci scalar is

$$R = -T + \mathbb{B} = -6(\frac{\ddot{a}}{a} + \frac{\dot{a}^2}{a^2}). \tag{37}$$

In this present article, we observe that Noether symmetry of teleparallel $f(T)$ theory of gravity in matter dominated only yields $f(T) \propto f_0 T^{\frac{3}{2}}$. Such a form of $f(T)$ admits a solution of the cosmological scale factor, $a(t) = a_1 t + a_0$, in a flat R–W metric. This particular solution implies $\ddot{a} = 0$. On the contrary, $f(R) \propto R^{\frac{3}{2}}$ yields a cosmological solution $a(t) = \sqrt{a_4 t^4 + a_3 t^3 + a_2 t^2 + a_1 t + a_0}$. So, in general the two differs. However, when t is small enough, i.e. in the early matter dominated era, the two match. In particular, under the condition $\ddot{a} = 0$, teleparallel $f(T)$ gravity becomes equivalent to metric $f(R)$ gravity, since $\mathbb{B} = 2T$ and $R = T$, in view of equation (37). Essentially, Noether symmetry puts up a limit under which the two theories become equivalent. This clearly demonstrates that at least in the context of Noether symmetry it is practically of no use to consider teleparallel gravity over $f(R)$ theory of gravity.

5. Concluding remarks

In the present work we studied teleparallel gravity and explored the form of $f(T)$ invoking Noether symmetry in the background of isotropic and homogeneous R–W metric. Both the canonical point Lagrangians obtained following Lagrange multiplier method and the scalar–tensor equivalent one, have been found to admit the only symmetry $f(T) = f_0 T^{\frac{3}{2}}$ in the matter dominated era. This reveals the fact that Noether symmetry, when applied to explore the form of an unknown parameter, is independent of the choice of the configuration space variables. We have also noticed that in R–W metric, Noether symmetry yields identical form of the cosmic scale factor ($a(t) = a_1 t + a_0$) both in teleparallel $f(T)$ theory of gravity and Palatini $f(\mathfrak{R})$ theory of gravity [32], in the matter dominated era. This establishes a sort of equivalence between the two. It has also been demonstrated that in the context of Noether symmetry teleparallel gravity turns out to be a special case of $f(R)$ theory of gravity.

It is clear that the form of $f(T)$ so obtained is not much appreciable. This is because, the coasting solution so obtained although fits SnIa data perfectly in the matter dominated era [33] fails to fit other available cosmological data. Particularly, it does not admit a long Friedmann-like matter dominated era, prior to the recent accelerated expansion of the universe. So application of Noether symmetry to choose a form of $f(T)$ becomes useless. In this context, we would like to mention that recently it has been observed that indeed Noether symmetry of $f(R)$ theory of gravity yields forms other than $f(R) \propto R^{\frac{3}{2}}$ [39]. In particular the other forms are $f(R) \propto R^2, \frac{1}{R}, R^{\frac{7}{5}}$. However, this requires a new symmetry generator, which includes the $\binom{0}{0}$ equation of Einstein in the form $\pounds_X L - \eta H = XL - \eta H = 0$, where, H is the Hamiltonian constraint of the theory being expressed in terms of configuration space variables (the $^0_0$ equation of Einstein) and η is a function of the coordinates in general. There is also possibility of finding other forms of $f(R)$ under proper investigation. Likewise, we do expect that several other forms of $f(T)$ might also emerge in view of the above mentioned symmetry generator. This we pose in a future communication.

Acknowledgements

I would like to thank my supervisor Dr. A.K. Sanyal for useful discussions and language editing. I also thank to reviewer's for their comments to improve this manuscript.

References

[1] H. Wei, X.J. Guo, L.F. Wang, Phys. Lett. B 707 (2012) 298, arXiv:1112.2270 [gr-qc].

[2] D.N. Riess, et al., Astron. J. 116 (1998) 1009;
D.N. Riess, et al., Astron. J. 730 (2011) 119.

[3] S. Perlmutter, et al., Astrophys. J. 517 (1999) 565.

[4] G.R. Bengochea, Phys. Lett. B 695 (2011) 405, arXiv:1008.3188;
S.H. Chen, J.B. Dent, S. Dutta, E.N. Saridakis, Phys. Rev. D 83 (2011) 023508, arXiv:1008.1250;
J.B. Dent, S. Dutta, E.N. Saridakis, J. Cosmol. Astropart. Phys. 1101 (2011) 009, arXiv:1010.2215;
R. Ferraro, F. Fiorini, Phys. Lett. B 702 (2011) 75, arXiv:1103.0824;
R. Ferraro, F. Fiorini, arXiv:1106.6349 [gr-qc];
C.G. Boehmer, A. Mussa, N. Tamanini, arXiv:1107.4455 [gr-qc].

[5] A. Einstein, Sitzungsber. Preuss. Akad. Wiss. Phys. Math. Kl. 217 (1928) 224;
A. Einstein, Math. Ann. 102 (1930) 685;
see A. Unzicker, T. Case, arXiv:physics/0503046 for English translation.

[6] K. Hayashi, T. Shirafuji, Phys. Rev. D 19 (1979) 3524, Addendum: Phys. Rev. D 24 (1981) 3312.

[7] R. De Ritis, et al., Phys. Rev. D 42 (1990) 1091.

[8] R. De Ritis, et al., Phys. Lett. A 161 (1991) 230.

[9] S. Capozziello, et al., Phys. Rev. D 80 (2009) 104030, arXiv:0908.2362;
C. Rubano, et al., Phys. Rev. D 69 (2004) 103510, arXiv:astro-ph/0311537;
G. Esposito, R. Roychowdhury, C. Rubano, P. Scudellaro, arXiv:1009.2887 [hep-th];
M. Tsamparlis, A. Paliathanasis, arXiv:1111.5567 [astro-ph.CO];
S. Basilakos, M. Tsamparlis, A. Paliathanasis, Phys. Rev. D 83 (2011) 103512, arXiv:1104.2980.

[10] S. Capozziello, R. De Ritis, P. Scudellaro, Int. J. Mod. Phys. D 2 (1993) 463;
S. Capozziello, R. De Ritis, Class. Quantum Gravity 11 (1994) 107;
R. De Ritis, C. Rubano, P. Scudellaro, Europhys. Lett. 32 (1995) 185;
R. De Ritis, A.A. Marino, C. Rubano, P. Scudellaro, Phys. Rev. D 62 (2000) 043506, arXiv:hep-th/9907198;
A.K. Sanyal, C. Rubano, E. Piedipalumbo, Gen. Relativ. Gravit. 35 (2003) 1617, arXiv:astro-ph/0210063;
E. Piedipalumbo, P. Scudellaro, G. Esposito, C. Rubano, arXiv:1112.0502 [astro-ph.CO].

[11] S. Capozziello, R. De Ritis, C. Rubano, M. Demianski, Phys. Rev. D 52 (1995) 3288;
S. Capozziello, R. De Ritis, A.A. Marino, Class. Quantum Gravity 14 (1998) 3259;
B. Modak, S. Kamilya, Int. J. Mod. Phys. A 13 (1998) 3915;
B. Modak, S. Kamilya, S. Biswas, Gen. Relativ. Gravit. 32 (2000) 1615, arXiv:gr-qc/9909046;
A.K. Sanyal, B. Modak, Class. Quantum Gravity 18 (2001) 3767, arXiv:gr-qc/0107052;
S. Fay, Class. Quantum Gravity 18 (2001) 4863, arXiv:gr-qc/0309088;
S. Kamilya, B. Modak, Gen. Relativ. Gravit. 36 (2004) 673;
H. Motavali, S. Capozziello, M. Rowshan Almeh Jog, Phys. Lett. B 666 (2008) 10.

[12] S. Capozziello, R. De Ritis, P. Scudellaro, Nuovo Cimento B 109 (1994) 159.

[13] S. Capozziello, R. De Ritis, C. Rubano, P. Scudellaro, Int. J. Mod. Phys. D 5 (1996) 85.

[14] K.G. Zloshchastiev, Phys. Rev. D 64 (2001) 084026, arXiv:hep-th/0101075.

[15] S. Capozziello, R. De Ritis, Nuovo Cimento B 109 (1994) 795;
S. Capozziello, G. Lambiase, Gen. Relativ. Gravit. 32 (2000) 295, arXiv:gr-qc/9912084;
A.K. Sanyal, B. Modak, C. Rubano, E. Piedipalumbo, Gen. Relativ. Gravit. 37 (2005) 407, arXiv:astro-ph/0310610;
B. Modak, A. Ghose, R.N. Bose, Gen. Relativ. Gravit. 37 (2005) 985;
K. Sarkar, N. Sk, R. Mandal, A.K. Sanyal, Int. J. Geom. Methods Mod. Phys. 14 (1750), arXiv:1507.03444 [hep-th];
S. Ruz, K. Sarkar, N. Sk, A.K. Sanyal, Mod. Phys. Lett. A 30 (2015) 1550119, arXiv:1505.00740 [gr-qc].

[16] S. Capozziello, P. M-Moruno, C. Rubano, Phys. Lett. B 664 (2008) 12.

[17] B. Vakili, Phys. Lett. B 664 (2008) 16;
B. Vakili, Phys. Lett. B 669 (2008) 206.

[18] K. Sarkar, N. Sk, S. Ruz, S. Debnath, A.K. Sanyal, Int. J. Theor. Phys. 52 (2013) 1515, arXiv:1207.3219 [astro-ph.CO].

[19] N. Sk, A.K. Sanyal, Int. J. Mod. Phys. D 26 (2017) 1750162, https://doi.org/10.1142/S0218271817501620, arXiv:1609.01824 [gr-qc].

[20] S. Capozziello, G. Marmo, C. Rubano, P. Scudellaro, Int. J. Mod. Phys. D 6 (1997) 491, arXiv:gr-qc/9606050;
E. Di Grezia, G. Mangano, G. Miele, Mod. Phys. Lett. A 20 (2005) 605, arXiv:hep-th/0407257.

[21] S. Kamilya, B. Modak, S. Biswas, Gen. Relativ. Gravit. 36 (2004) 661.

[22] A.K. Sanyal, C. Rubano, E. Piedipalumbo, Gen. Relativ. Gravit. 43 (2011) 2807, arXiv:1107.0560.

[23] S. Capozziello, R. De Ritis, P. Scudellaro, Int. J. Mod. Phys. D 3 (1994) 609;
S. Capozziello, V.I. Manko, G. Marmo, C. Stornaiolo, Gen. Relativ. Gravit. 40 (2008) 2627, arXiv:0706.3018;
B. Vakili, F. Khazaie, arXiv:1109.3352 [gr-qc];
B. Vakili, Phys. Lett. B 669 (2008) 206, arXiv:0809.4591.

[24] Y.F. Cai, S. Capozziello, M. De Laurentis, E.N. Saridakis, arXiv:1511.07586 [gr-qc].

[25] I. Hussain, M. Jamil, F.M. Mahomed, Astrophys. Space Sci. 337 (2012) 373.

[26] M. Jamil, F.M. Mahomed, D. Momeni, Phys. Lett. B 315 (2011).

[27] N. Sk, A.K. Sanyal, Astrophys. Space Sci. 342 (2012) 549, arXiv:1302.0411 [astro-ph.CO].

[28] N. Sk, A.K. Sanyal, J. Astrophys. 2013 (2013) 590171, arXiv:1311.2539 [gr-qc].

[29] M.F. Shamir, A. Jhangeer, A.A. Bhatti, Chin. Phys. Lett. 29 (2012) 080402, arXiv:1207.1008 [gr-gc].

[30] N. Sk, A.K. Sanyal, Chin. Phys. Lett. 30 (2013) 020401, arXiv:1302.0411 [astro-ph.CO].

[31] M. Roshan, F. Shojai, Phys. Lett. B 668 (2008) 238, arXiv:0809.1272 [gr-qc].

[32] N. Sk, arXiv:1610.07445 [gr-qc].

[33] K. Sarkar, N. Sk, S. Debnath, A.K. Sanyal, Int. J. Theor. Phys. 52 (2012) 1194, arXiv:1201.2987 [astro-ph.CO].

[34] R.M. Wald, A. Zoupas, Phys. Rev. D 61 (2000) 084027, arXiv:gr-qc/9911095.

[35] S. Capozziello, R. De Ritis, C. Rubano, P. Scudellaro, Riv. Nuovo Cimento 19 (1996) 1.

[36] S. Basilakos, S. Capozziello, M. De Laurentis, A. Paliathanasis, M. Tsamparlis, Phys. Rev. D 88 (2013) 103526, arXiv:1311.2173 [gr-qc].

[37] S. Bahamonde, C.G. Boehmer, M. Wright, Phys. Rev. D 92 (10) (2015) 104042.

[38] S. Bahamonde, S. Capozziello, Eur. Phys. J. C 77 (2017) 107.

[39] N. Sk, A.K. Sanyal, arXiv:1708.02624v1 [gr-qc].

18

Novel vacuum conditions in inflationary collapse models

Gabriel R. Bengochea [a,*], Gabriel León [b]

[a] Instituto de Astronomía y Física del Espacio (IAFE), CONICET – Universidad de Buenos Aires, (1428) Buenos Aires, Argentina
[b] Grupo de Astrofísica, Relatividad y Cosmología, Facultad de Ciencias Astronómicas y Geofísicas, Universidad Nacional de La Plata, Paseo del Bosque S/N (1900) La Plata, Argentina

ARTICLE INFO

Editor: M. Trodden

Keywords:
Cosmology
Inflation
Quantum cosmology

ABSTRACT

Within the framework of inflationary models that incorporate a spontaneous reduction of the wave function for the emergence of the seeds of cosmic structure, we study the effects on the primordial scalar power spectrum by choosing a novel initial quantum state that characterizes the perturbations of the inflaton. Specifically, we investigate under which conditions one can recover an essentially scale free spectrum of primordial inhomogeneities when the standard Bunch–Davies vacuum is replaced by another one that minimizes the renormalized stress–energy tensor via a Hadamard procedure. We think that this new prescription for selecting the vacuum state is better suited for the self-induced collapse proposal than the traditional one in the semiclassical gravity picture. We show that the parametrization for the time of collapse, considered in previous works, is maintained. Also, we obtain an angular spectrum for the CMB temperature anisotropies consistent with the one that best fits the observational data. Therefore, we conclude that the collapse mechanism might be of a more fundamental character than previously suspected.

1. Introduction

Inflation is considered as a fundamental component of the standard ΛCDM cosmological model characterizing the initial stages of the universe [1–4]. Essentially, according to the inflationary paradigm, the early universe underwent an accelerated expansion induced by a scalar field named the inflaton. In addition, it is widely accepted that the quantum fluctuations of the inflaton gave birth to the primordial curvature perturbation, which in turn, generated the primeval density perturbations [5–9]. These primordial perturbations are thus responsible for the origin of all the observed structure in the universe. The predicted properties of such perturbations are consistent with recent observational data from the cosmic microwave background (CMB) [10–12]. In particular, the data are consistent with a nearly scale invariant spectrum associated to the perturbations, which also favors the simplest inflationary models [12,13].

According to the standard inflationary picture, the dynamical expansion of the early universe is governed by Einstein equations which are symmetry preserving; the symmetry being the homogeneity and isotropy. Another important aspect is that, when considering the quantum description of the fields, the vacuum state associated to the inflaton is also homogeneous and isotropic, i.e. it is an eigenstate of the operator generating spatial translations and rotations. Furthermore, the dynamical evolution of the vacuum satisfies the Schrödinger equation, which does not break translational and rotational invariance. As a consequence, we arrive at an important conundrum: it is not clear how from a perfect symmetric initial situation (both in the background spacetime and in the quantum state that characterizes the inflaton), and based on dynamics that preserves the symmetries (the homogeneity and isotropy), one ends up with a final state that is inhomogeneous and anisotropic describing the late observed universe. The aforementioned problem was originally introduced in [14] (and extensively discussed in [15,16]) together with a possible solution: the self-induced collapse hypothesis. The collapse proposal consists that at some point, during the inflationary epoch, a spontaneous change occurs, transforming the original quantum state of the inflaton (the vacuum) into a new quantum state lacking the symmetries of the initial state.

It is worthwhile to mention that the situation we are facing is connected with the so called quantum measurement problem. Sometimes in the literature, the problem is presented as the quantum-to-classical transition of the primordial quantum fluctua-

* Corresponding author.
E-mail addresses: gabriel@iafe.uba.ar (G.R. Bengochea), gleon@fcaglp.unlp.edu.ar (G. León).

tions, and then decoherence is introduced into the picture [17,18]. Although, decoherence can provide a partial understanding of the issue, it does not fully addresses the problem mainly because decoherence does not solve the quantum measurement problem. We will not dwell in all the conceptual aspects regarding the appeal of decoherence during inflation, neither the perceived advantages that the objective reduction models could offer when applied to the early universe. Instead, we referred the interested reader to Refs. [14,15] for a more in depth analysis.

The collapse hypothesis during inflation has been analyzed using two approaches: In the first, one characterizes the post-collapse state phenomenologically through the expectation values and quantum uncertainties of the field, and its conjugated momentum, evaluated at the time of collapse [14,19–21]. In the second approach, one employs a particular collapse mechanism called the continuous spontaneous localization model, where a modification of the Schrödinger equation is proposed, resulting in an objective dynamical reduction of the wave function [22–26]. In both approaches, one obtains a prediction for the scalar and tensor power spectra that, in principle, is different from the standard prediction [27,28]. The first approach has been tested using the most recent data provided by the Planck collaboration, and, under certain circumstances, provides the same Bayesian evidence of the minimal standard cosmological model ΛCDM [29]. Therefore, we will follow the first approach to characterize the self-induced collapse, but the framework exposed in the present work can be extended to the second approach.

Another important feature of the collapse proposal is the adoption of semiclassical gravity [30], which serves to relate the spacetime description in terms of the metric and the degrees of freedom of the inflaton. In the semiclassical picture, gravity is always classical while the matter fields are treated quantum mechanically. We assume such a framework to be a valid approximation during the inflationary era, which is well after the full quantum gravity regime has ended. This approach is different from the standards models of inflation in which metric and matter fields are quantized simultaneously. We should mention that there are many arguments suggesting that the spacetime geometry might emerge from deeper, non-geometrical and fundamentally quantum mechanical degrees of freedom [31–35]. Therefore, in this work, we will employ Einstein semiclassical equations $G_{ab} = 8\pi G\langle \hat{T}_{ab}\rangle$.

On the other hand, the selection of the pre-collapse state, i.e. the vacuum state, which is perfectly homogeneous and isotropic, is not generic. It is known that since we are dealing with a theory of a scalar field (the inflaton) in a curved spacetime, the choice of the vacuum state is not unique [30,36]. Traditionally, the Bunch–Davies (BD) vacuum is selected when considering the quantum theory of the inflaton. The criterion used for the BD vacuum is based on finding a state $|0\rangle$ such that it minimizes the expectation value $\langle 0|\hat{H}(\eta_i)|0\rangle$ at some initial time η_i, with $\hat{H}$ the Hamiltonian associated to the perturbations [37,38]; this prescription is also called Hamiltonian diagonalization. On the other hand, there are known unresolved issues with such procedure. One is that $\langle 0|\hat{H}(\eta_i)|0\rangle$ can be minimized only at an instant η_i; at some other time $\eta_1 > \eta_i$, the BD vacuum does not achieve the sought minimization of the expectation value. In other words, the zero "particle" state is only defined at the time η_i, and as inflation unfolds, the state $|0\rangle$ contains "particles" at other time η_1. Another related issue is that usual renormalization methods, which make $\langle 0|\hat{H}(\eta_i)|0\rangle$ finite, can only be defined at $\eta_i \to -\infty$, that is, at the very early stages of inflation. Some authors consider those arguments sufficient to find alternatives to the Hamiltonian diagonalization method [39–41]. Here it is also important to mention that different choices other than the BD vacuum state have been analyzed previously. For example in Refs. [42–44] it is presented an analysis regarding the observable effects of trans-Planckian physics in the CMB and its relation with a non-BD vacuum. In addition, a non-BD vacuum is usually associated with large non-Gaussianities in the CMB [45,46].

One of the possible alternatives is proposed in Ref. [41]. Those authors suggest that, instead of minimizing $\langle 0|\hat{H}|0\rangle$, one should focus on minimizing the renormalized $\langle \hat{T}_{00}(x)\rangle$. Specifically, the vacuum $|\tilde{0}\rangle$ (which is not the same as the BD vacuum), is such that it minimizes the 0–0 component of the renormalized expectation value of the energy–momentum tensor, which can be considered as a local energy density. Moreover, the vacuum $|\tilde{0}\rangle$ only minimizes the renormalized $\langle \tilde{0}|\hat{T}_{00}(x)|\tilde{0}\rangle$ at some particular time η_0. However, conceptually, it is easier to handle a notion of an instantaneous local energy density minimum than dealing with a notion of "particle" that changes with time. Also, the time η_0 does not need to be taken in the limit $\eta_0 \to -\infty$, although, if one chooses to set η_0 at such early times, then $|\tilde{0}\rangle$ coincides with the prescription of the BD vacuum, but not with its physical interpretation of a "particle-less" state.

All previous works regarding the self-induced collapse proposal, when applied to the inflationary scenario, have been based on selecting the BD vacuum, which is the usual choice in traditional models of inflation as well. Nonetheless, one of the key objects in the inflationary collapse proposal, based on the semiclassical gravity framework, is the expectation value $\langle \hat{T}_{ab}(x)\rangle$. In our approach, if the post-collapse state does not share the same symmetries as the initial-vacuum-state then $\langle \hat{T}_{ab}\rangle$, evaluated in the post-collapse state, will result in a geometry that is no longer homogeneous and isotropic, thus, providing the primordial perturbations for cosmic structure. Therefore, a criterion based on selecting a vacuum state that minimizes the renormalized expectation value of $\hat{T}_{00}$ seems better suited for our picture than one based on choosing a zero "particle" state at a particular time. Furthermore, after the collapse, clearly $\langle \hat{T}_{00}\rangle$ will no longer be the same as the one evaluated in the vacuum state. Hence, if one thinks the collapse as a dynamical process, changing continuously from $|\tilde{0}\rangle$ to the post-collapse state, then it is clear to picture the expectation value of $\hat{T}_{00}$ also changing continuously. In particular, the value $\langle \tilde{0}|\hat{T}_{00}|\tilde{0}\rangle$ will transform from a minimum, which produces a perfectly symmetric spacetime, into a different value generating the perturbations of the geometry.

From discussion above the motivation for the present work is established. That is, we are interested in analyzing the possible effects on the primordial power spectrum generated by choosing the novel prescription based on minimizing the renormalized $\langle \hat{T}_{00}\rangle$. In particular, we are interested in analyzing which aspects of the collapse proposal are modified when the initial conditions are also changed. As we will show, one of our findings indicate that the parametrization of the time of collapse, for each mode of the field, surprisingly remains the same. This led us to think that the physics behind the self-induced collapse of the wave function should be studied in more detail.

The article is organized as follows: in Sect. 2, we review some basics about inflation in the semiclassical gravity framework; in Sect. 3, we analyze the quantization of perturbations, the vacuum choice and present the emergence of curvature perturbation within the collapse hypothesis. Then, we show our prediction for the scalar power spectrum. In Sect. 4 we make a discussion of our results, and finally in Sect. 5 we summarize our conclusions.

Regarding conventions and notation, we will be using a $(-,+,+,+)$ signature for the spacetime metric, and we will use units where $c = 1 = \hbar$.

2. Inflation in the semiclassical picture

In this section, we will summarize some basic concepts regarding the inflationary model in the framework of semiclassical gravity. Extra details can be consulted in previous works (e.g. [14, 19,47,48]).

In the inflationary regime, the dominant type of matter is modeled by a scalar field ϕ, called the inflaton, with a potential V responsible for the accelerating expansion. At the end of the inflationary epoch, the universe follows the standard Big Bang evolution whose transition mechanism is provided by a reheating period.

We begin describing the inflationary universe by the action of a scalar field minimally coupled to gravity,

$$S[\phi, g_{ab}] = \int d^4x\sqrt{-g}\left[\frac{1}{16\pi G}R[g] - \frac{1}{2}\nabla_a\phi\nabla_b\phi g^{ab} - V[\phi]\right]. \tag{1}$$

Varying this last equation with respect to the metric yields the Einstein field equations $G_{ab} = 8\pi G T_{ab}$, with G_{ab} the Einstein tensor.

We will use conformal coordinates and, as usual, we will split the metric and the scalar field into a background perfectly homogeneous and isotropic, plus small perturbations. That is, we write the metric as $g_{ab} = g_{ab}^{(0)} + \delta g_{ab}$, and $\phi = \phi_0(\eta) + \delta\phi(\mathbf{x}, \eta)$, where the background will be represented by a spatially flat FLRW spacetime and the homogeneous part of the scalar field (in the slow-roll regime) by $\phi_0(\eta)$. From Einstein equations for the background, it follows that $G_{00}^{(0)} = 8\pi G T_{00}^{(0)} = 8\pi G a^2\rho$, so the Friedmann equation is $3\mathcal{H}^2 = 8\pi G a^2\rho$ where $\mathcal{H} \equiv a'(\eta)/a(\eta)$ is the conformal Hubble parameter, and $a(\eta)$ is the scale factor. As is customary, the scale factor will be set to $a = 1$ at the present time. Remember that the inflationary phase extends between $-\infty < \eta < \eta_r$, where $\eta_r \approx -10^{-22}$ Mpc is the conformal time when inflation comes to an end. From here on, primes over functions will denote derivatives with respect to the conformal time η. During the inflationary phase, the potential V is the major contribution to the energy density ρ.

In the slow-roll inflationary model, the conformal Hubble parameter is expressed by $\mathcal{H} \simeq -1/[\eta(1-\epsilon_1)]$, with $\epsilon_1 \equiv 1 - \mathcal{H}'/\mathcal{H}^2$ the Hubble slow-roll parameter, which during inflation $1 \gg \epsilon_1 \simeq$ constant.

We will only focus on first-order scalar perturbations; hence, the FLRW perturbed metric can be written as

$$ds^2 = a^2(\eta)\{-(1-2\varphi)d\eta^2 + 2(\partial_i B)dx^i d\eta + \\ + [(1-2\psi)\delta_{ij} + 2\partial_i\partial_j E]dx^i dx^j\}. \tag{2}$$

Within the semiclassical framework, it is convenient to work with the well known gauge-invariant Bardeen potentials. They are defined as $\Phi \equiv \varphi + \frac{1}{a}[a(B - E')]'$ and $\Psi \equiv \psi + \mathcal{H}(E' - B)$. On the other hand, the inflaton perturbation can be modeled by the gauge-invariant fluctuation of the scalar field $\delta\phi^{(\mathrm{GI})}(\eta, \mathbf{x}) = \delta\phi + \phi_0'(B - E')$.

Working with the perturbed Einstein equations (in the absence of anisotropic stress), it can be found that $\Psi = \Phi$. Also, these perturbed equations, along with the Friedmann equation and the equation of motion for ϕ_0 in the slow-roll approximation, i.e. $3\mathcal{H}\phi_0' + a^2\partial_\phi V \approx 0$, imply that (see for instance Appendix A of [49]):

$$\nabla^2\Psi + \mu\Psi = 4\pi G\phi_0'\delta\phi'^{(\mathrm{GI})}, \tag{3}$$

where $\mu \equiv \mathcal{H}^2 - \mathcal{H}' = \epsilon_1\mathcal{H}^2$. In Fourier space, Eq. (3) results,

$$\Psi_{\mathbf{k}}(\eta) = \sqrt{\frac{\epsilon_1}{2}}\frac{H}{M_P(k^2-\mu)}a\delta\phi'_{\mathbf{k}}(\eta)^{(\mathrm{GI})}, \tag{4}$$

with H the Hubble parameter, $M_P \equiv \sqrt{1/(8\pi G)}$ is the reduced Planck mass, and we have also used the definition of ϵ_1. Notice that during most of the inflationary phase, the inequality $k^2 \gg \mu$ is satisfied (both when $|k\eta| \gg 1$ and $|k\eta| \ll 1$). Only when ϵ_1 starts departing from being a constant (i.e. when $\epsilon_1 \to 1$ which means that inflation is ending) that inequality is violated. Given that modes of observational interest are bigger than the Hubble radius ($|k\eta| \ll 1$) while the inflationary phase is still going on, the approximation $k^2 \gg \mu$ remains valid. Therefore, Eq. (4) can be approximated by

$$\Psi_{\mathbf{k}}(\eta) \simeq \sqrt{\frac{\epsilon_1}{2}}\frac{H}{M_P k^2}a\delta\phi'_{\mathbf{k}}(\eta)^{(\mathrm{GI})}. \tag{5}$$

In the semiclassical framework, Eq. (5) can be generalized to

$$\Psi_{\mathbf{k}}(\eta) \simeq \sqrt{\frac{\epsilon_1}{2}}\frac{H}{M_P k^2}a\langle\hat{\delta\phi}'_{\mathbf{k}}(\eta)^{(\mathrm{GI})}\rangle. \tag{6}$$

Last equation is expressed in terms of gauge-invariant quantities $\Psi_{\mathbf{k}}(\eta)$ and $\hat{\delta\phi}'_{\mathbf{k}}(\eta)^{(\mathrm{GI})}$. Note that, in our approach, the metric perturbation will be always a classical quantity.

3. Quantum perturbations, vacuum choice and collapse hypothesis

In this section, we perform the quantization of the perturbations. However, before proceeding with the quantization, we will first briefly address the subject of gauge and its relation with the metric and field perturbations.

In the following, we will choose to work with a fixed gauge and not in terms of the so-called gauge-invariant combinations. We are forced to do so because, in our approach, the adoption of the semiclassical gravity framework leads to consider a classical metric perturbation and a quantum field perturbation, i.e. the metric and field perturbations are treated on a different footing. This contrasts with the standard treatment in which, normally, one chooses to work with gauge invariant quantities which mix matter and geometry degrees of freedom. Then, the quantization results essentially the same for both types of perturbations (matter and geometry).

On the other hand, the choice of gauge implies that the time coordinate is attached to some specific slicing of the perturbed spacetime. And thus, our identification of the corresponding hypersurfaces, those of constant time as the ones associated with the occurrence of collapses–something deemed as an actual physical change–turns what is normally a simple choice of gauge into a choice of the distinguished hypersurfaces, tied to the putative physical process behind the collapse. This naturally leads to tensions with the expected general covariance of a fundamental theory, a problem that afflicts all known collapse models, and which in the non-gravitational settings becomes the issue of compatibility with Lorentz or Poincare invariance of the proposals. We must acknowledge that this generic problem of collapse models is indeed an open issue for the present approach. One would expect that its resolution would be tied to the uncovering of the actual physics behind what we treat here as the collapse of the wave function (which we view as a merely effective description). As it has been argued in related works, and in ideas originally exposed by Penrose [50], we hold that the physics that lies behind all this links the quantum treatment of gravitation with the foundational issues afflicting quantum theory in general; and in particular, those with connection to the so-called "measurement problem".

The gauge we choose is the *longitudinal gauge* ($B = E = 0$). The advantage of working with this gauge is that the action at second order involving the matter and metric perturbations is mathematically the same as the one using gauge invariant quantities, i.e. the Bardeen potentials and $\delta\phi^{(GI)}$. Also, note that Ψ represents the curvature perturbation in this gauge, and it is related to $\delta\phi$ in the exact same way as in Eq. (6) [51]. Therefore, we can be certain that the field perturbations are actual physical degrees of freedom and not pure gauge. Additionally, in our approach, before the collapse (i.e. in the vacuum state) there are no metric perturbations. Hence, the resulting action is the one involving only $\delta\phi$. After the collapse, when the metric perturbations are indeed present, the quantum theory should be modified as presented in [20]. However, we will not consider such backreaction, mainly because we are interested in describing the quantum theory using a non-BD vacuum.

Next, we will present the quantum theory for the field $\delta\phi(\mathbf{x},\eta)$, which will be carried out by choosing a vacuum state different from the BD vacuum. We point out that the criterion used is physically different from the usual BD vacuum. Then, we will characterize the collapse scheme, calculate the curvature perturbation, and finally we will show our expression for the primordial scalar power spectrum.

We start (for simplicity) re-scaling the field variable as $y = a\delta\phi$. Then, we proceed by expanding the action (1) up to second order in the scalar field perturbation y. This results in:

$$\delta S^{(2)} = \int d^4x \frac{1}{2}\bigg[y'^2 - (\nabla y)^2 + \left(\frac{a'}{a}\right)^2 y^2 - 2\left(\frac{a'}{a}\right) y y' - y^2 a^2 \partial^2_{\phi\phi} V \bigg]. \tag{7}$$

Therefore, the canonical momentum conjugated to y is $\pi \equiv \partial \delta\mathcal{L}^{(2)}/\partial y' = y' - (a'/a) y = a\delta\phi'$.

In order to facilitate the calculations, we will neglect the slow roll parameters ϵ_1 and $\epsilon_2 \equiv \epsilon_1'/(\mathcal{H}\epsilon_1)$ in the quantization procedure. At the end of the computations, we will argue how we can generalize our result to the quasi-de Sitter case, in which the slow roll parameters are considered.

Now, the field and momentum variables are promoted to operators satisfying the equal time commutator relations $[\hat{y}(\mathbf{x},\eta), \hat{\pi}(\mathbf{x}',\eta)] = i\delta(\mathbf{x}-\mathbf{x}')$ and $[\hat{y}(\mathbf{x},\eta), \hat{y}(\mathbf{x}',\eta)] = 0 = [\hat{\pi}(\mathbf{x},\eta), \hat{\pi}(\mathbf{x}',\eta)]$. Expanding the fields operators in Fourier modes yields

$$\hat{y}(\eta,\mathbf{x}) = \frac{1}{L^3}\sum_{\mathbf{k}} \hat{y}_{\mathbf{k}}(\eta) e^{i\mathbf{k}\cdot\mathbf{x}} \tag{8}$$

$$\hat{\pi}(\eta,\mathbf{x}) = \frac{1}{L^3}\sum_{\mathbf{k}} \hat{\pi}_{\mathbf{k}}(\eta) e^{i\mathbf{k}\cdot\mathbf{x}} \tag{9}$$

where the sums are over the wave vectors $\mathbf{k}$, satisfying $k_i L = 2\pi n_i$ for $i = 1,2,3$ with n_i integer. Also, we have defined $\hat{y}_{\mathbf{k}}(\eta) \equiv y_k(\eta)\hat{a}_{\mathbf{k}} + y_k^*(\eta)\hat{a}^\dagger_{-\mathbf{k}}$ and $\hat{\pi}_{\mathbf{k}}(\eta) \equiv g_k(\eta)\hat{a}_{\mathbf{k}} + g_k^*(\eta)\hat{a}^\dagger_{-\mathbf{k}}$, with $g_k(\eta) = y_k'(\eta) - \mathcal{H} y_k(\eta)$ and $\hat{a}_{\mathbf{k}}, \hat{a}^\dagger_{\mathbf{k}}$ being the usual annihilation/creator operators, respectively. Note that the quantization is on a finite cubic box of length L, and at the end of the calculations we will take the continuum limit ($L \to \infty$, $\mathbf{k} \to$ cont.).

From action (7), the equation of motion for $y_k(\eta)$ results in

$$y_k''(\eta) + \left(k^2 - \frac{a''}{a}\right) y_k(\eta) = 0 \tag{10}$$

with $a''/a = 2/\eta^2$. The general solution is,

$$y_k(\eta) = A_k\left(1 - \frac{1}{k\eta}\right)e^{-ik\eta} + B_k\left(1 + \frac{1}{k\eta}\right)e^{ik\eta} \tag{11}$$

where A_k and B_k are two constants (dependent on k) that will be fixed by the initial conditions at some η_0.

Therefore, to complete the quantization, we have to specify the solutions $y_k(\eta)$, through constants A_k and B_k. This choice is not completely free; to insure that canonical commutation relations between $\hat{y}$ and $\hat{\pi}$ give $[\hat{a}_{\mathbf{k}}, \hat{a}^\dagger_{\mathbf{k}'}] = L^3\delta_{\mathbf{k},\mathbf{k}'}$, they must satisfy:

$$y_k g_k^* - y_k^* g_k = i \tag{12}$$

for all k at some (and hence any) time η.

The choice of the $y_k(\eta)$ corresponds to the choice of a vacuum state $|0\rangle$ for the field, defined by $\hat{a}_{\mathbf{k}}|0\rangle = 0$ for all $\mathbf{k}$. In the present case, as on any non stationary spacetime, it is not unique. Condition (12) is not sufficient to fully determine $y_k(\eta)$. The traditional approach in inflationary models is to consider the (homogeneous and isotropic) so-called BD vacuum. In this case, the choice corresponds to the situation in which, when $k\eta_0 \to -\infty$, the solution $y_k(\eta) \to \frac{1}{\sqrt{2k}} e^{-ik\eta}$; this is, the solutions are the same as the ones with positive frequencies in the flat Minkowski spacetime. In the case of inflation in a quasi-de Sitter background, this last condition together with (12) correspond to fix $B_k = 0$ and $|A_k| = \sqrt{\frac{\pi}{4k}}$. Readers interested in how the quasi-de Sitter case is analyzed within collapse schemes, when the BD vacuum is chosen as the initial condition (and where the prediction for the scalar spectral index is $n_s \neq 1$), are invited to see the work [49].

At this point, we must make a short digression regarding our conceptual approach and its differences with the standard picture. Any selection of a vacuum (made through the choice of the $y_k(\eta)$ that we take as positive energy modes), would be a spatially homogeneous and isotropic state of the field, as it can be seen by evaluating directly the action of a translation or rotation operators (associated with the hypersurfaces η = constant of the background spacetime) on the state $|0\rangle$. A formal proof of this can be found, for instance, in Appendix A of [16]. As the dynamical evolution (through Schrödinger equation) preserves such symmetries, the state of the system will be symmetric (homogeneous and isotropic) at all times. In fact, there is nothing, given the standard unitary evolution of the quantum theory, that could be invoked to avoid such conclusion. The issue is then: How do we account for a universe with seeds of cosmic structure, starting from an isotropic and homogeneous background spacetime and an equally symmetric vacuum state? Note that this is an open issue in all current models of inflation relying in the traditional treatment of the primordial perturbations.

As we mentioned in the Introduction, one possible solution to the aforementioned problem relies on supplementing the standard inflationary model with an hypothesis involving the modification of quantum theory so as to include a spontaneous dynamical reduction of the quantum state (sometimes referred as the self-induced collapse of the wave function) [14,15]. The dynamical reduction can be considered as an actual physical process taking place independently of observers or measuring devices. Therefore, our approach regarding the origin of the primordial perturbations can be summarize as follows: a few e-folds after inflation has started, the universe finds itself in an homogeneous and isotropic quantum state. Then, during the inflationary regime, a quantum collapse of the wave function is triggered (by novel physics that could possibly be related to quantum gravitational effects), breaking in the process the unitary evolution of quantum mechanics and also, in general, the symmetries of the original state. That is, the post-collapse state will not be, in general, isotropic nor homogeneous. Also, the collapse mechanism functions as a generator of the metric perturbations, as will become clear below.

Readers familiar with the subject might take the posture that the problem we are characterizing is equivalent to the quantum-to-classical transition of the primordial perturbations. Several works in the literature, based on decoherence or evolution of the vacuum state into a squeezed state, have dealt with such a problem (see e.g. [17,18]). On the other hand, in Refs. [14–16] it is exposed why such arguments are not entirely convincing. Nevertheless, in the standard approach, at some point during inflation occurs the transition $\hat{\Psi}_{\mathbf{k}} \to \Psi_{\mathbf{k}} = Ae^{i\alpha_{\mathbf{k}}}$, with $\alpha_{\mathbf{k}}$ a random phase (recall that $\Psi_{\mathbf{k}}$ represents the metric perturbation). The amplitude A is identified with the quantum uncertainty of $\hat{\Psi}_{\mathbf{k}}$, i.e. $A^2 = \langle 0|\hat{\Psi}_{\mathbf{k}}^2|0\rangle$. Moreover, quantum expectation values are identified with ensemble averages of classical stochastic fields based on postulate, and the theoretical predictions agree with the observational data. Finally, note that in our approach, because our reliance on semiclassical gravity, the primordial curvature perturbation is always a classical quantity.

In the next subsections, we are going to analyze whether the replacement of the BD vacuum state by another one (motivated by different physical criteria) can affect the primordial scalar power spectrum, under the incorporation of the collapse hypothesis. As we will see, under certain conditions, one can recover a scale free spectrum for scalar perturbations, but generically there would be some characteristic deviations thereof. Note that neglecting the slow roll parameters indicates that our prediction for the primordial scalar power spectrum should be an essentially scale free spectrum, i.e. $\mathcal{P}(k) \propto \frac{1}{k^3}$. On the other hand, the observations (e.g. CMB [12]) suggest that $\mathcal{P}(k) \propto k^{-3+\mathcal{O}(\epsilon_1,\epsilon_2)}$. In other words, the scalar spectral index is such that $n_s \neq 1$. We think that when reincorporating the slow-roll parameters in the equation of motion for the field variable, we would obtain a prediction for n_s consistent with the observational data. However, the modification of the $\mathcal{P}(k)$ induced by the collapse hypothesis, would be practically the same as the one obtained using the mode functions, Eq. (11), which neglects the slow roll parameters. In fact, one is led to a similar conclusion in Refs. [14,21], in which the BD vacuum was chosen.

3.1. Novel vacuum conditions

As it is well known, the choice of a vacuum state is not unique in spacetimes that do not possess a time-like Killing field. This is precisely the case when, for example, we try to describe the inflationary phase of the early universe. There are several ways to choose the initial conditions; some of which can be seen in [52,53].

Traditionally, quantum initial conditions for perturbations in inflation are set using the BD vacuum.

The typical selection of the BD vacuum, described previously, can be deduced, for example, looking for the conditions that modes $y_k(\eta)$ must satisfy to achieve the diagonalization of the Hamiltonian of perturbations. However, those conditions are satisfied for a given initial time η_0; because as is known, in a curved spacetime the vacuum is a time-dependent notion. Hamiltonian diagonalization is the simplest approach for setting quantum initial conditions in a general spacetime, and derives the vacuum from the minimization of the Hamiltonian density. However, this approach has been criticised in the past [36,39].

In order to avoid the issues raised against Hamiltonian diagonalization, the authors in [41] motivated different initial conditions from the minimization of the *renormalized* stress–energy density. The authors in [41], start from the action for a scalar field ϕ with mass m,

$$S = \int d^4x\sqrt{-g}\Big[-\frac{1}{2}\nabla_a\phi\nabla_b\phi g^{ab} - \frac{1}{2}m^2\phi^2\Big]. \tag{13}$$

By expanding the field ϕ in Fourier modes in the context of a FLRW spacetime as

$$\phi(x) = \int \frac{d^3k}{(2\pi)^3 a(\eta)}\Big[\hat{a}_{\mathbf{k}}\chi_{\mathbf{k}}(\eta)e^{i\mathbf{k}\cdot\mathbf{x}} + \hat{a}_{\mathbf{k}}^\dagger\chi_{\mathbf{k}}^*(\eta)e^{-i\mathbf{k}\cdot\mathbf{x}}\Big], \tag{14}$$

if the field satisfies its equation of motion, and the commutation relation between $\hat{a}_{\mathbf{k}}$ and $\hat{a}_{\mathbf{k}}^\dagger$ is the standard one, then it is well known that the mode functions $\chi_{\mathbf{k}}$ must satisfy:

$$\chi_{\mathbf{k}}'' + \Big[k^2 + m^2a^2 - \frac{a''}{a}\Big]\chi_{\mathbf{k}} = 0 \tag{15}$$

$$\chi_{\mathbf{k}}\chi_{\mathbf{k}}'^{*} - \chi_{\mathbf{k}}^{*}\chi_{\mathbf{k}}' = i \tag{16}$$

Later, the authors computed a renormalized stress–energy tensor, $\langle 0|\hat{T}_{ab}|0\rangle_{\text{ren}}$, via a Hadamard point splitting procedure. To do that, they build the stress-tensor $\langle 0|\hat{T}_{ab}|0\rangle_{\text{ren}}$ using the Hadamard Green function with the mode expansion (14), and subtracting off de-Witt–Schwinger geometrical terms to obtain a non-divergent quantity. Then, finally they write:

$$\langle 0|\hat{T}_{00}(x)|0\rangle_{\text{ren}} = \frac{1}{2}\int \frac{d^3k}{(2\pi)^3a^2}(\chi_{\mathbf{k}}' - \frac{a'}{a}\chi_{\mathbf{k}})(\chi_{\mathbf{k}}'^{*} - \frac{a'}{a}\chi_{\mathbf{k}}^{*}) + (k^2 + m^2a^2)\chi_{\mathbf{k}}\chi_{\mathbf{k}}^{*} + \tilde{T} \tag{17}$$

where $\tilde{T}$ signifies additional terms arising from the renormalization process that have no dependence on the variables $\Sigma = \{\chi_{\mathbf{k}}, \chi_{\mathbf{k}}^{*}, \chi_{\mathbf{k}}', \chi_{\mathbf{k}}'^{*}\}$. Minimizing (17) with respect to Σ, subject to the normalization (16), yields the relations [41]:

$$|\chi_{\mathbf{k}}|^2 = \frac{1}{2\sqrt{k^2 + m^2a^2}} \tag{18}$$

$$\chi_{\mathbf{k}}' = \Big(-i\sqrt{k^2 + m^2a^2} + \frac{a'}{a}\Big)\chi_{\mathbf{k}} \tag{19}$$

Conditions (18) and (19) will be our guide to determine the novel vacuum conditions in the present work.

Now, let us return to our particular situation. Quantum field theory in curved spacetime describes the effects of gravity upon the quantum fields. The semiclassical Einstein equation describes how quantum fields act as the source of gravity. This equation is usually taken to be the classical Einstein equation, with the source as the quantum expectation value of the matter field stress–energy tensor operator $\hat{T}_{ab}$, that is,

$$G_{ab} = 8\pi G\langle \hat{T}_{ab}\rangle. \tag{20}$$

But, this expectation value is only defined after suitable regularization and renormalization.

As already mentioned, since we are not interested in the effects on the power spectrum (and its scalar spectral index) coming from slow-roll parameters, we assume $m = 0$ in Eq. (15), which is equivalent to neglect the second slow roll parameter ϵ_2. Moreover, note that the equation of motion for $y_k(\eta)$, Eq. (10), is identical to Eq. (15), which is the one obtained by the authors of [41] when $m = 0$. In particular, it involves the quantity $\frac{a''}{a}$. This contrasts with the traditional procedure involving the Mukhanov–Sasaki variable, which results in an equation of motion similar in structure to Eq. (15), but replacing $\frac{a''}{a} \to \frac{z''}{z}$, where $z \equiv \sqrt{2\epsilon_1}aM_P$. In other words, the quantum theory proposed by the authors of [41] is better suited for the field $y_k(\eta)$, than for the Mukhanov–Sasaki variable because strictly if $\epsilon_1' \neq 0$, then $\frac{a''}{a} \neq \frac{z''}{z}$.

Therefore, we consider once again, the general solution (11), which is

$$y_k(\eta) = A_k\Big(1 - \frac{1}{k\eta}\Big)e^{-ik\eta} + B_k\Big(1 + \frac{1}{k\eta}\Big)e^{ik\eta}. \tag{21}$$

Here, without loss of generality, we will assume $A_k \in \mathbb{R}$ and $B_k \in \mathbb{C}$. That is, only B_k will carry a complex phase. Normalization (12) imposes that

$$A_k^2 - |B_k|^2 = \frac{1}{2k}. \tag{22}$$

On the other hand, identifying $\chi_{\mathbf{k}}$ with $y_k(\eta)$, conditions (18) and (19) yield

$$A_k = +\sqrt{\frac{4z_0^2+1}{8kz_0^2}} \quad (23)$$

$$B_k = \frac{1}{+\sqrt{8k}\,|z_0|}\, e^{i\beta} \quad (24)$$

where we have defined $z_0 \equiv k\eta_0$ and $\beta \equiv -2z_0 + \arctan(2z_0) + \pi$. Note that when $z_0 \to -\infty$, the BD vacuum is recovered; this is, $A_k = \frac{1}{\sqrt{2k}}$ and $B_k = 0$.

Equation (21) together with Eqs. (23) and (24) constitute our choice of initial vacuum conditions at time η_0. In the next subsection, we will introduce the specific collapse scheme and calculate the primordial curvature perturbation.

3.2. Emergence of curvature perturbation within a collapse scheme

In this subsection, we are going to consider a modification to the standard inflationary proposal, designed to account for breaking the symmetries of the initial quantum state, leading to the generation of the primordial inhomogeneities.

As we have claimed, when considering a quantum description for the early universe, one must face the situation in which a completely homogeneous and isotropic stage must nevertheless lead, after some time, to a universe containing actual inhomogeneities and anisotropies. This issue has been considered at length in other works, including detailed discussions of the shortcomings of the most popular attempts to address the problem, and we will not repeat such extensive discussions here. It is clear that such transition from a symmetric situation to one that is not, cannot be simply the result of quantum unitary evolution, since, as we noted, the dynamics does not break these initial symmetries of the system. As discussed in [15], and despite multiple claims to the contrary (e.g. [18]), there is no satisfactory solution to this problem within the standard physical paradigms.

The proposal to handle this shortcoming was considered for the first time in [14]. There, the problem was addressed by introducing a new ingredient into the inflationary account of the origin of the seeds of cosmic structure: the self-induced collapse hypothesis. The basic idea is that an internally induced spontaneous collapse of the wave function of the inflaton field is the mechanism by which inhomogeneities and anisotropies arise at each particular scale. That proposal was inspired on early ones for the resolution of the measurement problem in quantum theory [54–58], which regarded the collapse of the wave function as an actual physical process taking place spontaneously. Also, on the ideas by R. Penrose and L. Diosi [50,59,60] who assumed that such process should be connected to quantum aspects of gravitation.

A *collapse scheme* [14,19] is a recipe to characterize and select the state into which each of the modes of the scalar field jumps at the corresponding time of collapse. The collapse itself is described in a purely phenomenological manner, without reference to any particular mechanism. As reported in, for instance, [14,29,48,61], the different collapse schemes generally give rise to different characteristic departures from the conventional Harrison–Zel'dovich flat primordial spectrum. There are, of course, more sophisticated theories describing the collapse dynamics, such as those in [22–25,55–58,62]. However, we will not consider those in the present study, which is meant a first exploration of such ideas in the context of different choices of the initial quantum state.

The self-induced collapse hypothesis is based on assuming that the collapse acts similar to a "measurement" (in an early universe where, clearly, there are no external observers or measuring devices), this lead us to consider Hermitian operators, which in ordinary quantum mechanics are the ones susceptible of direct measurement. Therefore, we will separate $\hat{y}_{\mathbf{k}}(\eta)$ and $\hat{\pi}_{\mathbf{k}}(\eta)$ into their real and imaginary parts: $\hat{y}_{\mathbf{k}}(\eta) = \hat{y}_{\mathbf{k}}{}^R(\eta) + i\hat{y}_{\mathbf{k}}{}^I(\eta)$ and $\hat{\pi}_{\mathbf{k}}(\eta) = \hat{\pi}_{\mathbf{k}}{}^R(\eta) + i\hat{\pi}_{\mathbf{k}}{}^I(\eta)$. In this way, the operators $\hat{y}_{\mathbf{k}}^{R,I}(\eta)$ and $\hat{\pi}_{\mathbf{k}}^{R,I}(\eta)$ are Hermitian operators and then they can be written as,

$$\hat{y}_{\mathbf{k}}^{R,I}(\eta) = \sqrt{2}\mathrm{Re}[y_k(\eta)\hat{a}_{\mathbf{k}}^{R,I}] \quad (25a)$$

$$\hat{\pi}_{\mathbf{k}}^{R,I}(\eta) = \sqrt{2}\mathrm{Re}[g_k(\eta)\hat{a}_{\mathbf{k}}^{R,I}] \quad (25b)$$

where $\hat{a}_{\mathbf{k}}^R \equiv \frac{1}{\sqrt{2}}(\hat{a}_{\mathbf{k}} + \hat{a}_{-\mathbf{k}})$, and $\hat{a}_{\mathbf{k}}^I \equiv \frac{-i}{\sqrt{2}}(\hat{a}_{\mathbf{k}} - \hat{a}_{-\mathbf{k}})$; and where the non-standard commutation relations for the $\hat{a}_{\mathbf{k}}^{R,I}$ are,

$$[\hat{a}_{\mathbf{k}}^{R,I}, \hat{a}_{\mathbf{k}'}^{R,I\dagger}] = L^3(\delta_{\mathbf{k},\mathbf{k}'} \pm \delta_{\mathbf{k},-\mathbf{k}'}). \quad (26)$$

In the last equation, the + and − signs correspond to the commutators with the R and I labels respectively; and all other commutators vanish.

Next, we will show how in our approach the quantum theory of the inflaton perturbations can be connected with the primordial curvature perturbation. Moreover, we will illustrate how the collapse process generates the seeds of cosmic structure. Here, we will proceed by choosing to work in the longitudinal gauge, and then, since $\hat{\pi}_{\mathbf{k}} = a\delta\hat{\phi}'_{\mathbf{k}}$, we will express Eq. (6) in terms of the expectation value of the conjugated momentum. Thus,

$$\Psi_{\mathbf{k}}(\eta) \simeq \sqrt{\frac{\epsilon_1}{2}}\frac{H}{M_P k^2}\langle\hat{\pi}_{\mathbf{k}}(\eta)\rangle. \quad (27)$$

At the initial conformal time η_0, the state $|0\rangle$ is perfectly symmetric, which implies that $\langle\hat{\pi}_{\mathbf{k}}(\eta)\rangle = 0$ and so, $\Psi_{\mathbf{k}} = 0$; i.e. there are no perturbations of the symmetric background spacetime. Afterwards, under the self-induced collapse hypothesis, at some later time η_k^c, called the time of collapse, a transition to a new state $|0\rangle \to |\Theta\rangle$ is produced, which does not have the initial symmetries. And in this new state, we will have that $\langle\hat{\pi}_{\mathbf{k}}(\eta)\rangle_\Theta \neq 0$ for all $\eta \geq \eta_k^c$, and $\Psi_{\mathbf{k}} \neq 0$. From Eq. (27), which was provided by the semiclassical framework, and given that all modes of the inflaton field are now in the post-collapse state $|\Theta\rangle$, we can clearly see that the expectation value $\langle\hat{\pi}_{\mathbf{k}}(\eta)\rangle$ serves as a source for $\Psi_{\mathbf{k}}$ for all $\mathbf{k}$. These collapses will be assumed to take place according to certain collapse scheme which we will describe in detail below.

Taking into account Eq. (27), and that the collapse is somehow analogous to an imprecise measurement of the operators $\hat{y}_{\mathbf{k}}^{R,I}(\eta)$ and $\hat{\pi}_{\mathbf{k}}^{R,I}(\eta)$, our next objective is to find an equation for the dynamics of the expectation values of $\langle\hat{\pi}_{\mathbf{k}}^{R,I}(\eta)\rangle$, evaluated in the post-collapse state. This equations, as we shall see, will be related to the values $\langle\hat{y}_{\mathbf{k}}^{R,I}(\eta_k^c)\rangle$ and $\langle\hat{\pi}_{\mathbf{k}}^{R,I}(\eta_k^c)\rangle$, through the proposed collapse scheme.

In the vacuum state, $\hat{y}_{\mathbf{k}}$ and $\hat{\pi}_{\mathbf{k}}$ individually are distributed according to Gaussian wave functions centered at zero with spread $(\Delta\hat{y}_{\mathbf{k}})_0^2$ and $(\Delta\hat{\pi}_{\mathbf{k}})_0^2$, respectively. Our assumption is that the effect of the collapse on a state is analogous to some sort of approximate measurement. Therefore, after the collapse the expectation values of the field and momentum operators, in each mode, will be related to the uncertainties of the initial state.

We will adopt a collapse scheme, where it is assumed that the expectation values of the field mode $\hat{y}_{\mathbf{k}}^{R,I}$ and their conjugate momentum $\hat{\pi}_{\mathbf{k}}^{R,I}$ acquire independent values randomly, and where the expectation (in the new state $|\Theta\rangle$) at the time of collapse is given by:

$$\langle\hat{y}_{\mathbf{k}}^{R,I}(\eta_k^c)\rangle_\Theta = \lambda_1\, x_{\mathbf{k},1}^{R,I}\sqrt{\left(\Delta\hat{y}_{\mathbf{k}}^{R,I}(\eta_k^c)\right)_0^2} \quad (28a)$$

$$\langle\hat{\pi}_{\mathbf{k}}^{R,I}(\eta_k^c)\rangle_\Theta = \lambda_2\, x_{\mathbf{k},2}^{R,I}\sqrt{\left(\Delta\hat{\pi}_{\mathbf{k}}^{R,I}(\eta_k^c)\right)_0^2} \quad (28b)$$

The parameters λ_1 and λ_2 are viewed as "switch-off/on" parameters. This is, they can only take the values 0 or 1 depending on which variable $\hat{y}_{\mathbf{k}}^{R,I}$, $\hat{\pi}_{\mathbf{k}}^{R,I}$ or both is affected by the collapse. For instance, in past works [14,63], the name *independent scheme* was coined for the case $\lambda_1 = 1 = \lambda_2$, i.e. $\hat{y}_{\mathbf{k}}^{R,I}$ and $\hat{\pi}_{\mathbf{k}}^{R,I}$ are both affected independently by the collapse. In this scheme the expectation value jumps to a random value $x_{\mathbf{k}}^{(R,I)}$ multiplied by the uncertainty of the vacuum state of the field. The random variables $x_{\mathbf{k},1}^{(R)}, x_{\mathbf{k},1}^{(I)}, x_{\mathbf{k},2}^{(R)}$ and $x_{\mathbf{k},2}^{(I)}$ are selected from a Gaussian distribution centered at zero, with unity spread, and all of them are assumed statistically uncorrelated. In Appendix A we show the quantum uncertainties $\left(\Delta\hat{y}_{\mathbf{k}}^{R,I}(\eta_k^c)\right)_0^2$ and $\left(\Delta\hat{\pi}_{\mathbf{k}}^{R,I}(\eta_k^c)\right)_0^2$ in the vacuum state within this collapse scheme.

Having established the relation between the curvature perturbation and the quantum matter fields [see Eq. (27)], as well as the characterization of the collapse, the next aim is to present an explicit expression for the curvature perturbation in terms of the parameters characterizing the collapse scheme. In order to achieve that goal, we must first find an expression for the evolution of the expectation values of the fields. In fact, as can be seen from Eq. (27), we will only be concerned with the expectation value of the conjugated momentum $\langle\hat{\pi}_{\mathbf{k}}(\eta)\rangle$.

In Appendix A, we show that:

$$\langle\hat{\pi}_{\mathbf{k}}(\eta)\rangle_\Theta = -L^{3/2}\Big[F(k\eta, z_k)\,|y_k(\eta_k^c)|\,\lambda_1\,X_{\mathbf{k},1} + \\ + G(k\eta, z_k)\,|g_k(\eta_k^c)|\,\lambda_2\,X_{\mathbf{k},2}\Big] \tag{29}$$

where $X_{\mathbf{k},1} \equiv x_{\mathbf{k},1}^R + i\,x_{\mathbf{k},1}^I$, $X_{\mathbf{k},2} \equiv x_{\mathbf{k},2}^R + i\,x_{\mathbf{k},2}^I$, and the functions $F(k\eta, z_k)$, $G(k\eta, z_k)$, $|y_k(\eta_k^c)|$ and $|g_k(\eta_k^c)|$ are also defined in Appendix A. Notice that the constants A_k and B_k appear in all these functions, and is in such constants that the information about the initial conditions is found, through its dependence with $z_0 \equiv k\eta_0$. Also, the parameter z_k is defined as $z_k \equiv k\eta_k^c$; thus, z_k is directly associated to the time of collapse η_k^c.

Finally, substituting Eq. (29) in Eq. (27), we find the expression for the curvature perturbation (in the longitudinal gauge):

$$\Psi_{\mathbf{k}}(\eta) = -\sqrt{\frac{\epsilon_1}{2}}\frac{H\,L^{3/2}}{M_P\,k^2}\Big[F(k\eta, z_k)\,|y_k(\eta_k^c)|\,\lambda_1\,X_{\mathbf{k},1} + \\ + G(k\eta, z_k)\,|g_k(\eta_k^c)|\,\lambda_2\,X_{\mathbf{k},2}\Big] \tag{30}$$

The curvature perturbation Ψ in the longitudinal gauge, is a constant quantity for modes "outside the horizon" during any given cosmological epoch, but not during the transition between epochs. In fact, during the transition from the inflationary stage to the radiation dominated stage, Ψ is amplified by a factor of $1/\epsilon_1$ [51,64]. On the other hand, an useful gauge-invariant quantity often encountered in the literature is the variable $\mathcal{R}(x)$. The field $\mathcal{R}(x)$ is a field representing the curvature perturbation in the *comoving gauge.* Its Fourier transform, represented by $\mathcal{R}_{\mathbf{k}}$, is constant for modes greater than the Hubble radius (irrespectively of the cosmological epoch), i.e. for modes with $k \ll \mathcal{H} = aH$ (and assuming adiabatic perturbations). This is, the value of $\mathcal{R}_{\mathbf{k}}$ during inflation (in the limit $k \ll \mathcal{H}$) will remain unchanged during the post-inflationary evolution, until the mode "re-enters the horizon", namely when $k \simeq \mathcal{H}$, at the later radiation/matter dominatted epochs. The curvature perturbation in the comoving gauge $\mathcal{R}$ and the curvature perturbation in the longitudinal gauge Ψ are related by $\mathcal{R} \equiv \Psi + (2/3)(\mathcal{H}^{-1}\Psi' + \Psi)/(1+\omega)$, with $\omega \equiv p/\rho$. Therefore, for modes such that $k \ll \mathcal{H}$, during the inflationary epoch $\omega + 1 \simeq 2\epsilon_1/3$, it is found that

$$\lim_{k\ll\mathcal{H}} \mathcal{R}_{\mathbf{k}}(\eta) \simeq \lim_{k\ll\mathcal{H}} \frac{\Psi_{\mathbf{k}}(\eta)}{\epsilon_1} \tag{31}$$

with $\Psi_{\mathbf{k}}(\eta)$, calculated during inflation, in the limit such that the modes are well outside the horizon.

In the next subsection, we are going to find an expression for the scalar power spectrum of the perturbations $\mathcal{R}_{\mathbf{k}}$, namely $P(k)$. The power spectrum serves as the initial condition to obtain the angular spectrum, denoted usually in the literature as C_l, which is the theoretical prediction that is contrasted with the observations directly. Henceforth, we will obtain an expression for $\mathcal{R}_{\mathbf{k}}$ during inflation explicitly, for the observationally relevant modes. Specifically, we take the limit $|k\eta| \to 0$ in the functions $F(k\eta, z_k)$ and $G(k\eta, z_k)$, and we define $M(z_0, z_k) \equiv \lim_{|k\eta|\to 0} F(k\eta, z_k)\,|y_k(\eta_k^c)|$ and $N(z_0, z_k) \equiv \lim_{|k\eta|\to 0} G(k\eta, z_k)\,|g_k(\eta_k^c)|$. Thus,

$$\mathcal{R}_{\mathbf{k}} = \frac{-HL^{3/2}}{\sqrt{2\epsilon_1}M_P k^2}\Big[M(z_0, z_k)\,\lambda_1\,X_{\mathbf{k},1} + N(z_0, z_k)\,\lambda_2\,X_{\mathbf{k},2}\Big] \tag{32}$$

Note the explicit dependence on the initial time η_0 through the quantity $z_0 \equiv k\eta_0$. Equation (32) is the main result of this section.

We strongly remark that the random variables $X_{\mathbf{k}}$ corresponding to the collapse scheme are fixed after the collapse of the wave function has occurred. In other words, if we somehow knew their exact value, we would be able to predict the exact value for $\mathcal{R}_{\mathbf{k}}$. We will do make use of the statistical properties of these random variables to be able to make theoretical predictions for the observational quantities. For a more detailed comparison between our approach and the traditional inflationary picture see Ref. [61].

Next, we will consider whether, and under what circumstances, one can obtain a prediction for the power spectrum of the scalar perturbations $\mathcal{R}_{\mathbf{k}}$, for the case of the observationally relevant modes, which have wavelengths greater than the Hubble radius at the time of inflation.

3.3. *Primordial scalar power spectrum*

In this subsection, we will calculate the primordial power spectrum of the scalar perturbations $\mathcal{R}_{\mathbf{k}}$, and analyze its relation with the CMB observations under our approach.

The temperature anisotropies of the CMB, $\delta T/T_0$, are the most direct observational quantity available, with T_0 the mean temperature today. Expanding $\delta T/T_0$ using spherical harmonics, the coefficients a_{lm} are

$$a_{lm} = \int \Theta(\hat{n})Y_{lm}^*(\theta, \varphi)d\Omega, \tag{33}$$

with $\hat{n} = (\sin\theta\sin\varphi, \sin\theta\cos\varphi, \cos\theta)$ and θ, φ the coordinates on the celestial two-sphere. By defining $\Theta(\hat{n}) \equiv \delta T(\hat{n})/T_0$ and assuming instantaneous recombination, the relation between the primordial perturbations and the observed CMB temperature anisotropies is

$$\Theta(\hat{n}) = [\Psi + \frac{1}{4}\delta_\gamma](\eta_D) + \hat{n}\cdot\vec{v}_\gamma(\eta_D) + 2\int_{\eta_D}^{\eta_0}\Psi'(\eta)d\eta, \tag{34}$$

where η_D is the time of decoupling; δ_γ and $\vec{v}_\gamma$ are the density perturbations and velocity of the radiation fluid.

On the other hand, the temperature anisotropies in Fourier modes is

$$\Theta(\hat{n}) = \sum_{\mathbf{k}} \frac{\Theta(\mathbf{k})}{L^3} e^{i\mathbf{k}\cdot R_D\hat{n}} \tag{35}$$

being R_D the radius of the last scattering surface. Then, the fluid motion equations can be solved with the initial condition provided by the curvature perturbation during inflation. Furthermore, using that $e^{i\mathbf{k}\cdot R_D\hat{n}} = 4\pi \sum_{lm} i^l j_l(kR_D) Y_{lm}(\theta, \varphi) Y^*_{lm}(\hat{k})$, expression (33) can be rewritten as

$$a_{lm} = \frac{4\pi i^l}{L^3} \sum_{\mathbf{k}} j_l(kR_D) Y^*_{lm}(\hat{k}) \Theta(\mathbf{k}), \tag{36}$$

with $j_l(kR_D)$ the spherical Bessel function of order l. To incorporate the linear evolution that relates the initial curvature perturbation $\mathcal{R}_{\mathbf{k}}$ and the anisotropies $\Theta(\mathbf{k})$, is defined a transfer function $T(k)$. This function results of solving the fluid motion equations (for each mode) with the initial condition provided by the curvature perturbation $\mathcal{R}_{\mathbf{k}}$, and then make use of Eq. (34) to relate it with the temperature anisotropies. In this way, $\Theta(\mathbf{k}) = T(k)\mathcal{R}_{\mathbf{k}}$. Therefore, the coefficients a_{lm}, in terms of the modes $\mathcal{R}_{\mathbf{k}}$, are given by

$$a_{lm} = \frac{4\pi i^l}{L^3} \sum_{\mathbf{k}} j_l(kR_D) Y^*_{lm}(\hat{k}) T(k) \mathcal{R}_{\mathbf{k}} \tag{37}$$

with $\mathcal{R}_{\mathbf{k}}$ during inflation, and in the limit $k \ll \mathcal{H}$.

Now, substituting the explicit form of $\mathcal{R}_{\mathbf{k}}$ given by Eq. (32) in Eq. (37), the coefficients a_{lm} are directly related to the random variables $X_{\mathbf{k}}$. Notice that the coefficients a_{lm} are a sum of random complex numbers (i.e. a sum over $\mathbf{k}$), where each term is characterized by the random complex variables $X_{\mathbf{k}}$. This leads to what can be considered effectively as a two-dimensional random walk. As it is well known, one cannot give a perfect estimate for the direction of the final displacement resulting from the random walk. However, it is possible to give an estimate for the length of the total displacement. In the present case, such length can be naturally associated with the magnitude $|a_{lm}|^2$. Hence, the most likely value of $|a_{lm}|^2$ can be estimated, and thus interpret it as our theoretical prediction for the observed $|a_{lm}|^2$. Moreover, since the collapse is being modeled by a random process, we can consider a set of possible realizations of such a process characterizing the universe in an unique manner, i.e., through the random variables $X_{\mathbf{k}}$. If the probability distribution function of $X_{\mathbf{k}}$ is Gaussian, then we can identify the most likely value $|a_{lm}|^2_{\text{ML}}$ with the mean value $\overline{|a_{lm}|^2}$ of all possible realizations. This is, $|a_{lm}|^2_{\text{ML}} = \overline{|a_{lm}|^2}$. The most likely value $|a_{lm}|^2_{\text{ML}}$ in our collapse scheme is explicitly given in Appendix B.

Since we are assuming that the $x^{R,I}_{\mathbf{k}}$ variables are uncorrelated, the ensemble average of the product of these random variables satisfies

$$\overline{x^R_{\mathbf{k}} x^R_{\mathbf{k}'}} = \delta_{\mathbf{k},\mathbf{k}'} + \delta_{\mathbf{k},-\mathbf{k}'} \quad \overline{x^I_{\mathbf{k}} x^I_{\mathbf{k}'}} = \delta_{\mathbf{k},\mathbf{k}'} - \delta_{\mathbf{k},-\mathbf{k}'} \tag{38}$$

We have also considered the correlation between the modes $\mathbf{k}$ and $-\mathbf{k}$ in accordance with the commutation relation given by $[\hat{a}^R_{\mathbf{k}}, \hat{a}^{R\dagger}_{\mathbf{k}'}]$ and $[\hat{a}^I_{\mathbf{k}}, \hat{a}^{I\dagger}_{\mathbf{k}'}]$.

Typically, the observational CMB data is presented in terms of the angular power spectrum, C_l. The definition of C_l is given in terms of the coefficients a_{lm} as $C_l = (2l+1)^{-1} \sum_m |a_{lm}|^2$. Therefore, we can use the prediction for $|a_{lm}|^2_{\text{ML}}$ for our collapse scheme considered, and give a theoretical prediction for C_l. Thus, form Eqs. (38) and using our values for the $|a_{lm}|^2_{\text{ML}}$ we can write,

$$C_l = 4\pi \int_0^\infty \frac{dk}{k} j_l^2(kR_D) T(k)^2 A\, Q(z_0, z_k) \tag{39}$$

where the explicit form of the function $Q(z_0, z_k)$ is shown in Appendix B, and A is:

$$A = \frac{H^2}{2\pi^2 M_P^2 \epsilon_1} \tag{40}$$

Also, we have taken the limit $L \to \infty$ and $\mathbf{k} \to$ continuum in order to go from sums over discrete $\mathbf{k}$ to integrals over $\mathbf{k}$.

In the standard inflationary paradigm, a well-known result is that the dimensionless power spectrum $P(k)$ for the perturbation $\mathcal{R}_{\mathbf{k}}$ and the C_l are related by

$$C_l = 4\pi \int_0^\infty \frac{dk}{k} j_l^2(kR_D) T(k)^2 P(k). \tag{41}$$

Therefore, by comparing Eq. (39) with Eq. (41) we can extract an equivalent power spectrum,[1] which finally turns out to be:

$$P(k) = \frac{H^2}{2\pi^2 M_P^2 \epsilon_1} Q(z_0, z_k). \tag{42}$$

Equation (42) is the main result of this work. Notice that because of $z_k = k\eta^c_k$ and $z_0 = k\eta_0$, the function $Q(z_0, z_k)$ depends on k explicitly.

In the next section, we will discuss the results, compare them with previous works, and analyze under which conditions one can recover an essentially scale free spectrum of primordial inhomogeneities, as suggested by the observations.

We would like to end this section by making some comments about our prediction for the power spectrum. Our model gives a direct theoretical prediction for the observed C_l, Eq. (39), and then from such expression we have read what can be identified as the "power spectrum" in the traditional approach of inflation. However, note that this is conceptually different from the traditional approach [65] in which the power spectrum is obtained from the two-point correlation function $\langle 0|\hat{\mathcal{R}}_{\mathbf{k}}\hat{\mathcal{R}}^*_{\mathbf{k}'}|0\rangle$. In contrast, our power spectrum is obtained from $\overline{\langle\hat{\pi}_{\mathbf{k}}\rangle\langle\hat{\pi}_{\mathbf{k}'}\rangle^*}$, where the expectation values are evaluated at the post-collapse state. In Appendix C, we show in detail the calculation of the CMB temperature angular spectrum and its relation with the scalar power spectrum. The interested reader can find there an explanation on how to calculate the power spectrum within the collapse framework, and why our proposal does not rely on the quantum two-point correlation function.

4. Results and discussion

Let us summarize briefly the results obtained in the present manuscript. We started by choosing a novel initial quantum vacuum state for the perturbations of the inflaton [whose mathematical description is given by Eqs. (21), (23) and (24)]. Then, we included the collapse hypothesis and finally arrived at Eq. (42) for the primordial scalar power spectrum. Note that, as already mentioned, the vacuum $y_k(\eta)$ in Eq. (21) includes the initial condition of the BD vacuum if $z_0 \to -\infty$. However, notice that the physical criteria for the choice of both vacuum states are very different.

Now, from our result shown in Eq. (42) for $P(k)$, we are going to analyze under which conditions a scale free spectrum can be obtained (this is, when the function $Q(z_0, z_k)$ does not depend on k and results in a constant). Also, we will analyze those cases where $P(k)$ shows small deviations from a scale invariant spectrum, but are still consistent with observational data. Here, we will adopt $\lambda_1 = 1 = \lambda_2$ (i.e. the independent collapse scheme).

We consider three cases, according to $|z_0|$ values:

[1] Bear in mind that there are two power spectrum in the literature: the dimensional power spectrum $\mathcal{P}(k)$ and the dimensionless power spectrum $P(k)$. The latter is defined in terms of the former by $P(k) \equiv (k^3/2\pi^2)\mathcal{P}(k)$. We are expressing our main result as the $P(k)$.

4.1. $|z_0| \to 0$

In this case, the (dimensional) scalar power spectrum results in $\mathcal{P}(k) \propto \frac{1}{k^5}$, both in the standard scenario and in ours having included the hypothesis of collapse. In our approach, if $|z_0| \to 0$, then the time of collapse must satisfy $|z_k| \to 0$ since the collapse always occurs for $\eta_k^c > \eta_0$. There is no possible parametrization for η_k^c, such that the spectrum is scale invariant; the proof can be seen in the Appendix D. We stress that, the loss of scale invariance in the resulting power spectrum, is not due to the physics behind the collapses, but because the novel vacuum choice is not the best option for smaller values of $|z_0|$.

4.2. $|z_0| \to \infty$

In this case, we observe that taking the limit $|z_0| \to \infty$, implies that the function $Q(z_0, z_k)$ results in exactly the same function as the one shown in Eq. (88) of Ref. [14], i.e.

$$Q(z_0, z_k) \to C(z_k) = \frac{1}{8}\Big[1 + \frac{2}{z_k^2}\sin^2(z_k) - \frac{1}{z_k}\sin(2z_k)\Big] \tag{43}$$

Note that the aforementioned result of Ref. [14] was obtained using the BD vacuum state. Equation (43) is expected because if $|z_0| \to \infty$, the initial condition of the BD vacuum is recovered.

In Ref. [14], it was shown that if $|z_k| \to \infty$ or $|z_k| \to 0$ then the function $C(z_k)$ is a constant. Thus, leading to an exactly scale invariant power spectrum. Also, if $\eta_k^c = \frac{\mathcal{A}}{k}$ is assumed (with $\mathcal{A}$ a constant), whatever the value of z_k, then the resulting spectrum is scale free, and its observational analysis can be consulted, for instance, in Refs. [29,61].

4.3. Other $|z_0|$ cases

For intermediate $|z_0|$ values, that is, values not included in the cases A and B described previously, in Fig. 1 we plot $Q(k)$ vs. k having chosen the time of collapse as $\eta_k^c = \frac{\mathcal{A}}{k}$ ($\mathcal{A}$ a constant), with $|\mathcal{A}| = 10^{-2}$ and $|\eta_0| = 10^4$ Mpc. Note that when a parametrization for η_k^c is chosen and the time η_0 is set, the function $Q(k)$ is only dependent on k. As it can be seen, the resulting function $Q(k)$ is constant for large values of k, while departures from a constant behavior for lower k. Therefore, we expect that the (dimensional) scalar power spectrum results in $\mathcal{P}(k) \sim \frac{1}{k^3}$ for large k values, while departures from the standard prediction affect the smallest ones. Since the observational relevant modes are such that $k \in [10^{-6}, 10^{-1}]$ Mpc^{-1}, only for observationally relevant small k values (i.e. low multipoles l), a difference is expected between our prediction for the C_l and that corresponding to a perfectly scale invariant spectrum. Also, note that Fig. 1 includes a wide range of intermediate $|z_0|$ values and, in addition, the graph is representative for $|z_k| \to 0$ and intermediate $|z_k|$ values.

Although here we will not perform a complete statistical analysis with the observational data, in Fig. 2 we show our predicted angular spectrum C_l and compare it with a fiducial model.

In order to perform our analysis, we modified the public available CAMB code [66]. The cosmological parameters of our fiducial flat ΛCDM model considered are: baryon density in units of the critical density $\Omega_b h^2 = 0.02225$, dark matter density in units of the critical density $\Omega_{cdm} h^2 = 0.1198$, Hubble constant $H_0 = 67.27$ km s^{-1} Mpc^{-1}, reionization optical depth $\tau = 0.079$, and the scalar spectral index, $n_s = 0.96$. Those are the best-fit values presented by the Planck Collaboration [12]. Recall that, we have neglected the effects on the power spectrum and the scalar spectral index coming from the slow-roll parameters. Consequently, our angular spectrum should be compared with a canonical scale free spectrum.

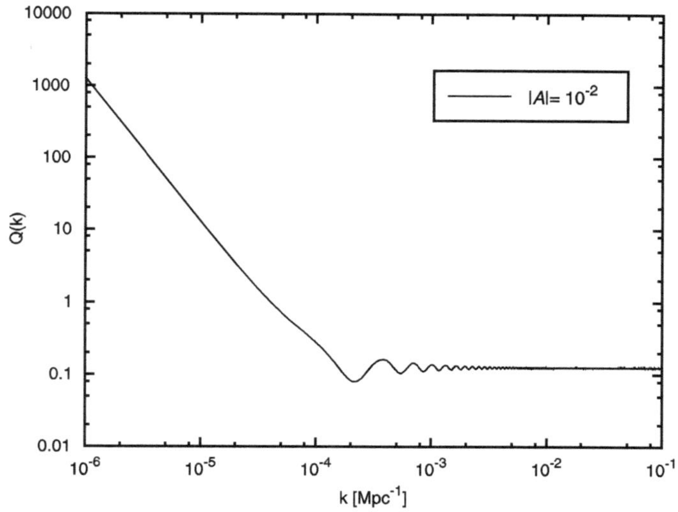

Fig. 1. The function $Q(k)$ vs. k, for intermediate $|z_0|$ values and when the parametrization $\eta_k^c = \frac{\mathcal{A}}{k}$ is assumed. The values considered are $|\mathcal{A}| = 10^{-2}$ and $|\eta_0| = 10^4$ Mpc. The behavior of Q is nearly constant except for large scales (i.e. small k values), for which, small deviations from a scale free spectrum are expected in our prediction for the C_l.

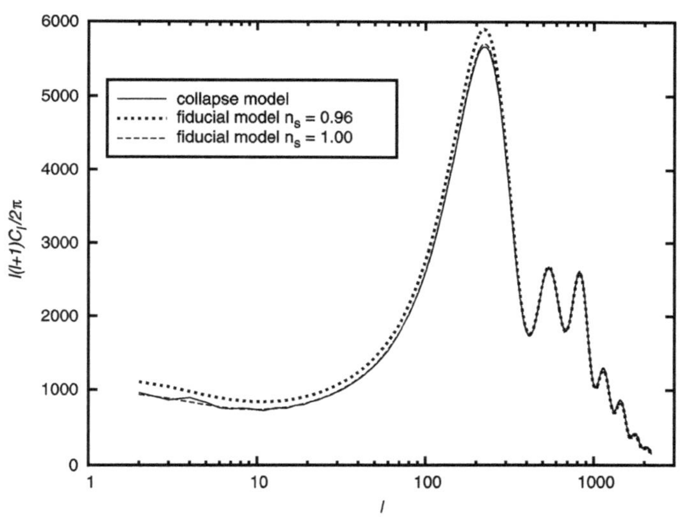

Fig. 2. Our prediction for the CMB angular spectrum within an inflationary collapse model, with $|z_k| = |\mathcal{A}| = 10^{-2}$, and having chosen a different vacuum state than the traditional one, being $|\eta_0| = 10^4$ Mpc. As a reference, we also show a fiducial model coming from the best-fit to the Planck data [12], with scalar spectral index $n_s = 0.96$ (dotted line) and $n_s = 1$ (dashed line). See the text for details.

In Fig. 2, we present three plots: One, is the fiducial model described previously. Another one is a quasi fiducial model with the best-fit values from Planck, except for the spectral index, for which $n_s = 1$. And, the remaining plot, corresponds to the predicted curve in our model, also with $n_s = 1$.

As it can be seen, our prediction agrees very well with the standard prediction curve plotted with the best-fit values from the Planck data. As anticipated, only small differences for low multipole values appear, where the cosmic variance is dominant.

We conclude this section with a final remark: the fact of having chosen a vacuum different from that which is typically chosen as the initial condition for inflation, led us to a function $Q(z_0, z_k)$ that is generically different from that of the authors in [14]. Then, one could think that it would be very difficult to find a parametrization for η_k^c, such that the shape of the power spectrum would be compatible with the CMB observations. However, as inferred from the plots, the parametrization $\eta_k^c = \frac{\mathcal{A}}{k}$ is still valid for a wide range of η_0 values.

5. Conclusions

In this work, we have analyzed under which conditions one can recover an essentially scale free spectrum of primordial inhomogeneities, when the standard BD vacuum is replaced by another one that minimizes the renormalized stress–energy tensor via a Hadamard procedure. This new prescription for selecting the initial vacuum state is better suited for cosmological models built to give a description of the early universe, particularly those that include the self-induced collapse proposal within the semiclassical gravity picture.

We found that a scale invariant scalar power spectrum can be obtained (and compatible with the CMB observations), for a wide range of initial times η_0. By choosing for the collapse time the parametrization $\eta_k^c = \mathcal{A}/k$ (being $\mathcal{A}$ a constant), in Fig. 2, we show that our predicted angular spectrum C_l agrees very well with the standard prediction curve plotted with the best-fit values from the recent Planck data. As anticipated in our analysis, only small differences for low multipole l values appear, where the cosmic variance is dominant.

For $|z_0| = k|\eta_0| \to 0$ the choice of this novel vacuum does not lead to a scale invariant scalar power spectrum, but not due to the collapse hypothesis. As a matter of fact, this is generically true for the standard inflationary model using the new vacuum choice. In other words, small values of $|k\eta_0|$ are not allowed by the observations using the new vacuum state either within the standard inflation or using the self-induced collapse hypothesis.

On the other hand, for values $|z_0| \gtrsim 1$, it is possible to obtain a scale free spectrum concordant with observations. In particular, for $|z_0| \gg 1$, the initial conditions are the same as the one provided by the BD vacuum.

The fact of having chosen a vacuum different from that which is typically chosen as the initial condition for inflation, led us to the function $Q(z_0, z_k)$, shown explicitly in Eq. (B.4). The obtained function is generically different from that of the authors in [14], who also considered the collapse proposal but using the standard BD vacuum. Therefore, one could think that it would be very difficult to find a parametrization for η_k^c, such that the power spectrum would become scale free, and, as a consequence, compatible with the CMB data. However, we have found that the parametrization $\eta_k^c = \frac{\mathcal{A}}{k}$, which is the same as the one originally proposed in all the previous works based on the collapse hypothesis, is still valid for a wide range of η_0 values. Thus, we conclude that the collapse mechanism might be of a more fundamental character than previously suspected.

Note that the model considered here involved some phenomenological characterization of the self-induced collapse proposal. For instance, the dependence of the time of collapse on each mode k, and the Gaussian distribution in the random variables. However, this characterization can be taken as ansatz modifications of the standard inflationary scenario, inspired by collapse schemes proposals based on spontaneous individual collapses (e.g. the GRW model [58]). The GRW objective reduction model has been originally proposed to deal with the quantum measurement problem, independent of any cosmological context. In this work, we used a GRW-inspired collapse scheme, by incorporating some of its generic features. Nevertheless, we think that a dynamical reduction mechanism (which can be seen as less *ad hoc* than the one considered here), such as the Continuous Spontaneous Localization (CSL) model [57,67], can be subjected to the same analysis presented in the present paper.

Finally, it is also important to mention some previous results regarding other observables; specifically, the primordial bispectrum [61,68] and tensor modes [27,28]. Those results were obtained under our self-induced collapse proposal but maintaining the usual choice of the BD vacuum. In respect to the bispectrum, we have obtained a completely different shape than the usual one. As a matter of fact, the characterization is not based on the usual quantum three-point function. Meanwhile, a possible detection of primordial gravitational waves would be considered as the "smoking-gun" between our proposal, based on semiclassical gravity, and the traditional one. Our framework predicts a strong suppression of the tensor modes amplitude; essentially undetectable by any present or future experiments. Based on the results obtained in this paper, we think that the aforementioned predictions will remain unchanged.

Acknowledgements

G.R.B. is supported by CONICET (Argentina). G.R.B. acknowledges support from the PIP 112-2012-0100540 of CONICET (Argentina). G.L. acknowledges financial support from CONICET (Argentina). We thank the anonymous referee for her/his comments and objective suggestions that contributed to the clarity of the work.

Appendix A. Explicit equations of Sect. 3.2

In this Appendix, we are going to show some steps to arrive at Eq. (29). We will use the collapse scheme chosen in Eq. (28), where appear the expectation values of $\langle \hat{y}_{\mathbf{k}}^{R,I}(\eta_k^c)\rangle_\Theta$ and $\langle \hat{\pi}_{\mathbf{k}}^{R,I}(\eta_k^c)\rangle_\Theta$, evaluated at the collapse time, and related to the quantum uncertainties of the vacuum state.

From Eqs. (25), together with definitions of $\hat{a}_{\mathbf{k}}^R$ and $\hat{a}_{\mathbf{k}}^I$ and the non-standard commutation relations for the $\hat{a}_{\mathbf{k}}^{R,I}$ in Eq. (26), we can rewrite the quantum uncertainties of the vacuum state $y_k(\eta)$ of Eq. (21) (at collapse time) as

$$\left(\Delta \hat{y}_{\mathbf{k}}^{R,I}(\eta_k^c)\right)_0^2 = \frac{L^3}{4}|y_k(\eta_k^c)|^2$$

$$\left(\Delta \hat{\pi}_{\mathbf{k}}^{R,I}(\eta_k^c)\right)_0^2 = \frac{L^3}{4}|g_k(\eta_k^c)|^2$$

where $g_k = y_k' - \mathcal{H}y_k$. Then, the collapse scheme can be written as:

$$\langle \hat{y}_{\mathbf{k}}^{R,I}(\eta_k^c)\rangle_\Theta = \lambda_1\, x_{\mathbf{k},1}^{R,I}\, \frac{L^{\frac{3}{2}}}{2}|y_k(\eta_k^c)| \tag{A.1a}$$

$$\langle \hat{\pi}_{\mathbf{k}}^{R,I}(\eta_k^c)\rangle_\Theta = \lambda_2\, x_{\mathbf{k},2}^{R,I}\, \frac{L^{\frac{3}{2}}}{2}|g_k(\eta_k^c)| \tag{A.1b}$$

Now, we need the values of $\hat{y}_{\mathbf{k}}(\eta)$ and $\hat{\pi}_{\mathbf{k}}(\eta)$ for $\eta \geq \eta_k^c$, in the post-collapse state. To do this, we introduce the quantity $d_{\mathbf{k}}^{R,I} \equiv \langle\Theta|\hat{a}_{\mathbf{k}}^{R,I}|\Theta\rangle$, that determines the expectation value of the field and momentum operator for the mode **k** at all times after the collapse. That is, from Eq. (25), we have

$$\langle \hat{y}_{\mathbf{k}}^{R,I}(\eta)\rangle_\Theta = \sqrt{2}\mathrm{Re}[y_k(\eta)d_{\mathbf{k}}^{R,I}] \tag{A.2}$$

$$\langle \hat{\pi}_{\mathbf{k}}^{R,I}(\eta)\rangle_\Theta = \sqrt{2}\mathrm{Re}[g_k(\eta)d_{\mathbf{k}}^{R,I}] \tag{A.3}$$

which corresponds to expectation values at any time after the collapse in the post-collapse state $|\Theta\rangle$. One can then relate the value of $d_{\mathbf{k}}^{R,I}$ with the value of the expectation value of the fields operators at the time of collapse $\langle \hat{y}_{\mathbf{k}}^{R,I}(\eta_k^c)\rangle_\Theta = \sqrt{2}\mathrm{Re}[y_k(\eta_k^c)d_{\mathbf{k}}^{R,I}]$,

$\langle \hat{\pi}^{R,I}_{\mathbf{k}}(\eta^c_k)\rangle_\Theta = \sqrt{2}\text{Re}[g_k(\eta^c_k)d^{R,I}_{\mathbf{k}}]$. Using the latter relations to express $d^{R,I}_{\mathbf{k}}$ in terms of the expectation values at the time of collapse, and substituting it in Eq. (A.3), we obtain an expression for the expectation value of the momentum field operator in terms of the expectation value at the time of collapse. Since $\langle \hat{\pi}_{\mathbf{k}}(\eta)\rangle_\Theta = \langle \hat{\pi}^R_{\mathbf{k}}(\eta)\rangle_\Theta + i\,\langle \hat{\pi}^I_{\mathbf{k}}(\eta)\rangle_\Theta$, we write

$$\langle \hat{\pi}_{\mathbf{k}}(\eta)\rangle_\Theta = \frac{1}{D(z_k)}\Big\{ F(k\eta, z_k)\Big[\langle \hat{y}^R_{\mathbf{k}}(\eta^c_k)\rangle_\Theta + i\langle \hat{y}^I_{\mathbf{k}}(\eta^c_k)\rangle_\Theta\Big] + + G(k\eta, z_k)\Big[\langle \hat{\pi}^R_{\mathbf{k}}(\eta^c_k)\rangle_\Theta + i\langle \hat{\pi}^I_{\mathbf{k}}(\eta^c_k)\rangle_\Theta\Big]\Big\} \tag{A.4}$$

where $z_k \equiv k\eta^c_k$. On the other hand, functions D, F and G are defined by:

$$\begin{aligned} D(z_k) &\equiv \text{Im}[g_k(\eta^c_k)]\text{Re}[y_k(\eta^c_k)] - \text{Im}[y_k(\eta^c_k)]\text{Re}[g_k(\eta^c_k)] \\ F(k\eta, z_k) &\equiv \text{Re}[g_k(\eta)]\text{Im}[g_k(\eta^c_k)] - \text{Im}[g_k(\eta)]\text{Re}[g_k(\eta^c_k)] \\ G(k\eta, z_k) &\equiv \text{Im}[g_k(\eta)]\text{Re}[y_k(\eta^c_k)] - \text{Re}[g_k(\eta)]\text{Im}[y_k(\eta^c_k)] \end{aligned} \tag{A.5}$$

Substituting in Eq. (A.4) what was found in Eqs. (A.1), we can finally write,

$$\langle \hat{\pi}_{\mathbf{k}}(\eta)\rangle_\Theta = \frac{L^{\frac{3}{2}}}{2D(z_k)}\Big[F(k\eta, z_k)\,|y_k(\eta^c_k)|\,\lambda_1\, x_{\mathbf{k},1} + + G(k\eta, z_k)\,|g_k(\eta^c_k)|\,\lambda_2\, x_{\mathbf{k},2}\Big] \tag{A.6}$$

with $X_{\mathbf{k},1} \equiv x^R_{\mathbf{k},1} + i x^I_{\mathbf{k},1}$ and $X_{\mathbf{k},2} \equiv x^R_{\mathbf{k},2} + i x^I_{\mathbf{k},2}$. From the equation for $y_k(\eta)$, Eq. (21) at η^c_k, together with the relation between A_k and B_k given by Eq. (22), we obtain that $D = -1/2$. The explicit forms for F and G functions turn out to be:

$$\begin{aligned} F(k\eta, z_k) &= (A_k^2 - |B_k|^2)\,k^2 \sin(k\eta - z_k) \\ G(k\eta, z_k) &= -\frac{k(A_k^2 - |B_k|^2)}{z_k} \\ &\quad\times \Big[z_k\cos(k\eta - z_k) + \sin(k\eta - z_k)\Big] \end{aligned} \tag{A.7}$$

On the other hand,

$$\begin{aligned} |y_k(\eta^c_k)|^2 &= \frac{(A_k^2 + |B_k|^2)(1+z_k^2)}{z_k^2} + 2|A_k||B_k|(z_k^2 - 1) \\ &\quad\times \cos(2z_k + \beta) - 4|A_k||B_k| z_k \sin(2z_k + \beta) \\ |g_k(\eta^c_k)|^2 &= k^2\Big[A_k^2 + |B_k|^2 - 2|A_k||B_k|\,\times \\ &\quad\times \cos(2z_k + \beta)\Big] \end{aligned} \tag{A.8}$$

Remember that A_k and B_k [Eqs. (23) and (24)] have dependence on $z_0 \equiv k\eta_0$, the initial condition of the vacuum state chosen.

Appendix B. The $|a_{lm}|^2_{ML}$ and the explicit form of $Q\,(z_0, z_k)$

Since we are interested in the observational relevant modes, which have wavelengths greater than the Hubble radius during inflation, as we mentioned in section 3.2, we will take the limit $|k\eta| \to 0$ in the functions F and G given in Eqs. (A.7), and we will define M and N as:

$$\begin{aligned} M(z_0, z_k) &\equiv \lim_{|k\eta|\to 0} F(k\eta, z_k)\,|y_k(\eta^c_k)| \\ N(z_0, z_k) &\equiv \lim_{|k\eta|\to 0} G(k\eta, z_k)\,|g_k(\eta^c_k)| \end{aligned} \tag{B.1}$$

Note that we have explicitly written that there is a dependence on $z_0 = k\eta_0$ in the functions M and N.

The explicit expressions for $|a_{lm}|^2_{\text{ML}}$ can be found by substituting $\mathcal{R}_{\mathbf{k}}$, given in Eq. (32), into Eq. (37) and then making the identification $|a_{lm}|^2_{\text{ML}} = \overline{|a_{lm}|^2}$. This is,

$$|a_{lm}|^2_{\text{ML}} = \frac{16\pi^2}{L^6}\sum_{\mathbf{k},\mathbf{k}'} j_l(kR_D)j_l(k'R_D) \times Y^*_{lm}(\hat{k})Y_{lm}(\hat{k}')T(k)T(k')\overline{\mathcal{R}_{\mathbf{k}}\mathcal{R}^*_{\mathbf{k}'}} \tag{B.2}$$

Therefore, it is

$$\begin{aligned} |a_{lm}|^2_{\text{ML}} &= \frac{8\pi^2 H^2}{L^3\epsilon_1 M_P^2}\sum_{\mathbf{k},\mathbf{k}'} \frac{j_l(kR_D)j_l(k'R_D)}{k^2\,k'^2} \\ &\times Y^*_{lm}(\hat{k})Y_{lm}(\hat{k}')T(k)T(k')\Big[\lambda_1^2 M(z_0, z_k)M(z_0, z_{k'}) \\ &\times \overline{X_{\mathbf{k},1}X^*_{\mathbf{k}',1}} + \lambda_2^2\, N(z_0, z_k)N(z_0, z_{k'})\overline{X_{\mathbf{k},2}X^*_{\mathbf{k}',2}}\Big] \end{aligned} \tag{B.3}$$

where we have used that the random variables $X_{\mathbf{k},1}$ and $X_{\mathbf{k},2}$ are uncorrelated, so $\overline{X_{\mathbf{k},1}X^*_{\mathbf{k}',2}} = 0 = \overline{X_{\mathbf{k},2}X^*_{\mathbf{k}',1}}$.

From Eq. (B.3), and since we are assuming that the $x^{R,I}_{\mathbf{k}}$ variables are uncorrelated, and thus the ensemble average of the product of these random variables satisfies Eq. (38), we can finally identify the function $Q\,(z_0, z_k)$ with

$$Q\,(z_0, z_k) \equiv \frac{1}{k}\Big[\lambda_1^2\, M^2(z_0, z_k) + \lambda_2^2\, N^2(z_0, z_k)\Big]$$

which explicitly turns out to be:

$$\begin{aligned} Q\,(z_0, z_k) &= \frac{1}{16\,z_k^2\,z_0^2}\Bigg\{\lambda_2^2\Bigg[1 + 2z_0^2 - z_0\sqrt{4 + \frac{1}{z_0^2}} \\ &\times \cos[2z_k - 2z_0 + \arctan(2z_0)]\Bigg][z_k\cos(-z_k) + \sin(-z_k)]^2 \\ &+ \lambda_1^2\,\sin^2(-z_k)\Bigg[(1+z_k^2)(1+2z_0^2) + z_0\sqrt{4 + \frac{1}{z_0^2}} \\ &\times \Big[(z_k^2 - 1)\cos[2z_k - 2z_0 + \arctan(2z_0)] \\ &- 2z_k\sin[2z_k - 2z_0 + \arctan(2z_0)]\Big]\Bigg]\Bigg\} \end{aligned} \tag{B.4}$$

Appendix C. Equivalent power spectra $P(k)$

In the traditional inflationary scenario, the power spectrum $P(k)$ is obtained by computing the quantum two-point correlation function. That is, if $\hat{\mathcal{R}}$ represents the quantum field associated to the scalar metric perturbation, then the power spectrum is taken to be

$$\langle 0|\hat{\mathcal{R}}_{\mathbf{k}}\hat{\mathcal{R}}^*_{\mathbf{k}'}|0\rangle = \frac{2\pi^2}{k^3}P(k)\delta(\mathbf{k} - \mathbf{k}') \tag{C.1}$$

On the other hand, let us recall that in general, the definition of the power spectrum is given in terms of $\mathcal{R}_{\mathbf{k}}$, i.e. a classical stochastic field and not a quantum field. Therefore, the standard approach is based on the identification:

$$\langle 0|\hat{\mathcal{R}}_{\mathbf{k}}\hat{\mathcal{R}}^*_{\mathbf{k}'}|0\rangle = \overline{\mathcal{R}_{\mathbf{k}}\mathcal{R}^*_{\mathbf{k}'}} \tag{C.2}$$

with $\overline{\mathcal{R}_{\mathbf{k}}\mathcal{R}^*_{\mathbf{k}'}}$ denoting an average over an ensemble of classical stochastic fields. The justification for the relation above relies on arguments based on decoherence and the squeezing nature of the evolved vacuum state [18,69] (although we do not subscribe to such arguments for the reasons exposed in [15,16]). It is also important to mention that based on the above hypotheses, there is a strong matching between the predictions based on the standard approach and the observational data.

On the other hand, in our proposal, the procedure to obtain an equivalent power spectrum is different from the traditional approach. We start by focusing on the temperature anisotropies of the CMB observed today on the celestial two-sphere and its relation to the scalar metric perturbation $\mathcal{R}$. In Fourier space, this relation can be written as (see Sect. 3.3),

$$\Theta(\mathbf{k}) = T(k)\mathcal{R}_{\mathbf{k}} \tag{C.3}$$

where $T(k)$ is known as the transfer function, which contains the physics between the beginning of the radiation-dominated era and the present, i.e the modifications associated with late-time physics. [A well known result (the *Sachs–Wolfe effect*) is, for instance, that $T(k) \simeq 1/5$ for very large scales].

On the other hand, the observational data are described in terms of the coefficients a_{lm} of the multipolar series expansion

$$\frac{\delta T}{T_0}(\theta,\varphi) = \sum_{lm} a_{lm} Y_{lm}(\theta,\varphi), \qquad a_{lm} = \int \frac{\delta T}{T_0}(\theta,\varphi) Y^*_{lm}(\theta,\varphi) d\Omega, \tag{C.4}$$

here θ and φ are the coordinates on the celestial two-sphere, with $Y_{lm}(\theta,\varphi)$ as the spherical harmonics.

The values for the quantities a_{lm} are then given by

$$a_{lm} = \frac{4\pi i^l}{3} \int \frac{d^3k}{(2\pi)^3} j_l(kR_D) Y^*_{lm}(\hat{k}) T(k) \mathcal{R}_{\mathbf{k}} \tag{C.5}$$

with $j_l(kR_D)$ being the spherical Bessel function of order l, and R_D is the comoving radius of the last scattering surface. The metric perturbation $\mathcal{R}_{\mathbf{k}}$ is the primordial curvature perturbation (in the comoving gauge).

By using Eq. (27) (with $\mathcal{R}_{\mathbf{k}} \simeq \Psi_{\mathbf{k}}/\epsilon_1$) into Eq. (C.5) we obtain

$$a_{lm} = \frac{4\pi i^l}{3} \frac{H}{\sqrt{2\epsilon_1} M_P} \int \frac{d^3k}{(2\pi)^3} j_l(kR_D) Y^*_{lm}(\hat{k}) T(k) \frac{\langle \hat{\pi}_{\mathbf{k}} \rangle}{k^2}. \tag{C.6}$$

The previous expression shows how the expectation value of the momentum field in the post-collapse state acts as a source for the coefficients a_{lm}.

Furthermore, the angular power spectrum is defined by

$$C_l = \frac{1}{2l+1} \sum_m |a_{lm}|^2. \tag{C.7}$$

For the reasons presented in Sect. 3.3, we can identify the observed value $|a_{lm}|^2$ with the most likely value of $|a_{lm}|^2_{ML}$ and in turn, assume that the most likely value coincides approximately with the average $\overline{|a_{lm}|^2}$.

Thus, in our approach, the observed C_l coincides with

$$C_l \simeq \frac{1}{2l+1} \sum_m \overline{|a_{lm}|^2}. \tag{C.8}$$

From Eq. (C.6) we obtain

$$\begin{aligned} \overline{|a_{lm}|^2} &= \left(\frac{4\pi}{3}\right)^2 \int \frac{d^3k d^3k'}{(2\pi)^6} j_l(kR_D) j_l(k'R_D) \\ &\times Y^*_{lm}(\hat{k}) Y_{lm}(\hat{k}') T(k) T(k') \overline{\mathcal{R}_{\mathbf{k}} \mathcal{R}^*_{\mathbf{k}'}} \\ &= \left(\frac{4\pi}{3}\right)^2 \int \frac{d^3k d^3k'}{(2\pi)^6} j_l(kR_D) j_l(k'R_D) \\ &\times Y^*_{lm}(\hat{k}) Y_{lm}(\hat{k}') T(k) T(k') \left[\frac{H^2}{2\epsilon_1 M_P^2} \frac{\overline{\langle \hat{\pi}_{\mathbf{k}} \rangle \langle \hat{\pi}_{\mathbf{k}'} \rangle^*}}{k^2 k'^2}\right]. \end{aligned} \tag{C.9}$$

Consequently using the generic definition of the power spectrum,

$$\overline{\mathcal{R}_{\mathbf{k}} \mathcal{R}^*_{\mathbf{k}'}} \equiv \frac{2\pi^2}{k^3} P(k) \delta(\mathbf{k} - \mathbf{k}') \tag{C.10}$$

and also using Eq. (C.9), the power spectrum, associated to $\mathcal{R}_{\mathbf{k}}$, in our approach is given by

$$P(k) = \frac{H^2}{4\pi^2 \epsilon_1 M_P^2} \frac{\overline{\langle \hat{\pi}_{\mathbf{k}}(\eta) \rangle \langle \hat{\pi}_{\mathbf{k}}(\eta) \rangle^*}}{k}. \tag{C.11}$$

The quantity $\overline{\langle \hat{\pi}_{\mathbf{k}}(\eta) \rangle \langle \hat{\pi}_{\mathbf{k}}(\eta) \rangle^*}$ is obtained by using Eq. (29) in the limit $-k\eta \to 0$, i.e. when the proper wavelength of the modes of interest are bigger than the Hubble radius.

Appendix D. A non-scale invariant power spectrum for $|z_0| \to 0$

In this Appendix, we will show that if $|z_0| \to 0$, in both approaches, i.e. in the standard inflationary model and in our picture with the additional collapse hypothesis, then the resulting shape of the scalar power spectrum $\mathcal{P}(k)$ is not consistent with the observational data.

In the standard approach, the modes v_k of the Mukhanov–Sasaki variable satisfy [37,65],

$$v_k'' + \left(k^2 - \frac{z''}{z}\right) v_k = 0 \tag{D.1}$$

Note that the equation of motion for $y_k(\eta)$, Eq. (10), is identical to Eq. (D.1). That is, when neglecting the slow roll parameters $\frac{z''}{z} \simeq \frac{a''}{a} = \frac{2}{\eta^2}$. As we know, a general solution in such a case will be,

$$v_k(\eta) = A_k\left(1 - \frac{1}{k\eta}\right) e^{-ik\eta} + B_k\left(1 + \frac{1}{k\eta}\right) e^{ik\eta} \tag{D.2}$$

where we assumed $A_k \in \mathbb{R}$ and $B_k \in \mathbb{C}$.

On the other hand, in the standard scenario, the (dimensional) scalar power spectrum is obtained from

$$\mathcal{P}(k) \simeq \lim_{-k\eta \to 0} \frac{|v_k(\eta)|^2}{M_P^2 \epsilon_1 a^2(\eta)} \tag{D.3}$$

By using Eq. (D.2) into Eq. (D.3) one obtains

$$\mathcal{P}(k) \simeq \frac{H^2}{M_P^2 \epsilon_1 k^2} \left[A_k^2 + |B_k|^2 - 2A_k \mathrm{Re}(B_k)\right] \tag{D.4}$$

From Eqs. (23) and (24) we have seen that if $z_0 \to -\infty$, the initial conditions provided by the BD vacuum are recovered, which implies that, $A_k = \frac{1}{\sqrt{2k}}$ and $B_k = 0$. In that case, Eq. (D.4) implies $\mathcal{P}(k) \propto \frac{1}{k^3}$, so a scale invariant power spectrum is obtained.

However, if $|z_0| \to 0$ then we can make the following approximations in Eqs. (23) and (24):

$$|A_k|^2 \simeq \frac{1}{8k|z_0|^2} \qquad |B_k|^2 = \frac{1}{8k|z_0|^2} \tag{D.5}$$

$$A_k \simeq \frac{1}{\sqrt{8k}|z_0|} \qquad \mathrm{Re}(B_k) \simeq -\frac{1}{\sqrt{8k}|z_0|} \tag{D.6}$$

and since $z_0 \equiv k\eta_0$, we obtain from Eq. (D.4),

$$\mathcal{P}(k) \simeq \frac{H^2}{8k^3} \frac{1}{|z_0|^2} \propto \frac{1}{k^5} \tag{D.7}$$

Therefore, if $|z_0| \to 0$, then the shape of the power spectrum, which resulted from choosing a vacuum state such that it minimizes the renormalized stress–energy tensor via a Hadamard procedure, is not compatible with the observational data. The initial conditions obtained from the vacuum choice fix the value of A_k and B_k given by Eqs. (23) and (24).

Now, let us show that under our approach, with the collapse hypothesis included, we arrive at the same result. For simplicity, we will assume $\lambda_1 = 1 = \lambda_2$.

For $|z_0| \to 0$, and taking into account that $|z_k| < |z_0|$, we perform a Taylor expansion in Eq. (B.4). At the leading order in $|z_0|$ and $|z_k|$, we obtain:

$$Q(z_0, z_k) \simeq \frac{1}{8|z_0|^2} + \frac{|z_k|^2}{12|z_0|^2} + \mathcal{O}(|z_0|, |z_k|^4) \tag{D.8}$$

Again, keeping the first relevant term in Eq. (D.8), and since $z_0 = k\eta_0$ we finally arrive at

$$\mathcal{P}(k) \propto \frac{1}{k^5} \tag{D.9}$$

We observe that the result is independent of the parametrization for η_k^c. Therefore, as in the standard case, the shape of the power spectrum is not consistent with the data. We attained this negative result not due to the collapse hypothesis but because of the initial conditions, provided by the novel choice of the vacuum state, when $|z_0| \to 0$.

References

[1] A.A. Starobinsky, Phys. Lett. B 91 (1980) 99.
[2] A.H. Guth, Phys. Rev. D 23 (1981) 347.
[3] A.D. Linde, Phys. Lett. B 108 (1982) 389.
[4] A. Albrecht, P.J. Steinhardt, Phys. Rev. Lett. 48 (1982) 1220.
[5] V.F. Mukhanov, G.V. Chibisov, JETP Lett. 33 (1981) 532.
[6] V.F. Mukhanov, G. Chibisov, Sov. Phys. JETP 56 (1982) 258.
[7] A.A. Starobinsky, Sov. Astron. Lett. 9 (1983) 302.
[8] S.W. Hawking, I.G. Moss, Nucl. Phys. B 224 (1983) 180.
[9] S.W. Hawking, Phys. Lett. B 115 (1982) 295.
[10] P.A.R. Ade, et al., Planck Collaboration, Astron. Astrophys. 594 (2016) A13.
[11] N. Aghanim, et al., Planck Collaboration, Astron. Astrophys. 594 (2016) A11.
[12] P.A.R. Ade, et al., Planck Collaboration, Astron. Astrophys. 594 (2016) A20, arXiv:1502.02114.
[13] J. Martin, C. Ringeval, R. Trotta, V. Vennin, J. Cosmol. Astropart. Phys. 1403 (2014) 039.
[14] A. Perez, H. Sahlmann, D. Sudarsky, Class. Quantum Gravity 23 (2317) (2005), arXiv:gr-qc/0508100.
[15] D. Sudarsky, Int. J. Mod. Phys. D 20 (2011) 509, arXiv:0906.0315.
[16] S. Landau, G. León, D. Sudarsky, Phys. Rev. D 88 (2013) 023526, arXiv:1107.3054.
[17] D. Polarski, A.A. Starobinsky, Class. Quantum Gravity 13 (377) (1996), arXiv:gr-qc/9504030.
[18] C. Kiefer, D. Polarski, Adv. Sci. Lett. 2 (2009) 164, arXiv:0810.0087.
[19] A. de Unánue, D. Sudarsky, Phys. Rev. D 78 (2008) 043510, arXiv:0801.4702.
[20] A. Diez-Tejedor, D. Sudarsky, J. Cosmol. Astropart. Phys. 7 (2012) 045, arXiv:1108.4928.
[21] G. León, S.J. Landau, M.P. Piccirilli, Phys. Rev. D 90 (2014) 083525, arXiv:1410.1562.
[22] P. Cañate, P. Pearle, D. Sudarsky, Phys. Rev. D 87 (2013) 104024, arXiv:1211.3463.
[23] J. Martin, V. Vennin, P. Peter, Phys. Rev. D 86 (2012) 103524, arXiv:1207.2086.
[24] S. Das, K. Lochan, S. Sahu, T. Singh, Phys. Rev. D 88 (2013) 085020, arXiv:1304.5094.
[25] G. Leon, G.R. Bengochea, Eur. Phys. J. C 76 (2016) 29, arXiv:1502.04907.
[26] S. Alexander, D. Jyoti, J. Magueijo, Phys. Rev. D 94 (2016) 043502, arXiv:1602.01216.
[27] G. León, L. Kraiselburd, S.J. Landau, Phys. Rev. D 92 (2015) 083516, arXiv:1509.08399.
[28] G. León, A. Majhi, E. Okon, D. Sudarsky, Reassessing the link between B-modes and inflation, arXiv:1607.03523, 2016.
[29] M. Benetti, S.J. Landau, J.S. Alcaniz, J. Cosmol. Astropart. Phys. 1612 (2016) 035, arXiv:1610.03091.
[30] S. Hollands, R.M. Wald, Phys. Rep. 574 (2015) 1, arXiv:1401.2026.
[31] J.A. Wheeler, in: B.S. DeWitt, C. DeWitt (Eds.), Relativity, Groups, and Topology, Gordon and Breach, 1964.
[32] D. Finkelstein, Phys. Rev. 184 (1969) 1261.
[33] T. Konopka, F. Markopoulou, S. Severini, Phys. Rev. D 77 (2008) 104029.
[34] D. Oriti, in: D. Oriti (Ed.), Approaches to Quantum Gravity, Cambridge University Press, 2009.
[35] H. Steinacker, Class. Quantum Gravity 27 (2010) 133001.
[36] S. Fulling, Aspects of Quantum Field Theory in Curved Spacetime, London Mathematical Society St, Cambridge University Press, ISBN 9780521377683, 1989.
[37] V.F. Mukhanov, H.A. Feldman, R.H. Brandenberger, Phys. Rep. 215 (1992) 203.
[38] V. Mukhanov, S. Winitzki, Introduction to Quantum Effects in Gravity, Cambridge University Press, ISBN 0521868343, 2007.
[39] S.A. Fulling, Gen. Relativ. Gravit. 10 (1979) 807.
[40] C. Armendariz-Picon, J. Cosmol. Astropart. Phys. 0702 (2007) 031, arXiv:astro-ph/0612288.
[41] W. Handley, A. Lasenby, M. Hobson, Phys. Rev. D 94 (2016) 024041, arXiv:1607.04148.
[42] J. Martin, R.H. Brandenberger, Phys. Rev. D 63 (2001) 123501, arXiv:hep-th/0005209.
[43] U.H. Danielsson, Phys. Rev. D 66 (2002) 023511, arXiv:hep-th/0203198.
[44] J. Martin, C. Ringeval, Phys. Rev. D 69 (2004) 083515, arXiv:astro-ph/0310382.
[45] E. Komatsu, Class. Quantum Gravity 27 (2010) 124010, arXiv:1003.6097.
[46] I. Agullo, J. Navarro-Salas, L. Parker, J. Cosmol. Astropart. Phys. 1205 (2012) 019, arXiv:1112.1581.
[47] G. León, D. Sudarsky, Class. Quantum Gravity 27 (2010) 225017, arXiv:1003.5950.
[48] A. Diez-Tejedor, G. Leon, D. Sudarsky, Gen. Relativ. Gravit. 44 (2012) 2965, arXiv:1106.1176.
[49] G. Leon, S. Landau, M.P. Piccirilli, Eur. Phys. J. C 75 (2015) 393, arXiv:1502.00921.
[50] R. Penrose, Gen. Relativ. Gravit. 28 (1996) 581.
[51] R.H. Brandenberger, H. Feldman, V.F. Mukhanov, pp. 19–30, arXiv:astro-ph/9307016, 1993.
[52] C. Armendariz-Picon, E.A. Lim, J. Cosmol. Astropart. Phys. 0312 (2003) 006, arXiv:hep-th/0303103.
[53] U.H. Danielsson, Phys. Rev. D 66 (2002) 023511, arXiv:hep-th/0203198.
[54] D. Bohm, J. Bub, Rev. Mod. Phys. 38 (1966) 453.
[55] P.M. Pearle, Phys. Rev. D 13 (1976) 857.
[56] P. Pearle, Int. J. Theor. Phys. 18 (1979) 489.
[57] P.M. Pearle, Phys. Rev. A 39 (1989) 2277.
[58] G. Ghirardi, A. Rimini, T. Weber, Phys. Rev. D 34 (1986) 470.
[59] L. Diosi, Phys. Lett. A 120 (1987) 377.
[60] L. Diosi, Phys. Rev. A 40 (1989) 1165.
[61] S.J. Landau, C.G. Scóccola, D. Sudarsky, Phys. Rev. D 85 (2012) 123001, arXiv:1112.1830.
[62] S. Das, S. Sahu, S. Banerjee, T. Singh, Phys. Rev. D 90 (2014) 043503, arXiv:1404.5740.
[63] G. León, A. De Unánue, D. Sudarsky, Class. Quantum Gravity 28 (2011) 155010, arXiv:1012.2419.
[64] N. Deruelle, V.F. Mukhanov, Phys. Rev. D 52 (5549) (1995), arXiv:gr-qc/9503050.
[65] V. Mukhanov, Physical Foundations of Cosmology, Cambridge University Press, New York, 2005.
[66] A. Lewis, A. Challinor, A. Lasenby, Astrophys. J. 538 (2000) 473, arXiv:astro-ph/9911177.
[67] A. Bassi, G.C. Ghirardi, Phys. Rep. 379 (257) (2003), arXiv:quant-ph/0302164.
[68] G. León, D. Sudarsky, J. Cosmol. Astropart. Phys. 1506 (2015) 020, arXiv:1503.01417.
[69] L. Grishchuk, Y. Sidorov, Phys. Rev. D 42 (1990) 3413.

19

On Minkowski Functionals of CMB polarization

Pravabati Chingangbam [a], Vidhya Ganesan [a,b,*], K.P. Yogendran [c,1], Changbom Park [d]

[a] Indian Institute of Astrophysics, Koramangala II Block, Bangalore 560 034, India
[b] Department of Physics, Indian Institute of Science, Bangalore 560 012, India
[c] IISER Tirupati, Karakambadi Road, Tirupati 517 507, India
[d] Korea Institute for Advanced Study, 85 Hoegiro, Dongdaemun-gu, Seoul 02455, Republic of Korea

ARTICLE INFO

Editor: H. Peiris

ABSTRACT

CMB polarization data is usually analyzed using E and B modes because they are scalar quantities under rotations along the lines of sight and have distinct physical origins. We explore the possibility of using the Stokes parameters Q and U for complementary analysis and consistency checks in the context of searches for non-Gaussianity. We show that the Minkowski Functionals (MFs) of Q, U are invariant under local rotations along the lines of sight even though Q, U are spin-2 variables, for full sky analysis. The invariance does not hold for incomplete sky. For local type primordial non-Gaussianity, when we compare the non-Gaussian deviations of MFs for Q, U to what is obtained for E mode or temperature fluctuations, we find that the amplitude is about an order of magnitude lower and the shapes of the deviations are different. This finding can be useful in distinguishing local type non-Gaussianity from other origins of non-Gaussianity in the observed data. Lastly, we analyze the sensitivity of the amplitudes of the MFs for Q, U and the number density of singularities of the total polarization intensity to the tensor-to-scalar ratio, r, and find that all of them decrease as r increases.

1. Introduction

Towards the last stages of the epoch of recombination quadrupolar anisotropies must have been present in the intensity of photons. This anisotropy would lead to a net linear polarization of the Cosmic Microwave Background (CMB) photons [1–6] as a result of Thompson scattering with free electrons. In addition to the temperature fluctuations, the CMB polarization is a vital repository of clues about the physical properties, origin of primordial fluctuations and history of the Universe. In the standard inflationary [7–11] ΛCDM cosmology the quadrupolar anisotropies can be traced back to two physical origins, namely, anisotropies in the scalar density fluctuations of the plasma, and tensor perturbations of the metric.

Observations of polarized CMB photons measure the Stokes parameters Q and U along each line of sight. They transform as spin-2 objects under rotations about the line of sight. Using spin two spherical harmonics they can be re-expressed as the so-called E and B modes [12,13]. To first order in perturbations, E mode is sourced by scalar density perturbations while B mode is sourced by tensor perturbations. Due to polarization taking place only during the last stages of decoupling the rms of E mode is about one order of magnitude lower than that of temperature fluctuations. E mode has been measured [14] and used for cosmological analysis [15–17]. Generic inflationary models predict that the ratio of the amplitudes of the primordial tensor and scalar perturbations, denoted by r, is less than one, with the precise value being model dependent. Currently, the detection of B mode sourced by primordial tensor perturbations is one of the foremost goals of observational cosmology. The precise knowledge of its rms value, which translates into knowledge of r, will strongly constrain inflation models. From observations by BICEP2 and KECK Array the latest constraint on r is < 0.07 at 95% CL [18].

Inflation predicts that the fluctuations in the energy density and metric during the very early stages of the Universe are random variables with a nearly Gaussian probability distribution function. The statistical properties of these fluctuations are inherited by the temperature fluctuations and polarization of the CMB. One of the important tools to analyze the statistical properties of these random fields are the Minkowski Functionals (MFs) [19–26]. They are quantities that characterize the geometrical and topological

* Corresponding author.
E-mail addresses: prava@iiap.res.in (P. Chingangbam), vidhya@iiap.res.in (V. Ganesan), yogendran@iisertirupati.ac.in (K.P. Yogendran), cbp@kias.re.kr (C. Park).
[1] On leave from IISER Mohali, Sector 81, Mohali, India.

properties of excursion sets of the CMB fields. They have been applied to temperature fluctuations to constrain primordial non-Gaussianity [27,28] and more recently on the E mode polarization [17,29]. They have also been used to identify traces of residual foreground contamination in WMAP data [30]. In [31], they show that MFs can be used to detect the presence of tiny foreground residuals in lensed B mode.

In this paper we focus attention on MFs for CMB polarization. Most analysis of polarization data focus on using E and B modes since they are scalar quantities under rotations along the lines of sight and the clean separation of their physical origins. Their invariance under rotations means that their MFs are invariant under such rotations. When dealing with observed data, E, B are obtained from the directly observed Q, U variables using spin-2 spherical harmonic functions. This step is complicated by the fact that we need to work with incomplete sky coverage of the data. In the work by Santos et al. [32], they show that the process of decomposition into E, B results in contamination in their morphological properties. Given that MFs are real space quantities it is then a natural question to ask whether it is possible to extract the physical information that we seek from MFs of E, B equally well from Q, U.

An early work that uses the genus, which is one of the MFs, for Q and U can be found in [33]. Because of their spin-2 nature it is not immediately obvious whether their MFs measured by different observers (or experiments) can be meaningfully compared. We clarify this issue and show analytically that for full sky coverage their MFs are invariant under rotations along the line of sight. We then confirm the invariance by performing numerical calculations of MFs. The invariance breaks down for incomplete sky.

Next, we investigate how non-Gaussian deviations of primordial density fluctuations manifest in MFs of Q, U. We restrict our analysis to local type primordial non-Gaussianity. We find, in comparison to what is obtained for E mode and temperature fluctuations, the non-Gaussian deviations of the MFs for Q, U have corresponding amplitudes that are about an order of magnitude lower, but the deviation shapes are distinct. Analytic expressions for MFs and the number density of singularities of the total polarization intensity, P, for Gaussian primordial perturbation, are derived in [34,35]. The amplitudes of the MFs and the number density of singularities are expressed in terms of the variances of Q, U. Their invariance under rotations along the lines of sight relies on the above result. We analyze the effect of tensor perturbations on the amplitudes of the MFs and the number density of singularities and find that both decrease as the amplitude of tensor perturbations is increased. This result is useful for searches for B mode and consistency checks of the CMB data. We would like to mention that we do not take into account observational effects, such as beam effect and instrument noise, in this paper since our goal is to elucidate the theoretical issues. It is expected that when such realistic effects are included the statistical significance of the results on the non-Gaussian deviations and the sensitivity to the presence of primordial tensor perturbations will weaken.

This paper is organized as follows. In Section 2 we briefly describe the CMB polarization simulation that we use for our analysis. In Section 3 we analyze the effect of rotations of the coordinate axes along the line of sight on the variances of Q and U. In Section 4 we introduce MFs, discuss their numerical computation and demonstrate their invariance under global rotations of the coordinate axes along the lines of sights. Further, we calculate the MFs for Q and U containing input primordial non-Gaussianity. In Section 5 we present the effect of including B mode on the variances and the amplitude of MFs of Q and U. We also show the effect on the number of singularities of the total polarization intensity. We end by summarizing the results along with a discussion of their implications in Section 6.

2. CMB polarization fields and their simulations

The Stokes parameters Q and U transform as spin-two objects under a rotation by angle α about each line of sight, given by

$$\begin{pmatrix} Q' \\ U' \end{pmatrix} = R(2\alpha)\begin{pmatrix} Q \\ U \end{pmatrix} = \begin{pmatrix} \cos 2\alpha\ Q + \sin 2\alpha\ U \\ -\sin 2\alpha\ Q + \cos 2\alpha\ U \end{pmatrix}. \tag{1}$$

Equivalently, Eq. (1) can be written as $(Q' \pm iU') = e^{\mp i2\alpha}(Q \pm iU)$. They are related to the total polarization intensity as, $P \equiv \sqrt{Q^2+U^2}$ and the polarization angle as, $\varphi \equiv \frac{1}{2}\tan^{-1} U/Q$. P is invariant under rotations about the line of sight while φ is not. Further, Q and U can be expressed as E and B modes [12] by expanding $Q \pm iU$ in terms of spin-two spherical harmonics, $Y_{\pm 2,\ell m}$,

$$Q \pm iU = \sum_{\ell m} a_{\pm 2,\ell m} Y_{\pm 2,\ell m}, \tag{2}$$

and defining

$$a_{E,\ell m} = -\frac{1}{2}\left(a_{2,\ell m} + a_{-2,\ell m}\right)$$
$$a_{B,\ell m} = \frac{i}{2}\left(a_{2,\ell m} - a_{-2,\ell m}\right), \tag{3}$$

and

$$E(\hat{n}) = \sum\left(\frac{\ell+2}{\ell-2}\right)^{1/2} a_{E,\ell m} Y_{\ell m}$$
$$B(\hat{n}) = \sum\left(\frac{\ell+2}{\ell-2}\right)^{1/2} a_{B,\ell m} Y_{\ell m}. \tag{4}$$

It is useful for our subsequent analysis to invert Eq. (3). Inserting $a_{\pm 2,\ell m}$ into Eq. (2) gives

$$Q = -\frac{1}{2}\sum_{\ell m}\left\{a_{E,\ell m}\left(Y_{2,\ell m} + Y_{-2,\ell m}\right) + i a_{B,\ell m}\left(Y_{2,\ell m} - Y_{-2,\ell m}\right)\right\}$$
$$U = \frac{i}{2}\sum_{\ell m}\left\{a_{E,\ell m}\left(Y_{2,\ell m} - Y_{-2,\ell m}\right) + i a_{B,\ell m}\left(Y_{2,\ell m} + Y_{-2,\ell m}\right)\right\}. \tag{5}$$

For our analysis we produce simulations of E and B mode with Gaussian statistics for primordial scalar and tensor perturbations and corresponding input angular power spectra. The ΛCDM cosmological parameters values used are $\Omega_c h^2 = 0.1198$, $\Omega_b h^2 = 0.02225$, $H_0 = 67.27$, $n_s = 0.9645$, $\ln(10^{10} A_s) = 3.094$, $\tau = 0.079$, taken from the 2015 PLANCK data [16]. The input angular power spectra were obtained using the `CAMB` package [36,37] and the map simulations were made using the `HEALPIX` package [38,39]. The map resolution corresponds to HEALPIX parameter NSIDE value 1024. The amplitudes of B mode maps are fixed by choosing values of the tensor-to-scalar ratio, r. Maps of Q, U are then made using the E and B maps.

3. Variances of Q, U and ∇Q, ∇U

In this section we examine the transformation of the variances of Q, U and their gradients ∇Q, ∇U, under rotations along the line of sight. Let us denote the following variances of field X by

$$\Sigma_0^X \equiv \langle XX \rangle, \quad \Sigma_1^X \equiv \langle \nabla X \cdot \nabla X \rangle, \tag{6}$$

where X can be either Q or U. Note that $\langle\rangle$ here means averaging over the surface of the sphere.

For simplicity, we first consider rotation by the same angle α along every line of sight. Then from Eq. (1) we get

$$\langle Q'Q'\rangle = \cos^2(2\alpha)\langle Q\,Q\rangle + \sin^2(2\alpha)\langle UU\rangle + \sin 4\alpha\,\langle Q\,U\rangle, \quad (7)$$

$$\langle U'U'\rangle = \sin^2(2\alpha)\langle Q\,Q\rangle + \cos^2(2\alpha)\langle UU\rangle - \sin 4\alpha\,\langle Q\,U\rangle, \quad (8)$$

$$\langle \nabla Q'\cdot\nabla Q'\rangle = \cos^2(2\alpha)\langle \nabla Q\cdot\nabla Q\rangle + \sin^2(2\alpha)\langle \nabla U\cdot\nabla U\rangle + \sin 4\alpha\,\langle \nabla Q\cdot\nabla U\rangle, \quad (9)$$

$$\langle \nabla U'\cdot\nabla U'\rangle = \sin^2(2\alpha)\langle \nabla Q\cdot\nabla Q\rangle + \cos^2(2\alpha)\langle \nabla U\cdot\nabla U\rangle - \sin 4\alpha\,\langle \nabla Q\cdot\nabla U\rangle, \quad (10)$$

Eqs. (7)–(10) imply that $\langle Q\,Q\rangle \neq \langle Q'Q'\rangle$ and $\langle \nabla Q\cdot\nabla Q\rangle \neq \langle \nabla Q'\cdot\nabla Q'\rangle$ if the cross-correlations $\langle Q\,U\rangle$ and $\langle \nabla Q\cdot\nabla U\rangle$ are non-zero. Using Eq. (5) we get

$$\langle Q\,U\rangle = \frac{i}{4}\sum_{\ell m\ell' m'}\Big\{a_{E,\ell m}a^*_{E,\ell' m'}\int d\Omega\,(Y_{2,\ell m}+Y_{-2,\ell m})\times\left(Y^*_{2,\ell' m'}-Y^*_{-2,\ell' m'}\right) + a_{B,\ell m}a^*_{B,\ell' m'}\int d\Omega\,(Y_{2,\ell m}-Y_{-2,\ell m})\times\left(Y^*_{2,\ell' m'}+Y^*_{-2,\ell' m'}\right) + \text{cross terms}\Big\} \quad (11)$$

Let $\langle\rangle_{\text{ens}}$ denote averaging over an ensemble of Universes. Using the following relations which follow from isotropy of the linear perturbations,

$$\langle a_{E,\ell m}a^*_{E,\ell' m'}\rangle_{\text{ens}} = \delta_{\ell\ell'}\delta_{mm'}\langle |a_{E,\ell m}|^2\rangle_{\text{ens}} \quad (12)$$

$$\langle a_{B,\ell m}a^*_{B,\ell' m'}\rangle_{\text{ens}} = \delta_{\ell\ell'}\delta_{mm'}\langle |a_{B,\ell m}|^2\rangle_{\text{ens}} \quad (13)$$

$$\langle a_{E,\ell m}a^*_{B,\ell' m'}\rangle_{\text{ens}} = 0, \quad (14)$$

the ensemble average of $\langle Q\,U\rangle$ becomes

$$\langle\langle Q\,U\rangle\rangle_{\text{ens}} = \frac{i}{4}\sum_{\ell m}\Big\{\langle |a_{E,\ell m}|^2\rangle_{\text{ens}}\int d\Omega\,(Y_{2,\ell m}+Y_{-2,\ell m})\left(Y^*_{2,\ell m}-Y^*_{-2,\ell m}\right) + \langle |a_{B,\ell m}|^2\rangle_{\text{ens}}\int d\Omega\,(Y_{2,\ell m}-Y_{-2,\ell m})\left(Y^*_{2,\ell m}+Y^*_{-2,\ell m}\right)\Big\}. \quad (15)$$

$Y_{s,\ell m}$ satisfy the conjugacy relation

$$Y^*_{s,\ell m} = (-1)^{m+s}Y_{-s,\ell-m}, \quad (16)$$

where s is the spin index. Using the conjugacy relation and the reality condition for $a^*_{E,\ell m}$, we get

$$\langle\langle Q\,U\rangle\rangle_{\text{ens}} = \frac{i}{4}\sum_{\ell m}\Big\{\langle |a_{E,\ell m}|^2\rangle_{\text{ens}} - \langle |a_{B,\ell m}|^2\rangle_{\text{ens}}\Big\}\times\int d\Omega\big(-Y_{2,\ell m}Y^*_{-2,\ell m} + Y_{-2,\ell m}Y^*_{2,\ell m}\big). \quad (17)$$

The two terms in the integrand above are complex conjugates. Again using the conjugacy relation we get

$$Y_{2,\ell m}Y^*_{-2,\ell m} = (-1)^{m-2}\,Y_{2,\ell m}Y_{2,\ell-m} \quad (18)$$

Since the dependence of $Y_{2,\ell m}$ on the coordinates θ, ϕ is like $Y_{2,\ell m} = f(\theta)e^{im\phi}$ and $Y_{2,\ell -m} = g(\theta)e^{-im\phi}$, each term in the integrand of Eq. (17) is real. Therefore, the relative sign in the integrand leads to the two terms canceling. Thus we get

$$\langle\langle Q\,U\rangle\rangle_{\text{ens}} = 0. \quad (19)$$

Since the cancellation occurs before we have integrated over the sky, we actually have

$$\langle Q\,U\rangle_{\text{ens}} = 0, \quad (20)$$

which holds at every point (θ,ϕ). Using Eq. (19) in Eq. (7) we get

$$\langle \Sigma_0^{Q'}\rangle_{\text{ens}} = \cos^2(2\alpha)\langle\langle Q\,Q\rangle\rangle_{\text{ens}} + \sin^2(2\alpha)\langle\langle UU\rangle\rangle_{\text{ens}}. \quad (21)$$

Using $\langle Q\,Q\rangle_{\text{ens}} = \langle UU\rangle_{\text{ens}}$ we get

$$\langle \Sigma_0^{Q'}\rangle_{\text{ens}} = \langle \Sigma_0^{Q}\rangle_{\text{ens}}. \quad (22)$$

And similarly for U.

Next, to calculate $\langle \nabla Q\cdot\nabla U\rangle$ we need to simplify the factor containing gradients of the $Y_{\pm 2,\ell m}$'s, given below,

$$\nabla\left(Y_{2,\ell m}+Y_{-2,\ell m}\right)\cdot\nabla\left(Y^*_{2,\ell m}-Y^*_{-2,\ell m}\right) = \nabla Y_{2,\ell m}\cdot\nabla Y^*_{2,\ell m} - \nabla Y_{-2,\ell m}\cdot\nabla Y^*_{-2,\ell m} - \nabla Y_{2,\ell m}\cdot\nabla Y^*_{-2,\ell m} + \nabla Y_{-2,\ell m}\cdot\nabla Y^*_{2,\ell m}. \quad (23)$$

The first two terms cancel using the conjugacy relation in Eq. (16). So we are left with

$$\langle\langle \nabla Q\cdot\nabla U\rangle\rangle_{\text{ens}} = \frac{i}{4}\sum_{\ell m}\Big\{\langle |a_{E,\ell m}|^2\rangle_{\text{ens}} - \langle |a_{B,\ell m}|^2\rangle_{\text{ens}}\Big\}\int d\Omega\left(-\nabla Y_{2,\ell m}\cdot\nabla Y^*_{-2,\ell m} + \nabla Y_{-2,\ell m}\cdot\nabla Y^*_{2,\ell m}\right) \quad (24)$$

Again using the conjugacy relation, and using the fact that the dependence of $Y_{2,\ell m}$ on ϕ is $e^{im\phi}$ while that of $Y_{2,\ell -m}$ is $e^{-im\phi}$, we can show that each term inside the integrand is a real function and hence the two terms cancel. Therefore,

$$\langle\langle \nabla Q\cdot\nabla U\rangle\rangle_{\text{ens}} = 0. \quad (25)$$

Again, the zero correlation holds at every (θ,ϕ).

We have shown that the variances are invariant under a *global* rotation by the same angle about every line of sight. If we allow the rotation angle to vary for different lines of sight and retrace the above calculation, then the rotation factors will be part of the integrand over the sphere. However, in order to prove Eqs. (19) and (25) we only use the properties of $Y_{\pm 2,\ell m}$ and do not need to carry out the integration at any step. Therefore, the invariance holds for direction dependent rotations also.

In the case of incomplete sky due to Galactic and point sources masks, the relations (12) and (13) no longer hold because isotropy is broken. Therefore, in this case

$$\langle\langle Q\,U\rangle\rangle_{\text{ens}} \neq 0, \quad \langle\langle \nabla Q\cdot\nabla U\rangle\rangle_{\text{ens}} \neq 0,$$

which implies that Σ_0^X and Σ_1^X are not invariant under rotations along the lines of sight.

4. Minkowski Functionals

A useful way to study the statistical properties of a random field is to choose suitable threshold values of the field and analyze regions that have field values above each threshold. Such regions are called *excursion sets*. The morphological properties of the excursion sets and their variation with the threshold value can reveal the statistical nature of the field. The shapes of the excursion sets can be quantified in terms of geometrical and topological quantities, namely, the Minkowski Functionals (MFs). There are three MFs for two-dimensional manifolds such as the excursion sets of the CMB. The first, denoted by V_0, is the area fraction of the excursion set. The second, denoted by V_1, is the total length of iso-temperature contours or boundaries of the excursion set. The third, denoted by V_2, is the genus which is the difference between the numbers of hot spots and cold spots. Let f denote a generic random field, $\sigma_0 = \sqrt{\Sigma_0}$ is the rms of f, and let $u \equiv f/\sigma_0$. Let ν be the threshold value chosen from the range of u. Then, the MFs are defined mathematically as follows:

$$V_0(\nu) \equiv \int da, \quad V_1(\nu) \equiv \frac{1}{4}\int_C dl, \quad V_2(\nu) \equiv \frac{1}{2\pi}\int_C K\, dl, \qquad (26)$$

where da is the area element of the excursion set, C denotes contours that form the boundaries of the excursion sets, dl is the line element on C and K is the curvature of the contours. Closely related to the MFs are the two Betti numbers [40,41], of which the first is the number of the connected regions and the second is the number of holes in the connected regions. The difference of the first and second Betti numbers gives the genus.

In this section, we study the effect of rotations along the line of sight on the MFs of Q and U. Further, we analyze how primordial non-Gaussianity will show up in the MFs of Q and U. For all calculations shown in this section we set the B mode to be zero.

4.1. Numerical computation of Minkowski Functionals for random fields

In general, for a given random field we may not know the analytic form for the MFs or we may need to test whether an observed random field has the statistical property that we expect from theory. In such situations we need to calculate MFs using numerical methods. For the numerical calculation of MFs we employ the method due to Schmalzing and Gorski [24], which relies on expressing Eq. (26) as a discrete sum involving first and second order covariant derivatives on the sphere and a δ-functional of the field. The discretization of the δ-function introduces numerical inaccuracies [42]. For a given random field, let V_i and $V_i^{\rm an}$ denote the numerically calculated MFs and the exact value, respectively. Then we can write $V_i = V_i^{\rm an} + R_i^{\Delta\nu}$, where $R_i^{\Delta\nu}$ denotes the residual numerical error. The superscript $\Delta\nu$ is the threshold bin size and it indicates that the residual error is dependent on it. $R_i^{\Delta\nu}$ is given by

$$R_i^{\Delta\nu}(\nu) = \frac{1}{\Delta\nu}\int_{\nu-\Delta\nu/2}^{\nu+\Delta\nu/2} du\, V_i(u) \;-\; V_i^{\rm an}(\nu). \qquad (27)$$

4.2. Minkowski functionals for Gaussian Q and U

For a Gaussian random field, f, the MFs as functions of the threshold ν, are given by

$$V_k(\nu) = A_k\, H_{k-1}(\nu)\, e^{-\nu^2/2}, \quad k = 0, 1, 2. \qquad (28)$$

Here $H_k(\nu)$ is the k-th Hermite polynomial and the amplitudes are

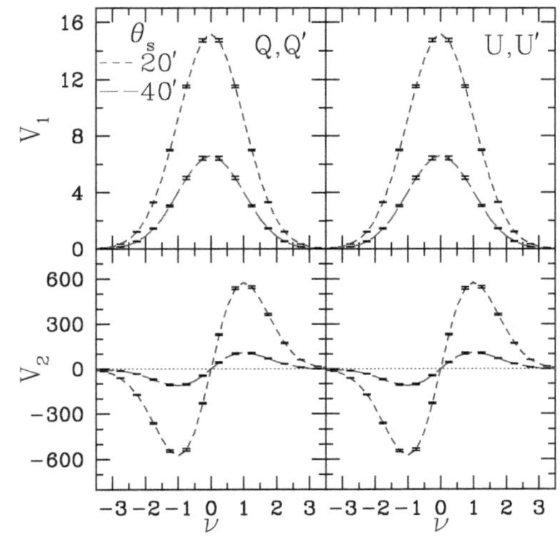

Fig. 1. Upper panels show plots of the contour length, V_1, for Gaussian Q, Q' (left) and U, U' (right), for smoothing angles $\theta_s = 20'$ (red) and $\theta_s = 40'$ (blue). Q', U' have been obtained by rotating Q, U about each line of sight by angle $\alpha = 45°$. The two plots in each panel are indistinguishable, demonstrating numerically that the amplitudes are invariant under global rotations along the line of sight. All plots are average over 1000 simulations and the error bars are the sample variances. Lower panels show the genus, V_2. We have repeated the calculations for other rotation angles and the results remain the same. (For interpretation of the references to color in this figure legend, the reader is referred to the web version of this article.)

$$A_0 = \frac{1}{\sqrt{2\pi}}, \quad A_1 = \frac{1}{8}\frac{\sigma_1}{\sqrt{2}\sigma_0}, \quad A_2 = \frac{1}{(2\pi)^{3/2}}\left(\frac{\sigma_1}{\sqrt{2}\sigma_0}\right)^2, \qquad (29)$$

where $\sigma_1 = \sqrt{\Sigma_1}$ is the rms of the gradient of the field.

If the primordial scalar and tensor fluctuations are Gaussian, we expect Q and U to be Gaussian for linear perturbations and their MFs should be of the form given in Eq. (28). The ratio $r_c \equiv \sigma_0/\sigma_1$ is usually referred to as the correlation length of hot and cold structures in a given random field. The amplitudes of the MFs depend on powers of r_c. As shown in Section 3, σ_i^X, and hence r_c, are invariant under rotations by the same angle about every line of sight. Hence the MFs should be invariant. However, for incomplete sky they are not invariant. Our subsequent analyses focus on full sky calculations.

We have computed the MFs for simulated Gaussian Q, U and their corresponding Q', U' obtained by rotating Q, U by angle α. In Fig. 1 we plots of the contour length and the genus for Gaussian Q, Q' and U, U' for rotation angle $\alpha = 45°$, demonstrating their invariance. We show results for two smoothing angles $\theta_s = 20', 40'$. We have obtained the same result for other choices of α. B mode has been set to zero for these calculation.

The residual error defined in Eq. (28) calculated numerically for Q and U (dashed lines) and the analytic form obtained from integrating the first term of Eq. (28) for a Gaussian *scalar* field (black dotted lines), are shown in Fig. 2. We chose two smoothing angles $\theta_s = 20', 40'$. The bin sizes used are $\Delta\nu = 0.25, 0.65, 1$. The analytic form of the residuals grow larger for larger bin sizes and are not affected much by changes in the smoothing scale. It is interesting to note that the residual errors for the contour length for Q and U are nearly the same and seems to agree with what is expected from Eq. (27) for Gaussian scalar fields (see Fig. (2) or (3) of [42]). If we zoom in the figure we find that at small bin sizes there is noticeable difference between the dashed and dotted lines. The difference gets more pronounced at larger smoothing angles. They agree very well at larger bin size, such as can be seen for the case $\Delta\nu = 1$. The genus residuals exhibit similar behavior but we find much stronger disagreement between the dashed lines and

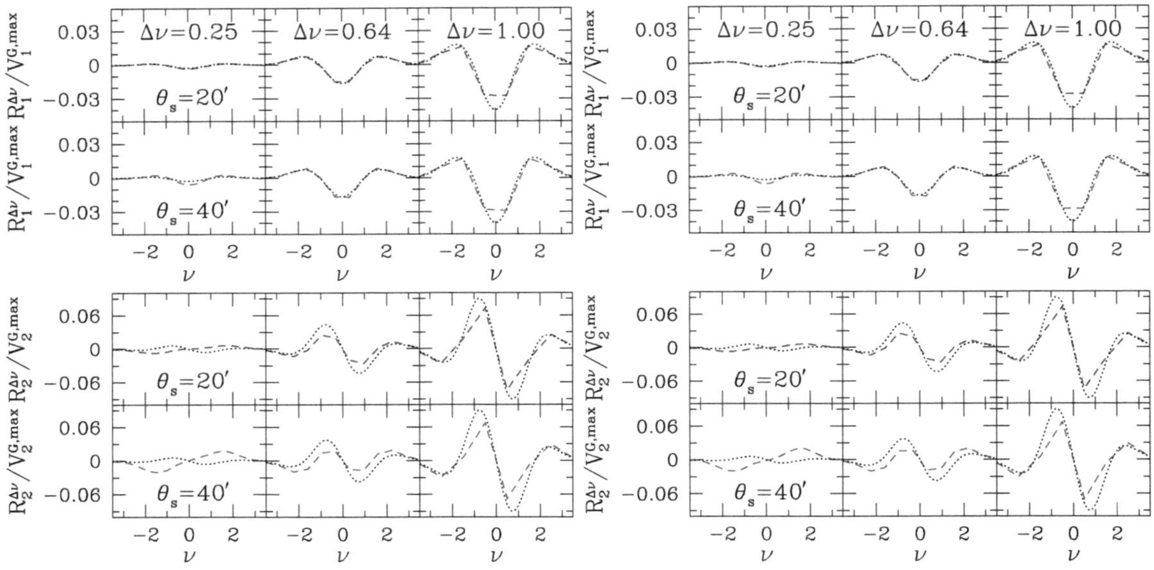

Fig. 2. *Left*: Upper panels show of the residuals defined in Eq. (27) of V_1, obtained numerically for Gaussian Q for smoothing angles $\theta_s = 20'$ (red) and $\theta_s = 40'$ (blue) for $\Delta\nu = 0.25, 0.64, 1$. All plots are average over 1000 maps. The black dotted lines are the analytic form of the residuals obtained by integrating the first term on the right hand side of Eq. (27) for a Gaussian *scalar* field. Lower panels show the residuals for V_2. *Right*: Same as left plots for U. (For interpretation of the references to color in this figure legend, the reader is referred to the web version of this article.)

dotted lines at small bin sizes. Moreover, there is small but noticeable difference between the residuals of Q and U.

We would like to mention that we have repeated the residual calculations for temperature maps and we have reproduced the results of [42] very well. Hence we rule out the possibility that the disagreement at small bin size arises due to mistakes in our numerical calculations. We think that this disagreement is due to the spin-two nature of Q and U. The derivatives that we have used to implement the Schmalzing and Gorski method are covariant derivatives for scalar fields on the sphere, and not what should be the appropriate covariant derivative for a $U(1)$ bundle on the sphere which should be relevant for spin-two variables such as Q, U. We do not pursue this question further since the mathematics that is relevant for clarifying it is beyond the scope of this paper.

Next, let the non-Gaussian deviations of the MFs be denoted by

$$\Delta V_i \equiv V_i^{NG} - V_i^{G}, \tag{30}$$

where $i = 0, 1, 2$. In the following we consider local type primordial non-Gaussianity parametrized by the variable $f_{\rm NL}$. Our calculations here are done for full sky since our aim is to bring out the non-Gaussian effects. Considering incomplete sky will decrease the statistical significance of our results. We use simulations of $a_{E,\ell m}$ that contain input local type primordial non-Gaussianity that have been made publicly available by Elsner and Wandelt [43]. The values of the input ΛCDM parameters are those obtained from the WMAP 5 years data [44]. The resolution is set by NSIDE = 512. Gaussian and non-Gaussian Q, U maps with our chosen $f_{\rm NL}$ values are constructed from the corresponding $a_{E,\ell m}$. These maps are then used to calculate the MFs from which ΔV_i are calculated. We have done the calculations using both the Schmalzing and Gorski method and a geometrical method described in [45,46] and the results agree with each other.

In Fig. 3 we show the MFs for Q and their non-Gaussian deviations. We have not shown the results for U since they are the same, as expected. The top panels show V_i for Gaussian (black) as well as non-Gaussian maps for $f_{\rm NL} = 1000$, for smoothing angle $\theta_s = 5'$. The plots are indistinguishable by eye. The lower panels show the non-Gaussian deviations rescaled by the corresponding $V_i^{G,\max}$. From the plots of ΔV_i we can make two main observations. Firstly, the amplitude of deviations for Q is much smaller than what is obtained for temperature fluctuations (compare Fig. 3 with Fig. (2) of [47]) and for E mode (compare with Fig. (5) of [29]). We have chosen unphysically large values of $f_{\rm NL}$ because the numerical calculation for realistic values become quite noisy. Secondly, the shape of deviations is different from what is seen for temperature fluctuations and E mode. The genus deviation is similar to what is seen for cubic order local primordial non-Gaussianity (see Fig. (4) of [48] and Fig. (1) of [49]).

5. Effect of primordial tensor perturbations on Minkowski Functionals

The probability distribution function (PDF) for the total polarization intensity, P, has the Rayleigh form $\frac{1}{\sigma_0} P\, e^{-P^2/2\sigma_0}$, where σ_0 is the rms of Q or U, under the assumption that they are equal. As shown in Section 3, this holds only for complete sky coverage. For the subsequent discussions we consider only complete sky coverage and will refer to σ_0, σ_1 without the field superscript. In [29] the authors have shown that the PDF of P varies significantly with the amplitude of B mode. Since the PDF of P is completely characterized by σ_0, for Gaussian Q, U, we can quantify the effect of including B mode by calculating how it affects σ_0. Here we take this observation further and study the effect on σ_0, σ_1, the amplitude of the MFs for Q, U, and the number density of singularities of P. We use the tensor-to-scalar ratio, r, to quantify the effect of primordial tensor perturbations, and study how the various quantities vary as r is varied.

In the left and middle panels of Fig. 4 we show how σ_0 and σ_1, computed using Eqn. (6) on simulated maps, vary with r between 0.05 to 0.2. We have used two smoothing scales, $10'$ and $90'$ to highlight the variation in the variances with the smoothing scale also. We find that inclusion of B mode increases the variances. Note that the slopes of the dependence on r vary with the smoothing scale. The dependence of σ_0 and σ_1 on r are not linear, even though over the small range that we have considered here they appear to be so on visual inspection.

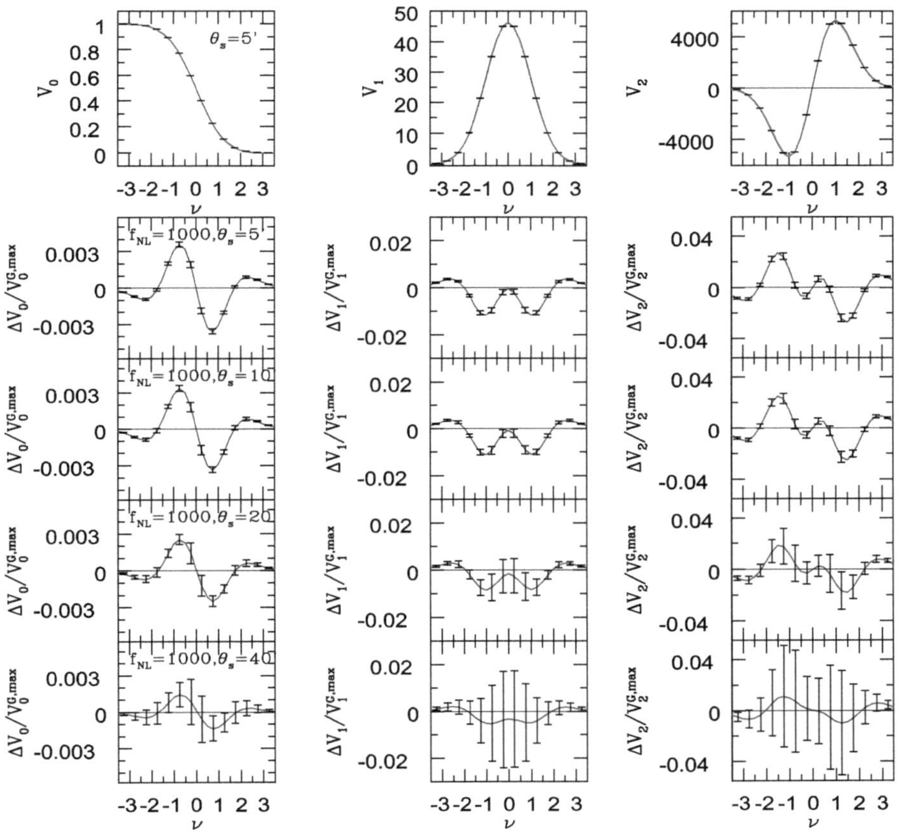

Fig. 3. *Top panels*: The three Minkowski Functionals for Gaussian and non-Gaussian cases with $f_{NL} = 1000$ are shown. The plots are not distinguishable by eye since the differences are small. *Lower panels*: Non-Gaussian deviations for the MFs for $f_{NL} = 1000$ for different smoothing angles. We have chosen unrealistically large value of f_{NL} because the plots become noisy for small values of f_{NL}. All plots are average over 1000 simulations and the error bars are their sample variances.

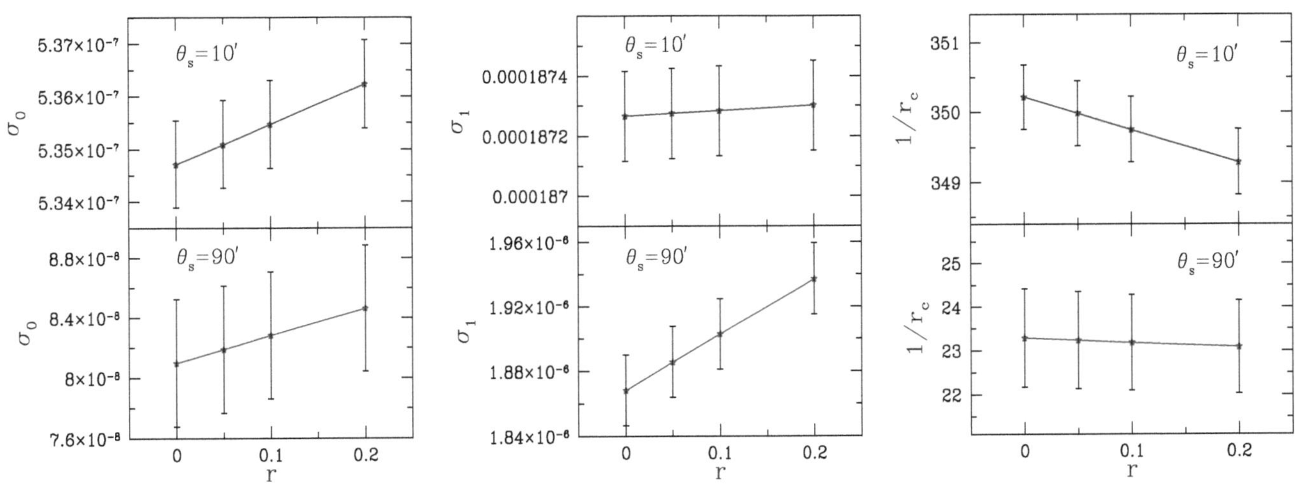

Fig. 4. The left panel shows how σ_0 varies as r is varied, for smoothing angles $\theta_s = 10', 90'$. The middle and right panels show σ_1 and r_c. The stars indicate values of r at which we have done the calculations. The superscript Q has been dropped. All plots are average over 1000 simulations and the error bars are the sample variances.

To study the effect of r on the amplitudes of the MFs of Q, U it suffices to find out how r_c varies with r. In the right panel of Fig. 4 we have plotted r_c^{-1} versus r between 0.05 to 0.2, for different smoothing scales. As seen in the plot, the presence of B mode with increasing amplitudes results in decrease of the amplitude of MFs. Note that the genus is more sensitive to r than the contour length.

For Gaussian P, it was shown by Naselsky and Novikov [34] that the amplitudes of the MFs are proportional to r_c^{-i}, where $i = 1, 2$. The behavior of r_c^{-1} noted above implies that the amplitudes of MFs for P also decrease as we increase r.

The points where $Q = 0 = U$ and hence $P = 0$, are referred to as singular points. Let us denote the number density of singularities by N_{sing}. In [34], it is shown to be given by $N_{sing} = 1/4\pi r_c^2$. Therefore, our calculation here shows that N_{sing} is sensitive to r and decreases as r is increased. This corroborates the result in [50] where the authors found that N_{sing} is sensitive to changes in r. Here we have quantified the nature of the dependence.

6. Conclusion

Analyses of CMB polarization data in the form of the Stokes parameters Q, U can provide information that complement the analyses using E, B and serve as consistency checks of the results. Q, U are what are measured in CMB polarization observations and hence if we use them for cosmological analysis we bypass the complications related to incomplete sky coverage that arise when we transform to E, B variables.

In this paper we address issues related to analyzing Q, U in the context of searches for primordial non-Gaussianity using MFs. It makes sense to use Q, U for cosmological analysis only if the observable quantities that we define using them are invariant under the transformations under which they transform as spin-two variables. We first show analytically that under rotations along the lines of sight the MFs are invariant. This implies that calculations of MFs for Q, U can be meaningfully compared between different observers (or different observing instruments) and the physical information obtained from them should be the same. However, we find that the invariance holds only when there is full sky coverage and under the assumption of statistical isotropy of the fluctuations. Thus the result is not immediately applicable to data from actual experiments where parts of the sky are masked due to our uncertain knowledge of Galactic emissions and point sources.

We have further calculated non-Gaussian deviations of the MFs that arise from local type primordial non-Gaussianity. We find the magnitudes of the deviations are about an order of magnitude lower in comparison to what is seen for E mode or temperature fluctuations and the shapes are distinct. For non-Gaussian analysis using masked observed data MFs an important step is to estimate the Gaussian component. This is usually done by using the Gaussian formulae given in Eq. (28), with the amplitudes calculated from the variances of the field obtained from the data. This is approximate and is reasonable only for very weakly non-Gaussian fields. The non-Gaussian deviation can then be calculated by subtracting the Gaussian estimate from the MFs calculated from the data. Note that this can be applied to MFs of Q, U for masked data. However, for the estimation of error bars, calculations using observed data and simulation can be compared only if the simulations use as input the $x - y$ coordinate choices along each line of sight that is used by the observational setup.

Lastly, we analyze the effect of the presence of primordial tensor perturbations on MFs for Q, U and P. We also discuss the effect on the number density of singularities in P. All these quantities can be expressed in terms of $r_c = \sigma_0/\sigma_1$. We show that r_c is sensitive to the presence of primordial tensor perturbations, and increases as r is increased. This result can potentially be used in analyzing polarization data and the search for B mode.

Acknowledgements

The computation required for this work was carried out on the Hydra cluster at the Indian Institute of Astrophysics. We acknowledge use of the CAMB [36,37] and HEALPIX packages [38,39] which were used to derive some of the results. We thank Anne Ducout and Dmitri Pogosyan for providing us the CND_REG2D source code for computing the Minkowski Functionals. We also thank F. Elsner for the use of his simulations of non-Gaussian E mode polarization.

References

[1] Martin Rees, Astrophys. J. 153 (1968) L1.
[2] J.R. Bond, G. Efstathiou, Mon. Not. R. Astron. Soc. 226 (1987) 655.
[3] R. Crittenden, R.L. Davis, P.J. Steinhardt, Astrophys. J. 417 (1993) L13.
[4] R.A. Frewin, A.G. Polnarev, P. Coles, Mon. Not. R. Astron. Soc. 266 (1994) L21.
[5] D. Coulson, R.G. Crittenden, N. Turok, Phys. Rev. Lett. 73 (1994) 2390.
[6] R.G. Crittenden, D. Coulson, N.G. Turok, Phys. Rev. D 52 (1995) 5402.
[7] A.A. Starobinsky, Pis'ma Zh. Eksp. Teor. Fiz. 30 (1979) 719; JETP Lett. 30 (1979) 682.
[8] A.H. Guth, Phys. Rev. D 23 (1981) 347.
[9] A.A. Starobinsky, Phys. Lett. B 117 (1982) 175.
[10] A.D. Linde, Phys. Lett. B 108 (1982) 175.
[11] A. Albrecht, P.J. Steinhardt, Phys. Rev. Lett. 48 (1982) 1220.
[12] M. Zaldarriaga, U. Seljak, Phys. Rev. D 55 (1997) 1830.
[13] M. Kamionkowski, A. Kosowsky, A. Stebbins, Phys. Rev. D 55 (1997) 7368.
[14] J. Kovac, E.M. Leitch, C. Pryke, J.E. Carlstrom, N.W. Halverson, W.L. Holzapfel, Nature 420 (2002) 772.
[15] A. Kogut, et al., WMAP Collaboration, Astrophys. J. Suppl. Ser. 148 (2003) 161.
[16] P.A.R. Ade, et al., Planck Collaboration, Planck 2015 results. XIII. Cosmological parameters, Astron. Astrophys. 594 (2016) A13.
[17] P.A.R. Ade, et al., Planck Collaboration, Planck 2015 results. XVII. Constraints on primordial non-Gaussianity, Astron. Astrophys. 594 (2016) A17.
[18] P.A.R. Ade, et al., BICEP2 and Keck Array Collaborations, Phys. Rev. Lett. 116 (3) (2016) 031302.
[19] H. Tomita, Prog. Theor. Phys. 76 (1986) 952.
[20] P. Coles, Mon. Not. R. Astron. Soc. 234 (1988) 509.
[21] J.R. Gott, C. Park, R. Juzkiewicz, W.E. Bies, F.R. Bouchet, A. Stebbins, Astrophys. J. 352 (1990) 1.
[22] K.R. Mecke, T. Buchert, H. Wagner, Astron. Astrophys. 288 (1994) 697.
[23] J. Schmalzing, T. Buchert, Astrophys. J. 482 (1997) L1.
[24] J. Schmalzing, K.M. Gorski, Mon. Not. R. Astron. Soc. 297 (1998) 355.
[25] S. Winitzki, A. Kosowsky, New Astron. 3 (1998) 75.
[26] T. Matsubara, Astrophys. J. 584 (2003) 1.
[27] E. Komatsu, et al., Astrophys. J. Suppl. Ser. 192 (2011) 18.
[28] P.A.R. Ade, et al., Planck Collaboration, Astron. Astrophys. 571 (2014) A24.
[29] V. Ganesan, P. Chingangbam, K.P. Yogendran, C. Park, J. Cosmol. Astropart. Phys. 02 (2015) 028.
[30] P. Chingangbam, C. Park, J. Cosmol. Astropart. Phys. 1302 (2013) 031.
[31] L. Santos, K. Wang, W. Zhao, J. Cosmol. Astropart. Phys. 07 (2016) 029.
[32] L. Santos, K. Wang, Y. Hu, W. Fang, W. Zhao, J. Cosmol. Astropart. Phys. 01 (2017) 043.
[33] C-G. Park, C. Park, J. Korean Astron. Soc. 35 (2002) 67.
[34] P.D. Naselsky, D.I. Novikov, Astrophys. J. 507 (1998) 31.
[35] A.D. Dolgov, A.G. Doroshkevich, D.I. Novikov, I.D. Novikov, Int. J. Mod. Phys. D 8 (1999) 189.
[36] A. Lewis, A. Challinor, A. Lasenby, Astrophys. J. 538 (2000) 473.
[37] http://camb.info/.
[38] K.M. Gorski, E. Hivon, A.J. Banday, B.B. Wandelt, F.K. Hansen, M. Reinecke, M. Bartelmann, Astrophys. J. 622 (2005) 759.
[39] http://healpix.sourceforge.net.
[40] P. Chingangbam, C. Park, K.P. Yogendran, R. van de Weygaert, Astrophys. J. 755 (2012) 122.
[41] C. Park, P. Pranav, P. Chingangbam, R. van de Weygaert, B. Jones, G. Vegter, I. Kim, J. Hidding, et al., J. Korean Astron. Soc. 46 (2013) 125.
[42] E.A. Lim, D. Simon, J. Cosmol. Astropart. Phys. 1 (2012) 48.
[43] F. Elsner, B.D. Wandelt, Astrophys. J. Suppl. Ser. 184 (2009) 264.
[44] E. Komatsu, et al., Astrophys. J. Suppl. Ser. 180 (2009) 330–376.
[45] C. Gay, C. Pichon, D. Pogosyan, Phys. Rev. D 85 (2012) 023011.
[46] A. Ducout, F. Bouchet, S. Colombi, D. Pogosyan, S. Prunet, Mon. Not. R. Astron. Soc. 429 (2013) 2104.
[47] C. Hikage, E. Komatsu, T. Matsubara, Astrophys. J. 653 (2006) 11.
[48] P. Chingangbam, C. Park, J. Cosmol. Astropart. Phys. 0912 (2009) 019.
[49] T. Matsubara, Phys. Rev. D 81 (2010) 083505.
[50] D. Huterer, T. Vachaspati, Phys. Rev. D 72 (2005) 043004.

Pre-inflation: Origin of the Universe from a topological phase transition

Mauricio Bellini [a,b,*]

[a] Departamento de Física, Facultad de Ciencias Exactas y Naturales, Universidad Nacional de Mar del Plata, Funes 3350, C.P. 7600, Mar del Plata, Argentina
[b] Instituto de Investigaciones Físicas de Mar del Plata (IFIMAR), Consejo Nacional de Investigaciones Científicas y Técnicas (CONICET), Mar del Plata, Argentina

ARTICLE INFO

Editor: M. Trodden

ABSTRACT

I study a model which describes the birth of the universe using a global topological phase transition with a complex manifold where the time, τ, is considered as a complex variable. Before the big bang τ is a purely imaginary variable so that the space can be considered as Euclidean. The phase transition from a pre-inflation to inflation is examined by studying the dynamical rotation of the time on the complex plane. Back-reaction effects are exactly calculated using Relativistic Quantum Geometry.

1. Introduction

One of the most important paradigm in modern cosmology is the explanation of how the universe reached the inflationary epoch. Inflation [1] is the more serious candidate to describe the expansion of the universe since the Planckian time to about 10^{-35} sec. This theory describes a quasi-exponential expansion that can resolve the flatness, horizon and monopole problems (among others). This theory has been very intensively tested [2] and provides a physical mechanism to explain the generation of primordial energy density fluctuations on super Hubble scales [3]. The most conservative assumption is that the energy density $\rho = P/\omega$ is due to a cosmological parameter which is constant and the equation of state is given by a constant $\omega = -1$, describing a vacuum dominated universe with pressure P and energy density ρ.

The theory that describes the earlier evolution of the universe is called pre-inflation [4]. The existence of a pre-inflationary epoch with fast-roll of the inflaton field would introduce an infrared depression in the primordial power spectrum. This depression might have left an imprint in the CMB anisotropy [5]. It is supposed that during pre-inflation the universe begun to expand from some Planckian-size initial volume to thereafter pass to an inflationary epoch. Some models consider the possibility of an pre-inflationary epoch in which the universe is dominated by radiation [6] In this framework Relativistic Quantum Geometry (RQG) [7], should be very useful when we try to study the evolution of the geometrical back-reaction effects given that we are dealing with Planckian energetic scales, and back-reaction effects should be very intense at these scales [8].

On the other hand, scalar fluctuations of the metric on cosmological scales can be studied in a non-perturbative formalism, describing not only small fluctuations, but the larger ones [9]. However, the metric fluctuations can be studied as a more profound phenomena in which the scalar metric fluctuations appear as a geometric response to the scalar field fluctuations by means of geometrical displacement from a Riemann manifold to a Weylian one, through RQG. The dynamics of the geometrical scalar field is defined on a Weyl-integrable manifold that preserves the gauge-invariance under the transformations of the Einstein's equations, that involves the cosmological constant. Our approach is different to quantum gravity. The natural way to construct quantum gravity models is to apply quantum field theory methods to the theories of classical gravitational fields interacting with matter. Our approach is different to quantum gravity because our subject of study is the dynamics of the geometrical quantum fields. This dynamics is obtained from the Einstein–Hilbert action, and not by using the standard effective action used in various models of quantum gravity [10]. It is supposed that during pre-inflation the universe began to expand from some Planckian-size initial volume to thereafter pass to an inflationary epoch. In this framework RQG should be very useful when we try to study the evolution of the geometrical back-reaction effects given that we are dealing with Planckian energetic scales, and back-reaction effects should be very intense at these scales.

In this letter we explore the idea that the universe could be created through a topological global phase transition, from a Eu-

* Correspondence to: Departamento de Física, Facultad de Ciencias Exactas y Naturales, Universidad Nacional de Mar del Plata, Funes 3350, C.P. 7600, Mar del Plata, Argentina.

E-mail address: mbellini@mdp.edu.ar.

clidean to an hyperbolic manifold, in which the time rotates in the complex plane from the imaginary axis to the real one. When the time reaches the real axis, the universe begins its inflationary expansion. The back-reaction effects are studied in detail.

2. The model

With the aim to describe a big-bang theory we shall consider a complex manifold, in terms of which the universe describes a background semi-Riemannian expansion. The line element for this case is

$$d\hat{S}^2 = \hat{g}_{\mu\nu} d\hat{x}^\mu d\hat{x}^\nu = e^{2i\theta(t)} d\hat{t}^2 + a^2(t)\hat{\eta}_{ij} d\hat{x}^i d\hat{x}^j, \tag{1}$$

with the signature: $(+,-,-,-)$. Here $\theta(t) = \frac{\pi}{2}\frac{a_0}{a}$, with $a \geq a_0$, t is a real parameter time and $H_0 = \pi/(2a_0) = 1/t_p$, such that $t_p = 5.4 \times 10^{-44}$ sec is the Planckian time. Notice that the metric (1) describes a complex manifold such that, at $t = 0$ the spacetime is Euclidean, but after many Planckian times, when $\theta \to 0$, it becomes hyperbolic. We shall define the background action $\mathcal{I}$ on this manifold, so that it describes the expansion driven by a scalar field, which is minimally coupled to gravity

$$\mathcal{I} = \int d^4x \sqrt{\hat{g}} \left[\frac{\hat{\mathcal{R}}}{16\pi G} + \left[\frac{1}{2}\dot{\phi}^2 - V(\phi) \right] \right], \tag{2}$$

where $\sqrt{\hat{g}} = ia^3 e^{i\theta}$. After some algebraic work, we obtain that the relevant (complex) Einstein equations are

$$3H^2(t) e^{-i\hat{\theta}} = 8\pi G \left(\frac{\dot{\phi}^2}{2} e^{-i\hat{\theta}} + e^{i\hat{\theta}} V(\phi) \right), \tag{3}$$

$$\left(3H^2 + 2\dot{H} - 2i\dot{\hat{\theta}} H \right) e^{-i\hat{\theta}} = 8\pi G \left(\frac{\dot{\phi}^2}{2} e^{-i\hat{\theta}} - e^{i\hat{\theta}} V(\phi) \right), \tag{4}$$

so that we can calculate the equation of state

$$\omega = P/\rho = -1 - \frac{2\dot{H}}{3H^2} + \frac{2i\dot{\hat{\theta}}}{3H}. \tag{5}$$

Here, the (complex) background pressure P and the energy density ρ, are

$$P = \frac{\dot{\phi}^2}{2} e^{-i\hat{\theta}} - e^{i\hat{\theta}} V(\phi), \tag{6}$$

$$\rho = \frac{\dot{\phi}^2}{2} e^{-i\hat{\theta}} + e^{i\hat{\theta}} V(\phi). \tag{7}$$

On the other hand, from the action (1) we obtain

$$\ddot{\phi} + \left(3\frac{\dot{a}}{a} - i\dot{\hat{\theta}} \right) \dot{\phi} + e^{2i\hat{\theta}} \frac{\delta V(\phi)}{\delta\phi} = 0, \tag{8}$$

that describes the dynamics of the background field $\phi(t)$. The metric (1) is not sufficiently explicit to describe the transition to an inflationary universe from a topological phase transition, because t is not exactly the dynamical coordinate that describes this transition. The correct dynamical variable in (1) is: $\tau = \int e^{i\hat{\theta}(t)} dt$, which describes the time in the complex plane. The idea is that τ becomes a space-like coordinate before the big bang, so that it can be considered as a reversal variable. However, after a phase transition we must require that it changes its signature and then can be considered as a causal variable. This dynamical change of signature describes a topological phase transition of the universe from an initial global Euclidean 4D space, to a final hyperbolic 4D spacetime.

3. An example: asymptotic de Sitter expansion

We shall consider a scale factor, related to a de Sitter expansion in the t-dynamical scale: $H_0 = \dot{a}/a(t)$, such that $\mathcal{H}(\tau) = \frac{1}{a(\tau)}\frac{da(\tau)}{d\tau} = H_0\, e^{-i\hat{\theta}(\tau)}$, is

$$a(\tau) = a_0\, e^{\mathrm{Ei}\left[1, i\frac{\pi}{2} a_0 e^{-H_0 \tau}\right]}. \tag{9}$$

Notice that we have used the fact that $\hat{\theta}(\tau) = \frac{\pi}{2}\frac{a_0}{a(\tau)}$. The expression (9) for the scale factor written in terms of τ makes it very difficult to describe the cosmological dynamics of the universe. For this reason, we shall search for another variable to describe the dynamics of this cosmological phase transition. A good candidate is the phase $\hat{\theta}$. Since $\hat{\theta}(t) = \frac{\pi}{2} e^{-H_0 t}$, we can rewrite the metric (1), as

$$d\hat{S}^2 = \left(\frac{\pi a_0}{2} \right)^2 \frac{1}{\hat{\theta}^2} \left[d\hat{\theta}^2 + \hat{\eta}_{ij} d\hat{x}^i d\hat{x}^j \right]. \tag{10}$$

If we desire to describe an initially Euclidean 4D universe, that thereafter evolves to an asymptotic value $\hat{\theta} \to 0$, we must require that $\hat{\theta}$ to have an initial value $\hat{\theta}_0 = \frac{\pi}{2}$. Furthermore, the nonzero components of the Einstein tensor, are

$$G_{00} = -\frac{3}{\hat{\theta}^2}, \qquad G_{ij} = \frac{3}{\hat{\theta}^2}\delta ij, \tag{11}$$

so that the radiation energy density and pressure, are respectively given in this representation by

$$\rho(\hat{\theta}) = \frac{1}{2\pi G}\frac{3}{(\pi a_0)^2}, \qquad P(\hat{\theta}) = -\frac{1}{4\pi G}\frac{3}{(\pi a_0)^2}. \tag{12}$$

The equation of state for the metric (10), is

$$\omega(\hat{\theta}) = -1. \tag{13}$$

We shall describe the case where the asymptotic evolution of the Universe is described by a vacuum expansion. In this case the asymptotic scale factor, Hubble parameter and the potential are respectively given by

$$a(t) = a_0\, e^{H_0 t}, \qquad \frac{\dot{a}}{a} = H_0, \qquad V = \frac{3}{8\pi G} H_0^2, \tag{14}$$

so that the background field in (8), is

$$\phi(t) = \phi_0. \tag{15}$$

This solution describes the background solution of the field that drives a phase transition of the global geometry from a 4D Euclidean space to a 4D hyperbolic spacetime.

In order to describe the exact back-reaction effects, we shall consider Relativistic Quantum Geometry (RQG), introduced in [7]. In this formalism the manifold is defined with a connection

$$\Gamma^\alpha_{\beta\gamma} = \left\{ \begin{array}{c} \alpha \\ \beta\,\gamma \end{array} \right\} + \sigma^\alpha \hat{g}_{\beta\gamma}, \tag{16}$$

such that the covariant derivative of the metric tensor in the Riemannian background manifold is null (we denote with a semicolon the Riemannian-covariant derivative): $\Delta g_{\alpha\beta} = g_{\alpha\beta;\gamma}\, dx^\gamma = 0$, so that the Weylian covariant derivative [11] on the manifold generated by (16) is nonzero: $g_{\alpha\beta|\gamma} = \sigma_\gamma\, g_{\alpha\beta}$. To simplify the notation we denote $\sigma_\alpha \equiv \sigma_{,\alpha}$. From the action's point of view, the scalar field $\sigma(x^\alpha)$ is a generic geometrical transformation that leaves the action invariant [7]

$$\begin{aligned} \mathcal{I} &= \int d^4\hat{x} \sqrt{-\hat{g}} \left[\frac{\hat{R}}{2\kappa} + \hat{\mathcal{L}} \right] \\ &= \int d^4\hat{x} \left[\sqrt{-\hat{g}} e^{-2\sigma} \right] \left\{ \left[\frac{\hat{R}}{2\kappa} + \hat{\mathcal{L}} \right] e^{2\sigma} \right\}, \end{aligned} \tag{17}$$

Hence, Weylian quantities will be varied over these quantities in a semi-Riemannian manifold so that the dynamics of the system preserves the action: $\delta\mathcal{I}=0$, and we obtain

$$-\frac{\delta V}{V}=\frac{\delta\left[\frac{\hat{R}}{2\kappa}+\hat{\mathcal{L}}\right]}{\left[\frac{\hat{R}}{2\kappa}+\hat{\mathcal{L}}\right]}=2\,\delta\sigma, \tag{18}$$

where $\delta\sigma=\sigma_\mu dx^\mu$ is an exact differential and $V=\sqrt{-\hat{g}}$ is the volume of the Riemannian manifold. Of course, all the variations are in the Weylian geometrical representation, and assure us gauge invariance because $\delta\mathcal{I}=0$.

$$\frac{1}{\hat{\rho}}\frac{\delta\hat{\rho}}{\delta S}=-2\left(\frac{\pi}{2a_0}\right)\hat{\theta}\sigma', \tag{19}$$

such that [12], for $\sigma'=\left\langle(\sigma')^2\right\rangle^{1/2}$

$$\left\langle(\sigma')^2\right\rangle=\frac{1}{(2\pi)^3}\int d^3k(\xi_k)'\,(\xi_k^*)', \tag{20}$$

where the modes ξ_k must be restricted to

$$(\xi_k^*)'\xi_k-(\xi_k)'\xi_k^*=i\hat{\theta}^2\left(\frac{2}{\pi a_0}\right)^2, \tag{21}$$

in order to the field σ to be quantized [7]

$$\left[\sigma(x),\sigma_\mu(y)\right]=i\hbar\Theta_\mu\delta^{(4)}(x-y). \tag{22}$$

Here, $\Theta_\mu=\left[\hat{\theta}^2\left(\frac{2}{\pi a_0}\right)^2,0,0,0\right]$ are the components of the background relativistic tetra-vector on the Riemann manifold. The equation of motion for the modes of σ: $\xi_k(\hat{\theta})$, is

$$\xi_k''-\frac{2}{\hat{\theta}}\xi_k'+k^2\,\xi_k(\hat{\theta})=0, \tag{23}$$

where the *prime* denotes the derivative with respect to $\hat{\theta}$. The exact solution for the modes $\xi_k(\hat{\theta})$ are

$$\xi_k(\hat{\theta})=C_1(k)\left[k\hat{\theta}\cos(k\hat{\theta})-\sin(k\hat{\theta})\right]+C_2(k)\left[k\hat{\theta}\sin(k\hat{\theta})+\cos(k\hat{\theta})\right]. \tag{24}$$

If we take $C_2(k)=i\,C_1(k)$, we obtain that:

$$C_1(k)=\frac{i}{2}\left(\frac{\pi}{2a_0}\right)k^{-3/2},\qquad C_2(k)=-\frac{1}{2}\left(\frac{\pi}{2a_0}\right)k^{-3/2}, \tag{25}$$

so that the quantized solution of (23) results to be

$$\xi_k(\theta)=\frac{i}{2}\left(\frac{\pi}{2a_0}\right)k^{-3/2}\,e^{-ik\hat{\theta}}\left[k\hat{\theta}-i\right]. \tag{26}$$

Therefore, the fluctuations (20), are

$$\left\langle(\sigma')^2\right\rangle=\frac{1}{8}\frac{\hat{\theta}^2}{(4a_0)^2}\epsilon^4k_0^4, \tag{27}$$

such that $\epsilon\ll 1$ and $k_0=\frac{\sqrt{2}}{\hat{\theta}}$. Hence, the amplitude of energy-density fluctuations on super Hubble scales, becomes

$$\left|\frac{1}{\hat{\rho}}\frac{\delta\hat{\rho}}{\delta S}\right|=\frac{\pi\epsilon^2}{4\sqrt{2}a_0^2}, \tag{28}$$

which is a constant.

4. Final comments

We have studied a model that describe the origin of the universe using a global topological phase transition from a 4D Euclidean manifold to an asymptotic 4D hyperbolic one. To develop this idea we have introduced a complex time, τ. The interesting of this idea is that τ is a space-like coordinated before the big bang, so that can be considered as a reversal variable. However, after the phase transition it changes its signature and then can be considered as a causal variable. Due to the fact that the description of the cosmological dynamics of the universe as a function of τ become complicated, we have expressed introduced the phase $\hat{\theta}$ as a dynamical variable.

As an example, we have explored the case where the universe evolves through an asymptotic de Sitter expansion. We have studied back-reaction effects in a pre-inflationary universe using RQG. This formalism makes possible the non-perturbative treatment of the vacuum fluctuations of the spacetime, by making a displacement from a semi-Riemannian to a Weylian one. In this description the dynamics of the geometrical field σ describes the geometrical quantum fluctuations with respect to the Riemannian (classical) background. In the example here studied, we have found that back-reaction becomes frozen in time.

Finally, the field σ has a geometrical origin, but must be interpreted as a primordial gravitational quantum potential, which generates a distortion of the metric. After pre-inflation, σ could decay in different kinds of fields, one of which should be the inflaton field.

Acknowledgements

M. Bellini acknowledges UNMdP grant No. EXA651/14 and CONICET (Argentina) grant No. PIP (CONICET) 112-201101-00325CO for financial support.

References

[1] A.A. Starobinsky, Phys. Lett. B 91 (1980) 99;
A.H. Guth, Phys. Rev. D 23 (1981) 347;
A.D. Linde, Phys. Lett. B 129 (1983) 177.
[2] R.L. Smoot, et al., Astrophys. J. 396 (1992) L1.
[3] M. Bellini, H. Casini, R. Montemayor, P. Sisterna, Phys. Rev. D 54 (1996) 7172.
[4] H. Kurki-Suonio, F. Graziani, G.J. Mathews, Phys. Rev. D 44 (1991) 3072;
S. Hirai, Class. Quantum Gravity 22 (2005) 1239;
G.J. Mathews, N.Q. Lan, T. Kajino, I.S. Suh, Phys. Rev. D 92 (2015) 123514.
[5] A. Gruppuso, N. Kitazawa, N. Mandolesi, P. Natoli, A. Sagnotti, Phys. Dark Universe 11 (2016) 68.
[6] I-Chin Wang, Kin-Wang Ng, Phys. Rev. D 77 (2008) 083501.
[7] L.S. Ridao, M. Bellini, Phys. Lett. B 751 (2015) 565;
L.S. Ridao, M. Bellini, Astrophys. Space Sci. 357 (94) (2015).
[8] M.R.A. Arcodía, M. Bellini, Eur. Phys. J. C 76 (6) (2016) 326.
[9] M. Anabitarte, M. Bellini, Eur. Phys. J. C 60 (2009) 297.
[10] I.L. Buchbinder, S.D. Odintsov, I.L. Shapiro, Effective Action in Quantum Gravity, Institute of Physics Publishing, Bristol–Philadelphia, 1992.
[11] H. Weyl, Philosophy of Mathematics and Natural Science, Princeton University Press, 1949, english version.
[12] M. Bellini, Phys. Dark Universe 11 (2016) 64.

21

Primordial gravitational waves induced by magnetic fields in an ekpyrotic scenario

Asuka Ito*, Jiro Soda

Department of Physics, Kobe University, Kobe 657-8501, Japan

ARTICLE INFO

Editor: H. Peiris

ABSTRACT

Both inflationary and ekpyrotic scenarios can account for the origin of the large scale structure of the universe. It is often said that detecting primordial gravitational waves is the key to distinguish both scenarios. We show that this is not true if the gauge kinetic function is present in the ekpyrotic scenario. In fact, primordial gravitational waves sourced by the gauge field can be produced in an ekpyrotic universe. We also study scalar fluctuations sourced by the gauge field and show that it is negligible compared to primordial gravitational waves. This comes from the fact that the fast roll condition holds in ekpyrotic models.

1. Introduction

Inflation has succeeded in solving several issues in big bang cosmology and explaining the temperature anisotropy of the cosmic microwave background radiation (CMB) and the large scale structure of the universe. However, it is known that bouncing universe models [1] such as the ekpyrotic scenario [2] based on superstring theory [3] can do the same job [4].[1] Therefore, it is important to clarify which scenario is actually realized in the early stage of the universe.

In the ekpyrotic scenario, the primordial fluctuations are produced in a slowly contracting (ekpyrotic) phase. The spectrum of the scalar and tensor vacuum fluctuations become blue-tilted at the end of the ekpyrotic phase. We therefore need an additional scalar field to explain the temperature anisotropy of the CMB [6]. In the ekpyrotic scenario, the amplitudes of primordial gravitational waves [7] are quite small and practically unobservable [8]. Hence, it is often said that, if we could detect the primordial gravitational waves, we would be able to disprove the ekpyrotic scenario. However, if there could exist another mechanism for producing gravitational waves in the ekpyrotic scenario, the story would be completely different. Indeed, we show that there exists a mechanism for producing abundant gravitational waves in the ekpyrotic phase.

The key is the presence of magnetic fields in the early universe. Observationally, there are several evidences for magnetic fields to exist on various cosmological scales [9]. Although the origin of primordial magnetic fields is unknown, the presence of magnetic fields on extra galactic scales [10] implies that the seed of magnetic fields must be produced in the early universe. Notably, there are attempts to make primordial magnetic fields with the gauge kinetic function in an inflationary universe [11] or in a bouncing universe [12].

In this paper, we first show that scale invariant magnetic fields can be produced in the ekpyrotic phase in the presence of the gauge kinetic function. Next, we show that the magnetic fields can become a source of abundant gravitational waves (such mechanism works also in inflation [13]). It turns out that the gravitational wave spectrum is nearly scale invariant (slightly blue) at the end of the ekpyrotic phase. Hence, it is difficult to discriminate between inflation and the ekpyrotic scenario by merely detecting primordial gravitational waves. We also study scalar fluctuations induced by the magnetic fields and show that the sourced tensor to scalar ratio should be more than unity, which implies that scalar fluctuations in the CMB should be dominated by quantum fluctuations produced by an additional scalar field as is often assumed in the ekpyrotic scenario.

The paper is organized as follows. In section 2, we review the ekpyrotic scenario briefly and explain background evolution in the ekpyrotic phase. In section 3, we derive the mode functions of the gauge field and show that scale invariant magnetic fields can be produced in the ekpyrotic scenario. In section 4, we demonstrate that abundant gravitational waves with scale invariance are pro-

* Corresponding author.
E-mail address: asuka-ito@stu.kobe-u.ac.jp (A. Ito).

[1] The pre-big bang scenario is also a kind of the models [5]. Our conclusion could apply to it too.

duced from the scale invariant magnetic fields. In section 5, we show that scalar fluctuations are also produced by the scale invariant magnetic fields. It turns out that the tensor to scalar ratio should be larger than unity in the ekpyrotic scenario. The final section is devoted to the conclusion.

2. Ekpyrotic phase

In the ekpyrotic scenario, two branes residing in an extra dimension approach, collide and bounce off to each other. From the four-dimensional point of view, they correspond to a contracting universe and an expanding universe, respectively. The ekpyrotic scenario can be described by a four dimensional effective theory with a scalar field ϕ moving in an effective potential $V(\phi)$ specified below. The action reads

$$S = \int d^4x\sqrt{-g}\left[\frac{M_{pl}^2}{2}R - \frac{1}{2}(\partial_\mu\phi)(\partial^\mu\phi) - V(\phi)\right], \tag{1}$$

where M_{pl} represents the reduced Plank mass, g is the determinant of the metric $g_{\mu\nu}$, and R is the Ricci scalar. The scalar field represents the separation l between two branes $l \sim e^{\phi}$. The contracting universe ($\dot{\phi} < 0$) is connected to the expanding universe ($\dot{\phi} > 0$) through a bounce (a collision of two branes). The scalar and tensor vacuum fluctuations are produced in the contracting phase where the scalar field rolls down a negative steep potential

$$V(\phi) \simeq V_0 e^{\lambda\frac{\phi}{M_{pl}}}, \tag{2}$$

where V_0 is a negative constant. Note that λ is also negative and satisfies the fast roll condition $|\lambda| \gg 1$ to keep isotropy of the universe. Thus, we can take an isotropic metric ansatz in this phase as

$$ds^2 = a(\tau)\left[-d\tau^2 + dx^2 + dy^2 + dz^2\right], \tag{3}$$

where we used a conformal time τ. It is straightforward to derive scaling solutions from Eqs. (1)~(3)

$$a(\tau) = a_{end}\left(\frac{-\tau}{-\tau_{end}}\right)^{\frac{2}{\lambda^2-2}}, \quad \frac{\phi(\tau)}{M_{pl}} = \phi_0 - \frac{2\lambda}{\lambda^2-2}\ln(-M_{pl}\tau), \tag{4}$$

where τ_{end} (< 0) and a_{end} represent the moment and the scale factor at the end of the ekpyrotic phase, respectively. The obtained vacuum scalar and tensor power spectrums are blue-tilted, so that we need an additional scalar field to explain the CMB observation [6]. Then, the ekpyrotic scenario predicts the nearly scale invariant scalar power spectrum and the blue-tilted tensor power spectrum. The situation is different from inflation where both spectra are nearly scale invariant.

3. Scale invariant magnetic fields

In this section, we show that scale invariant magnetic fields can be produced from quantum fluctuations due to interaction between a scalar field and a gauge field in the ekpyrotic phase. We consider the action

$$S = \int d^4x\sqrt{-g}\left[\frac{M_{pl}^2}{2}R - \frac{1}{2}(\partial_\mu\phi)(\partial^\mu\phi) - V(\phi) - \frac{1}{4}f^2(\phi)F_{\mu\nu}F^{\mu\nu}\right], \tag{5}$$

where $F_{\mu\nu} = \partial_\mu A_\nu - \partial_\nu A_\mu$ is the field strength of the gauge field coupled to the scalar field ϕ and $f(\phi)$ represents the gauge kinetic function. Now, let us take the gauge kinetic function as exponential type functional form which is ubiquitous in models obtained from dimensional reduction

$$f(\phi) = f_0 e^{\rho\frac{\phi}{M_{pl}}}, \tag{6}$$

where it has been set to be unity at the end of the ekpyrotic phase, and then there is no strong coupling problem. We treat the gauge field as a test field and ignore the back reaction from the gauge field in the background. Thus, using the background solution (4), we can express the gauge kinetic function as

$$f(\phi) \propto (-\tau)^{-\frac{2\rho\lambda}{\lambda^2-2}}. \tag{7}$$

Let us expand the gauge field in Fourier space as

$$\vec{A}(\tau,\vec{x}) = \int \frac{d^3k}{(2\pi)^{3/2}}\boldsymbol{A_k}(\tau)e^{i\boldsymbol{k}\cdot\boldsymbol{x}}. \tag{8}$$

Then the part for the gauge field in the action (5) can be rewritten as

$$S_{gauge} = \frac{1}{2}\int d\tau d^3k f^2(\phi)\left[\boldsymbol{A}'_{\boldsymbol{k}}\boldsymbol{A}'_{-\boldsymbol{k}} - k^2\boldsymbol{A_k}\boldsymbol{A_{-k}}\right], \tag{9}$$

where a prime represents a derivative with respect to the conformal time. The Fourier mode of the gauge field can be promoted into the operator and expanded by the creation and annihilation operators satisfying commutation relations $\left[\hat{a}_{\boldsymbol{k}}^{(\sigma)}, \hat{a}_{\boldsymbol{k}'}^{(\rho)\dagger}\right] = \delta_{\sigma\rho}\delta(\boldsymbol{k}-\boldsymbol{k}')$ as

$$\hat{\boldsymbol{A}}_{\boldsymbol{k}}(\tau) = \sum_{\sigma=+,-}\vec{e}^{(\sigma)}(\hat{k})\left[U_k(\tau)\hat{a}_{\boldsymbol{k}}^{(\sigma)} + U_k^*(\tau)\hat{a}_{-\boldsymbol{k}}^{(\sigma)\dagger}\right], \tag{10}$$

where σ represents the two polarization degrees of freedom of the gauge field. The circular polarization vectors $\vec{e}^{\,(\sigma)}$ satisfy the relations

$$\begin{aligned}
&\vec{k}\cdot\vec{e}^{\,(\pm)}(\hat{k}) = 0, \\
&\vec{k}\times\vec{e}^{\,(\pm)}(\hat{k}) = \mp ik\vec{e}^{\,(\pm)}(\hat{k}), \\
&\left(\vec{e}^{\,(\sigma)}(\hat{k})\right)^* = \vec{e}^{\,(\sigma)}(-\hat{k}), \quad \left(\vec{e}^{\,(\sigma)}(\hat{k})\right)^*\cdot\vec{e}^{\,(\rho)}(\hat{k}) = \delta_{\sigma\rho}.
\end{aligned} \tag{11}$$

The mode functions obey the equations derived from the action (9)

$$U_k'' + 2\frac{f'}{f}U_k' + k^2U_k = 0. \tag{12}$$

Using new variables $u_k \equiv fU_k$, we get

$$u_k'' + \left(k^2 - \frac{f''}{f}\right)u_k = 0. \tag{13}$$

Substituting Eq. (7) into Eq. (13) and solving it with the Bunch–Davies initial condition, we get the mode functions

$$u_k(\tau) = \frac{1}{\sqrt{2k}}\sqrt{\frac{-k\tau\pi}{2}}H^{(1)}_{-\frac{1}{2}\frac{\lambda^2+4\rho\lambda-2}{\lambda^2-2}}(-k\tau), \tag{14}$$

where $H_\gamma^{(1)}(x)$ is the Hankel function of the first kind. Now, we define the electric and magnetic fields as

$$\vec{E}(\tau,\boldsymbol{x}) \equiv -\frac{f}{a^2}\partial_\tau\vec{A}(\tau,\boldsymbol{x}), \quad \vec{B}(\tau,\boldsymbol{x}) \equiv \frac{f}{a^2}\left(\nabla\times\vec{A}(\tau,\boldsymbol{x})\right). \tag{15}$$

They can be expanded in Fourier space as

$$\vec{E}(\tau,\boldsymbol{x}) = \int \frac{d^3k}{(2\pi)^{3/2}} \hat{\boldsymbol{E}}_{\boldsymbol{k}}(\tau) e^{i\boldsymbol{k}\cdot\boldsymbol{x}} , \tag{16}$$

$$\vec{B}(\tau,\boldsymbol{x}) = \int \frac{d^3k}{(2\pi)^{3/2}} \hat{\boldsymbol{B}}_{\boldsymbol{k}}(\tau) e^{i\boldsymbol{k}\cdot\boldsymbol{x}} , \tag{17}$$

where

$$\hat{\boldsymbol{E}}_{\boldsymbol{k}}(\tau) = \sum_{\sigma=+,-} \vec{e}^{(\sigma)}(\hat{k}) \left[\mathcal{E}_k \hat{a}_{\boldsymbol{k}}^{(\sigma)} + \mathcal{E}_k^* \hat{a}_{-\boldsymbol{k}}^{(\sigma)\dagger} \right] , \tag{18}$$

$$\hat{\boldsymbol{B}}_{\boldsymbol{k}}(\tau) = \sum_{\sigma=+,-} \sigma \vec{e}^{(\sigma)}(\hat{k}) \left[\mathcal{B}_k \hat{a}_{\boldsymbol{k}}^{(\sigma)} + \mathcal{B}_k^* \hat{a}_{-\boldsymbol{k}}^{(\sigma)\dagger} \right] . \tag{19}$$

Here, we defined

$$\mathcal{E}_k(\tau) = -\frac{f}{a^2} \partial_\tau U_k(\tau) , \quad \mathcal{B}_k(\tau) = \frac{f}{a^2} k U_k(\tau) . \tag{20}$$

Using the mode functions (14), we obtain

$$\mathcal{E}_k(\tau) = -\frac{\sqrt{\pi}}{2} k^{1/2} (-k\tau)^{1/2} a_{end}^{-2} \left(\frac{-\tau}{-\tau_{end}} \right)^{-\frac{4}{\lambda^2-2}} \times H^{(1)}_{\frac{1}{2}\frac{\lambda^2-4\rho\lambda-2}{\lambda^2-2}}(-k\tau) , \tag{21}$$

$$\mathcal{B}_k(\tau) = \frac{\sqrt{\pi}}{2} k^{1/2} (-k\tau)^{1/2} a_{end}^{-2} \left(\frac{-\tau}{-\tau_{end}} \right)^{-\frac{4}{\lambda^2-2}} \times H^{(1)}_{-\frac{1}{2}\frac{\lambda^2+4\rho\lambda-2}{\lambda^2-2}}(-k\tau) . \tag{22}$$

In the superhorizon limit $|k\tau| \to 0$, we can use an approximation

$$H_\gamma^{(1)}(x) \simeq -\frac{i\Gamma(-\gamma)}{\pi} e^{-i\pi\gamma} \left(\frac{x}{2} \right)^\gamma . \tag{23}$$

Then the magnetic fields are given by

$$\mathcal{B}_k(\tau) = -\frac{(\lambda^2-2)^2}{\sqrt{\pi}} i 2^{-\frac{1}{2}\frac{5\lambda^2-4\rho\lambda-10}{\lambda^2-2}} \Gamma\left(\frac{1}{2}\frac{\lambda^2+4\rho\lambda-2}{\lambda^2-2} \right) \times e^{\frac{\pi i}{2}\frac{\lambda^2+4\rho\lambda-2}{\lambda^2-2}} k^{\frac{1}{2}\frac{\lambda^2-4\rho\lambda-2}{\lambda^2-2}} (-\tau)^{-\frac{2(\rho\lambda+2)}{\lambda^2-2}} (-\tau_{end})^{\frac{2\lambda^2}{\lambda^2-2}} H_{end}^2 , \tag{24}$$

where we used a relation

$$a = \left(\frac{2}{\lambda^2-2} \right) \frac{1}{\tau H} . \tag{25}$$

In order to obtain the scale invariant magnetic fields, we require

$$\rho = \frac{\lambda^2-2}{\lambda} . \tag{26}$$

In this case, the electric fields are always subdominant compared with the magnetic fields. So we only consider the magnetic fields as the source of gravitational waves. Substituting Eq. (26) into Eq. (24), we get

$$\mathcal{B}_k(\tau) = \frac{3\sqrt{2}}{8} (\lambda^2-2)^2 k^{-3/2} \left(\frac{-\tau}{-\tau_{end}} \right)^{-\frac{2\lambda^2}{\lambda^2-2}} H_{end}^2 . \tag{27}$$

We find that the steeper the potential becomes, namely $|\lambda|$ is bigger, the more magnetic field is amplified.

To avoid destroying the background evolution by back reaction from the magnetic fields, at least, we need the condition that the energy density of the electromagnetic field does not exceed that of the scalar field at the end of the ekpyrotic phase

$$\begin{aligned} \langle \rho_{em} \rangle &= \frac{1}{2} \frac{1}{(2\pi)^3} \int_{k_{in}}^{k_{end}} 2 \times \mathcal{B}_k(\tau_{end})^2 d^3k \\ &= \frac{9}{64\pi^2} (\lambda^2-2)^4 H_{end}^4 \ln\left(\frac{k_{end}}{k_{in}} \right) \\ &< 3M_{pl}^2 H_{end}^2 , \end{aligned} \tag{28}$$

where k_{in} and k_{end} represent the scales where a mode exits the Hubble horizon at the beginning and at the end of the ekpyrotic phase, respectively. Let us also check the back reaction problem in the equation for the scalar field. The hamiltonian constraint and the equation for the scalar field are given as

$$H^2 = \frac{1}{3M_{pl}^2} \left(\frac{1}{2}\dot{\phi}^2 + V(\phi) + <\rho_{em}> \right) , \tag{29}$$

$$\ddot{\phi} = -3H\dot{\phi} - V_{,\phi} + \frac{2\rho}{M_{pl}} <\rho_{em}> , \tag{30}$$

where an overdot denotes a derivative with respect to the cosmic time. Asuuming that the gauge field is negligible in Eq. (29), the fast roll condition let Eq. (29) be

$$\frac{1}{2}\dot{\phi}^2 \simeq -V(\phi) . \tag{31}$$

Differentiating the both sides of Eq. (31) with respect to the time, we find that the first term is negligible compared with the second term in the right-hand side of Eq. (30). Thus, we have the relation

$$<\rho_{em}> \simeq \frac{M_{pl}}{2\rho} V_{,\phi} , \tag{32}$$

when the gauge field is significant in Eq. (30). Then the ratio of the energy density of the gauge field to that of the scalar field is

$$\frac{<\rho_{em}>}{3M_{pl}^2 H^2} \gg \frac{<\rho_{em}>}{V} \simeq \frac{M_{pl}}{2\rho} \frac{V_{,\phi}}{V} = \frac{1}{2}\frac{\lambda}{\rho} . \tag{33}$$

The most right term is order unity in our scenario since λ and ρ are same order from Eq. (26). Therefore, as far as the ratio of the energy density of the gauge field to that of the scalar field is small, the gauge field can be treated as a test field in any equations.

Taking a look at Eq. (28), for example, if we set $H_{end} = 10^{-5} M_{pl}$, we obtain the minimum value of λ about -17. Thus, the cosmological magnetic fields observed at present can be produced in the ekpyrotic scenario [9]. Remarkably, such magnetic fields can also induce abundant primordial gravitational waves. We will see it in the next section.

4. Gravitational waves from magnetic fields

In this section, we calculate the gravitational waves induced by the magnetic fields studied in the previous section. The method is similar to that used in inflationary universe [13,14]. One can get the tensor sector of the action (5) as

$$S_{GW} = \int d\tau d^3x \left[\frac{M_{pl}^2}{8} a^2 \left(h'_{ij} h'^{ij} - \partial_k h_{ij} \partial_k h^{ij} \right) + \frac{1}{2} a^4 \left(E_i E_j + B_i B_j \right) h^{ij} \right] , \tag{34}$$

where h_{ij} is the transverse traceless tensor and we used the definition of the electric and magnetic fields (15). The tensor fluctuation can be expanded in Fourier space as

$$h_{ij} = \sum_{\sigma=+,-} \int \frac{d^3k}{(2\pi)^{3/2}} \hat{h}^{(\sigma)}_{\boldsymbol{k}} e^{i\boldsymbol{kx}} \Pi^{(\sigma)}_{ij} , \tag{35}$$

$$\hat{h}^{(\sigma)}_{\boldsymbol{k}}(\tau) = V_k(\tau)\hat{a}^{(\sigma)}_{\boldsymbol{k}} + V^*_k(\tau)\hat{a}^{(\sigma)\dagger}_{-\boldsymbol{k}} , \tag{36}$$

where $\Pi^{(\sigma)}_{ij}$ are polarization tensors constructed by circular polarization vectors as $\Pi^{(\sigma)}_{ij} \equiv e^{(\sigma)}_i e^{(\sigma)}_j$ and we have used creation and annihilation operators. Substituting Eqs. (16)~(19) and (35) into Eq. (34), we obtain

$$\begin{aligned} S_{GW} = \sum_{\sigma=+,-} \int d\tau d^3k \Bigg[& \frac{M^2_{pl}}{4} a^2 \left(\hat{h}'^{(\sigma)}_{\boldsymbol{k}} \hat{h}'^{(\sigma)}_{-\boldsymbol{k}} - k^2 \hat{h}^{(\sigma)}_{\boldsymbol{k}} \hat{h}^{(\sigma)}_{-\boldsymbol{k}} \right) \\ & - \frac{a^4}{2} \int \frac{d^3p}{(2\pi)^{3/2}} \left(\hat{E}_{i,\boldsymbol{p}} \hat{E}_{j,\boldsymbol{k}-\boldsymbol{p}} + \hat{B}_{i,\boldsymbol{p}} \hat{B}_{j,\boldsymbol{k}-\boldsymbol{p}} \right) \\ & \times e^{*(\sigma)}_i(\hat{k}) e^{*(\sigma)}_j(\hat{k})\, \hat{h}^{(\sigma)}_{-\boldsymbol{k}} \Bigg] . \end{aligned} \tag{37}$$

Using the variable $v_k \equiv \frac{M_{pl}}{2} a V_k$, we can get the equation for the mode function of the gravitational waves as

$$v''_k(\tau) + \left(k^2 - \frac{a''}{a} \right) v_k(\tau) = S^{(\sigma)}(\tau, \boldsymbol{k}) , \tag{38}$$

where the source term is defined by

$$\begin{aligned} S^{(\sigma)}(\tau, \boldsymbol{k}) = & -\frac{a^3}{M_{pl}} \int \frac{d^3p}{(2\pi)^{3/2}} \left(\hat{E}_{i,\boldsymbol{p}} \hat{E}_{j,\boldsymbol{k}-\boldsymbol{p}} + \hat{B}_{i,\boldsymbol{p}} \hat{B}_{j,\boldsymbol{k}-\boldsymbol{p}} \right) \\ & \times e^{*(\sigma)}_i(\hat{k}) e^{*(\sigma)}_j(\hat{k}) . \end{aligned} \tag{39}$$

We define the power spectrum of tensor fluctuations as

$$\left\langle h^{(\sigma)}_{\boldsymbol{k}} h^{(\sigma)}_{\boldsymbol{k}'} \right\rangle = \frac{2\pi^2}{k^3} P^{(\sigma)}(k) \delta^{(3)}(\boldsymbol{k} + \boldsymbol{k}') . \tag{40}$$

Let us divide the tensor fluctuations into the two parts. The one comes from vacuum fluctuations and the other comes from the gauge field. Since they are uncorrelated to each other, we can write the tensor power spectrum as the sum

$$P^{(\sigma)}(k) = P^{(\sigma)}_v(k) + P^{(\sigma)}_s(k) . \tag{41}$$

From Eqs. (38)~(41), we can deduce

$$\begin{aligned} P^{(\sigma)}_s(k) = & \frac{k^3}{\pi^2 M^4_{pl} a^2} \int \frac{d^3p}{(2\pi)^3} \left(1 + (\hat{k} \cdot \hat{p})^2 \right) \left(1 + (\hat{k} \cdot \widehat{\boldsymbol{k} - \boldsymbol{p}})^2 \right) \\ & \times \left| \int d\tau' a^3(\tau') G_k(\tau, \tau') \mathcal{B}_p(\tau') \mathcal{B}_{|\boldsymbol{k}-\boldsymbol{p}|}(\tau') \right|^2 , \end{aligned} \tag{42}$$

where we ignored the subdominant contribution of the electric fields and used an identity $\left| \vec{e}^{(\alpha)}(\hat{k}) \cdot \vec{e}^{(\beta)}(\hat{k}') \right|^2 = \frac{1}{4} \left(1 - \alpha\beta(\hat{k} \cdot \hat{k}') \right)^2$. Let us define the Green's function $G_k(\tau, \tau')$ for Eq. (38). Substituting the scale factor in Eq. (4) into the homogeneous part of Eq. (38), we obtain

$$v''_k(\tau) + \left(k^2 + \frac{2(\lambda^2 - 4)}{(\lambda^2 - 2)^2 \tau^2} \right) v_k(\tau) = 0 . \tag{43}$$

In the fast roll limit ($\lambda \to \infty$), the Green's function obtained from Eq. (43) becomes

$$\begin{aligned} G_k(\tau, \tau') \equiv & \frac{\cos(k\tau)\sin(k\tau') - \sin(k\tau)\cos(k\tau')}{k} \\ \simeq & \tau' \qquad \left(|k\tau|, \left| k\tau' \right| \ll 1 \right) , \end{aligned} \tag{44}$$

where we took the superhorizon limit since the Green's function just oscillates and does not contribute the time integration of Eq. (42) in the subhorizon regime. Substituting Eqs. (27) and (44) into Eq. (42) and using the new variables $\vec{q} \equiv \frac{\vec{p}}{k}$, $\vec{q}' \equiv \frac{\vec{p}-\vec{k}}{k}$ and $z \equiv -k\tau$, we get the power spectrum at the end of the ekpyrotic phase as

$$\begin{aligned} P_s(k) = & 2 \times P^{(\sigma)}_s(k) \\ = & \frac{81}{256\pi^5} (\lambda^2 - 2)^4 \left(\frac{H_{end}}{M_{pl}} \right)^4 z_{end}^{\frac{4\lambda^2-4}{\lambda^2-2}} \\ & \times \int d^3q q^{-3} q'^{-3} \left(1 + (\hat{k} \cdot \hat{q})^2 \right) \left(1 + (\hat{k} \cdot \hat{q}')^2 \right) \\ & \times \left| \int dz' z'^{-\frac{3\lambda^2-4}{\lambda^2-2}} \right|^2 , \end{aligned} \tag{45}$$

where we used the fact that there is no polarization of gravitational waves. Since the gauge field becomes relevant as the source of the gravitational waves after the ekpyrotic phase starts, we consider the region

$$\lambda^2 |\tau_{in}| \gtrsim \frac{1}{p} , \frac{1}{\left| \vec{p} - \vec{k} \right|} \tag{46}$$

(see Eq. (25)). Here, τ_{in} is the time when the ekpyrotic phase starts. Note that we take the limit $\lambda \gg 1$ hereafter. Multiplying it by k, we get

$$\frac{1}{q_{in}} > \frac{1}{q} , \frac{1}{q'} , \tag{47}$$

where $q_{in} \equiv \left| \lambda^2 k \tau_{in} \right|^{-1}$ represents the infrared cut off of the momentum integral. Then the momentum integral is calculated as

$$\begin{aligned} & \int_{q_{in}}^{1} d^3q q^{-3} q'^{-3} \left(1 + (\hat{k} \cdot \hat{q})^2 \right) \left(1 + (\hat{k} \cdot \hat{q}')^2 \right) \\ & = 2\pi \int_{q_{in}}^{1} dq q^{-1} \int d\theta \sin\theta \frac{(1 + \cos^2\theta) \left(1 + \left(\sqrt{\frac{(1-q\cos\theta)^2}{1+q^2-2q\cos\theta}} \right)^2 \right)}{\left(1 + q^2 - 2q\cos\theta \right)^{3/2}} \\ & = 2\pi \int_{q_{in}}^{1} \frac{16}{15} \frac{q^4 - q^2 - 5}{q(q+1)(q-1)} dq , \end{aligned} \tag{48}$$

where we defined $\cos\theta \equiv \hat{k} \cdot \hat{q}$ and we approximately evaluated the integral in the range $q_{in} < q < 1$. One can see that there are two poles at $q = q_{in} \ll 1$ and $q = 1$. The later one is corresponding to the $q' = q_{in} \ll 1$. From the symmetry between q and q', we can evaluate the integral (48) at $q = q_{in}$ by multiplying it by 2

$$\frac{64\pi}{3} \ln q_{in}^{-1} . \tag{49}$$

On the other hand, the time integral in Eq. (45) can be evaluated at $z' = z_{end}$ approximately. We therefore obtain

$$P_s(k) \simeq \frac{27}{16\pi^4} \lambda^8 \left(\frac{H_{end}}{M_{pl}} \right)^4 \ln \left[\left(\frac{k}{k_{in}} \right) \right] . \tag{50}$$

There is a factor $\left(\frac{H_{end}}{M_{pl}} \right)^4$ in the spectrum (50) because of the non-linear contribution of the magnetic fields (27). One can see that sourced gravitational waves have a nearly scale invariant spectrum. This conclusion is different from the well-known blue-tilted spectrum in the ekpyrotic scenario [8]. Most importantly, there appears a factor λ^8 in $P_s(k)$. For example, if we set $H_{end} = 10^{-5} M_{pl}$, $\lambda = -17$, the amplitude of the power spectrum is about 10^{-11}. This is comparable with the gravitational waves in the inflationary universe $\sim \left(\frac{H_{end}}{\pi M_{pl}}\right)^2$. Therefore, we can not discriminate between inflation and the ekpyrotic scenario just by detecting primordial gravitational waves. In the next section, let us calculate scalar fluctuations sourced by the gauge field and discuss if the ekpyrotic model with a gauge field is compatible with the CMB data.

5. Scalar fluctuations from magnetic fields

Now, we calculate scalar fluctuations sourced by the gauge field in the present scenario. Too much production of scalar fluctuations implies incompatibility with the CMB data. Fortunately, we will soon see that the sourced scalar fluctuations are smaller than the tensor fluctuations. Let us show it by repeating the same procedure we used in the previous section. As is discussed in [14], the equation for linear perturbation of the scalar field $\delta\phi$ in the flat slicing gauge is given by

$$s'' - \left(\nabla^2 + \frac{z''}{z}\right) s \simeq -a^3 \left(\frac{f_{,\phi}}{f} + \frac{\phi'}{4M_{pl}^2 \mathcal{H}}\right) \vec{B}^2 , \tag{51}$$

where we ignored the subdominant contributions of the electric fields and used $s = a\delta\phi$, $z \equiv \frac{a\phi'}{\mathcal{H}}$. From Eqs. (4) and (6), we obtain

$$\frac{\phi'}{\mathcal{H}} = -M_{pl}\lambda , \quad \frac{f_{,\phi}}{f} = \frac{\rho}{M_{pl}} . \tag{52}$$

Since we are considering the scale invariant magnetic fields as the source of the scalar fluctuations, ρ satisfies Eq. (26). Hence, working in Fourier space, we can rewrite Eq. (51) as

$$s''_{\boldsymbol{k}}(\tau) + \left(k^2 - \frac{a''}{a}\right) s_{\boldsymbol{k}}(\tau) \simeq S(\tau, \boldsymbol{k}) , \tag{53}$$

where the source term is defined by

$$S(\tau, \boldsymbol{k}) = -\frac{3a^3}{4M_{pl}^2} \frac{\lambda^2 - 8/3}{\lambda} \int \frac{d^3p}{(2\pi)^{3/2}} \left(\hat{B}_{i,\boldsymbol{p}} \hat{B}_{i,\boldsymbol{k}-\boldsymbol{p}}\right) . \tag{54}$$

We define the power spectrum of scalar fluctuations as

$$\left\langle \left(\frac{\mathcal{H}}{\phi'}\right) \delta\phi_{\boldsymbol{k}} \left(\frac{\mathcal{H}}{\phi'}\right) \delta\phi_{\boldsymbol{k}'} \right\rangle = \frac{2\pi^2}{k^3} \mathcal{P}(k) \delta^{(3)}(\boldsymbol{k} + \boldsymbol{k}') . \tag{55}$$

One can see that the fast roll condition suppresses the scalar power spectrum by the factor $\left(\frac{\mathcal{H}}{\phi'}\right)^2 = \frac{1}{M_{pl}^2 \lambda^2}$. The power spectrum can be divided into two parts like the tensor power spectrum as

$$\mathcal{P}(k) = \mathcal{P}_v(k) + \mathcal{P}_s(k) . \tag{56}$$

From Eqs. (53)~(56), we can deduce

$$\mathcal{P}_s(k) = \frac{9k^3}{16\pi^2 M_{pl}^4 a^2} \frac{(\lambda^2 - 8/3)^2}{\lambda^4} \int \frac{d^3p}{(2\pi)^3} \left(1 + (\hat{p} \cdot \widehat{\boldsymbol{k} - \boldsymbol{p}})^2\right) \times \left| \int d\tau' a^3(\tau') G_k(\tau, \tau') \mathcal{B}_p(\tau') \mathcal{B}_{|\boldsymbol{k}-\boldsymbol{p}|}(\tau') \right|^2 , \tag{57}$$

where the Green's function $G_k(\tau, \tau')$ for Eq. (53) is the same as that for Eq. (44). Substituting Eqs. (27) and (44) into Eq. (57) and using the variables $\vec{q} \equiv \frac{\vec{p}}{k}$, $\vec{q}' \equiv \frac{\vec{p} - \vec{k}}{k}$ and $z \equiv -k\tau$, we get the scalar power spectrum at the end of the ekpyrotic phase as

$$\mathcal{P}_s(k) = \frac{729}{8192\pi^5} \frac{(\lambda^2 - 2)^4 (\lambda^2 - 8/3)^2}{\lambda^4} \left(\frac{H_{end}}{M_{pl}}\right)^4 z_{end}^{\frac{4\lambda^2 - 4}{\lambda^2 - 2}} \times \int d^3q q^{-3} q'^{-3} \left(1 + (\hat{q} \cdot \hat{q}')^2\right) \left| \int dz' z'^{-\frac{3\lambda^2 - 4}{\lambda^2 - 2}} \right|^2 . \tag{58}$$

The momentum integral is carried out as

$$\begin{aligned} &\int_{q_{in}}^{1} d^3q q^{-3} q'^{-3} \left(1 + (\hat{q} \cdot \hat{q}')^2\right) \\ &= 2\pi \int_{q_{in}}^{1} dq q^{-1} \int d\theta \sin\theta \frac{\left(1 + \left(\frac{(q - \cos\theta)}{\sqrt{1 + q^2 - 2q\cos\theta}}\right)^2\right)}{(1 + q^2 - 2q\cos\theta)^{3/2}} \\ &= -2\pi \int_{q_{in}}^{1} \frac{8}{3} \frac{1}{q(q+1)(q-1)} dq , \end{aligned} \tag{59}$$

where $\cos\theta \equiv \hat{k} \cdot \hat{q}$ and we approximately evaluated the integral in the range $q_{in} < q < 1$. From the symmetry between q and q', we can calculate Eq. (59) at $q = q_{in}$ by multiplying it by 2 as is done for tensor fluctuations. The result reads

$$\frac{32\pi}{3} \ln q_{in}^{-1} . \tag{60}$$

The time integral is same as the case of tensor fluctuations and we can obtain the scalar power spectrum sourced by the scale invariant magnetic fields as

$$\mathcal{P}_s(k) \simeq \frac{243}{1024\pi^4} \lambda^8 \left(\frac{H_{end}}{M_{pl}}\right)^4 \ln\left[\left(\frac{k}{k_{in}}\right)\right] . \tag{61}$$

From Eqs. (50) and (61), the tensor to scalar ratio r_{source} is given by

$$r_{source} \simeq 7 . \tag{62}$$

This result is different from that in the inflationary universe, where the scalar fluctuations are enhanced by the inverse square of a slow roll parameter [14]. Taking a look at terms in the parenthesis of the right-hand side of Eq. (51), we see that it gives rise to a factor λ^2 in the scalar power spectrum in contrast to the case of tensor fluctuations. On the other hand, from Eq. (55), we see the scalar power spectrum is suppressed by $\frac{1}{\lambda^2}$ in contrast to the case of tensor fluctuations. These two factors have been canceled out. The numerical value (62) comes from accumulation of several factors such as the polarization degrees of freedom. Since the tensor to scalar ratio becomes larger than unity, we can say that the scalar fluctuations sourced by the scale invariant magnetic field are negligible in the ekpyrotic scenario due to the fast roll condition.

6. Conclusion

We studied the role of the gauge kinetic function in the ekpyrotic scenario and showed that abundant gravitational waves sourced by the gauge field can be produced. As a demonstration, we first showed that scale invariant magnetic fields can be produced in the ekpyrotic phase. It turned out that the magnetic fields induce nearly scale invariant gravitational waves (slightly blue) and the amplitude could be comparable with that of the inflationary universe. It turned out that it is difficult to disprove the ekpyrotic scenario by detecting primordial gravitational waves. In order to distinguish both scenarios, it is necessary to look at the details of the spectrum such as the tilt of the spectrum. Observing the distinction of higher order scalar perturbations is also important [15]. We should mention that the idea of finding an ekpyrotic model with observable gravitational waves on CMB scales using sourced fluctuations was put forward for the first time in [16] by investigating a different model with explicit parity violation. Our model has no explicit parity violation. Moreover, we also showed that the scalar fluctuations induced by the magnetic field are smaller than the sourced gravitational waves. Generally, as far as the fast roll condition is satisfied, the tensor to scalar ratio becomes more than unity in any ekpyrotic models with the gauge kinetic function. Therefore, our scenario would be compatible with the CMB data provided that nearly scale invariant scalar fluctuations are produced in a standard way with an additional scalar field [6].

It should be noted that we must check the non-gaussianity of the primordial scalar fluctuations in the present model [17]. Moreover, we should consider a bounce process from contracting to expanding to connect the spectrum at the end of the ekpyrotic phase with observables. We have not looked into this issue in this paper since the mechanism is model dependent and the detailed analysis is beyond the scope of this paper [1]. However, actually, although we fixed the parameters such as ρ, λ, H_{end} for simplicity in this paper, we can tune these parameters in our scenario so that our conclusion becomes valid for any ekpyrotic bouncing models. Therefore, our conclusion is robust.

Acknowledgements

This work was in part supported by MEXT KAKENHI Grant Number 15H05895.

References

[1] D. Battefeld, P. Peter, Phys. Rep. 571 (2015) 1, arXiv:1406.2790 [astro-ph.CO];
R. Brandenberger, P. Peter, arXiv:1603.05834 [hep-th].

[2] J. Khoury, B.A. Ovrut, P.J. Steinhardt, N. Turok, Phys. Rev. D 64 (2001) 123522, arXiv:hep-th/0103239.

[3] P. Horava, E. Witten, Nucl. Phys. B 475 (1996) 94, arXiv:hep-th/9603142;
P. Horava, E. Witten, Nucl. Phys. B 460 (1996) 506, arXiv:hep-th/9510209.

[4] A. Ijjas, P.J. Steinhardt, Class. Quantum Gravity 33 (4) (2016) 044001, arXiv:1512.09010 [astro-ph.CO].

[5] M. Gasperini, G. Veneziano, Phys. Rep. 373 (2003) 1, arXiv:hep-th/0207130;
M. Gasperini, G. Veneziano, Astropart. Phys. 1 (1993) 317, arXiv:hep-th/9211021;
R. Brustein, M. Gasperini, M. Giovannini, G. Veneziano, Phys. Lett. B 361 (1995) 45, arXiv:hep-th/9507017.

[6] M. Li, Phys. Lett. B 724 (2013) 192, arXiv:1306.0191 [hep-th];
A.M. Levy, A. Ijjas, P.J. Steinhardt, Phys. Rev. D 92 (6) (2015) 063524, arXiv:1506.01011 [astro-ph.CO];
A. Notari, A. Riotto, Nucl. Phys. B 644 (2002) 371, arXiv:hep-th/0205019.

[7] A.A. Starobinsky, JETP Lett. 30 (1979) 682, Pisma Zh. Eksp. Teor. Fiz. 30 (1979) 719.

[8] L.A. Boyle, P.J. Steinhardt, N. Turok, Phys. Rev. D 69 (2004) 127302, arXiv:hep-th/0307170.

[9] K. Subramanian, arXiv:1504.02311 [astro-ph.CO];
R. Durrer, A. Neronov, Astron. Astrophys. Rev. 21 (2013) 62, arXiv:1303.7121 [astro-ph.CO].

[10] W. Chen, J.H. Buckley, F. Ferrer, Phys. Rev. Lett. 115 (2015) 211103, arXiv:1410.7717 [astro-ph.HE];
W. Chen, B.D. Chowdhury, F. Ferrer, H. Tashiro, T. Vachaspati, Mon. Not. R. Astron. Soc. 450 (4) (2015) 3371, arXiv:1412.3171 [astro-ph.CO];
K. Takahashi, M. Mori, K. Ichiki, S. Inoue, H. Takami, Astrophys. J. 771 (2013) L42, arXiv:1303.3069 [astro-ph.CO];
A.M. Taylor, I. Vovk, A. Neronov, Astron. Astrophys. 529 (2011) A144, arXiv:1101.0932 [astro-ph.HE];
W. Essey, S. Ando, A. Kusenko, Astropart. Phys. 35 (2011) 135, arXiv:1012.5313 [astro-ph.HE];
K. Dolag, M. Kachelriess, S. Ostapchenko, R. Tomas, Astrophys. J. 727 (2011) L4, arXiv:1009.1782 [astro-ph.HE];
S. Ando, A. Kusenko, Astrophys. J. 722 (2010) L39, arXiv:1005.1924 [astro-ph.HE];
F. Tavecchio, G. Ghisellini, L. Foschini, G. Bonnoli, G. Ghirlanda, P. Coppi, Mon. Not. R. Astron. Soc. 406 (2010) L70, arXiv:1004.1329 [astro-ph.CO];
A. Neronov, I. Vovk, Science 328 (2010) 73, arXiv:1006.3504 [astro-ph.HE].

[11] B. Ratra, Astrophys. J. 391 (1992) L1;
M. Giovannini, Phys. Lett. B 659 (2008) 661, arXiv:0711.3273 [astro-ph];
K. Bamba, M. Sasaki, J. Cosmol. Astropart. Phys. 0702 (2007) 030, arXiv:astro-ph/0611701;
J. Martin, J. Yokoyama, J. Cosmol. Astropart. Phys. 0801 (2008) 025, arXiv:0711.4307 [astro-ph];
V. Demozzi, V. Mukhanov, H. Rubinstein, J. Cosmol. Astropart. Phys. 0908 (2009) 025, arXiv:0907.1030 [astro-ph.CO];
S. Kanno, J. Soda, M.a. Watanabe, J. Cosmol. Astropart. Phys. 0912 (2009) 009, arXiv:0908.3509 [astro-ph.CO];
R.J.Z. Ferreira, R.K. Jain, M.S. Sloth, J. Cosmol. Astropart. Phys. 1310 (2013) 004, arXiv:1305.7151 [astro-ph.CO];
T. Kobayashi, J. Cosmol. Astropart. Phys. 1405 (2014) 040, arXiv:1403.5168 [astro-ph.CO];
T. Fujita, R. Namba, arXiv:1602.05673 [astro-ph.CO].

[12] F.A. Membiela, Nucl. Phys. B 885 (2014) 196, arXiv:1312.2162 [astro-ph.CO];
L. Sriramkumar, K. Atmjeet, R.K. Jain, J. Cosmol. Astropart. Phys. 1509 (09) (2015) 010, arXiv:1504.06853 [astro-ph.CO];
D. Chowdhury, L. Sriramkumar, R.K. Jain, arXiv:1604.02143 [gr-qc].

[13] A. Ito, J. Soda, J. Cosmol. Astropart. Phys. 1604 (04) (2016) 035, arXiv:1603.00602 [hep-th].

[14] N. Barnaby, R. Namba, M. Peloso, Phys. Rev. D 85 (2012) 123523, arXiv:1202.1469 [astro-ph.CO].

[15] X. Chen, M.H. Namjoo, Y. Wang, arXiv:1601.06228 [hep-th];
X. Chen, M.H. Namjoo, Y. Wang, J. Cosmol. Astropart. Phys. 1602 (02) (2016) 013, arXiv:1509.03930 [astro-ph.CO].

[16] I. Ben-Dayan, arXiv:1604.07899 [astro-ph.CO].

[17] A. Fertig, J.L. Lehners, E. Mallwitz, Phys. Rev. D 89 (10) (2014) 103537, arXiv:1310.8133 [hep-th];
A. Ijjas, J.L. Lehners, P.J. Steinhardt, Phys. Rev. D 89 (12) (2014) 123520, arXiv:1404.1265 [astro-ph.CO].

22

Radiation and energy release in a background field of axion-like dark matter

Wei Liao

Institute of Modern Physics, School of Science, East China University of Science and Technology, 130 Meilong Road, Shanghai 200237, PR China

ARTICLE INFO

Editor: J. Hisano

ABSTRACT

We find that a fuzzy dark matter background and the mG scale magnetic field in the galactic center can give rise to a radiation with a very large energy release. The frequency of the radiation field is the same as the frequency of the oscillating axion-like background field. We show that there is an energy transfer between the fuzzy dark matter sector and the electromagnetic sector because of the presence of the generated radiation field and the galactic magnetic field. The energy release rate of radiation is found to be very slow in comparison with the energy of fuzzy dark matter but could be significant comparing with the energy of galactic magnetic field in the source region. Using this example, we show that the fuzzy dark matter together with a large scale magnetic field is possible to give rise to fruitful physics.

1. Introduction

An interesting hypothesis of dark matter (DM) in the universe is that the DM is composed of axion-like particles (ALPs). For ALPs in different mass range, they may have very different behaviors in evolution of universe. For example, for a mass of ALP with $m_a \sim 10^{-5}$ eV, ALPs can form boson stars [1]. For ultra-light ALP with $m_a \sim 10^{-23} - 10^{-21}$ eV, known as Fuzzy DM (FDM) [2], ALPs can have very large de Broglie wavelengths up to $1 \sim 10$ pc and avoid typical problems associated with Cold DM [2–4]. In this case, FDM is possible to form a diffuse DM background as a galactic halo [5,6]. In both of these cases, a very large number of light ALPs are concentrated in a volume of the scale of de Broglie wavelength. Hence, physics of these ALPs can be described by a classical scalar field and quantum fluctuations around this classical field are small [1].

Because interaction of ALP with ordinary matter is very weak, detection of it in laboratory or observation of its cosmological or astrophysical signature is very difficult [7]. Detecting the signatures of FDM is difficult in particular because of its ultra-light mass scale. To date, there are very few ways known to us which can possibly constrain FDM [8,9]. In this Letter we discuss a possible physical prediction of FDM in the presence of a large scale magnetic field. Although our discussions are for FDM, the discussion can be easily extended to ALP boson stars or condensates of other types of ALPs if there are magnetic fields with them.

E-mail address: liaow@ecust.edu.cn.

2. Electromagnetism in background of axion-like field

Considering coupling of ALP field $\phi(\vec{x},t)$ with electromagnetic (EM) field, the Lagrangian can be written as

$$\mathcal{L} = -\frac{1}{4}F_{\mu\nu}F^{\mu\nu} + \frac{\phi}{2F}g_\phi \widetilde{F}_{\mu\nu}F^{\mu\nu} - A_\mu J^\mu \tag{1}$$

where $\widetilde{F}_{\mu\nu} = \frac{1}{2}\varepsilon_{\mu\nu\rho\sigma}F^{\rho\sigma}$, J^μ being the 4-vector of an external current. F in the denominator, known as the decay constant, is in the range $10^{16} \sim 10^{18}$ GeV for FDM [4]. g_ϕ is a model-dependent parameter denoting the strength of the coupling of ϕ with EM field. Its magnitude may vary from 10^{-3} to 10^{-2}. A modified set of equations of motion can be found using (1). A $\phi F\tilde{F}$ term with a constant ϕ does not contribute to the equations of motion. However, in the presence of a background field $\phi(\vec{x},t)$ which depends on space and time, the equation of motion with electric source becomes

$$\partial_\nu F^{\nu\mu} = J^\mu - g_\phi \frac{(\partial_\nu \phi)}{F}\varepsilon^{\mu\nu\rho\sigma}F_{\rho\sigma}. \tag{2}$$

Together with the equations with no magnetic source, we get a modified set of equations:

$$\vec{\nabla}\cdot\vec{E} = J^0 + g_\phi \frac{2}{F}(\vec{\nabla}\phi)\cdot\vec{B} \tag{3}$$

$$\vec{\nabla}\times\vec{B} - \frac{\partial \vec{E}}{\partial t} = \vec{J} - g_\phi \frac{2}{F}[(\partial_t\phi)\vec{B} + (\vec{\nabla}\phi)\times\vec{E}] \tag{4}$$

$$\vec{\nabla}\cdot\vec{B}=0 \tag{5}$$

$$\vec{\nabla}\times\vec{E}+\frac{\partial\vec{B}}{\partial t}=0. \tag{6}$$

For FDM in a galaxy, e.g. in Milky Way, we can further simplify these equations. Since the typical velocity in our galaxy is $\sim 10^{-3}$, FDM in the halo of a galaxy can be considered non-relativistic and be described by a classical field as [9].

$$\phi(\vec{x},t)=A(\vec{x})\cos(m_a t+\beta(\vec{x})) \tag{7}$$

The de Broglie wavelength of this non-relativistic FDM is found to be

$$\lambda_a=\frac{1}{m_a v_a}=63.6\ \text{pc}\ \frac{10^{-22}\ \text{eV}}{m_a}\frac{10^{-3}}{v_a}. \tag{8}$$

This means that the spatial derivative of ϕ should be proportional to $1/\lambda_a$ and should be small comparing with $\partial_t\phi$, i.e. $|\vec{v}|=|(\vec{\nabla}\phi)/(\partial_t\phi)|\ll 1$.

So for FDM, we can neglect terms with $\vec{\nabla}\phi$ in Eqs. (3) and (4) as a first approximation, as long as we do not consider the case with strong electric field. We get

$$\vec{\nabla}\cdot\vec{E}=J^0 \tag{9}$$

$$\vec{\nabla}\times\vec{B}-\frac{\partial\vec{E}}{\partial t}=\vec{J}-g_\phi\frac{2}{F}(\partial_t\phi)\vec{B} \tag{10}$$

$$\vec{\nabla}\cdot\vec{B}=0 \tag{11}$$

$$\vec{\nabla}\times\vec{E}+\frac{\partial\vec{B}}{\partial t}=0. \tag{12}$$

Note that in the approximation that $\vec{\nabla}\phi$ is neglected, the right-handed side of (10) is indeed divergence free and can be considered as an effective current of a neutral source with $J^0=0$: $\vec{J}_{eff}=\vec{J}-g_\phi\frac{2}{F}(\partial_t\phi)\vec{B}$ and $\vec{\nabla}\cdot\vec{J}_{eff}=0$.

Neglecting the gravitational potential and the potential energy caused by axion interaction, the energy–momentum tensor of FDM can be estimated using a free field Lagrangian of ϕ [9]. One can find that the leading term in energy density is time independent

$$\rho_{DM}(\vec{x})=\frac{1}{2}m_a^2A^2(\vec{x}) \tag{13}$$

and the oscillating part in energy density is proportional to $(\vec{\nabla}\phi)^2\propto\vec{v}^2\rho_{DM}$ which can be neglected. The energy density should vary slowly within a de Broglie wavelength (8) and be taken as a constant within a distance smaller than the de Broglie wavelength λ_a. In the following, we will use (13) and replace $m_aA(\vec{x})$ with $\sqrt{2\rho_{DM}(\vec{x})}$.

Apparently, the modified set of equations (3)–(6) may have very rich physical consequences. For example, one would expect that the cosmic or galactic ALP field may affect the evolution and development of cosmic or galactic magnetic field. In the present Letter we are not going to discuss this complicated problem. We assume that galactic magnetic field has been generated and serves as an external B field, and then discuss the possible radiation caused by the effective current coming from the galactic FDM field and external galactic B field.

3. Energy release in oscillating FDM field and external B field

Magnetic field in our galaxy, in particular in galactic center, has been measured by astrophysical observations [10–13]. It was found that inside a ring-like Central Molecular Zone (CMZ) the magnetic field in the galactic center is mainly perpendicular to the galactic plane [13], in particular in regions close to non-thermal filaments (NTFs). The magnetic field in the CMZ is mainly toroidal, that is oriented parallel to the galactic plane. The strength of magnetic field in the galactic center is quite uncertain. A global picture is that a pervasive magnetic field with a strength of mG exists in a central region with a radius no less than 150 pc. Challenges to this picture of pervasive magnetic field exists [10,13]. An alternative picture is that magnetic field in local regions of NTFs is of strength around mG and it is of strength of tens μG in the diffuse inter-cloud region. In this Letter, we do not discuss the configuration and strength of the magnetic field in the galactic center, rather show that this large scale magnetic field together with the oscillating FDM field is very possible to give rise to radiation and large energy release. In particular, we focus on effects caused by poloidal magnetic field inside the CMZ.

We use simplified equations (9)–(12) to present our result. We denote the external B field as $\vec{B}_{ex}$ and the electric and magnetic fields of radiation as $\vec{E}_r$ and $\vec{B}_r$. We take the interaction of FDM with external B field as a perturbation, and the radiation field also as a perturbation to the external magnetic field. Assuming the external B field satisfying a set of Maxwell equations without interaction with FDM, we can get a set of equations for the radiation field to the first order as

$$\vec{\nabla}\cdot\vec{E}_r=0 \tag{14}$$

$$\vec{\nabla}\times\vec{B}_r-\frac{\partial\vec{E}_r}{\partial t}=-g_\phi\frac{2}{F}(\partial_t\phi)\vec{B}_{ex} \tag{15}$$

$$\vec{\nabla}\cdot\vec{B}_r=0 \tag{16}$$

$$\vec{\nabla}\times\vec{E}_r+\frac{\partial\vec{B}_r}{\partial t}=0. \tag{17}$$

Writing $\vec{B}_r=\vec{\nabla}\times\vec{A}_r$ and $\vec{E}_r=-\frac{\partial}{\partial t}\vec{A}_r$, we can get the propagation equation of $\vec{A}_r$ in radiation gauge which can be solved using a complex vector field $\vec{\mathcal{A}}_r$ [14]

$$\vec{A}_r(\vec{x},t)=\text{Re}(\vec{\mathcal{A}}_r), \tag{18}$$

$$\vec{\mathcal{A}}_r(\vec{x},t)=\frac{1}{4\pi}e^{-im_at}\int d^3y\,\frac{\vec{\mathcal{J}}(\vec{y})}{|\vec{x}-\vec{y}|}\,e^{ik|\vec{x}-\vec{y}|} \tag{19}$$

where $\vec{\mathcal{J}}(\vec{y})=g_\phi\frac{2i}{F}m_aA(\vec{y})\vec{B}_{ex}(\vec{y})e^{-i\beta(\vec{y})}$. k in (19) is the wave number of radiation and we have $k=m_a$ in the present case with an oscillating FDM field. Introducing $\lambda_k=2\pi/k$, we can find

$$\lambda_k=0.4\ \text{pc}\ \frac{10^{-22}\ \text{eV}}{m_a}. \tag{20}$$

λ_k is much smaller than the spatial scale of the pervasive magnetic field in galactic center. (18) and (19) suggest that there are periodic EM fields with a period

$$T=\frac{2\pi}{m_a}\approx 1.3\ \text{year}\ \frac{10^{-22}\ \text{eV}}{m_a} \tag{21}$$

in a fixed position in our galaxy. Effects of such a periodic EM field are very interesting subjects to study.

A real evaluation of (19) is very complicated and difficult, not only because we do not really know the detail of the magnetic field in galactic center, but also because $A(\vec{y})$ and $\beta(\vec{y})$ are uncertain. Instead, we make a rough estimate of the radiation field and the radiation power. In particular, if there are special regions in galactic center in which the magnetic field is much stronger than surrounding regions, e.g. in regions close to NTFs, we can evaluate the integration (19) in these special regions independently. So we can evaluate the contribution of a particular source region to the radiation field. The total contribution of all these source regions can be obtained by summing their contributions. This is the strategy of the rough estimate in the present Letter.

To simplify calculation, we can evaluate radiation field at a distance far away from the source region and with $|\vec{x}|k \gg 1$, e.g. at the solar distance from the center of galaxy $R_\circ = 8$ kpc. For such a distance far away from a particular source region in the galactic center, we can find

$$\vec{B}_r = \mathrm{Re}(\vec{\mathcal{B}}_r\, e^{-im_a t+ikr}),\ \vec{E}_r = \mathrm{Re}(\vec{\mathcal{E}}_r\, e^{-im_a t+ikr}), \tag{22}$$

$$\vec{\mathcal{B}}_r \approx -\frac{k}{2\pi}\frac{1}{r}g_\phi\frac{m_a A_S}{F}e^{-i\beta_S}\int_S d^3y\,\vec{n}\times\vec{B}_{ex}(\vec{y})e^{-ik\vec{n}\cdot\vec{y}}, \tag{23}$$

where $\vec{n} = \vec{x}/|\vec{x}|$ and $r = |\vec{x}|$. Electric field of radiation in the vacuum can be estimated as $\vec{\mathcal{E}}_r = -\vec{n}\times\vec{\mathcal{B}}_r$. A_S and β_S in (22) and (23) are the values of $A(\vec{y})$ and $\beta(\vec{y})$ taken at the source region. This replacement can be done because $e^{-ik\vec{n}\cdot\vec{y}}$ in (23) is usually a function changing much faster than $A(\vec{y})$ and $\beta(\vec{y})$. In our estimate, we restrict the integration to a source region with a scale no more than λ_k. Writing $|\int_S d^3y\ \vec{B}_{ex}(\vec{y})e^{-ik\vec{n}\cdot\vec{y}}| = |\vec{B}_{ex}|\Omega_S$ where Ω_S is the effective volume, we can make an estimate of the order of magnitude. Suppressing angular dependence which does not affect the estimate of the order of magnitude, we find

$$|\vec{B}_r| \sim 6.7\times10^{-9}\ \mathrm{mG}\left(\frac{10^{-22}\ \mathrm{eV}}{m_a}\right)^2\frac{R_\circ}{r}\frac{g_\phi}{10^{-2}}\frac{10^{15}\ \mathrm{GeV}}{F} \times\left(\frac{\rho_{DM}(S)}{10^4\rho_\odot}\right)^{1/2}\frac{|\vec{B}_{ex}|\Omega_S}{\mathrm{mG}\,\lambda_k^3}. \tag{24}$$

where $\rho_\odot = 0.3$ GeV cm^{-3} is the DM energy density at the position of solar system. We have taken $10^4\rho_\odot$ as a reference energy density in calculation because the DM energy density can increase to $10^3 \sim 10^4$ of $\rho_\odot$ in the galactic center according to the popular Navarro–Frenk–White (NFW) density profile of DM [15]. For the estimate of $|E_r|$, the numerical factor in (24) is replaced by 2.0×10^{-7} V m^{-1}.

The time averaged radiation power is estimated using (24) as

$$\frac{dP}{d\Omega} = \frac{1}{2}\mathrm{Re}[r^2\vec{n}\cdot(\vec{E}_r\times\vec{B}_r)] \sim 1.0\times10^{39}\ \mathrm{erg\ year^{-1}}\left(\frac{10^{-22}\ \mathrm{eV}}{m_a}\right)^4\left(\frac{g_\phi}{10^{-2}}\right)^2 \times\left(\frac{10^{15}\ \mathrm{GeV}}{F}\right)^2\frac{\rho_{DM}(S)}{10^4\rho_\odot}\left(\frac{|\vec{B}_{ex}|\Omega_S}{\mathrm{mG}\,\lambda_k^3}\right)^2. \tag{25}$$

One can compare (25) with E_{DM} and E_B, the typical total energy of DM and magnetic field in the source region of a volume λ_k^3. We can find

$$\frac{P}{E_{DM}} \sim 10^{-16}\ \mathrm{year}^{-1}\frac{10^{-22}\ \mathrm{eV}}{m_a}\left(\frac{g_\phi}{10^{-2}}\right)^2\left(\frac{10^{15}\ \mathrm{GeV}}{F}\right)^2 \times\left(\frac{|\vec{B}_{ex}|}{\mathrm{mG}}\right)^2\left(\frac{\Omega_S}{\lambda_k^3}\right)^2 \tag{26}$$

and

$$\frac{P}{E_B} \sim 10^{-8}\ \mathrm{year}^{-1}\frac{10^{-22}\ \mathrm{eV}}{m_a}\left(\frac{g_\phi}{10^{-2}}\right)^2\left(\frac{10^{15}\ \mathrm{GeV}}{F}\right)^2 \times\frac{\rho_{DM}(S)}{10^4\rho_\odot}\left(\frac{\Omega_S}{\lambda_k^3}\right)^2. \tag{27}$$

We see that the energy release rate is slow but may not be negligible. The energy release rate is really slow in comparison with the energy of DM in the source region. However, the energy release rate could still be significant comparing with the energy of the galactic magnetic field in the source region, in particular if g_ϕ could be larger than 10^{-2}.

To know in more detail about the energy transfer in energy budget, we can use (1) to get the equation of motion of ϕ in the presence of $\phi F\tilde{F}$ interaction and find

$$\partial_\mu T_\phi^{\mu\nu} = g_\phi\frac{1}{2F}(\partial^\nu\phi)\tilde{F}^{\rho\sigma}F_{\rho\sigma} = -g_\phi\frac{2}{F}(\partial^\nu\phi)\vec{E}\cdot\vec{B} \tag{28}$$

where $T_\phi^{\mu\nu} = (\partial^\mu\phi)(\partial^\nu\phi) - \eta^{\mu\nu}\frac{1}{2}[(\partial\phi)^2 - m^2\phi^2]$ is the energy-momentum tensor of ϕ with a free field Lagrangian. Similarly, we can also use the modified Maxwell equation (2) to get the property of the energy–momentum tensor for EM field. Setting $J^\mu = 0$ in (2) and with a bit of algebra we find

$$\partial_\mu\Theta_{em}^{\mu\nu} = -g_\phi\frac{1}{2F}(\partial^\nu\phi)\tilde{F}^{\rho\sigma}F_{\rho\sigma} = g_\phi\frac{2}{F}(\partial^\nu\phi)\vec{E}\cdot\vec{B}, \tag{29}$$

where $\Theta_{em}^{\mu\nu} = -F^\mu{}_\lambda F^{\nu\lambda} + \frac{1}{4}\eta^{\mu\nu}F^{\alpha\beta}F_{\alpha\beta}$ [14]. It is clear that there is an energy transfer between the FDM sector and EM sector. The rate of energy transfer would be proportional to $(\partial_t\phi)\vec{E}_r\cdot\vec{B}_{ex}$ in the present case. One can show that there are an oscillating term with frequency $2m_a$ and a constant term in $(\partial_t\phi)\vec{E}_r\cdot\vec{B}_{ex}$. The constant term in $(\partial_t\phi)\vec{E}_r\cdot\vec{B}_{ex}$ would give rise to a flow of energy between the FDM sector and the EM sector. However, the direction of energy flow depends on the sign of the imaginary part of $\int d^3xd^3y\ \vec{B}_{ex}(\vec{x})\cdot\vec{B}_{ex}(\vec{y})e^{ik|\vec{x}-\vec{y}|}/|\vec{x}-\vec{y}|$ which depends on the detailed distribution of the magnetic field strength in the source region. If this double integration is positive in the source region, energy is transferred from the FDM sector to the EM sector, and a rough estimate shows that the rate of energy injection to EM sector would be of the same order of magnitude of the energy loss rate of radiation shown in (26) or (27). Apparently, energy transfer between FDM sector and EM sector and the energy loss in radiation should all be taken into account, e.g. in the development and evolution of galactic magnetic field. A detailed study of this topic is beyond the scope of the present Letter.

As a comparison, we can also estimate the radiation caused by the Earth, the Sun and a magnetic neutron star. For these stellar objects with scale much smaller than λ_k, factor $e^{-ik\vec{n}\cdot k}$ in (23) can be taken as one and we get $\int d^3y\ \vec{B}_{ex}(\vec{y}) = \frac{2}{3}\vec{m}$ where $\vec{m}$ is the magnetic moment of the source object. For magnetic moments of the Earth and the Sun at order of 10^{22} A m^2 and 10^{29} A m^2, the radiation powers are $P \sim 10^{-14}$ erg year^{-1} for the Earth and $P \sim 10^{-1}$ erg year^{-1} for the Sun separately. So they can be safely neglected. For compact object like neutron star, the magnetic field could be much more than 10^{10} G. However the magnetic moment of the neutron star is hard to be much larger than that of a star like the Sun. It is reasonable to expect that the power of radiation caused by a magnetic neutron star might be as large as that caused by the Sun but should not be much larger. We can see that the rate of energy release caused by these stellar objects are all very small. On the other hand, the energy release caused by the large scale magnetic field in galactic center could be huge.

We emphasize that results given in (25), (26) and (27) are for particular source regions in galactic center, not for the whole galactic center. The total energy release given by the whole galactic center could be a sum of many such kind of sources with strong magnetic field. This is in particular true for the possible picture that mG scale magnetic field comes only with NTFs. So the rate of total energy release in the galactic center could be much larger than that given in (25). One should also notice that our estimate is very uncertain because our knowledge about the magnetic field in the galactic center is very uncertain.

4. Conclusion

In summary, we have shown that the $\phi F\tilde{F}$ interaction term in an oscillating background field of axion-like DM and a background of magnetic field gives rise to an effective oscillating current in the Maxwell equation, so that it can give rise to EM radiation and energy transfer between FDM and EM sectors. The frequency of the radiation field equals to the frequency of the oscillating background field of axion-like DM. In other words, the energy of the radiation photon equals to the energy of the non-relativistic axion-like DM, i.e. the mass of axion-like DM. In the case of FDM, the energy of the radiation photon is $10^{-23} \sim 10^{-21}$ eV and the period of the EM wave of radiation is $0.1 \sim 10$ year. We have estimated the strength of radiation and find that the FDM background and the large scale magnetic field in the galactic center can give rise to a radiation with a very large energy release. We found that the energy release rate of radiation is very slow in comparison with the energy of FDM but it could be significant comparing with the energy of galactic magnetic field in the source region. Needless to say, a very interesting question is how to directly detect this EM wave of a very long wavelength coming from the galactic center. Given the very weak strength of this radiation field at the solar distance from the galactic center, it should be a difficult and very challenging topic.

We note that the radiation can be absorbed by plasma in galaxy, in particular by plasma in galactic center, so that it can possibly affect the physics of the development and evolution of plasma in galactic center. Detailed study of the energy release rate and the impact on galactic plasma can be done using models of galactic magnetic field and models of plasma in galactic center. This detailed research is out of the scope of the present Letter. Other interesting physics include the energy transfer between FDM and EM sectors in models of galactic magnetic field, the effect of oscillating FDM field in the development and evolution of galactic, inter-galactic and cosmic magnetic fields, possible effects caused by the effective charge in $(\vec{\nabla}\phi) \cdot \vec{B}_{ex}$ term in (3), effects of the predicted diffuse EM radiation field on the physics of CMB polarization and propagation of cosmic ray. These topics are all of great interests for future study. In conclusion, we have pointed out that the FDM background together with a large scale magnetic field can give rise to fruitful physics.

Acknowledgements

This work is supported by National Science Foundation of China (NSFC), grant No. 11375065, and Shanghai Key Laboratory of Particle Physics and Cosmology, grant No. 15DZ2272100. I would like to thank ITP CAS Beijing for the hospitality during the course of this work.

References

[1] A.H. Guth, M.P. Hertzberg, C. Prescod-Weinstein, Phys. Rev. D 92 (2015) 103513.
[2] W. Hu, R. Barkana, A. Gruzinov, Phys. Rev. Lett. 85 (2000) 1158.
[3] T. Matos, L.A. Urena-Lopez, Gen. Relativ. Gravit. 39 (2007) 1279.
[4] For a recent detailed discussion, see e.g. L. Hui, J.P. Ostriker, S. Tremaine, E. Witten, arXiv:1610.08297.
[5] M.R. Baldeschi, Phys. Lett. B 122 (1983) 221.
[6] S.-J. Sin, Phys. Rev. D 50 (1994) 3650;
Jae-weon Lee, In-gyu Koh, Phys. Rev. D 53 (1996) 2236.
[7] For a recent review of axion cosmology, see D.J.E. Marsh, Phys. Rep. 643 (2016) 1.
[8] D.J.E. Marsh, P.G. Ferreira, Phys. Rev. D 82 (2010) 103528;
R. Hlozek, D. Grin, D.J.E. Marsh, P.G. Ferreira, Phys. Rev. D 91 (2015) 103512.
[9] A. Khmelnitsky, V. Rubakov, J. Cosmol. Astropart. Phys. 1402 (2014) 019.
[10] K. Ferriere, Astron. Astrophys. 505 (2009) 1183.
[11] K. Ferriere, in: Astrophysical Dynamics: From Stars to Galaxies, Proc. Int. Astron. Union 6 (S271) (2011) 170.
[12] M.R. Morris, J. Phys. Conf. Ser. 54 (2006) 1.
[13] M.R. Morris, arXiv:1406.7859.
[14] J.D. Jackson, Classical Electrodynamics, 3rd edition, John Wiley & Sons Inc., 1999.
[15] J.F. Navarro, C.S. Frenk, S.D. White, Astrophys. J. 462 (1996) 563.

23

Screening in perturbative approaches to LSS

Matteo Fasiello [a,*], Zvonimir Vlah [a,b]

[a] Stanford Institute for Theoretical Physics and Department of Physics, Stanford University, Stanford, CA 94306, United States
[b] Kavli Institute for Particle Astrophysics and Cosmology, Stanford University and SLAC, Menlo Park, CA 94025, United States

ARTICLE INFO

Editor: M. Trodden

ABSTRACT

A specific value for the cosmological constant Λ can account for late-time cosmic acceleration. However, motivated by the so-called cosmological constant problem(s), several alternative mechanisms have been explored. To date, a host of well-studied dynamical dark energy and modified gravity models exists. Going beyond ΛCDM often comes with additional degrees of freedom (dofs). For these to pass existing observational tests, an efficient screening mechanism must be in place. The linear and quasi-linear regimes of structure formation are ideal probes of such dofs and can capture the onset of screening. We propose here a semi-phenomenological "filter" to account for screening dynamics on LSS observables, with special emphasis on Vainshtein-type screening.

1. Introduction

The existence of a dynamical mechanism responsible for late-time cosmic acceleration often requires additional degrees of freedom (dofs) besides those of general relativity. On the other hand, the latter is, to an exquisite level of accuracy, a good description of the physics we see at "small" scales such as within the solar system. For the overall picture to be consistent, a screening mechanism must be in place. Screening is expected to be efficient in highly dense regions. Conversely, low-density environments make up the ideal settings to access the additional dynamics of beyond-ΛCDM models.

Large scale structure probes are an optimal case in point. The linear regime of structure formation is the environment where the additional dofs are most transparent and testable. These scales are well-described by perturbation theory. Crucially, the number of available modes grows approximately like the cube of the wavenumber, making any gain on the k-reach of the perturbative theory significant. An analytical description of the mildly-non-linear regime of structure formation [1–3] is highly desirable: these scales are a precious repository of information on both primordial physics (e.g. non-Gaussianities [4–6]) and late-time dynamics (see [7] and references therein). Our focus here will be on the latter: the mildly-non-linear regime can capture the onset of screening dynamics, which is central to dark energy and modified gravity models.

* Corresponding author.
E-mail address: matteorf@gmail.com (M. Fasiello).

2. A new scale

There has been considerable recent effort towards expanding the fluid description of dark matter to include an additional dynamical component (see e.g. [8,9] and [10–12] for earlier work on the same specific model). These works are based on the notion that the large hierarchy of scales in between the size of the observable universe $1/H_0$ and the highly non-linear-regime of structure formation $1/k_{\rm NL}$ allows for a clean perturbative treatment of the $k \ll k_{NL}$ modes. Naturally, the small expansion parameter is $k/k_{\rm NL}$. By employing a full-fledged effective theory approach [2,13], the microphysics of yet smaller scales can be encapsulated in a number of "UV" coefficients[1] to be determined by comparison with observations and/or simulations.

However, as argued above, in general screening will suppress the effects of the additional dofs in dark-energy (DE) and modified gravity (MG) at small scales i.e. in the highly-non-linear regime. We sketch in Fig. 1 the *total* power spectrum *vs.* the ΛCDM behaviour under one of the screening mechanisms that most clearly exemplifies this effect: Vainshtein screening (see e.g. [14] for a detailed N-body analysis).

There exists in other words a scale, we shall call it $k_{\rm V}$, at which screening becomes active. Any attempt at an accurate and general description of beyond-ΛCDM dynamics of structure formation in screened theories needs to take $k_{\rm V}$ into account (see [15] for interesting work that includes $k_{\rm V}$-related effects up to linear order).

[1] These multiply at each order all possible operators allowed by the symmetries of the theory (e.g. rotational invariance).

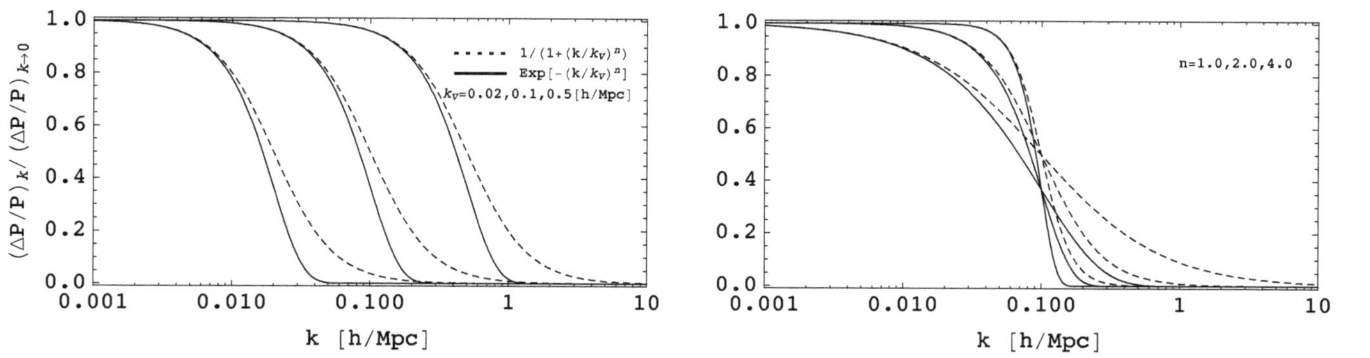

Fig. 1. Shown in this plot is a sketch of screening effects on the fractional difference between the power spectrum in nDGP [16] (here as a Vainshtein-screened theory and a precursor of Galileons and massive gravity theories) and ΛCDM. Two test functional forms are used: Lorentzian (dashed line) and Gaussian-like (solid line), for several different values of the parameter k_V and slope n. We refer the reader to the discussion around Eq. (9) for more details. It's clear that, according to the region of parameter space probed (different k_V values), screening can happen already at linear scales, at quasi-linear ones, or only deep in the non-linear regime. We refer the reader to Fig. (1) of [14] for the results, derived via N-body simulations. Note also that here, as opposed to [14], we have normalized the profiles by their value at low k.

In the specific case of Vainshtein-screened theories, some readers may be more familiar with the quantity in coordinate space related to k_V, the so-called *Vainshtein* radius, r_V. Models such as DGP [16] and non-linear massive gravity [17] exhibit an efficient implementation of such screening mechanism.

The definition of r_V typically depends on the specific configuration[2] under study. Most importantly, it depends on a set of defining parameters for the theory. For the above examples, r_V depends on the *cross-over scale* in DGP and on the *graviton mass* in massive gravity. It is then clear how the mildly-non-linear regime of structure formation can be used to set powerful bounds on DE and MG models.

Our ability to access screening depends crucially on the k_{NL} *vs.* k_V hierarchy:

- In the asymptotic region where $k_V \gg k_{NL}$ accounting for screening is hardly necessary: all dofs are manifest and the perturbative expansion breaks down (at $k \lesssim k_{NL}$) long before screening becomes relevant.
- Complementarily, for too small a k_V, $k_V \ll k_{NL}$, screening will be extremely efficient and for all intents and purposes our description will coincide with ΛCDM.

The interesting regime at hand corresponds to

$$k_V \lesssim k_{NL} \,. \tag{1}$$

Intriguingly, there exist several setups where this regime provides the most compelling cosmological solutions (see [18] for one such example).

2.1. Useful mismatch in the two expansions

As shown in Fig. 1, depending on the k_V *vs.* k_{NL} hierarchy, screening can become relevant already at linear scales, or only at n-loop order in the k/k_{NL} expansion, or in general "in between" loops. This happens because non-linearities can become important at very different scales on the dark matter and the dark energy side (see also Fig. (1) of [14]). It is relying on this very fact that one can hope to access an intrinsically non-linear (in k/k_V on the MG side) phenomenon such as screening already at quasi linear (in k/k_{NL} on the DM side) scales.

Depending on the value of k_V and the strength of screening, non-linearities on the MG side can start suppressing the gravitational coupling between MG and DM at very different scales. It is this allowed "mismatch" between k_V and k_{NL} that grants access to screening. Our screening model will be a phenomenological take on highly non-linear screening effects for the power spectrum (PS) of the total density contrast: it should be thought of as resulting from the resummation of the non-linearities in k/k_V all the while the perturbative expansion is kept for the k/k_{NL} parameter.

3. Setup

Let us show how the screening effects regulated by k_V come about in a typical setup. Consider a Lagrangian made up by the standard GR and matter content plus an additional scalar (split into standard kinetic term + interactions) directly coupled to matter:

$$\mathcal{L} \sim \mathcal{L}_{EH} + \mathcal{L}_m + (\partial_\mu \phi)^2 + \mathcal{L}_\phi^{\rm int} + \frac{\beta}{M_{\rm Pl}}\, \phi\, T_m \,. \tag{2}$$

Such a scenario naturally emerges in dark energy models as well as e.g. in the *decoupling* limit of modified gravity theories [19,20].

The existence of strong derivative ϕ self-interactions is the key to screening dynamics. As soon as the non-linearities in $\mathcal{L}_\phi^{\rm int}$ are important, they too will contribute a non-negligible kinetic term and affect the canonical normalization of $\delta\phi$. In other words, the kinetic term has the form $Z(\bar\phi)(\partial\delta\phi)^2$, with $Z \to 1$ only in the linear regime. Upon normalizing one finds $\delta\phi \sim \delta\phi_c/Z_{\rm int}$, where again $Z_{\rm int}$ depends on the background value of ϕ and the self-interactions coefficients (see e.g. [21]). For $Z_{\rm int} \gg 1$, the field ϕ coupling to matter is heavily suppressed. We identify the condition $Z_{\rm int} \gg 1$ with a strong screening regime where the presence of the additional dof will not be detectable.[3]

Let us schematically write the equation of motion for the dark matter + additional dof system in the Newtonian limit:

$$\begin{aligned}
&\frac{\partial \delta_m}{\partial \tau} + \partial_i[(1+\delta_m)v_m^i] = 0\,,\\
&\frac{\partial v_m^i}{\partial \tau} + \mathcal{H} v_m^i + v_m^j \partial_j v_m^i = -\nabla^i \Phi\,,\\
&\nabla^2 \Phi = \frac{3}{2}\mathcal{H}^2 \Omega_m \delta_m + F(\bar\phi)\nabla^2\delta\phi\\
&\nabla^2\delta\phi + \text{non linearities} = \frac{\beta}{M_{\rm Pl}}\,\delta_m\,,
\end{aligned} \tag{3}$$

where we have split the scalar background value from its fluctuations in $\phi = \bar\phi + \delta\phi$ and used the fact that the fifth force from

[2] For example, the effective dimensionality and symmetry of the source + test-particle system.

[3] Although see [22].

the extra field will affect dark matter dynamics via the Poisson equation. The function $F(\bar{\phi})$ tracks the screening strength and is therefore related to $Z_{\rm int}$.

Note that we have instead been deliberately agnostic about the equation of motion for $\delta\phi$: for stability, we require it be at most second order in time derivatives. A well-studied [23] example is the cubic Galileon:

$$\nabla^2\phi + \frac{1}{\Lambda^3}\left[(\nabla^2\phi)^2 - (\nabla_i\nabla_j\phi)^2\right] = \beta\frac{\rho}{M_{\rm Pl}}\,. \tag{4}$$

Galileon interactions are ubiquitous: one can think of them in this context as emerging in the decoupling limit of massive (bi)gravity or as a small subset of Horndeski-type interactions. Indeed, it has been shown [24] that the broader class of Horndeski theories exhibits Vainshtein screening. The presence of Λ in Eq. (4) identifies the threshold in momentum at which non-linearities become relevant. For example, in massive gravity $\Lambda = \Lambda_3 \equiv (m^2 M_{\rm Pl})^{1/3}$, proving how a small mass can in principle activate screening at arbitrarily small momentum scales. However, if massive gravity is enlisted to explain cosmic acceleration, the value of m cannot stray too far from the current value of the Hubble constant H_0.

Shifting the focus back on Eq. (3), we identify the two regimes in the DE/MG side via $Z_{\rm int}$, with $Z_{\rm int} \ll 1$ corresponding to the regime where the dynamics is well approximated by the linear solution, and $Z_{\rm int} \gtrsim 1$ requiring non-linearities to be taken into account.

4. Modelling screening

We want to model the observables resulting from the solution to Eq. (3) in a regime sensitive to screening effects. To this aim, we assume that the system has been solved up to a certain perturbative order "l" in $k/k_{\rm NL}$[4] and, in particular, that the solution is known for the *total* density contrast variable defined as the RHS of the Poisson equation, $\nabla^2\Phi \equiv \frac{3}{2}\mathcal{H}^2\Omega_m\delta_T$.

We model screening dynamics on crucial observables, such as the power spectrum of the total density, in the following way:

$$P_{\rm res}\big|_N(k,\tau) = \sum_{n=0}^{N} P^{(n)}_{\rm res}(k,\tau) = \sum_{n=0}^{N}\int \frac{d^3k'}{(2\pi)^3}\,\mathcal{K}^N_n(k',k,\tau)P^{(n)}(k',\tau), \tag{5}$$

where $P^{(n)}$ is the n'th perturbative solution for the power spectrum related to the system in Eq. (3), N stands for the perturbative order *up to* which the expression is valid and n signals instead a specific order in the expansion. We formally introduced here the kernels $\mathcal{K}^N_n(k',k,\tau)$ to describe the resummed dynamics of higher order contributions in k/k_V. In other words, kernels are to account for the part of the screening dynamics that is *not* captured by the perturbative expansion. Indeed, the non-linearities in the DE/MG sector play an increasing role at higher momenta; given the hierarchy, the k/k_V parameter becomes order one much sooner than $k/k_{\rm NL}$ and so needs to be resummed. This resummation affects in particular also observables at $k \ll k_V$. This further implies that kernels, in addition to varying according to the perturbative order index n, should also depend on the overall PT order N. The reason is that depending on the working PT order N, part of the screening is captured perturbatively, while the kernels are responsible for the resummation of the "residual" screening. As we go higher in perturbation theory, the kernels have indeed less screening to account for.

The structure of Eq. (5) is reminiscent of the recently proposed resummation schemes in the context of the baryon acoustic oscillations (BAO) [25–27] (see also [28]). The physics we are describing is of course quite different but the analogy stems from the fact that here too the kernels account for the effects from non-linear physics (in the k/k_V expansion) that we need to resum. More specifically, in our case the non-linearities to be resummed as k approaches k_V are those in the dark energy/modified gravity sector. These are propagated to the dark matter sector gravitationally as clear from Poisson's equation. The coupling between the two sectors is suppressed as DE/MG non-linearities become important and in particular the contribution from the DE/MG sector to the total density contrast becomes much weaker. As a result, the kernels in Eq. (5) are sensitive to the DE/MG contributions to observables and essentially blind to the purely dark matter ΛCDM-like sector.

A top-bottom exact derivation of the kernels, ideally via a Lagrangian formulation, is beyond the scope of this paper and we leave it to upcoming work. It is important to point out at this stage that we have chosen to apply the screening filter directly at the level of observables. As should be clear from *Section* 3, we could have instead opted to introduce kernels already at the level of the fields and then propagate theirs effects trough the equations of motion, all the way to observables. Our handling observables directly is certainly simpler but it relies on the fact that our parametrization for screening kernels is sufficiently general to cover for what would have instead been a convolution of kernels if applied at the field level. The fact that this is not necessarily always true is down to the phenomenological nature of our approach.

From here on instead, we proceed phenomenologically. Organizing the total power spectrum contribution in generalized cosmology, P, as a ΛCDM piece plus the remaining $\Delta P = P - P_{\Lambda{\rm CDM}}$, we can write, for the linear calculation,

$$P^{(0)}_{\rm res}(k,\tau) = P^{(0)}_{\Lambda{\rm CDM}}(k,\tau) + K_0(k,\tau)\Delta P^{(0)}(k,\tau), \tag{6}$$

where $P^{(0)}_{\Lambda{\rm CDM}}$ is the usual linear ΛCDM power spectra (the usual output of Boltzmann codes such as CAMB [29] or CLASS [30,31]), while $P^{(0)}$ is the linear solution in generalized cosmology (also linear or low-order in the k/k_V expansion). Note also that we shall refer to the total power spectrum also as $P_{\rm res}$, this to underscore the resummation of screening effects. The phenomenological nature of Eq. (6) is already evident from the fact that the kernels now act directly on the "external" observables, as opposed to the more general prescription in Eq. (5). The kernel K_0 is in Eq. (6) to capture screening effects much beyond the linear order in the k/k_V expansion, it is a resummation to all orders in PT. As one proceeds beyond linear order, the expression for the total power spectrum reads, after resumming the k/k_V expansion, as

$$P_{\rm res}\big|_N(k,\tau) = \sum_{n=0}^{N}\left[P^{(n)}_{\Lambda{\rm CDM}}(k,\tau) + K^N_n(k,\tau)\Delta P^{(n)}(k,\tau)\right]. \tag{7}$$

Here $P^{(n)}_{\Lambda{\rm CDM}}$ represents the n-th loop expansion of the power spectrum in the ΛCDM cosmology and $\Delta P^{(n)}$ perturbatively (both in the k/k_V and $k/k_{\rm NL}$) captures the dynamics beyond ΛCDM at every loop. Let us stress again the effect of the K^N_n factors, where the index N stands for the PT order we are working at and the index n stands for an expansion in k/k_V: as one goes higher in perturbation theory (increasing N), more of the screening dynamics is captured already perturbatively and so the N dependence of kernels K^N_n is there to ensure one does not "double count" the perturbative and

[4] The structure of the perturbative expansion is more complex in non-scaling universe but we nevertheless adopt it here for the sake of convenience.

the resummed screening contributions. We will illustrate this below with a specific example.

The discussion so far has been relatively general as the specific phenomenon we want to describe is encoded in the form of the kernels. We now specialize the analysis to the screening mechanism known as Vainshtein screening (VM).

4.1. Vainshtein screening – resummation

In the mechanism first studied by Vainshtein [32], the suppression of the coupling with matter originates from kinetic interactions in the DE/MG sector, such as the ones generating the second term in Eq. (4). We refer the reader to [33,34] for important early works in the context of structure formation and to e.g. [35–37] for more recent studies. Let us see how the framework we have outlined takes shape in the case of VM.

It is convenient at this stage to express the generic K_n^N in terms of the Taylor expansion of the generic reduced non-linear form K^N, i.e. we can write

$$K_n^N(k,\tau) = K(k,\tau)\big[K\big]^{-1}\Big|_{N-n}(k,\tau), \tag{8}$$

where the last term is the $(N-n)$-th order Taylor polynomial of the inverse of the reduced kernel K^N, typically a function of time and the k/k_V parameter. Note that in employing this form for the kernels there is already an element of choice. We now take on the form that the reduced kernels should have to account for Vainshtein screening. The most immediate constraints come from the asymptotic regimes:
- in the $k \to 0$, and for very low k in general ($k \ll k_V < k_{NL}$), these kernels (i.e. resummation) will not be necessary and must therefore reduce to unity.
- in the complementary regime, $k \gtrsim k_V$ kernels ought to screen very efficiently and should therefore render any non-ΛCDM feature in the spectrum negligible.

The most natural candidates as reduced kernels $K(k,\tau)$ to model the VM are:

$$K_G(k,\tau) = \exp\left(-\sum_m \alpha_m(\tau)(k/k_V)^{2m}\right),$$
$$K_L(k,\tau) = 1/\left(1+\sum_m \alpha_m(\tau)(k/k_V)^{2m}\right), \tag{9}$$

where subscripts G and L indicate respectively Gaussian and Lorentzian forms and the coefficients α_m have a time dependence of their own. The presence of only even powers of k is due to rotational invariance. Both expressions clearly satisfy the asymptotic requirements but the following considerations point to utilizing the Gaussian kernels. From Eq. (8) one can see how, in the case of the Lorentzian kernel, it is necessary for the sum over m to go up to $\lfloor N/2 \rfloor + 1$ in order to ensure that K_n^N gives the desired asymptotic behaviour in the high k limit. This in turn makes the reduced Lorentzian kernel K_L sensitive to the specific PT order one is working at. As a consequence, one should in principle write it with an N index as well. Note that this is not the case for the Gaussian kernels. Using the simplified form of Eq. (8), the formula in Eq. (7) becomes

$$P_{res}\big|_N(k,\tau) = P_{\Lambda CDM}\big|_N(k,\tau) + K(k,\tau)\sum_{n=0}^{N}\big[K\big]^{-1}\Big|_{N-n}(k,\tau)\Delta P^{(n)}(k,\tau), \tag{10}$$

where again the form we have chosen for the kernels in Eq. (9) guarantees the correct behaviour in the asymptotic regions, with the second line of Eq. (10) becoming negligible at sufficiently high k. We should note here that, besides the coefficients α_m the scale k_V itself should in principle have a time dependence. This is clearly seen, for example, at the level of the Poisson equation within Eq. (3) where the background evolution is stored in the function F. Indeed, the scale k_V is determined by a "universal" quantity such as Λ in Eq. (4) and also by background quantities sensitive to time-evolution. At this level of analysis, and given the freedom on the $\alpha_m(\tau)$ and the possible ensuing degeneracies between α_m and k_V time-dependence, we have opted not to place a time dependence directly on k_V. The merits of this choice are best tested within a specific model for which the exact screening kernels can be extracted following a full Lagrangian perturbation theory treatment.[5]

4.2. Vainshtein screening – perturbative build-up

The framework that we have setup so far will account for the "residual" screening effects, those that escape perturbation theory at the given working order. However, it is often the case the perturbative solutions themselves are hard to obtain without resorting to idealized configurations such as, for example, those endowed with spherical symmetry. On the other hand, an analytical handle on LSS dynamics is crucial in view of upcoming data from astronomical surveys. It is paramount that we develop analytical tools to complement the role of N-body simulations (see [43] for interesting recent developments) in the study of structure formation.

In this context, the use and extension of Einstein–Boltzmann solvers to include DE/MG is an important and timely development [38,39] (see also [40]). However, it is often the case that available codes account only for the dynamics up to quadratic order in the Lagrangian and therefore do not fully account for screening. Our framework has already been setup to include the resummed screening component and we will now extend it to model also the perturbative screening build-up. This of course with the ultimate goal to make contact with simulations.

As ever, the known behaviour in the asymptotics will act as our guiding principle. Let us proceed by assuming that we know the perturbative expression for the power spectrum up to order $n-1$ and would like to estimate the n-th order contribution. Since we organize our observables around the known ΛCDM result, the quantity to be determined at order n will be $\Delta P^{(n)}(k/k_{NL}, k/k_V, \tau)$. In the following, we propose two different ways to estimate $\Delta P^{(n)}$. The first will be particularly effective at very low perturbative orders, as close as possible to the linear solution, the other in the complementary regime. We first use the result (see e.g. [41]) valid for ΛCDM cosmology expansion at large scales:

$$P^{(n)}_{\Lambda CDM}(k,\tau)/k^2 P^{(0)}_{\Lambda CDM} \sim \text{const}_\Lambda\,, \quad k \to 0. \tag{11}$$

It has been shown that this approximation is reliable up to scales of almost 0.1 h/Mpc for the two-loop power spectrum and past 0.1 h/Mpc at higher orders (see Fig. 4 in [41]). We now extend this expansion beyond ΛCDM and write

$$\Delta P^{(n)}(k,\tau)/k^2\, P^{(0)} \sim \text{const} - \text{const}_\Lambda \frac{P^{(0)}_{\Lambda CDM}}{P^{(0)}}\,, \tag{12}$$

where this is valid for $k \ll k_V$. The above equation provides our first estimate of the difference between the (unknown) perturbative expression at n-th order of the total power spectrum and the

[5] We are grateful to the anonymous Referee for stressing this point as well as the related observation on the convolution (*vs.* direct application) of kernels in Eq. (7), (9).

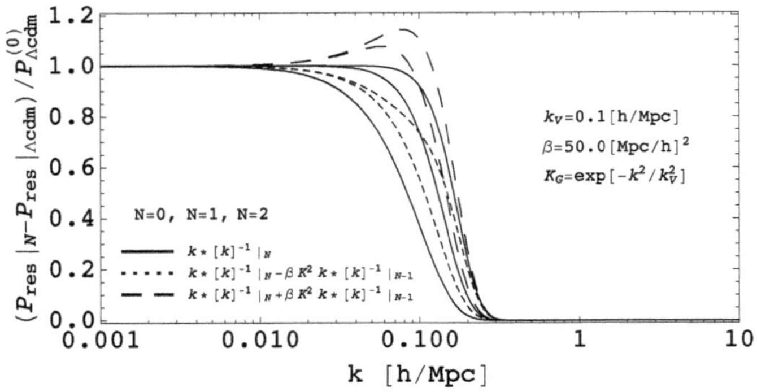

Fig. 2. The fractional difference between a fully screened total power spectrum and the ΛCDM result. Different colours indicate different perturbative orders. Continuous, dotted, and dashed lines stand for the action of the additional kernel $\tilde{K}$ in Eq (13), which in this approximation is regulated by the value of just one parameter, β. Note that for the $N=2$ case we have dropped the last contribution from Eq. (13) as it does not modify the physical picture in this specific example. (For interpretation of the colours in this figure, the reader is referred to the web version of this article.)

ΛCDM one. As such, this expression can be used in Eq. (10) in order to include also the residual screening.

In particular, whenever[6] one can write $\Delta P^{(n)}(k,\tau)/k^2\,\Delta P^{(0)} \sim$ const and k_V happens to be small, e.g. ~ 0.1 h/Mpc, there is a dramatic simplification of the overall results for Eq. (10), which in this case reads:

$$
\begin{aligned}
P_{\rm res}\big|_N(k,\tau) &= P_{\Lambda{\rm CDM}}\big|_N(k,\tau) \\
&+ K(k,\tau)\big[K\big]^{-1}\Big|_N(k,\tau)\Delta P^{(0)}(k,\tau) \\
&+ K(k,\tau)\,\big[K\big]^{-1}\Big|_{N-1}(k,\tau)\,\Delta P^{(1)}(k,\tau) \\
&+ k^2\Delta P^{(0)}(k,\tau)K(k,\tau)\sum_{n=2}^{N}\beta_n^0\big[\tilde{K}\big]^{-1}\Big|_{N-n}(k,\tau)\,,
\end{aligned} \tag{13}
$$

where we have used the fact that, for k_V in the vicinity of 0.1 h/Mpc, one need only have the exact perturbative solution up to one loop and can rely on the approximation for higher orders contributions (see e.g. [41] Fig 4). Note that β_n does in principle also depend on $k/k_{\rm NL}$. We stress that in this configuration the last term in the last line of Eq. (13) can be further simplified in favour of the usual K kernel times another compact kernel with no need for the sum over n. In Fig. 2 we illustrate how the fractional difference between a fully screened total power spectrum and the ΛCDM PS would look like whenever the relation in Eq. (12) can be simplified this one step further. In particular we assume the following $\Delta P^{(0)} \sim P^{(0)}_{\Lambda{\rm CDM}}$ and $\Delta P^{(n)} \sim k^2\Delta P^{(0)} \sim k^2 P^{(0)}_{\Lambda{\rm CDM}}$.

Let us now consider another way to estimate the $\Delta P^{(n)}$ and its embedding in Eq. (10):

$$
\begin{aligned}
P_{\rm res}\big|_N(k,\tau) &= P_{\Lambda{\rm CDM}}\big|_N(k,\tau) \\
&+ K(k,\tau)\sum_{n=0}^{N-1}\big[K\big]^{-1}\Big|_{N-n}(k,\tau)\Delta P^{(n)}(k,\tau) \\
&+ K(k,\tau)\,\bar{K}(k,k/k_{\rm NL},\tau)\,\bar{\Delta}P^{(N)}(k,\tau)\,,
\end{aligned} \tag{14}
$$

[6] Although this depends on the specific cosmology, one obvious parameter is the deviation of the equation of state from $w=-1$. More in general, one needs to establish to what extent going beyond-ΛCDM affects the variance of fluctuations and velocity dispersion. Once this is done at one loop order, the results of [41] point to a reliable extrapolation to higher orders as well.

where in the last line we have isolated the term $\Delta P^{(N)}$ to be estimated and written it as $\Delta P = \bar{K}\bar{\Delta}P$. The role of the new kernel $\bar{K}$ is to model the perturbative screening contribution and that is why it must depend also on $k/k_{\rm NL}$. More explicitly, in order to estimate the value of ΔP at higher perturbative orders, we propose the following:

$$
\Delta P^{(N)} = \bar{K}\bar{\Delta}P^{(N)} \equiv \bar{K}\Delta P^{(N-1)}, \tag{15}
$$

where we are using the fact that, at higher orders, the most reliable way to estimate $\Delta P^{(N)}$ is to employ the value of the known closest observable, $\Delta P^{(N-1)}$, and control it with $\bar{K}$. Let us then explore some of the properties we demand of the new kernel. First of all, at $k/k_{\rm NL}$ scales where the N-th order contribution in perturbation theory becomes relevant, we require that $\bar{K} \ll 1$ so that $\bar{K}\,\Delta P^{N-1}$ is effectively of order N.

The specific form of $\bar{K}$ is hard to pin down for a generic theory with a screening mechanism that could be either perturbatively very strong or very weak at order N in the expansion. However, the task becomes easier if the scales where the N-th order contribution is important are also the ones at which screening becomes *rapidly* strong. In such a scenario, even if $\bar{K}$ is modeling a perturbative contribution to screening, the rapid perturbative onset of screening will be well-approximated by the Gaussian or Lorentzian form in Eq. (9) and suitable α_n coefficients will readily account for an effect of order N (and not $N-1$) in a ΔP derived via Eq. (15). Note also that in the rapid perturbative screening limit $\bar{K}$ need not depend on $k/k_{\rm NL}$: the dynamics of the two expansions decouples in this limit and at the next perturbative order one may well use directly the ΛCDM result for the total power spectrum.

5. Embedding in the "EFT of LSS"

The modeling of screening we have proposed can be readily embedded within the effective approach to LSS dynamics. Let us consider the asymptotic regions. For very small k the shielding effect is negligible and the EFT prescription [13] will generate the appropriate *counterterms* for both the dark matter and dark energy component. We stress that at small perturbative order the counterterms can be common to both components [9] or, in other words, degenerate (see [42] for a derivation). At perturbative orders above the one where (strong) screening occurs observables coincide with their ΛCDM counterpart and so do counterterm operators. This is natural as our filter is nothing other than a phenomenological resummation in the k/k_V expansion; as such, it bypasses the need for counterterms. The coefficients α_n in our kernels will also vary depending on the different shielding strengths associated with different screening theories/interactions. Such difference can be found already within the same model: for example, the cubic, quartic and quintic Galileon interactions generate a different suppression of the coupling to matter.

6. Conclusions

The data from upcoming astronomical surveys (Euclid, LSST) will put to the test our best ideas on the mechanism responsible for the current acceleration of the Universe. There are intriguing proposals that go beyond the ΛCDM model: from dark energy to IR modifications of general relativity. The additional degrees of freedom that typically characterize beyond-ΛCDM models come with an associated scale, k_V, beyond which the corresponding fifth force is suppressed to the point of being currently undetectable. As we have seen, if the hierarchy between k_V and the scale of dark matter non-linearities $k_{\rm NL}$ is benevolent, $k_V \lesssim k_{\rm NL}$, screening dynamics will be accessible in LSS setups already at quasi-linear scales. This is precisely the regime where perturbative analytical tools, such as effective field theory, are most efficient.

In this work we have proposed a phenomenological ansatz to model screening dynamics. Our framework accounts for the "residual screening" that is not captured in perturbation theory but is crucial to obtain reliable predictions for LSS observables. We have further put forward a mechanism to estimate also the perturbative screening component whenever the exact result is not known. Our formalism can be readily adapted to several screening mechanisms and to different layers of approximation. In the second part of the text however, we have adopted to focus on one specific screening mechanism, Vainshtein screening, and provided the corresponding kernels K.

We applied our formalism to the total density power spectrum, for which we have provided a resummation scheme for higher order effects in k/k_V. We stress in particular the usefulness of Eq. (13): under certain assumptions it can model screening by relying on exact inputs solely from the linear theory.

Our approach here is phenomenological in nature. However, by enforcing a number of constraints from exact asymptotic solutions and from symmetries of the physical system, we have been able to identify very efficient kernels that account for screening dynamics. The most natural next step is to analyze screening via the Lagrangian PT formalism, which we address in upcoming work.

7. Acknowledgements

We are delighted to thank Diego Blas for illuminating conversations. We are very grateful to Martin White and Miguel Zumalacarregui for precious feedback on a draft version of the manuscript. MF is supported in part by NSF PHY-1068380; ZV is supported in part by DOE contract DEAC02-76SF00515.

References

[1] F. Bernardeau, S. Colombi, E. Gaztanaga, R. Scoccimarro, Phys. Rep. 367 (2002) 1, arXiv:astro-ph/0112551.
[2] D. Baumann, A. Nicolis, L. Senatore, M. Zaldarriaga, J. Cosmol. Astropart. Phys. 1207 (2012) 051, arXiv:1004.2488.
[3] R.A. Porto, L. Senatore, M. Zaldarriaga, J. Cosmol. Astropart. Phys. 1405 (2014) 022, arXiv:1311.2168.
[4] N. Dalal, O. Dore, D. Huterer, A. Shirokov, Phys. Rev. D 77 (2008) 123514, arXiv: 0710.4560.
[5] R. Angulo, M. Fasiello, L. Senatore, Z. Vlah, J. Cosmol. Astropart. Phys. 1509 (09) (2015) 029, arXiv:1503.08826.
[6] V. Assassi, D. Baumann, F. Schmidt, J. Cosmol. Astropart. Phys. 1512 (12) (2015) 043, arXiv:1510.03723.
[7] D. Huterer, et al., Astropart. Phys. 63 (2015) 23, arXiv:1309.5385.
[8] M. Fasiello, Z. Vlah, Phys. Rev. D 94 (6) (2016) 063516, arXiv:1604.04612.
[9] M. Fasiello, Z. Vlah, arXiv:1611.00542.
[10] E. Sefusatti, F. Vernizzi, J. Cosmol. Astropart. Phys. 1103 (2011) 047, arXiv:1101.1026.
[11] S. Anselmi, G. Ballesteros, M. Pietroni, J. Cosmol. Astropart. Phys. 1111 (2011) 014, arXiv:1106.0834.
[12] S. Anselmi, D. Lopez Nacir, E. Sefusatti, J. Cosmol. Astropart. Phys. 1407 (2014) 013, arXiv:1402.4269.
[13] J.J.M. Carrasco, M.P. Hertzberg, L. Senatore, J. High Energy Phys. 1209 (2012) 082, arXiv:1206.2926.
[14] B. Falck, K. Koyama, G.b. Zhao, B. Li, J. Cosmol. Astropart. Phys. 1407 (2014) 058, arXiv:1404.2206.
[15] D. Alonso, E. Bellini, P.G. Ferreira, M. Zumalacárregui, Phys. Rev. D 95 (6) (2017) 063502, arXiv:1610.09290.
[16] G.R. Dvali, G. Gabadadze, M. Porrati, Phys. Lett. B 485 (2000) 208, arXiv:hep-th/0005016.
[17] C. de Rham, G. Gabadadze, A.J. Tolley, Phys. Rev. Lett. 106 (2011) 231101, arXiv:1011.1232.
[18] G. D'Amico, C. de Rham, S. Dubovsky, G. Gabadadze, D. Pirtskhalava, A.J. Tolley, Phys. Rev. D 84 (2011) 124046, arXiv:1108.5231.
[19] N.A. Ondo, A.J. Tolley, J. High Energy Phys. 1311 (2013) 059, arXiv:1307.4769.
[20] M. Fasiello, A.J. Tolley, J. Cosmol. Astropart. Phys. 1312 (2013) 002, arXiv:1308.1647.
[21] C. de Rham, Living Rev. Relativ. 17 (2014) 7, arXiv:1401.4173.
[22] L. Hui, A. Nicolis, Phys. Rev. Lett. 109 (2012) 051304, arXiv:1201.1508.
[23] N. Bartolo, E. Bellini, D. Bertacca, S. Matarrese, J. Cosmol. Astropart. Phys. 1303 (2013) 034, arXiv:1301.4831.
[24] K. Koyama, G. Niz, G. Tasinato, Phys. Rev. D 88 (2013) 021502, arXiv:1305.0279.
[25] L. Senatore, M. Zaldarriaga, J. Cosmol. Astropart. Phys. 1502 (02) (2015) 013, arXiv:1404.5954.
[26] Z. Vlah, M. White, A. Aviles, J. Cosmol. Astropart. Phys. 1509 (09) (2015) 014, arXiv:1506.05264.
[27] D. Blas, M. Garny, M.M. Ivanov, S. Sibiryakov, J. Cosmol. Astropart. Phys. 1607 (07) (2016) 028, arXiv:1605.02149.
[28] L.F. de la Bella, D. Regan, D. Seery, S. Hotchkiss, arXiv:1704.05309.
[29] A. Lewis, A. Challinor, A. Lasenby, Astrophys. J. 538 (2000) 473, arXiv:astro-ph/9911177.
[30] J. Lesgourgues, arXiv:1104.2932.
[31] D. Blas, J. Lesgourgues, T. Tram, J. Cosmol. Astropart. Phys. 1107 (2011) 034, arXiv:1104.2933.
[32] A.I. Vainshtein, Phys. Lett. B 39 (1972) 393.
[33] A. Lue, R. Scoccimarro, G.D. Starkman, Phys. Rev. D 69 (2004) 124015, arXiv: astro-ph/0401515.
[34] K. Koyama, F.P. Silva, Phys. Rev. D 75 (2007) 084040, arXiv:hep-th/0702169.
[35] C. de Rham, L. Heisenberg, Phys. Rev. D 84 (2011) 043503, arXiv:1106.3312.
[36] A. Taruya, K. Koyama, T. Hiramatsu, A. Oka, Phys. Rev. D 89 (4) (2014) 043509, arXiv:1309.6783.
[37] A. Taruya, T. Nishimichi, F. Bernardeau, T. Hiramatsu, K. Koyama, Phys. Rev. D 90 (12) (2014) 123515, arXiv:1408.4232.
[38] B. Hu, M. Raveri, N. Frusciante, A. Silvestri, Phys. Rev. D 89 (10) (2014) 103530, arXiv:1312.5742.
[39] M. Zumalacarregui, E. Bellini, I. Sawicki, J. Lesgourgues, arXiv:1605.06102.
[40] M. Lagos, T. Baker, P.G. Ferreira, J. Noller, J. Cosmol. Astropart. Phys. 1608 (08) (2016) 007, arXiv:1604.01396.
[41] D. Blas, M. Garny, T. Konstandin, J. Cosmol. Astropart. Phys. 1401 (01) (2014) 010, arXiv:1309.3308.
[42] M. Lewandowski, A. Maleknejad, L. Senatore, arXiv:1611.07966.
[43] W. Fang, B. Li, G.B. Zhao, arXiv:1704.02325.

24

Secondary scintillation yield of xenon with sub-percent levels of CO_2 additive for rare-event detection

The NEXT Collaboration

C.A.O. Henriques [a], E.D.C. Freitas [a], C.D.R. Azevedo [c], D. González-Díaz [i], R.D.P. Mano [a], M.R. Jorge [a], L.M.P. Fernandes [a], C.M.B. Monteiro [a,*], J.J. Gómez-Cadenas [b,1], V. Álvarez [b], J.M. Benlloch-Rodríguez [b], F.I.G.M. Borges [d], A. Botas [b], S. Cárcel [b], J.V. Carríon [b], S. Cebrían [e], C.A.N. Conde [d], J. Díaz [b], M. Diesburg [f], R. Esteve [g], R. Felkai [b], P. Ferrario [b], A.L. Ferreira [c], A. Goldschmidt [h], R.M. Gutiérrez [j], J. Hauptman [k], A.I. Hernandez [j], J.A. Hernando Morata [i], V. Herrero [g], B.J.P. Jones [l], L. Labarga [m], A. Laing [b], P. Lebrun [f], I. Liubarsky [b], N. López-March [b], M. Losada [j], J. Martín-Albo [b,2], G. Martínez-Lema [l], A. Martínez [b], A.D. McDonald [l], F. Monrabal [l], F.J. Mora [g], L.M. Moutinho [c], J. Muñoz Vidal [b], M. Musti [b], M. Nebot-Guinot [b], P. Novella [b], D.R. Nygren [l,1], B. Palmeiro [b], A. Para [f], J. Pérez [b], M. Querol [b], J. Renner [b], L. Ripoll [n], J. Rodríguez [b], L. Rogers [l], F.P. Santos [d], J.M.F. dos Santos [a], A. Simón [b], C. Sofka [o,3], M. Sorel [b], T. Stiegler [o], J.F. Toledo [g], J. Torrent [b], Z. Tsamalaidze [p], J.F.C.A. Veloso [c], R. Webb [o], J.T. White [o,4], N. Yahlali [b]

[a] *LIBPhys, Physics Department, University of Coimbra, Rua Larga, 3004-516 Coimbra, Portugal*
[b] *Instituto de Física Corpuscular (IFIC), CSIC & Universitat de València, Calle Catedrático José Beltrán, 2, 46980 Paterna, Valencia, Spain*
[c] *Institute of Nanostructures, Nanomodelling and Nanofabrication (i3N), Universidade de Aveiro, Campus de Santiago, 3810-193 Aveiro, Portugal*
[d] *LIP, Departamento de Física, Universidade de Coimbra, Rua Larga, 3004-516 Coimbra, Portugal*
[e] *Laboratorio de Física Nuclear y Astropartículas, Universidad de Zaragoza, Calle Pedro Cerbuna, 12, 50009 Zaragoza, Spain*
[f] *Fermi National Accelerator Laboratory, Batavia, IL 60510, USA*
[g] *Instituto de Instrumentación para Imagen Molecular (I3M), Universitat Politècnica de València, Camino de Vera, s/n, Edificio 8B, 46022 Valencia, Spain*
[h] *Lawrence Berkeley National Laboratory (LBNL), 1 Cyclotron Road, Berkeley, CA 94720, USA*
[i] *Instituto Gallego de Física de Altas Energías, Univ. de Santiago de Compostela, Campus sur, Rúa Xosé María Suárez Núñez, s/n, 15782 Santiago de Compostela, Spain*
[j] *Centro de Investigación en Ciencias Básicas y Aplicadas, Universidad Antonio Nariño, Sede Circunvalar, Carretera 3 Este No. 47 A-15, Bogotá, Colombia*
[k] *Department of Physics and Astronomy, Iowa State University, 12 Physics Hall, Ames, IA 50011-3160, USA*
[l] *Department of Physics, University of Texas at Arlington, Arlington, TX 76019, USA*
[m] *Departamento de Física Teórica, Universidad Autónoma de Madrid, Campus de Cantoblanco, 28049 Madrid, Spain*
[n] *Escola Politècnica Superior, Universitat de Girona, Av. Montilivi, s/n, 17071 Girona, Spain*
[o] *Department of Physics and Astronomy, Texas A&M University, College Station, TX 77843-4242, USA*
[p] *Joint Institute for Nuclear Research (JINR), Joliot-Curie 6, 141980 Dubna, Russia*

ARTICLE INFO

Editor: W. Haxton

ABSTRACT

Xe–CO_2 mixtures are important alternatives to pure xenon in Time Projection Chambers (TPC) based on secondary scintillation (electroluminescence) signal amplification with applications in the important field of rare event detection such as directional dark matter, double electron capture and double beta decay detection. The addition of CO_2 to pure xenon at the level of 0.05–0.1% can reduce significantly the scale of electron diffusion from 10 mm/$\sqrt{m}$ to 2.5 mm/$\sqrt{m}$, with high impact on the discrimination

* Corresponding author.
E-mail address: cristina@gian.fis.uc.pt (C.M.B. Monteiro).

1 NEXT Co-spokesperson.
2 Now at University of Oxford, United Kingdom.
3 Now at University of Texas at Austin, USA.
4 Deceased.

Keywords:
Double beta decay
Neutrino
Rare event detection
Electroluminescence
Secondary scintillation
Xenon

efficiency of the events through pattern recognition of the topology of primary ionization trails. We have measured the electroluminescence (EL) yield of Xe–CO_2 mixtures, with sub-percent CO_2 concentrations. We demonstrate that the EL production is still high in these mixtures, 70% and 35% relative to that produced in pure xenon, for CO_2 concentrations around 0.05% and 0.1%, respectively. The contribution of the statistical fluctuations in EL production to the energy resolution increases with increasing CO_2 concentration, being smaller than the contribution of the Fano factor for concentrations below 0.1% CO_2.

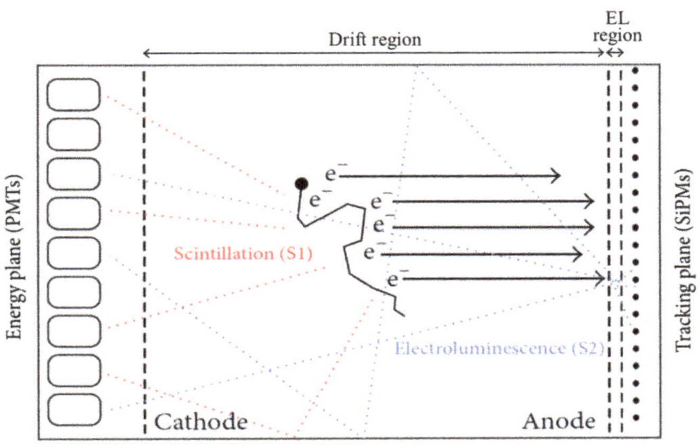

Fig. 1. Schematic of the EL-based TPC developed by the NEXT Collaboration for double-beta decay searches in ^{136}Xe.

1. Introduction

Many experiments aiming for rare event detection such as double beta decay (DBD) and double electron capture (DEC), with or without neutrino emission, as well as directional dark matter (DDM) use high-pressure xenon (HPXe) as the detection/target medium [1–7]. The physics behind these experiments is of paramount importance in contemporary particle physics and cosmology.

When compared to liquid xenon and double phase xenon TPCs [8–14], detection in the gas phase offers some important advantages. While the event detection in liquid TPCs allows for compactness and self-shielding, some features may be essential for the above experiments to succeed. The impact of background depends strongly on the achieved energy resolution, which is much better for event detection in gas than in liquid. Furthermore, event interaction in the gas will allow for discrimination of the rare event topological signature, as demonstrated for DBD and DEC detection [15,16,5], in contrast to the interaction in liquid, where the extremely reduced dimensions of the primary ionization trail rules out any possible trail pattern recognition.

In particular, optical TPCs based on secondary scintillation (electroluminescence) amplification of the primary ionization signal are the most competitive alternatives to those based on charge avalanche amplification. For the latter, the limited charge amplification at high pressure impacts the energy resolution, yielding at present a best value around 3% at 2.5 MeV for a 1 kg-scale prototype based on micromegas [17], to be compared to 0.7% obtained for an electroluminescence (EL) amplification prototype of similar dimensions [18]. In addition, when compared to conventional electronic readout of the charge avalanche, EL optical readout through a photosensor has the advantage of mechanically and electrically decoupling the amplification region, rendering more immunity to electronic noise, radiofrequency pickup and high voltage issues.

Fig. 1 depicts a schematic of a typical optical TPC. Most of the gas volume is occupied by the conversion/drift region where the radiation interaction takes place exciting or ionizing the gas atoms/molecules and leading to the emission of primary scintillation (the t_0 signal of the event) resulting from the gas de-excitation or electron/ion recombination. A low electric field, below the gas excitation threshold, is applied to the drift region to minimize recombination and to guide the primary electrons towards a shallow region with electric field intensity between the gas excitation and ionization thresholds, the scintillation region. Upon crossing this region, each electron gains from the electric field enough kinetic energy to excite the gas atoms/molecules by electron impact, leading to a large scintillation output upon gas de-excitation (electroluminescence). A pixelated photosensor plane enables to determine the x- and y-positions of the primary electrons arriving at the EL region, and the time interval between primary and EL scintillation pulses enables to determine the z-position of where the ionization takes place.

Absolute values of the EL light yield have been measured in uniform electric fields [19–21] and in the modern micropatterned electron multipliers, as GEM, THGEM, MHSP and micromegas [22–24]. The statistical fluctuations in the EL produced in charge avalanches are dominated by the statistical fluctuations in the total number of electrons produced in the avalanche, since all the electrons contribute to EL production. On the other hand, the statistical fluctuations in the EL produced for uniform electric fields below the gas ionization threshold are negligible when compared to those associated with the primary ionization formation [25]. The latter situation is most important when event to background discrimination is also based on the energy deposited in the gas, as is the case of DEC and neutrinoless double beta decay, where the best achievable detector energy resolution is important for efficient background rejection.

The effectiveness of event discrimination based on the topological signature of the ionization trail is related to the low electron drift velocity of xenon and, mainly, to its large electron diffusion. The large electron diffusion is determined by the inefficient electron energy loss in elastic collisions with the xenon atoms, in particular in the range of reduced electric fields of few tens of V/cm/bar used in the drift region. Diffusion hinders the finer details of the ionization trail, especially for large drift distances, and the discrimination based on the topological signature of the events becomes less effective [26].

The aforementioned problem can be mitigated by adding a molecular gas, like CO_2, CH_4 or CF_4, to pure xenon. With the addition of such molecules, new molecular degrees of freedom from vibrational and rotational states are made available for electron energy transfer in inelastic collisions. In this case, the energy distribution of the ionization electron cloud in the drift region tends to build up around the energy of the first vibrational level, typically at ~0.1 eV, even in the presence of minute concentrations of molecular additives.

Until recently, it was believed that the presence of molecular species in the noble gas would dramatically reduce the EL yield that could be achieved. Experimental studies performed for Ar [27] have shown that the presence of CO_2 and CH_4 in concentrations

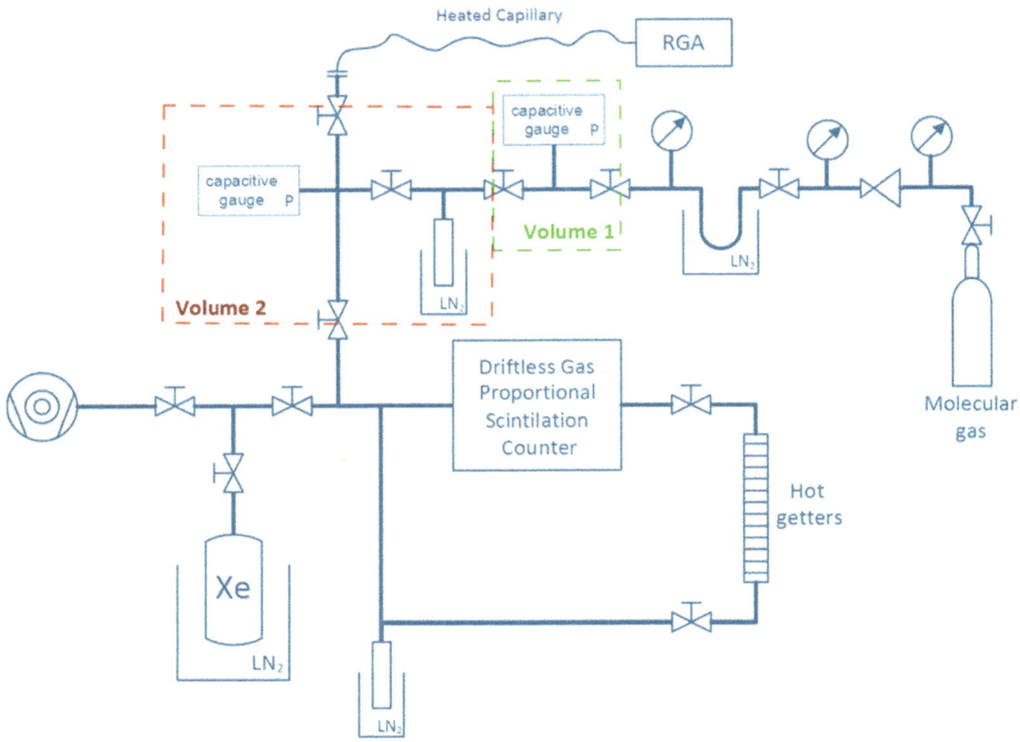

Fig. 2. Schematic of the experimental apparatus, including the gas proportional scintillation counter, the gas circulation and purifying system with SAES St-707 getters, the gas entrance and exit systems including the turbo-pump, two calibration volumes (volume 1 in green, volume 2 in red), the liquid nitrogen mixing vessels and the RGA connection through a heated capillary. (For interpretation of the references to colour in this figure legend, the reader is referred to the web version of this article.)

as low as ~50 ppm and ~200 ppm, respectively, resulted in an EL reduction above 70%. Detailed Monte Carlo simulation studies of electron drift in xenon at atmospheric pressure and room temperature [28] have shown that the average number of elastic collisions between successive inelastic collisions is very large, of the order of 10^4, for the typical reduced electric fields applied to the EL region. This fact partly explained the importance of gas purity for the EL yield of noble gases: if an electron has significant probability of colliding with a molecular impurity before it obtains from the electric field sufficient energy to excite a noble gas atom, it may lose part of its energy without leading to EL photon emission, resulting in a decrease in the EL production. Depending on the conditions, excimer quenching, photo-absorption and dissociative attachment can jeopardize performance as well.

Within the NEXT Collaboration [1], which has built a HPXe TPC for DBD studies with the ^{136}Xe isotope, we proposed to revisit the addition of molecular additives to xenon, at sub-percent level, to reduce electron diffusion in the TPC, hence improving the topological discrimination capabilities. Preliminary experimental studies and simulations for different concentrations of CO_2 and CH_4 gases that are common in TPCs and whose elementary cross-sections are well known, have shown encouraging results [29], leading to acceptable EL losses and only small degradation in the detector energy resolution. Simulation results obtained with Magboltz [38] have shown that Xe-CO_2 mixtures with concentrations of 0.05–0.1% of CO_2 would be sufficient to reduce the transversal and longitudinal diffusion coefficients to acceptable values (~ 2.5 mm/$\sqrt{\text{m}}$). These concentrations are almost one order of magnitude lower than those needed for CH_4, in order to obtain a similar diffusion reduction [29].

Those results led us to perform a detailed study on the effect of the addition of CO_2 to pure xenon both on the EL yield and on the energy resolution, for additive concentrations below 1% [29]. In this work we present those studies. CO_2 is a priori the most interesting option due to its low cost and easy handling, since it is non-flammable.

2. Experimental setup and methodology

The experimental setup especially projected for these studies includes a Gas Proportional Scintillation Counter (GPSC) [25], which is connected to a gas re-circulation system in order to continuously purify the gas or the mixture using SAES St-707 getters; a Residual Gas Analyser (RGA) that provides a real-time direct measurement of the molecular additive concentrations; a vacuum pumping system to maintain the RGA in continuous operation; associated electronics and suitable control and data-acquisition electronics for both systems, the RGA and the GPSC. The xenon gas is 4.8 grade from Messer containing less than 1 ppm of the main molecular gasses. The getter efficiency is very effective in removing the outgassing from the detector; the EL degradation is very slow after closing the gas circulation through the getters while presenting a much faster recovery when the gas circulation is restored. E.g., it takes two hours for the EL to be reduced by 20% after closing the getters, while it takes only 10 minutes to restore the original performance after gas circulation through the getters is resumed. Therefore, xenon purity in normal detector operation is similar or even better than that of the gas inside the original bottle. The main components of the experimental setup are illustrated in Fig. 2.

The GPSC used in this work is depicted in Fig. 3. It is of the 'driftless' type, i.e. without drift region, and has been already used in [30,31]. This design was chosen for the present work because it allows to study the influence of molecular additives on the EL parameters, minimizing the effects that may arise in the electron drift through the drift region and gas scintillation transparency.

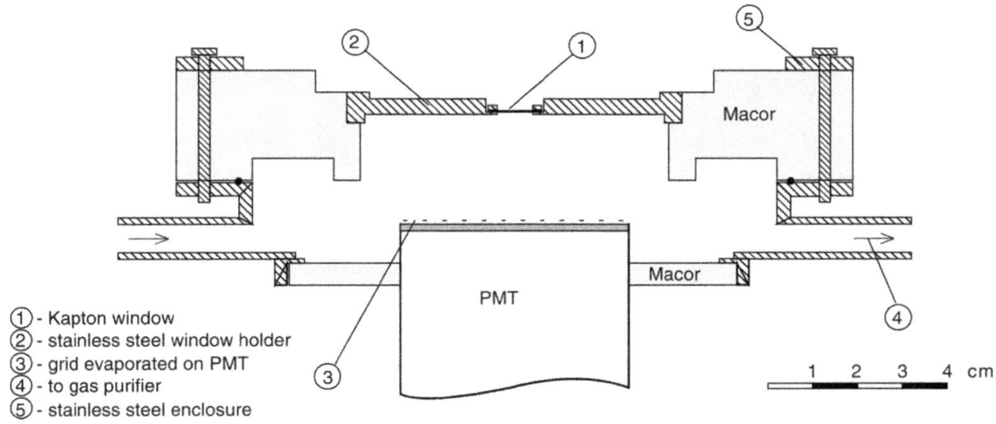

Fig. 3. Schematic of the driftless GPSC used in this work, including its principal components.

The EL region, 2.5 cm long, is delimited by a Kapton radiation window (8 mm in diameter, mounted on a stainless steel holder) aluminised on the inner side, and by the PMT quartz window, vacuum-evaporated with a chromium grid (100 μm width and 1000 μm spacing), electrically connected to the photocathode pin of the PMT. The EL electric field is established by applying a negative high voltage to the detector window and its holder (4 to 10.5 kV), which are insulated from the stainless steel detector body by a ceramic material (Macor), being the detector body, the chromium grid of the PMT window and its photocathode kept at 0 V. More information on this detector can be found in [30,31]. The reduced electric field inside the detector is set below the gas ionization threshold in order for EL to be produced without any charge multiplication in the EL region.

A 2-mm collimated, 5.9 keV x-ray beam from a ^{55}Fe radioactive source irradiated the detector window along the detector axis, being the 6.4-keV x-rays of the Mn K_β line absorbed by a chromium film. The 5.9 keV x-rays interact in the gas resulting in the release of electrons and photons. These ionization electrons are accelerated throughout the EL region exciting the noble gas atoms and inducing EL as a result of the atoms' de-excitation processes. The amount of EL is more than 3 orders of magnitude higher than primary scintillation. The EL pulse is collected by the PMT, whose output signal is subsequently shaped, amplified and, finally, digitized through a Multi-Channel Analyser (MCA). A typical pulse-height distribution obtained in the MCA for 5.9 keV x-rays is depicted in Fig. 4a.

In a driftless chamber, the amount of EL depends on the distance travelled by the primary electron cloud in the EL region and, therefore, on the x-ray interaction depth. Consequently, the pulse-height distribution generated by the MCA has the typical Gaussian shape (from a monoenergetic line) convoluted with an exponential tail towards the low-energy region, due to the exponential law of the x-ray attenuation. Since, for 5.9-keV x-rays, the absorption length in 1 bar of xenon is about 2.7 mm, very small when compared to the long EL region of 25 mm, the observed full absorption peak in the pulse-height distribution has an almost Gaussian-like shape.

The intrinsic response of the GPSC for 5.9 keV x-rays was obtained by deconvolution of the overall full absorption peak distribution into a sum of a large number, 250, of Gaussian functions corresponding to x-ray interactions at successive depths, $\Delta z =$ 0.01 cm, with areas decreasing according to the exponential absorption law for the 5.9 keV x-rays and with the same relative FWHM, which was left as a free parameter. The centroid of each Gaussian follows the integration of the solid angle subtended by the PMT photocathode along the path corresponding to each x-ray penetration, being the centroid of the rightmost Gaussian, i.e. the one having the highest centroid, left as a free parameter. Fig. 4b depicts an experimental pulse-height distribution for the full energy peak and the resulting fit obtained by the deconvolution procedure used, denoting a very good agreement. The GPSC pulse amplitude and energy resolution were taken from the centroid and FWHM of the Gaussian corresponding to x-ray interactions taking place just below the window, i.e. the rightmost one. The obtained amplitude is within 2% of the peak channel of the full energy peak, Fig. 4b. The obtained energy resolution is somewhat below 7%, instead of the ~8% obtained for a Gaussian fit to the right side. Directly analysing the PMT pulse waveforms with a given duration (i.e. corresponding to the same x-ray penetration) we obtained a Gaussian shape pulse-height distribution with energy resolution not higher than 7.3%, absolute value. The "true" energy resolution should be lower than this value, since the signal-to-noise ratio in this case is not negligible, contributing to higher statistical fluctuations.

The small volumes in Fig. 2, each one read by an accurate capacitive pressure gauge, were used to calibrate the RGA. Volume 2 is filled with pure xenon from the detector volume, while volume 1 is filled with CO_2. For the RGA calibration, only these volumes were used, being isolated from the detector volume before the reference mixtures were done avoiding, in this way, any error that might result from CO_2 adsorption to the inner surfaces of the GPSC and, mainly, to the getters. We consider that the amount of CO_2 adsorbed or released by the walls of the calibration volumes is negligible. The calibration process has shown a good linear correlation between the concentration measured in the RGA and the initial additive concentration based on the pressure gauge readings, within the studied concentration range. In order to avoid a pressure-dependent non-linearity of the RGA, calibration and detector operation have been carried out at the same total pressure of about 1.13 bar, for both pure xenon and its mixtures.

The EL studies were performed when the RGA partial pressures stabilized and, likewise, the additive concentration was calculated from an average over several measurements done during the time interval when the EL studies were performed.

Before setting each mixture, a measurement of the CO_2 background was performed in the GPSC filled with pure xenon, having the getters at 250 °C, in order to ensure maximum xenon purity. This background was, afterwards, subtracted from the RGA CO_2 reading once the mixture was done. For CO_2 only mass 44 peak was used, as the other peaks are superimposed on other molecular species, while for Xe all the peaks are considered.

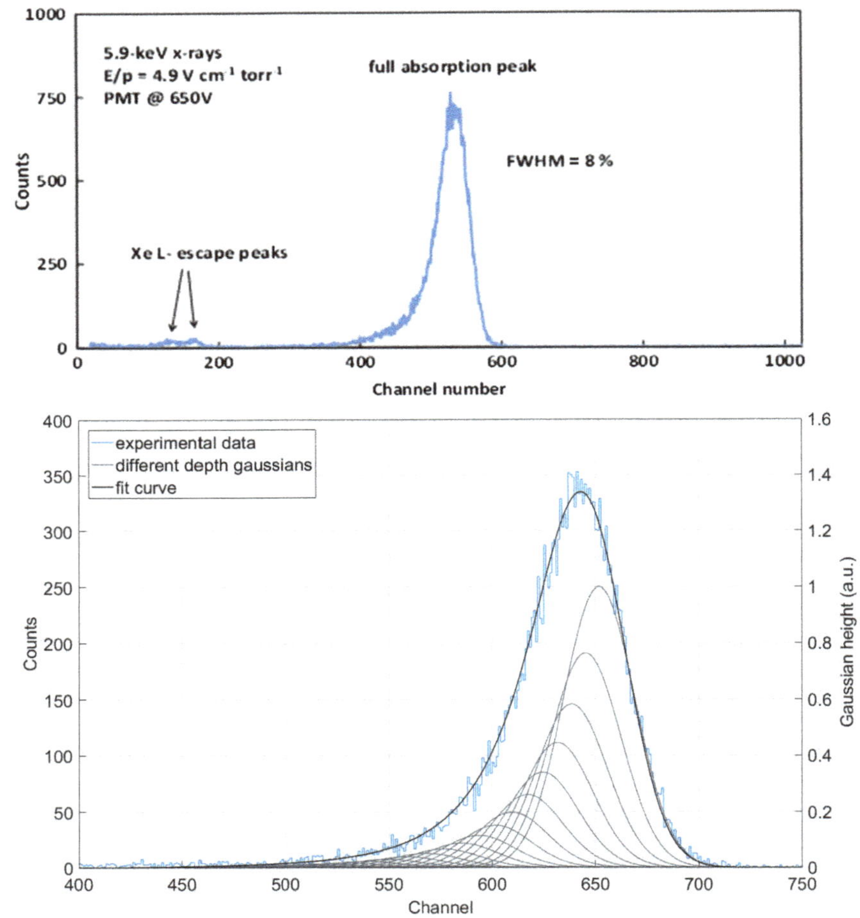

Fig. 4. a) Pulse-height distribution for 5.9-keV x-rays absorbed in the xenon driftless GPSC for a reduced electric field of 3.7 kV/cm/bar; b) detail of the full energy peak (blue histogram), the fit function (black solid line), and some of the Gaussians resulting from the deconvolution procedure used in this work (black and grey solid lines). (For interpretation of the references to colour in this figure legend, the reader is referred to the web version of this article.)

As the hot getters react with CO_2, the getter temperature was reduced to 80 °C before the mixture was done. At this getter temperature CO_2 is only slightly absorbed. On the other hand, the EL parameters in pure xenon were found to degrade only slightly and only after several days of operation when cooling down the getters from 250 °C to 80 °C. Furthermore, it was observed that, despite the CO_2 being absorbed in the getters, part of it is also transformed into CO that escapes to the gas phase, as observed by the correlated growth of the partial pressure at mass 28 (related with CO) as the concentration of CO_2 decreases. This effect increases with increasing getter temperature. Therefore, some CO is present in the Xe–CO_2 mixtures, with concentrations that are roughly constant for all the studied mixtures, for a getter operation temperature of 80 °C. Simulations have shown that the impact of the presence of CO on the yield is small, within a 10% effect at most, for the lowest CO_2 concentrations. The simulations were carried out using the same algorithms as in [29] and considering CO concentrations below 0.05%.

In this work only relative values were measured for the EL yield. Absolute values for the EL reduced yield, Y/N, N being the density of the molecules in the gas, were obtained normalizing the relative pulse amplitude as a function of reduced electric field, E/N, obtained in this work for pure xenon, to the absolute values of the EL reduced yield obtained in [19]. The same normalization constant was, then, used to normalize the remaining EL curves obtained for the different mixtures.

3. Experimental results

In Fig. 5, the reduced EL yield, number of photons produced per electron per cm of path divided by the atomic number density, Y/N, as a function of the reduced electric field, E/N, applied to the EL region[5] is shown for different CO_2 concentrations added to pure xenon. The two data sets for the 0.174% of CO_2 are related to two independent measurements. Interestingly, the reduced yield exhibits the typical approximate linear dependence of EL with reduced electric field even in the presence of CO_2. The solid lines present fits to the data, excluding the data points near the EL threshold where the EL response of GPSCs deviates from the linear trend [30]. As expected, the EL yield decreases as the CO_2 amount increases. Nevertheless, CO_2 concentrations of ~0.05%, which allow an overall electron diffusion around 2.5 mm/$\sqrt{\text{m}}$ [29], can be acceptable in terms of EL yield since, in spite of having an EL yield reduction up to 35% when compared to pure xenon, this reduction may be tolerable in the cases where the EL is large enough. For comparison, it must be recalled that such a reduction is achieved in Ar mixtures when CO_2 concentration reaches 10 ppm [27].

As anticipated, for the same reduced electric field intensity, the EL threshold increases with increasing CO_2 content since, upon

[5] At $T = 293$ K we have Y/N (ph/e/atom $\times$ 10^{-17} cm^2) $= 2.276 \times 10^3 Y/p$ (ph/e cm^{-1} bar^{-1}) and E/N (Td) $= 2.276E/p$ (kV cm^{-1} bar^{-1}); (1 Townsend $=$ 10^{-17} V cm^2).

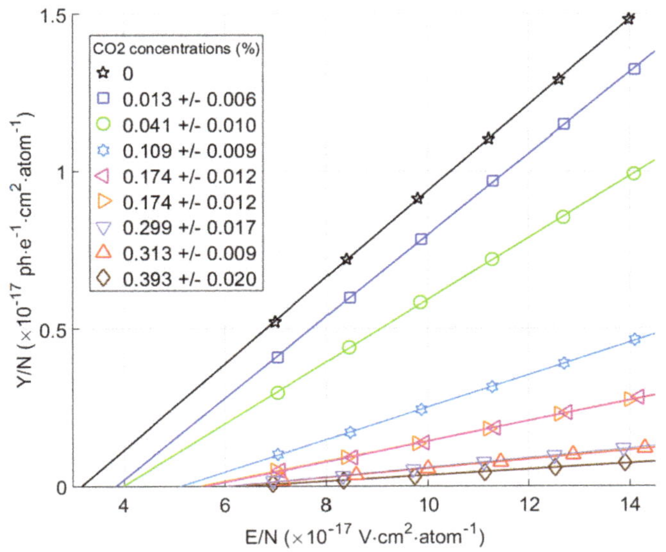

Fig. 5. EL reduced yield, Y/N, as a function of the reduced electric field, E/N, for different concentrations of CO_2 added to pure xenon. The errors are less than few percent and, hence, the error bars are within the symbols. The solid lines are linear fits to the data.

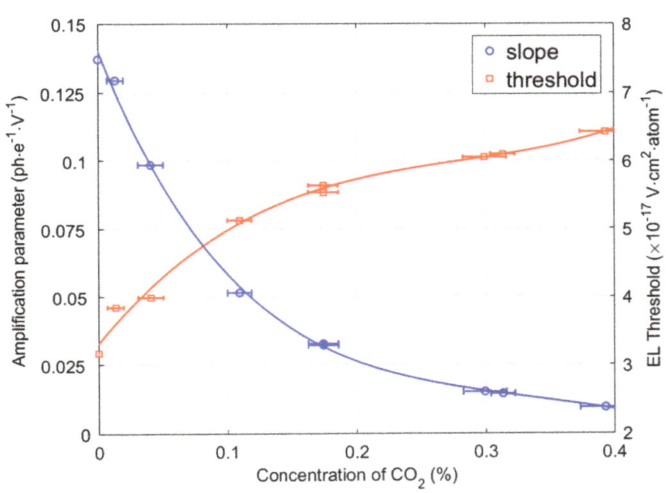

Fig. 6. Light amplification parameter and EL threshold of the lines fitted to the reduced EL yield (Fig. 5). The solid lines serve only to guide the eye.

colliding with a CO_2 molecule, the electron loses energy to rotational and vibrational excited states, reducing the average electron energy. Although qualitative in nature, the behaviour of the EL threshold shows how this cooling seems to be very efficient up to concentrations around 0.1% (as indicated by Magboltz simulations), values for which the EL loss remains acceptable, hinting that a compromise in terms of electron cooling/excimer scintillation does exist. Additional losses can be expected in CO_2 from dissociative attachment and excimer quenching, this last one being indeed the main source identified earlier in [27]. Fig. 6 summarizes the EL threshold and reduced yield slope, from Fig. 5 data, as a function of CO_2 concentration.

The impact of the molecular additive on the TPC energy resolution is an important parameter to be considered, in particular in double electron capture and in neutrinoless double beta decay detection, as it is a tool to effectively discriminate the rare events against background. In Fig. 7 we present the GPSC energy resolution (FWHM) for 5.9 keV x-rays as a function of reduced electric field, for the different CO_2 concentrations used in this work. The

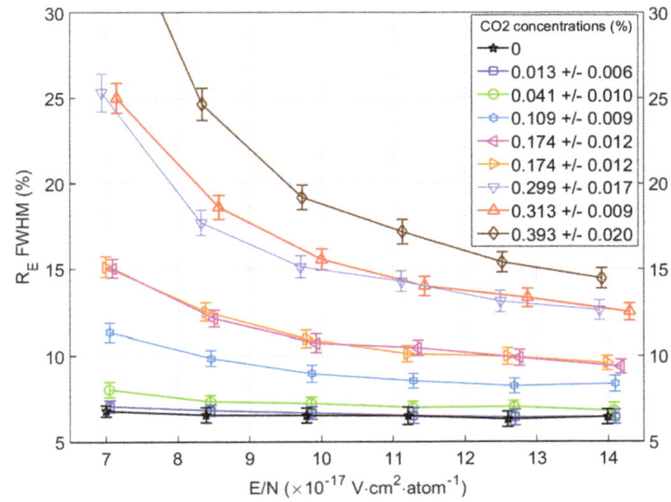

Fig. 7. Detector energy resolution as a function of E/N for 5.9 keV x-rays and for different concentrations of CO_2. The solid lines serve only to guide the eye. Error bars include both statistical and estimated systematic errors.

error bars include the uncertainties in the fit and the statistical error associated to multiple readings, added quadratically. The errors are overestimated as we do not know the correlation between both values. Nevertheless, we assumed these conservative errors.

From the experimental data, centroid and FWHM, it is possible to assess the fluctuations in the EL production assuming that the obtained pulse-height distributions are well-described by the mathematical model already discussed in the literature. The energy resolution (FWHM) of a GPSC, R_E, is given by [25]:

$$R_E = 2\sqrt{2\ln 2}\sqrt{\frac{\sigma_e^2}{\overline{N}_e^2} + \frac{1}{\overline{N}_e}\left(\frac{\sigma_{EL}^2}{\overline{N}_{EL}^2}\right) + \frac{\sigma_{pe}^2}{\overline{N}_{pe}^2} + \frac{1}{\overline{N}_{pe}}\left(\frac{\sigma_q^2}{\overline{G}_q^2}\right)} \tag{1}$$

where the first term under the square root describes the relative fluctuations in the number of ionization electrons induced by the interaction, N_e, the second term describes the relative fluctuations associated to the number of EL photons produced in the EL region per primary electron, N_{EL}, and the last two terms describe the relative fluctuations in the photosensor, namely the relative fluctuations in the number of photoelectrons released from the PMT photocathode by the EL burst, N_{pe}, and the relative fluctuations in the number of electrons collected in the PMT anode per photoelectron, i.e. the relative fluctuations in the gain of the electron avalanche in the PMT. The electronic noise is not included in Eq. (1) since it is negligible, as shown by the amplifier pulse waveforms in the oscilloscope. Since the process of photoelectron release from a photocathode by the incoming photons is described by a Poisson distribution, its variance is $\sigma_{pe}^2 = \overline{N}_{pe}$ and the relative fluctuations in the PMT are given by:

$$\frac{1+\left(\frac{\sigma_q^2}{\overline{G}_q^2}\right)}{c\overline{N}_e\overline{N}_{EL}} = \frac{k}{\overline{N}_{EL}}, \tag{2}$$

c represents the light collection efficiency, related to the anode grid transparency (Fig. 3), 81%, to the PMT quantum efficiency, 20% for 172 nm, and to the average solid angle subtended by the PMT photocathode relative to the primary electron path in the EL region, 30%. Therefore, k is a constant, which depends on the scintillation readout geometry and on the photosensor itself (for our PMT we measured a relative gain fluctuation of 0.55).

From the data for pure xenon, we can experimentally determine the contributions to the energy resolution from the statistical

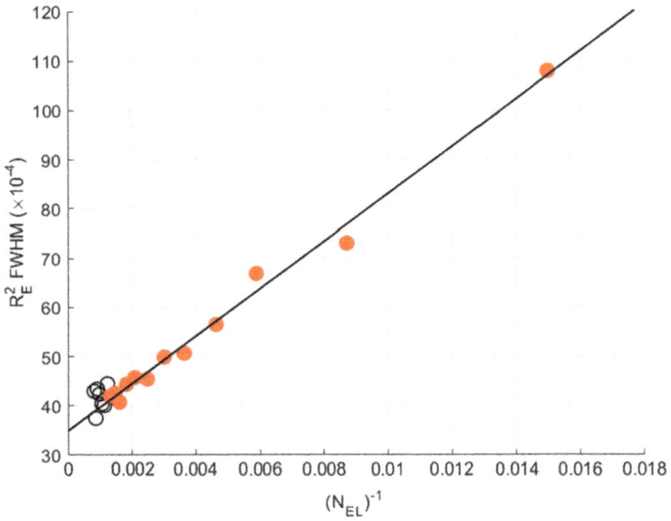

Fig. 8. The square of the energy resolution (R_E) as a function of the inverse of the average number of photons produced in the EL region per primary electron for 5.9 keV x-rays and for pure xenon. The solid line presents a linear fit to the data points in red. (For interpretation of the references to colour in this figure legend, the reader is referred to the web version of this article.)

fluctuations due to the primary ionization formation and due to the photosensor, since the contribution from the statistical fluctuations to the EL is negligible when compared to the other factors [25]. In Fig. 8 we depict R_E^2 as a function of N_{EL}^{-1} for pure xenon. A linear function, as imposed by Eq. (1), is fitted to the data points, excluding those with the highest N_{EL}, which depart from the linear trend. This behaviour is similar to the one observed in standard type GPSCs, with drift region [32,33]. The first term is obtained from the line interception with the vertical axis while $8\ln(2)k$ is the slope of the line. From Fig. 8 we obtain a Fano factor $F = \sigma_e^2/\overline{N}_e = 0.17 \pm 0.04$ for xenon, using a w-value of 22 eV [34], i.e. the average energy needed to produce one electron–ion pair in xenon ($N_e = E_x/w$, E_x being the x-ray energy; for x-rays the w-value does not depend on the applied electric field, above ~20 V/cm/bar, since recombination in this case is negligible). This result is in good agreement with the values normally found in the literature, between 0.13 and 0.25 [32,35–37]. In addition, the result we obtain for c from Eq. (2) (0.056) is similar to what is obtained from calculations based on the geometry (0.048). These agreements show the robustness of the present method.

The terms of Eq. (1) obtained from the fit in Fig. 8 (the 1st and the last 2), for the relative fluctuations in the primary ionization formation and for the relative fluctuations in the photosensor, are constant for all CO_2 concentrations studied in this work, since the Fano factor and the w-values of those mixtures are not expected to change significantly for these low additive concentrations, as shown by Degrad simulation [38], and k is constant for a given geometry and photosensor setup. Therefore, it is possible to determine the fluctuations associated to the EL production as a function of reduced electric field for the different mixtures, using the R_E data from Fig. 7, as these fluctuations are the only unknown variable in Eq. (1).

In Fig. 9 we present the square of the relative standard deviation in the number of EL photons produced in the EL region per primary electron,

$$Q = \left(\frac{\sigma_{EL}^2}{\overline{N}_{EL}^2}\right), \tag{3}$$

as a function of reduced electric field in the EL region, for the different concentrations of CO_2 added to pure xenon. A striking observation that can be made in Fig. 9 is that Q becomes non-negligible as the CO_2 concentration increases, largely independent of the reduced electric field. For a CO_2 concentration as low as 0.1% $Q \sim 0.08$, i.e. about half the value of the Fano factor, while for 0.2% CO_2 Q becomes comparable. For a CO_2 concentration of 0.05% Q is found to be around 0.02, a value that has a negligible impact on the energy resolution. For higher CO_2 concentrations and lower E/N, the signal-to-noise ratio decreases significantly, resulting in an artificially high energy resolution and, consequently, an over-estimated Q value obtained from Eq. (1), as the noise is not included in this equation. For that reason, those points are not included in Fig. 9. As the uncertainty in Q is dominated by the uncertainty in the energy resolution, the error bars are also over-estimated, as explained before.

The rise in the contribution from Q cannot be explained if we take only into account the effect of EL reduction with increasing CO_2, since even a reduction of one order of magnitude in EL still means a very high number of EL photons and, in addition, one would expect a decrease in Q with increasing electric field in the EL region, instead of an almost constant trend. We believe that this effect is due to dissociative electron attachment to

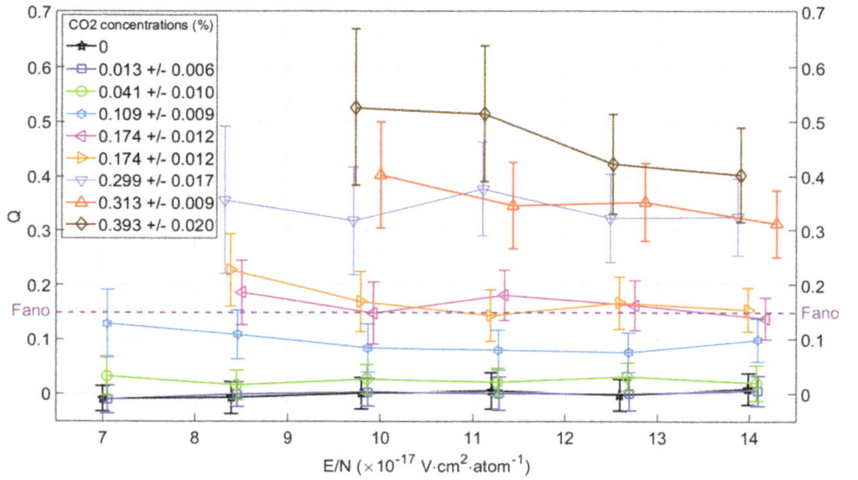

Fig. 9. Relative standard deviation in the number of EL photons per primary electron once squared (Q), as a function of the reduced electric field in the EL region (E/N) for different concentrations of CO_2. The solid lines serve only to guide the eye.

CO_2 molecules, for which we find a good quantitative agreement with simulations [39]; in the absence of this effect, simulations predict values for the Q-factor around 0.02, little dependent on concentration. Indeed, the cross-section for electron attachment is non-negligible and has narrow peaks for electron energies between 4-5 and 7–9 eV, e.g. see Fig. 1 in [40], and these are energies that the electrons will eventually reach in the EL region in order to be able to excite the xenon atoms. The simulated attachment implies for the highest CO_2 concentration a relatively modest 10's of %-level loss of ionization electrons; this effect is, therefore, not the only responsible for the EL reduction. For example, for the highest concentrations of CO_2, the electron attachment probability is about 65%, according to Magboltz simulation [41], which would lead to a maximum EL reduction of 65% if all attachment would happen right at the beginning of the electron path; however, experimental results show EL reductions above 90%. Most of the remaining EL reduction is explained by quenching. The presence of attachment becomes, nonetheless, the main source of fluctuations in the EL signal for concentrations already above 0.17% CO_2.

4. Discussion

CO_2 concentrations in the range of 0.05%–0.1% correspond to a characteristic size of the electron diffusion ellipsoid of $\sqrt[3]{\sigma_x\sigma_y\sigma_z} \cong 2.5\ \text{mm} \times \sqrt{\frac{10\ \text{bar}}{P}}$ after a 1 m drift through the TPC [29,39]. This value can be found for reduced drift fields in the range of $E/P =$ [20–30] V/cm/bar, by resorting to the latest Magboltz cross-section database. The drastic change experimentally observed in the EL threshold (Fig. 6) suggests that electron cooling is in fact strongly active even for these minute concentrations. Moreover, simulations indicate that a minimum of diffusion exists in the above field range; therefore, there is little to be gained by increasing (or decreasing) the drift field in the TPC. The presence of such minima can be found experimentally and theoretically for xenon mixtures and additives like CH_4 or TMA in [17], and it becomes narrower at low concentrations.

In pure xenon at 10 bar, electric fields in the aforementioned range pose no problems concerning charge recombination for primary electrons, as can be readily noticed, for instance, in [24]. For admixtures, however, the situation at high pressures is less clear. Some qualitative arguments can be drawn: in xenon at 5 bar and in a 0.22% TMA admixture, for example, the additional contribution to the Fano factor stemming from fluctuations in the charge recombination process is less than 0.1 [42]. Since the measured diffusion and drift coefficients are, in that case, similar to those simulated for the CO_2 optimum (0.05%–0.1%), and being the ionization density close to the one attempted in NEXT (10 bar), measurements performed in Xe–TMA can be used to estimate an upper bound to the effect expected in Xe–CO_2. Besides this initial charge recombination, it must be noted that the drift velocity will be reduced for optimum CO_2 concentrations by a factor of around 2, which is not expected to have a dramatic effect on the electron lifetime, according to the measurements performed in [17], again for TMA admixtures.

The above aspects will soon be evaluated for high pressures in NEXT-DEMO, a large prototype with a drift length of 30 cm and a hexagonal cross section of 8-cm apothem, operated with ~1.5 kg of natural xenon at a pressure of 10 bar [43]. Other relevant effects like the pressure-dependence of the EL yields and the light fluctuations will also be evaluated. Simulations performed in [39] indicate that, despite the anticipated deterioration at high pressure, both the Q factor and the finite-statistic term from the PMTs can be kept at the level of the Fano factor at 10 bar, as long as the CO_2 concentration remains in the range of 0.05–0.1%. Concerning the primary scintillation yields, a tolerable reduction within a factor of 5–10 is expected in the same concentration range.

A possible drawback that can arise from the use of CO_2 is related to the gas stability in the long term and the associated formation of CO. This can be handled using specific getters for CO_2. On the other hand, being these devices cold getters, one has to evaluate the radon emanation.

5. Conclusions

We have performed experimental studies on the reduced electroluminescence yield of Xe–CO_2 mixtures at room temperature. We have demonstrated that the addition of CO_2 to pure xenon, at concentration levels of few tenths of a percent, does not kill the proportional electroluminescence (EL) yield entirely, as it has been assumed during the last decades. CO_2 concentrations of 0.05% and 0.1% at around atmospheric pressure lead only to an EL reduction of 35% and 70%, respectively, relative to that produced in pure xenon at the same reduced electric field. Such a modest reduction seems tolerable, provided the number of photons produced per ionization electron is very large and also because it may be readily compensated by increasing the reduced electric field, since higher field can be applied to the EL region, as the ionization threshold increases with increasing CO_2 concentration.

On the other hand, the intrinsic energy resolution of xenon-based TPCs (i.e., photo-detection statistics neglected) degrades with increasing CO_2 concentration; for a concentration of 0.05% the contribution of the statistical fluctuations associated to EL production is a factor of 6 lower than the Fano factor, for 0.1% nearly half and for 0.2% it is slightly above it. This degradation in the energy resolution cannot be, however, compensated by an increase in the reduced electric field intensity. Based on both the approximate linear dependence of Q on the CO_2 concentration and the comparison with Magboltz simulations, these large fluctuations can be attributed to dissociative attachment of ionization electrons to CO_2 molecules. Seemingly, this process can only be mitigated by using shallower EL regions. Nevertheless, a compromise has to be found between the thickness of this region and the amount of EL produced.

The above findings are important for xenon-based TPCs relying on EL signal amplification, which are being increasingly used for rare-event detection such as directional dark matter, double electron capture and double beta decay detection. Particularly, the addition of CO_2 to pure xenon at the level of 0.05%–0.1% will reduce significantly the electron transverse diffusion from 10 mm/$\sqrt{\text{m}}$ to the level of few mm/$\sqrt{\text{m}}$, having a high impact on the discrimination of events through pattern recognition of the topology of primary ionization trails.

Other molecular additives, such as CH_4, do not present the drawback of having significant electron attachment but, on the other hand, higher concentrations will be needed to obtain similar electron transverse and longitudinal diffusions as in CO_2. Nevertheless, former work in [27] has shown that the addition of CH_4 to pure argon has less impact on the reduction in the mixture EL yield when compared to the addition of CO_2.

Acknowledgements

The NEXT Collaboration acknowledges support from the following agencies and institutions: the European Research Council (ERC) under the Advanced Grant 339787-NEXT; the Ministerio de Economía y Competitividad of Spain under grants FIS2014-53371-C04 and the Severo Ochoa Program SEV-2014-0398; the GVA of Spain under grant PROMETEO/2016/120; the Portuguese FCT under project PTDC/FIS-NUC/2525/2014; the U.S. Department of Energy under contracts number DE-AC02-07CH11359 (Fermi National Accelerator Laboratory) and DE-FG02-13ER42020 (Texas A&M); the University of Texas at Arlington. C.A.O.H., E.D.C.F., C.M.B.M. and C.D.R.A. acknowledge FCT under grants PD/BD/105921/2014, SFRH/BPD/109180/2015, SFRH/BPD/76842/2011 and SFRH/BPD/79163/2011, respectively.

References

[1] J. Martin-Albo, et al., NEXT Collaboration, J. High Energy Phys. 5 (2016) 159.
[2] T. Brunner, et al., Int. J. Mass Spectrom. 379 (2015) 110.
[3] J. Galan, J. Instrum. 11 (2016) P04024.
[4] D.Yu. Akimov, et al., arXiv:physics/9704021v1, 16 Apr. 1997.
[5] Yu.M. Gavrilyuk, et al., Phys. At. Nucl. 78 (2016) 1563.
[6] D. Nygren, J. Phys. Conf. Ser. 460 (2013) 012006.
[7] G. Mohlabeng, et al., J. High Energy Phys. 7 (2015) 092.
[8] E. Aprile, et al., XENON100 Collaboration, Phys. Rev. Lett. 109 (2012) 181301.
[9] D.S. Akerib, et al., LUX Collaboration, Phys. Rev. Lett. 112 (2014) 091303.
[10] Andi Tan, et al., PandaX-II Collaboration, Phys. Rev. Lett. 117 (2016) 121303.
[11] J.B. Albert, et al., EXO-200 Collaboration, Nature 510 (7504) (2014) 229.
[12] A. Gando, et al., KamLAND-Zen Collaboration, Phys. Rev. Lett. 117 (2016) 082503.
[13] K. Abe, et al., XMASS Collaboration, Phys. Lett. B 759 (2016) 64.
[14] E. Aprile, et al., XENON Collaboration, Phys. Rev. C 95 (2017) 024605.
[15] R. Luescher, et al., Phys. Lett. B 434 (1998) 407.
[16] P. Ferrario, et al., NEXT Collaboration, J. High Energy Phys. 1 (2016) 104.
[17] D. González-Díaz, et al., NEXT Collaboration, Nucl. Instrum. Methods 604 (2015) 8.
[18] D. Lorca, et al., NEXT Collaboration, J. Instrum. 9 (2014) P10007.
[19] C.M.B. Monteiro, et al., J. Instrum. 2 (2007) P05001 and references therein.
[20] C.M.B. Monteiro, et al., Phys. Lett. B 668 (2008) 167.
[21] E.D.C. Freitas, et al., Phys. Lett. B 684 (2010) 205.
[22] C.M.B. Monteiro, et al., Phys. Lett. B 677 (2009) 133.
[23] C.M.B. Monteiro, et al., Phys. Lett. B 714 (2012) 18.
[24] C. Balan, et al., J. Instrum. 6 (2011) P02006.
[25] J.M.F. dos Santos, et al., X-Ray Spectrom. 30 (2001) 373 and references therein.
[26] J. Renner, et al., NEXT Collaboration, J. Instrum. 12 (2017) T01004.
[27] T. Himi, et al., Nucl. Instrum. Methods 205 (1983) 591.
[28] F.P. Santos, et al., J. Phys. D, Appl. Phys. 27 (1994) 42.
[29] C.D.R. Azevedo, et al., J. Instrum. 11 (2016) C02007.
[30] P.C.P.S. Simões, et al., X-Ray Spectrom. 30 (2001) 342.
[31] P.C.P.S. Simões, et al., Nucl. Instrum. Methods A 505 (2003) 247.
[32] J.A.M. Lopes, et al., IEEE Trans. Nucl. Sci. 48 (2001) 312.
[33] C.M.B. Monteiro, et al., IEEE Trans. Nucl. Sci. 48 (2001) 1081.
[34] T.H.V.T. Dias, et al., J. Appl. Phys. 82 (1997) 2742.
[35] D.F. Anderson, et al., Nucl. Instrum. Methods 163 (1979) 125.
[36] T.Z. Kowalski, et al., Nucl. Instrum. Methods A 279 (1989) 567.
[37] S.J.C. do Carmo, et al., IEEE Trans. Nucl. Sci. 55 (2008) 2637.
[38] S. Biagi, Degrad, http://consult.cern.ch/writeup/magboltz/.
[39] C. Azevedo, D. Gonzalez-Diaz, S. Biagi, et al., FERMILAB-PUB-17-267-CD, arXiv: 1705.09481 [physics.ins-det], 2017.
[40] J. Escada, et al., J. Instrum. 4 (2009) P11025.
[41] S. Biagi, Magboltz, http://consult.cern.ch/writeup/magboltz/.
[42] E. Ruiz-Choliz, et al., Nucl. Instrum. Methods A 799 (2015) 137.
[43] V. Álvarez, et al., NEXT Collaboration, J. Instrum. 8 (2013) P09011.

25

Singularities in FLRW spacetimes

Huibert het Lam *, Tomislav Prokopec

Institute for Theoretical Physics, Spinoza Institute and EMMEΦ, Utrecht University, Postbus 80.195, 3508 TD Utrecht, The Netherlands

ARTICLE INFO

Editor: M. Trodden

Keywords:
FLRW spacetime
Incomplete geodesics
Singularity

ABSTRACT

We point out that past-incompleteness of geodesics in FLRW spacetimes does not necessarily imply that these spacetimes start from a singularity. Namely, if a test particle that follows such a trajectory has a non-vanishing velocity, its energy was super-Planckian at some time in the past if it kept following that geodesic. That indicates a breakdown of the particle's description, which is why we should not consider those trajectories for the definition of an initial singularity. When one only considers test particles that do not have this breakdown of their trajectory, it turns out that the only singular FLRW spacetimes are the ones that have a scale parameter that vanishes at some initial time.

1. Introduction

Hubble's law, the observed abundance of elements, the cosmic background radiation and the large scale structure formation in the universe are strong evidence that the universe expanded from an initial very high dense state to how we observe it now. However, what happened exactly during this hot density state is still an open problem. One of the questions that needs to be answered is whether there was a singularity at the beginning of spacetime. Such a singularity is in accordance with the very general theorems of Hawking and Penrose [1,2] defined as a non-spacelike geodesic that is incomplete in the past. One uses this definition because test particles move on these trajectories and thus have only traveled for a finite proper time.

The flatness, horizon and magnetic monopole problem can be solved with a period of exponential expansion in the very early universe [3,4]. To avoid a singularity before that period, it was suggested that one can have past-eternal inflation in which the universe starts from an almost static universe and flows towards a period of exponential expansion. This way the universe would not have a beginning. One of the characteristics of inflationary models is that the Hubble parameter H is positive. In [5] it was shown that when the average Hubble parameter along a geodesic H_{av} is positive, the geodesic is past-incomplete such that we would have a singularity. This is also applicable to models of eternal inflation in which the average Hubble parameter along geodesics does not go to zero sufficiently fast (i.e. such that we do not have that H_{av} is zero). In [6], a model of eternal inflation was given with all non-spacelike geodesics complete, but in [7] these kind of models were shown to be quantum mechanically unstable. Hence, this would imply that also models of eternal inflation start from a singularity.

In [8] it was pointed out that in De Sitter space the test particles that follow those past-incomplete trajectories and have a non-vanishing velocity, will have an energy that becomes arbitrarily large when going back in the past. This can be generalized to general Friedmann–Lemaître–Robertson–Walker (FLRW) spacetime and means that the energy of such a test particle can become super-Planckian at some initial time such that their description breaks down. This is the reason one should not consider those trajectories when defining a singularity. When one only considers the trajectories of test particles that do not have a breakdown of the description of their trajectory, one finds that the only FLRW spacetimes that start from a singularity are the ones with a scale factor that vanishes at some initial time. This implies that models of eternal inflation or bouncing models are singularity free provided one requires sub-Planckian test particles at all times.

In this paper we first consider the past-(in)completeness of geodesics in spacetimes with an FLRW metric. We review the general singularity theorems of [1,2] applied to these models and we review the more general (in the context of cosmology) argument of [5]. After that we consider how the energies of test particles change in time. We adopt units in which the velocity of light $c = 1$.

* Corresponding author.
E-mail addresses: h.hetlam@uu.nl (H. het Lam), t.prokopec@uu.nl (T. Prokopec).

2. Past-(in)completeness of geodesics in FLRW spacetimes

Consider a universe with an FLRW metric which describes a spatially homogeneous, isotropic spacetime:

$$ds^2 = -dt^2 + a(t)^2 \left[\frac{dr^2}{1-\kappa r^2} + r^2 \left(d\theta^2 + \sin^2(\theta) d\varphi^2 \right) \right], \quad (1)$$

where κ is the curvature of spacelike three-surfaces and the scale factor $a(t)$ is normalized such that $a(t_1) = 1$ for some time t_1. This metric is a good description of our universe, since from experiments as WMAP and Planck, it follows that our universe is spatially homogeneous and isotropic when averaged over large scales. Geodesics $\gamma(\tau)$, where τ is an affine parameter, satisfy

$$\frac{d\gamma^0}{d\tau} = \frac{\sqrt{|\vec{V}(t_1)|^2 - \epsilon a^2}}{a}, \quad (2)$$

where $|\vec{V}|^2 = g_{ij}\dot{\gamma}^i\dot{\gamma}^j$ and ϵ is the normalization of the geodesic: $\epsilon = 0$ for null geodesics and $\epsilon = -1$ for timelike geodesics. We thus have a past-incomplete geodesic when

$$\int d\tau = \int_{t_0}^{t} \frac{a}{\sqrt{|\vec{V}(t_1)|^2 - \epsilon a^2}} dt \quad (3)$$

for an initial velocity $|\vec{V}(t_1)|$ is finite. Here t_0 is $-\infty$ if $a(t) > 0$ for all t, otherwise $t_0 \in \mathbb{R}$ is taken such that $a(t_0) = 0$. Notice that when $a(t_0) = 0$ for some time t_0, all non-spacelike geodesics are past-incomplete. When $t_0 = -\infty$ and the integral (3) is converging, we cannot immediately conclude that geodesics are past-incomplete. It is possible that we only consider a part of the actual spacetime. An example is given by $\kappa = 0$, and the Hubble parameter $H = \dot{a}/a$ satisfying $\dot{H}/H^2 = 0$, in which case $a(t) = e^{Ht}$ with H constant. If the whole manifold would be covered by these coordinates, it would result in past-incomplete geodesics. However, this model only describes one half, known as the Poincaré patch, of the larger De Sitter space; the whole space is described by choosing $\kappa = 1$, $a(t) = \cosh(Ht)/H$ which yields complete geodesics. See also [9] and [10]. When the integral (3) is diverging one can conclude that geodesics in that specific coordinate patch are past-complete. Of course, one can also assume that a certain model with $t_0 = -\infty$ covers the whole spacetime. Then the past-(in)completeness of a geodesic is determined by the integral (3).

From (3) we see that in spacetimes with $a(t) > A \in \mathbb{R}_{>0}$ all non-spacelike geodesics are past-complete. Hence for a spacetime to have a non-spacelike geodesic that is past-incomplete, $a(t)$ needs to become arbitrarily small.

There are a few theorems that prove that a spacetime contains a (past-)incomplete geodesic. Hawking and Penrose, [1,2], proved theorems that state that when

$$R_{\mu\nu}\dot{\gamma}^\mu\dot{\gamma}^\nu \geq 0 \quad (4)$$

for all geodesics γ and the spacetime obeys a few other conditions such as containing a trapped surface, there is a non-spacelike geodesic that is incomplete. Condition (4) for the metric (1) yields

$$\left(\frac{\ddot{a}}{a} + 2\frac{\dot{a}^2}{a^2} + 2\frac{\kappa}{a^2} \right)\epsilon - 2\left[\frac{\ddot{a}}{a} - \frac{\dot{a}^2}{a^2} - \frac{\kappa}{a^2} \right]\left(\dot{\gamma}^0\right)^2 \geq 0. \quad (5)$$

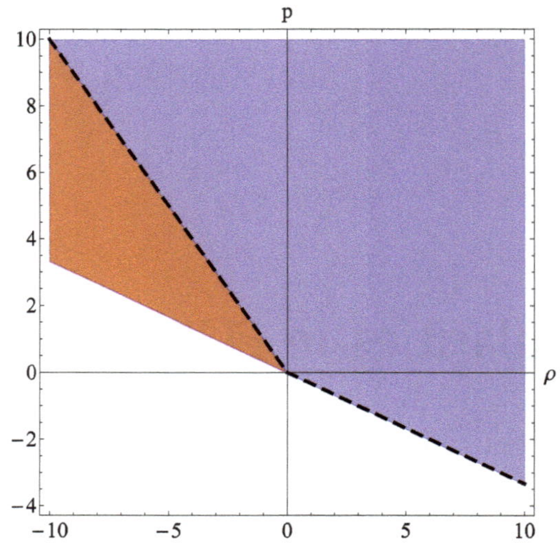

Fig. 1. Illustration of condition (8). For $\kappa < 0$ one needs (ρ, p) in the shaded area above the dashed line to apply the Hawking–Penrose singularity theorems. For $\kappa \geq 0$, we have less restrictions, the shaded area below the dashed line is also included, but it is impossible for an FLRW spacetime with non-negative spatial curvature to be in that area.

Using Eq. (2) one finds that condition (5) becomes

$$\kappa \geq 0 : \ddot{a} \leq 0;$$
$$\kappa < 0 : \begin{cases} \frac{\ddot{a}}{a} - \frac{\dot{a}^2}{a^2} - \frac{\kappa}{a^2} & \leq 0; \\ \ddot{a} & \leq 0. \end{cases} \quad (6)$$

In particular for all κ we need that $\ddot{a} \leq 0$ at all time, or that the spacetime is non-accelerating. Notice that when $\ddot{a} \leq 0$, a will always be zero at some time t_0 (this might be in the future), unless a is a positive constant ($H = 0$) in which case we do not have past-incomplete geodesics. Hence, when we want to use these theorems to say something about an initial singularity in an FLRW spacetime, we need a metric that has a scale parameter a that becomes zero at some time in the past. Describing the matter content of the universe by a perfect fluid

$$T_{\mu\nu} = (\rho + p)U_\mu U_\nu + p g_{\mu\nu}, \quad (7)$$

where p is the pressure, ρ the energy density and $U^\mu = (1, 0, 0, 0)$, the condition (6) translates via the Friedmann equations to

$$\kappa \geq 0 : \rho + 3p \geq 0;$$
$$\kappa < 0 : \begin{cases} \rho + p & \geq 0; \\ \rho + 3p & \geq 0. \end{cases} \quad (8)$$

Although it seems that we have less restrictions when $\kappa \geq 0$, it is impossible that $\rho + p < 0$ and $\rho + 3p \geq 0$ for non-negative spatial curvature. In Fig. 1 one finds an illustration of condition (8).

Another theorem that proves that a geodesic is past-incomplete was published in [5] and is also applicable to spacetimes that have $a(t) > 0$ for all t. It says that when the average Hubble parameter $H = \dot{a}/a$ along a non-spacelike geodesic, H_{av}, satisfies $H_{\text{av}} > 0$, the geodesic must be past-incomplete. For the metric (1), the argument is as follows. Consider a non-spacelike geodesic $\gamma(\tau)$ between an initial point $\gamma(\tau_i)$ and a final point $\gamma(\tau_f)$. We can integrate H along the geodesic, using Eq. (2):

$$\begin{aligned}\int_{\tau_i}^{\tau_f} H d\tau &= \int_{t_i}^{t_f} \frac{\dot{a}}{\sqrt{|\vec{V}(t_1)|^2 - \epsilon a^2}} dt \\ &= \int_{a(t_i)}^{a(t_f)} \frac{da}{\sqrt{|\vec{V}(t_1)|^2 - \epsilon a^2}} \\ &= \begin{cases} \frac{1}{|\vec{V}(t_1)|}\left[a(t_f) - a(t_i)\right], & \epsilon = 0 \\ \log\left(\frac{a(t_f)+\sqrt{|\vec{V}(t_1)|^2+a(t_f)^2}}{a(t_i)+\sqrt{|\vec{V}(t_1)|^2+a(t_i)^2}}\right) & \epsilon = -1 \end{cases} \\ &\leq \begin{cases} \frac{a(t_f)}{|\vec{V}(t_1)|}, & \epsilon = 0 \\ \log\left(\frac{a(t_f)+\sqrt{|\vec{V}(t_1)|^2+a(t_f)^2}}{|\vec{V}(t_1)|}\right) & \epsilon = -1. \end{cases}\end{aligned} \tag{9}$$

Notice that for the second equality sign, one should break up the integration domain into parts where $a = a(t)$ is injective, but that one will end up with the same result. Hence, this integral as function of the initial affine parameter τ_i is restricted by some fixed final τ_f. This means that when

$$H_{av} = \frac{1}{\tau_f - \tau_i} \int_{\tau_i}^{\tau_f} H d\tau > 0 \tag{10}$$

τ_i has to be some finite value such that the geodesic is past-incomplete. Notice that it is still possible to construct an FLRW spacetime that has $H > 0$ at all times and complete geodesics. For this we need that H_{av} must become zero when $\tau_i \to -\infty$. Examples are for instance given by spacetimes with $H > 0$ and $a \to a_0 > 0$ for $t \to -\infty$ (in this case we will have that $H \to 0$ as $t \to -\infty$).

3. Energy of test particles

As stated before, the definition of a singularity is based on the trajectories of massive test particles and massless particles. For cosmological spacetimes with an FLRW metric, we would like to study the energies of test particles over time. We will generalize the argument given in [8] for De Sitter space to a general FLRW spacetime.

Using Eq. (2) we find that for massive test particles

$$|\vec{V}|^2 = g_{ij}\dot{\gamma}^i\dot{\gamma}^j = \epsilon + \left(\dot{\gamma}^0\right)^2 = \frac{|\vec{V}(t_1)|^2}{a^2}. \tag{11}$$

We already saw that in order for a spacetime to have a past-incomplete non-spacelike geodesic, the scale parameter a needs to become arbitrarily small. With Eq. (11) this then implies that when the particle has a velocity $|\vec{V}(t_1)|$ at time t_1, the velocity and hence the energy $E^2 = m^2\left(1 + \frac{|\vec{V}(t_1)|^2}{a^2}\right)$ of a test particle with mass m become arbitrarily large when moving back to the past.

The statement above for massive test particles carries over to photons. In this case the angular frequency as observed by a co-moving observer is

$$\omega = \dot{\gamma}^0 = \frac{\omega(t_1)}{a}. \tag{12}$$

Thus also the energy of photons $E = \hbar\omega$ will become arbitrarily large when moving back to the past.

In [8] it was noted that one cannot have particles with arbitrarily high energies because if such a particle has a nonvanishing interaction cross section with any particle with a non-zero physical number density, then the particle will interact with an infinite number of them, breaking the Cosmological principle. However, the particle's energy cannot become arbitrarily high because it will reach the Planck energy $E_P = \sqrt{\frac{\hbar}{G}} \approx 1.22 \cdot 10^{19}$ GeV at some time t. With this energy, the particle's Compton wavelength is approximately equal to its Schwarzschild radius such that it will form a black hole. Therefore, the description of the particle's trajectory will break down. Scattering processes involving vacuum fluctuations may cause the test particle's energy to never reach the Planck energy. If these processes are significant the particle's trajectory is not a geodesic anymore. Near the Planck energy scattering processes are dominated by processes that involve the exchange of a graviton [11]. To estimate this effect we consider photon–photon scattering with the exchange of a graviton. We model the loss of energy of the photon when going back in time as

$$\frac{d}{dt}E = (-H - \sigma n)E, \tag{13}$$

where n is the number density of virtual photons and σ is the cross section of the scattering process. The particle gains energy from the expansion of the universe because $-H$ is positive (when going back in time) and it looses energy from the scattering with virtual photons. We estimate the density of virtual photons as one per Hubble volume:

$$n = \frac{1}{V_H} = -\frac{3H^3}{4\pi}. \tag{14}$$

The differential cross section for photon–photon scattering with the exchange of a graviton for unpolarized photons is [12]

$$\frac{d\sigma}{d\Omega} = \frac{\kappa^4}{8\pi^2}\frac{k^2}{\sin^2(\theta)}\left[1 + \cos^{16}\left(\frac{1}{2}\theta\right) + \sin^{16}\left(\frac{1}{2}\theta\right)\right] \tag{15}$$

where $\kappa = \sqrt{16\pi G}$, k is the momentum of the photon and θ is the scattering angle. Since we are primarily interested in large momentum exchange, we neglect small angle scatterings when calculating the total cross section of this process:

$$\begin{aligned}\sigma &= \int \frac{d\sigma}{d\Omega} d\Omega \\ &= \frac{\kappa^4}{\pi}\frac{k^2}{4}\int_{-1+\xi}^{1-\xi} \frac{1 + \frac{1}{256}(1+x)^8 + \frac{1}{256}(1-x)^8}{1-x^2} dx \\ &= \frac{\kappa^4}{\pi}\frac{k^2}{2}\int_{\xi}^{1} \frac{1 + \frac{1}{256}(2-y)^8 + \frac{1}{256}y^8}{y(2-y)} dy \\ &= \frac{\kappa^4}{\pi}\frac{k^2}{4}\left[2\log\frac{1}{\xi} - \frac{363}{140} + \log(4) + \mathcal{O}(\xi)\right],\end{aligned} \tag{16}$$

where we have the relation $\sin(\theta/2) = \sqrt{\xi/2}$. Taking only angles $.26\pi < \theta < .74\pi$ into account for the scattering, we have that $2\log\frac{1}{\xi} - \frac{363}{140} + \log(4) \approx 1$. With Eqs. (13), (14) and (16) we find that the energy of the test photon does not increase when

$$H \sim \sigma n = 48G^2E^2H^3, \tag{17}$$

where $E = k$ is the photon energy. Using the Hubble parameter of cosmic inflation which typically is about $-\hbar H \approx 10^{13}$ GeV, we find from (17) that the scattering process becomes significant when

$$\left(\frac{E}{E_P}\right)^2 \sim \frac{E_P^2}{48\hbar^2H^2} \approx 10^{10}. \tag{18}$$

Hence, processes involving gravitons will not cause the particle's energy to stay smaller than the Planck energy and a black hole will form. This implies that the description of the particle's trajectory (as a geodesic) breaks down, either because of interaction processes or by the formation of a black hole. The latter definitely happens when the initial energy is near the Planck energy.

Up to now, the maximum energy of a single particle that has been measured is of the order of 10^{20} eV [13] which is eight orders of magnitude smaller than the Planck scale. These particles were all cosmic ray particles, so their probable origin is a supernova, an active galactic nucleus, a quasar or a gamma ray-burst. Even when using this energy as an upper bound for the energy of test particles, we have that the description of the trajectories of non-comoving test particles breaks down at times that are certainly later than the Planck era, the period where we have to take quantum gravitational effects into account. In [8] the arbitrarily high energies of test particles were used to argue that these particles should be forbidden in De Sitter space. This can be done by using a different time arrow in the two patches of De Sitter space that one has in the flat slicing. That way the two coordinate patches become non-communicating and describe eternally inflating spacetimes. We will not look into these kind of constructions for general FLRW spacetimes but we want to use the arbitrarily high energies of test particles to give a consistent definition of a singularity. When the particle's description breaks down before it reaches the beginning of its trajectory, it is not very useful to use that particle as an indication for an initial singularity. That is the reason why we suggest to define a singularity in spacetimes with an FLRW metric that has a parameter a that becomes arbitrarily small, as a time-like geodesic with $|\vec{V}(t_1)| = 0$ that is past-incomplete. For such trajectories, we have that $dt = d\tau$ which means that a spacetime has no initial singularity when $a(t) > 0$ for all $t \in \mathbb{R}$. Hence, an FLRW spacetime starts from a singularity precisely when $a(t_0) = 0$ at some initial finite time t_0.

4. Conclusion

We pointed out that spacetimes with an FLRW metric such that $a(t) > 0$ for all $t \in \mathbb{R}$ have no initial singularity. This was done by first observing that in models that have $a(t) > A \in \mathbb{R}_{>0}$ all non-spacelike geodesics are past-complete. When a becomes arbitrarily small, it is possible that the spacetime contains a past-incomplete geodesic. With the usual definition of a singularity, this means that the spacetime has an initial singularity. However, that definition is based on a test particle that has that geodesic as trajectory. We pointed out that when this particle has an initial velocity, its energy will become super-Planckian at some time in the past if it kept following that geodesic. This means that the particle stops being a test particle and it does not matter that its trajectory is past-incomplete. For a model in which the scale factor becomes arbitrarily small, we should define an initial singularity as a trajectory of a comoving particle that is past-incomplete. This implies that the only FLRW spacetimes with an initial singularity are the ones such that $a(t_0) = 0$ at some initial time t_0. Hence, bouncing spacetimes and past-eternal inflationary models do not start from a singularity. One can use similar arguments to show that the only FLRW spacetimes that have a singularity in the future are the ones that have a scale factor such that $a(t)$ vanishes at some time in the future. It would be interesting to examine if similar results hold for universes that are obtained by perturbating an FLRW spacetime.

Acknowledgements

This work was supported in part by the D-ITP consortium, a program of the Netherlands Organization for Scientific Research (NWO) that is funded by the Dutch Ministry of Education, Culture and Science (OCW), and by the NWO Graduate Programme.

References

[1] S. Hawking, G. Ellis, The Large Scale Structure of Space–Time, Cambridge Monogr. Math. Phys., Cambridge University Press, 1973.
[2] S.W. Hawking, R. Penrose, The singularities of gravitational collapse and cosmology, Proc. R. Soc. Lond. A 314 (1970) 529–548.
[3] A.A. Starobinsky, A new type of isotropic cosmological models without singularity, Phys. Lett. B 91 (1980) 99–102.
[4] A.H. Guth, The inflationary universe: a possible solution to the horizon and flatness problems, Phys. Rev. D 23 (1981) 347–356.
[5] A. Borde, A.H. Guth, A. Vilenkin, Inflationary space-times are incomplete in past directions, Phys. Rev. Lett. 90 (2003) 151301, arXiv:gr-qc/0110012.
[6] G.F.R. Ellis, R. Maartens, The emergent universe: inflationary cosmology with no singularity, Class. Quantum Gravity 21 (2004) 223–232, arXiv:gr-qc/0211082.
[7] A. Mithani, A. Vilenkin, Did the universe have a beginning?, arXiv:1204.4658.
[8] A. Aguirre, S. Gratton, Steady state eternal inflation, Phys. Rev. D 65 (2002) 083507, arXiv:astro-ph/0111191.
[9] A. Aguirre, S. Gratton, Inflation without a beginning: a null boundary proposal, Phys. Rev. D 67 (2003) 083515, arXiv:gr-qc/0301042.
[10] A. Aguirre, Eternal inflation, past and future, arXiv:0712.0571.
[11] G. 't Hooft, Graviton dominance in ultrahigh-energy scattering, Phys. Lett. B 198 (1987) 61–63.
[12] D. Boccaletti, V. De Sabbata, P. Fortini, C. Gualdi, Photon–photon scattering and photon-scalar particle scattering via gravitational interaction (one-graviton exchange) and comparison of the processes between classical (general-relativistic) theory and the quantum linearized field theory, Nuovo Cimento B 60 (1969) 320–330.
[13] A. Aab, et al., Measurement of the cosmic ray spectrum above 4×10^{18} eV using inclined events detected with the Pierre Auger observatory, J. Cosmol. Astropart. Phys. 1508 (2015) 049, arXiv:1503.07786.

Strong deflection lensing by a Lee–Wick black hole

Shan-Shan Zhao [a,b], Yi Xie [a,b,*]

[a] School of Astronomy and Space Science, Nanjing University, Nanjing 210023, China
[b] Key Laboratory of Modern Astronomy and Astrophysics, Nanjing University, Ministry of Education, Nanjing 210093, China

ARTICLE INFO

Editor: M. Cvetič

ABSTRACT

We study strong deflection gravitational lensing by a Lee–Wick black hole, which is a non-singular black hole generated by a high derivative modification of Einstein–Hilbert action. The strong deflection lensing is expected to produce a set of relativistic images very closed to the event horizon of the black hole. We estimate its observables for the supermassive black hole in our Galactic center. It is found that the Lee–Wick black hole can be distinguished from the Schwarzschild black hole via such lensing effects when the UV scale is not very large, but the requiring resolution is much higher than current capability.

1. Introduction

A singularity is the point where Einstein's general relativity breaks down. It exists at the center of a black hole and appears at the beginning of the Big Bang. A quantum theory of gravity is believed to be able to remove these singularities. A non-singular black hole, also called a regular black hole, which was firstly discussed by Bardeen [1], refers to a sort of black holes without the singularities in their center.

In 1977, Stelle originally proposed a renormalizable and asymptotically free theory of higher-derivative quantum gravity [2], but it also suffers from existence of a massive ghost state. Weakly and non-locally modified theories of gravity [3–10] are self-consistent but likely have infinite complex conjugate poles [11]. A new local higher derivative theory without real poles was recently proposed [12–16], which is in agreement with the prescription of Lee–Wick model [17–19]. The Lee–Wick black hole is currently found out and it is proved to be non-singular [20].

While thermodynamics of the Lee–Wick black hole was discussed [20], investigations on its astrophysical properties and its resulting observability are still absent, especially for effects of its strong gravitational field. In this work, we will study strong deflection gravitational lensing by such a Lee–Wick black hole. Gravitational lensing in the strong gravitational field can produce an unique feature called relativistic images [21]. Relativistic images are a class of infinite discrete images existing on the two sides of the lens, due to photons winds several loops before escaping from the lens. If the source have timing signals, the time delay between different relativistic images can also be detected. These lensing effects belong to the strong deflection gravitational lensing (see [22] for a review).

Since the relativistic images are very close to the lens, direct observation of such lensing effects needs a very high angular resolution. The supermassive black hole in the center of the Milky Way, Sagittarius A* (Sgr A*), has the largest apparent angular diameter (i.e., shadow) ~ 50 microarcsecond (μas) among all the black holes known in the universe [23]. A global sub-mm very long baseline interferometry network, called Event Horizon Telescope (EHT), will firstly and soon give a direct image of Sgr A*, which can provide a new fundamental laboratory for testing black hole theories as well as gravity in the strong field regime [24–27].

In Sect. 2, the spacetime of a Lee–Wick black hole [20] will be briefly reviewed. We study its strong deflection gravitational lensing by using the method of strong deflection limit (SDL) [28] in Sect. 3. Taking Sgr A* as an example, we estimate numerical values of observables of the lensing in Sect. 4. In Sect. 5, conclusions will be represented.

2. Lee–Wick black hole

The action of the Lee–Wick theory is [12–16]

$$S=\frac{1}{8\pi G_N}\int d^4x\sqrt{|g|}\left[R+\Lambda^{-4}G_{\mu\nu}\Box R^{\mu\nu}\right], \tag{1}$$

where G_N is the gravitational constant, Λ refers to the UV scale and is expected but not necessary to be in the same order of Planck mass. By solving the approximated exact equations of motion, one can get a static and spherically symmetric black hole solution generated by a point-like mass source M. The spacetime is [20]

* Corresponding author.
E-mail address: yixie@nju.edu.cn (Y. Xie).

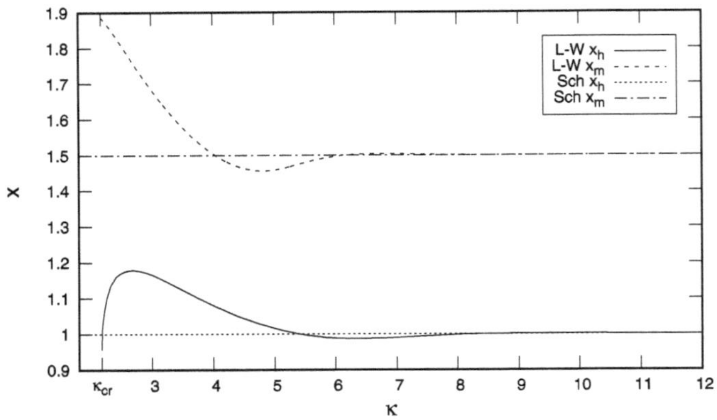

Fig. 1. Radii of event horizon (x_h) and photon sphere (x_m) of a Lee–Wick black hole and a Schwarzschild black hole. The horizontal axis is the model parameter κ starting from κ_{cr} and the vertical axis is the radial coordinate x. The solid line and the dashed line refer respectively to x_h and x_m of a Lee–Wick black hole, while $x_h = 1$ and $x_m = 1.5$ for a Schwarzschild black hole which are plotted with the dotted line and the dot-dashed line.

$$ds^2 = A(x)dt^2 - B(x)dx^2 - C(x)(d\theta^2 + \sin^2\theta d\phi^2), \tag{2}$$

where we use $2M$ as the unit of the length dimension and

$$A(x) = 1 - \frac{m(\chi)}{x}, \tag{3}$$

$$B(x) = \left[1 - \frac{m(\chi)}{x}\right]^{-1}, \tag{4}$$

$$C(x) = x^2. \tag{5}$$

The effective mass function $m(\chi)$ is

$$m(\chi) = 1 - e^{-\chi}\left[(1+\chi)\cos\chi + \chi\sin\chi\right], \tag{6}$$

where $\chi = \kappa x$ and $\kappa = \Lambda/\sqrt{2}$.

The model parameter κ should be larger than a critical value to ensure the existence of the event horizon(s). The critical point occurs when $A(x) = 0$ has only one real solution x_{cr} which can be numerically found as

$$\kappa_{cr} \approx 2.165 \quad \text{and} \quad x_{cr} \approx 0.953. \tag{7}$$

When $\kappa > \kappa_{cr}$, a Lee–Wick black hole have two event horizons: an outer one and an inner one. These two horizons merge to one as κ equals κ_{cr}. If $\kappa < \kappa_{cr}$, no horizon can survive. Since the spacetime metric (2) has no central singularity, the nonexistence of the event horizon does not break down the weak cosmic censorship conjecture. However, it was found [29] that destroying the event horizon of a regular black hole can violate the black hole's area theorem [30], causing the energy released in the collision of two black holes to exceed the Hawking bound; and such a violation is not supported by the current observations of gravitational waves [31,32]. Therefore, we only consider cases with $\kappa \geq \kappa_{cr}$ which were called Lee–Wick black hole solutions in [20]. The radius of the (outer) event horizon is plotted in Fig. 1 against different values of κ by the solid line. When κ becomes large enough, the Lee–Wick spacetime is approaching the Schwarzschild one.

3. Lensing in SDL

The strong deflection gravitational lensing under SDL can be described by two equations: an lens equation and a SDL equation for the deflection angle. The lens equation geometrically determines relationships among the observer, the lens and the light source. Assuming that the source and the observer are far from the lens and they are nearly aligned, we choose the asymptotically flat lens equation as [33]

$$\beta = \theta - \frac{D_{LS}}{D_{OS}}\Delta\alpha_n, \tag{8}$$

where $\Delta\alpha_n$ is the extra deflection angle of a photon looping over $2n\pi$; β is the angular separation between the source and the lens; θ is the angular separation between the image and the lens; the projected distances of the lens to the source and the observer to the lens are D_{LS} and D_{OS}.

The exact deflection angle for the spacetime (2) is given by [34, 35]

$$\alpha(x_0) = -\pi + \int_{x_0}^{\infty} \frac{2\sqrt{B(x)}}{\sqrt{C(x)}\sqrt{\frac{C(x)}{C_0}\frac{A_0}{A(x)} - 1}} dx, \tag{9}$$

where x_0 is the closest approach distance of the winding photon. A quantity with a subscript 0 means its corresponding value taking $x = x_0$. In order to deal with the divergence of Eq. (9) as x_0 approaching the photon sphere, we can expand the integral near the photon sphere by the method of SDL [28]. The photon sphere is the innermost orbit for a winding photon, which is defined by [36,37]

$$\frac{C'(x)}{C(x)} = \frac{A'(x)}{A(x)}. \tag{10}$$

The photon sphere of a Lee–Wick black hole with varying κ can be found in Fig. 1 and it is plotted with the dashed line.

The deflection angle in the SDL can be written as [28]

$$\alpha(\theta) = -\bar{a}\log\left(\frac{\theta D_{OL}}{u_m} - 1\right) + \bar{b} + \mathcal{O}[(u - u_m)\log(u - u_m)], \tag{11}$$

where $u = \sqrt{C_0/A_0}$ is the impact parameter and $u \approx \theta D_{OL}$. The subscript m means evaluating at $x = x_m$. $\bar{a}$ and $\bar{b}$ are the SDL coefficients expressed by [28]

$$\bar{a} = \frac{R_m}{2\sqrt{\beta_m}}, \tag{12}$$

$$\bar{b} = -\pi + b_R + \bar{a}\ln\frac{2\beta_m}{A_m}, \tag{13}$$

where

$$\beta_m = \frac{C_m(1 - A_m)^2\left(A_m C''_m - C_m A''_m\right)}{2A_m^2 C'^2_m}, \tag{14}$$

$$R_m = \frac{2(1 - A_m)\sqrt{A_m B_m}}{A'_m\sqrt{C_m}}, \tag{15}$$

$$b_R = \int_0^1 \left[\frac{2(1 - A_m)\sqrt{A(z)B(z)}}{A'(z)C(z)\sqrt{\frac{A_m}{C_m} - \frac{A(z)}{C(z)}}} - \frac{R_m}{z\sqrt{\beta_m}}\right] dz. \tag{16}$$

Here, z refers to a new variable $z = (A(x) - A_m)/(1 - A_m)$; $'$ and $''$ are the operators of taking once and twice derivatives against x.

We can also detect the time delay between different relativistic images from a time-varying light source. Supposing a photon traveling from the source to the observer, the total time span is [38]

$$T = \tilde{T}(x_0) - \int_{D_{OL}}^{\infty}\left|\frac{dt}{dx}\right| dx - \int_{D_{LS}}^{\infty}\left|\frac{dt}{dx}\right| dx, \tag{17}$$

where D_{OL} is the projected distance between the observer and the source. The second and third terms at the right hand side can be worked out since both of the observer and the source are far from the lens. The first term is given by [35,38,39]

$$\tilde{T}(x_0) = \int_{x_0}^{\infty} \frac{2\sqrt{B(x)C(x)A_0}}{A(x)\sqrt{\frac{C(x)}{C_0}\frac{A_0}{A(x)} - 1}} dx. \tag{18}$$

It can be also expanded in the SDL

$$\tilde{T}(u) = -\tilde{a}\ln\left(\frac{u}{u_m} - 1\right) + \tilde{b} + \mathcal{O}[(u - u_m)\log(u - u_m)], \tag{19}$$

where $\tilde{a}$ and $\tilde{b}$ are SDL coefficients and $\tilde{a} = \bar{a}\, u_m$ for the spacetime (2) [38].

Having the lens equation (8) and the deflection angle (11) as well as the time delay (19) in the SDL, we can calculate the observables of relativistic images including angular separation, brightness difference and time delay.

If the first image can be distinguished from others, three characteristic observables can be detected and related to SDL coefficients as [28]

$$\theta_\infty = \frac{u_m}{D_{\mathrm{OL}}}, \tag{20}$$

$$s = \theta_\infty \exp\left(\frac{\bar{b}}{\bar{a}} - \frac{2\pi}{\bar{a}}\right), \tag{21}$$

$$r = 2.5\log_{10}\left[\exp\left(2\pi/\bar{a}\right)\right], \tag{22}$$

where θ_∞ is the apparent radius of the photon sphere; s and r are the angular separation and brightness difference between the first image and other packed images.

If we can distinguish the time signals of the first image from those of the second image, the delay of the two signals $\Delta T_{2,1}$ can be also calculated by SDL coefficients as [38]

$$\Delta T_{2,1} = \Delta T^0_{2,1} + \Delta T^1_{2,1}, \tag{23}$$

where

$$\Delta T^0_{2,1} = 2\pi u_m, \tag{24}$$

$$\Delta T^1_{2,1} = 2\sqrt{\frac{B_m}{A_m}}\sqrt{\frac{u_m}{c_m}}\exp\left(\frac{\bar{b}}{\bar{a}}\right) \times \left[\exp\left(-\frac{\pi}{\bar{a}}\right) - \exp\left(-\frac{2\pi}{\bar{a}}\right)\right] \tag{25}$$

with

$$c_m = \beta_m\sqrt{\frac{A_m}{C_m^3}\frac{C_m'^2}{2(1 - A_m)^2}}. \tag{26}$$

4. Observables for Sgr A*

Supposing a Lee–Wick black hole with the same mass and the same distance of Sgr A*, we can numerically estimate all these observables in the SDL. From top to bottom, Fig. 2 respectively shows the apparent radius of the photon sphere θ_∞, the angular separation s between the first image and the other packed images and their brightness difference r against the variation of κ. The time delay between the first and the second images $\Delta T_{2,1}$, the corrected term $\Delta T^1_{2,1}$ and their ratio $\eta_{2,1} = \Delta T^1_{2,1}/\Delta T_{2,1}$ can be found in Fig. 3. For comparison, these observables for a Schwarzschild black hole are denoted by dashed lines in Figs. 2 and 3.

As we can see in Fig. 2, the smaller is κ, the larger is the deviation of the observables from their corresponding values for a Schwarzschild black hole. The curve of θ_∞ looks like a damped oscillation as κ increases, and it approaches its Schwarzschild value 26.54 µas very fast when κ is getting sufficiently big. Its maximum value of 29.96 µas occurs at κ_{cr} and its minimum is 26.18 µas at $\kappa = 4.2$. Supposing we have an angular resolution of 0.1 µas which is far beyond the current technology, a Lee–Wick black hole with $\kappa \lesssim 5.2$ can potentially be distinguished from a Schwarzschild black hole by measuring θ_∞. For a Lee–Wick black hole with a larger κ, the difference between its θ_∞ and the Schwarzschild one can hardly be detected with such a resolution. In this situation, three other observables, s, r and $\Delta T_{2,1}$, might be helpful for determining κ if the first image can be separated from other relativistic images, which demands that the angular resolution has

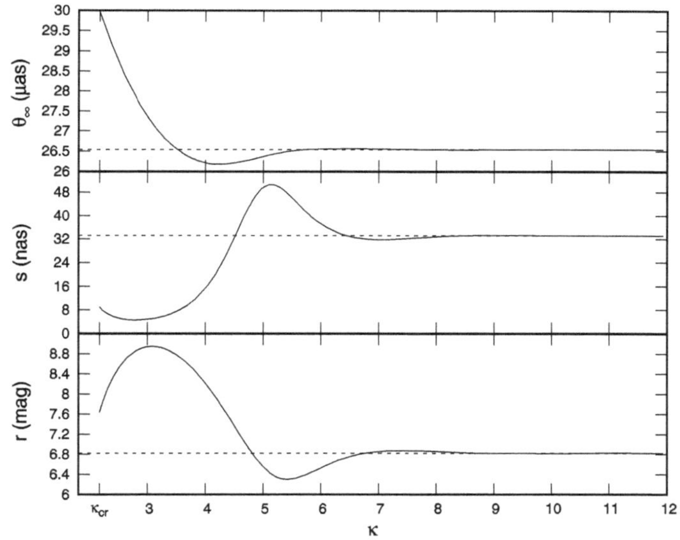

Fig. 2. Estimated θ_∞ (top panel), s (middle panel) and r (bottom panel) of a Lee–Wick black hole with the same mass and distance as Sgr A*: $D_{\mathrm{OL}} = 8.33$ kpc and $M_\bullet = 4.31 \times 10^6 M_\odot$ [40]. The solid lines show variation of these observables against κ, while the dashed lines refer to those of a Schwarzschild black hole: $\theta_\infty = 26.54$ µas, $s = 33.32$ nas, $r = 6.822$ mag.

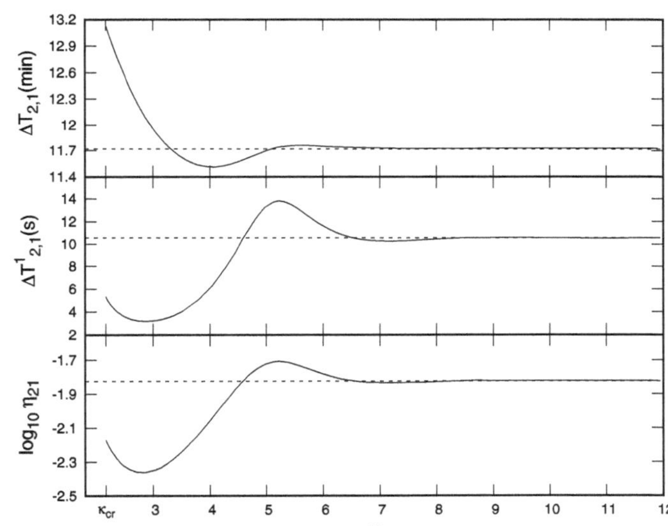

Fig. 3. Estimated time delays $\Delta T_{2,1}$ (top panel), corrected term $\Delta T^1_{2,1}$ (middle panel) and the ratio $\log_{10}\eta_{2,1}$ (bottom panel) of a Lee–Wick black hole with the same mass and distance as Sgr A*: $D_{\mathrm{OL}} = 8.33$ kpc and $M_\bullet = 4.31 \times 10^6 M_\odot$ [40]. The solid lines show variation of these observables against κ, while the dashed lines refer to those of a Schwarzschild black hole: $\Delta T_{2,1} = 11.72$ min, $\Delta T^1_{2,1} = 10.56$ s and $\log_{10}\eta_{2,1} = -1.8$.

to be improved to the level better than 10 nanoarcsecond (nas). The range of s varies from 4.681 nas to 50.65 nas and r changes from 6.302 mag to 8.947 mag. The main part of $\Delta T_{2,1}$ contributes no new information because it is proportional to θ_∞. But if the timing precision reaches 1 s, the corrected term $\Delta T^1_{2,1}$ can be resolved. Fig. 3 also shows that a larger $\Delta T^1_{2,1}$ has a higher ratio to the total time delay.

The Lee–Wick black hole we have discussed above is a non-rotating one. The profile of its shadow due to the strong deflection gravitational lensing is a circle with angular radius θ_∞. However, an astrophysical black hole is very likely spinning. The radius θ_∞ we obtained here might not be able to constrain possible deviations from Einstein's general relativity with the on-going experiments because strong deflection lensing by a highly rotating black hole [41] will be qualitatively different from the one by a non-rotating one [42,43]. In order to describe the spacetime of a rotating Lee–Wick black hole and its strong deflection lensing in a self-consistent way, its metric is indispensably needed but the solution of such a metric is absent for now. Nevertheless, based on strong deflection lensing by a Kerr black hole [41], we can intuitively expect that the angular momentum of a rotating Lee–Wick black hole would make the shadow distorted and displaced and the shape of such a shadow would also depend on κ. Furthermore, it would drift the caustics away from the optical axis, make the caustic with a finite extension and cause only one image visible instead of two sets of relativistic images [44]. We will leave the construction of the metric of a rotating Lee–Wick black hole and the detailed investigation on its strong deflection lensing for future works.

On the practical aspect, the determination of the model parameter κ through observation is another complicated work. For the mm-VLBI observation of Sgr A*, κ can be determined via comparing the scale and shape of the shadow between the observed image and the modeled image, which is constructed by general relativity magnetohydrodynamics simulation [45,46]. In this work, we just give a direct sense of what the relativistic images by a (non-rotating) Lee–Wick black hole will look like in an analytical approach; a fully numerical simulation of the physical process for a rotating Lee–Wick black hole and a sophisticated determination on the model parameter(s) by upcoming direct observation are beyond our scope.

5. Conclusions

We study the strong deflection gravitational lensing by a Lee–Wick black hole through the SDL method and estimate the observables of its relativistic images. By calculating the SDL coefficients numerically and making estimations by assuming that the black hole has the same mass and distance as Sgr A*, we show how its observables change with respect to the model parameter κ, which has a critical value $\kappa_{\rm cr} = 2.165$ for allowing existence of the event horizon. We find that, when $\kappa = \kappa_{\rm cr}$, the apparent radius of the photon sphere θ_∞ can reach $\sim$30 μas which is larger than the Schwarzschild one. But, as κ increases, the value of θ_∞ approaches the one of Schwarzschild black hole very fast. If the first relativistic image could be resolved from the others, their angular separation, brightness differences and possible time delay signals will be helpful for determining κ, although it demands technology far beyond current stage. In light of current projects such as EHT, these results might provide useful clues for detecting the Lee–Wick black hole.

Acknowledgement

This work is funded by the National Natural Science Foundation of China (Grant No. 11573015).

References

[1] J. Bardeen, Non-singular general-relativistic gravitational collapse, in: Proceedings of International Conference GR5, Tbilisi University Press, Tbilisi, USSR, 1968, p. 174.
[2] K.S. Stelle, Renormalization of higher-derivative quantum gravity, Phys. Rev. D 16 (1977) 953–969, http://dx.doi.org/10.1103/PhysRevD.16.953.
[3] L. Modesto, Super-renormalizable quantum gravity, Phys. Rev. D 86 (4) (2012) 044005, http://dx.doi.org/10.1103/PhysRevD.86.044005, arXiv:1107.2403.
[4] L. Modesto, L. Rachwał, Super-renormalizable and finite gravitational theories, Nucl. Phys. B 889 (2014) 228–248, http://dx.doi.org/10.1016/j.nuclphysb.2014.10.015, arXiv:1407.8036.
[5] C. Bambi, D. Malafarina, L. Modesto, Terminating black holes in asymptotically free quantum gravity, Eur. Phys. J. C 74 (2014) 2767, http://dx.doi.org/10.1140/epjc/s10052-014-2767-9, arXiv:1306.1668.
[6] Y. Zhang, Y. Zhu, L. Modesto, C. Bambi, Can static regular black holes form from gravitational collapse?, Eur. Phys. J. C 75 (2015) 96, http://dx.doi.org/10.1140/epjc/s10052-015-3311-2, arXiv:1404.4770.
[7] V.P. Frolov, Information loss problem and a 'black hole' model with a closed apparent horizon, J. High Energy Phys. 05 (2014) 49, http://dx.doi.org/10.1007/JHEP05(2014)049, arXiv:1402.5446.
[8] V.P. Frolov, A. Zelnikov, T. de Paula Netto, Spherical collapse of small masses in the ghost-free gravity, J. High Energy Phys. 06 (2015) 107, http://dx.doi.org/10.1007/JHEP06(2015)107, arXiv:1504.00412.
[9] E.T. Tomboulis, Renormalization and unitarity in higher derivative and nonlocal gravity theories, Mod. Phys. Lett. A 30 (2015) 1540005, http://dx.doi.org/10.1142/S0217732315400052.
[10] E.T. Tomboulis, Nonlocal and quasilocal field theories, Phys. Rev. D 92 (12) (2015) 125037, http://dx.doi.org/10.1103/PhysRevD.92.125037, arXiv:1507.00981.
[11] I.L. Shapiro, Counting ghosts in the "ghost-free" non-local gravity, Phys. Lett. B 744 (2015) 67–73, http://dx.doi.org/10.1016/j.physletb.2015.03.037, arXiv:1502.00106.
[12] L. Modesto, I.L. Shapiro, Superrenormalizable quantum gravity with complex ghosts, Phys. Lett. B 755 (2016) 279–284, http://dx.doi.org/10.1016/j.physletb.2016.02.021, arXiv:1512.07600.
[13] L. Modesto, Super-renormalizable or finite Lee–Wick quantum gravity, Nucl. Phys. B 909 (2016) 584–606, http://dx.doi.org/10.1016/j.nuclphysb.2016.06.004, arXiv:1602.02421.
[14] G.P. de Brito, P.I.C. Caneda, Y.M.P. Gomes, J.T. Guaitolini Junior, V. Nikoofard, Effective models of quantum gravity induced by Planck scale modifications in the covariant quantum algebra, arXiv:1610.01480.
[15] A. Accioly, B.L. Giacchini, I.L. Shapiro, Low-energy effects in a higher-derivative gravity model with real and complex massive poles, arXiv:1610.05260.
[16] B.L. Giacchini, On the cancellation of Newtonian singularities in higher-derivative gravity, Phys. Lett. B 766 (2017) 306–311, http://dx.doi.org/10.1016/j.physletb.2017.01.019, arXiv:1609.05432.
[17] T.D. Lee, G.C. Wick, Negative metric and the unitarity of the S-matrix, Nucl. Phys. B 9 (1969) 209–243, http://dx.doi.org/10.1016/0550-3213(69)90098-4.
[18] T.D. Lee, G.C. Wick, Finite theory of quantum electrodynamics, Phys. Rev. D 2 (1970) 1033–1048, http://dx.doi.org/10.1103/PhysRevD.2.1033.
[19] R.E. Cutkosky, P.V. Landshoff, D.I. Olive, J.C. Polkinghorne, A non-analytic S-matrix, Nucl. Phys. B 12 (1969) 281–300, http://dx.doi.org/10.1016/0550-3213(69)90169-2.
[20] C. Bambi, L. Modesto, Y. Wang, Lee–Wick black holes, Phys. Lett. B 764 (2017) 306–309, http://dx.doi.org/10.1016/j.physletb.2016.11.060, arXiv:1611.03650.
[21] C. Darwin, The gravity field of a particle, Proc. R. Soc. Lond. Ser. A 249 (1959) 180–194, http://dx.doi.org/10.1098/rspa.1959.0015.
[22] V. Bozza, Gravitational lensing by black holes, Gen. Relativ. Gravit. 42 (2010) 2269–2300, http://dx.doi.org/10.1007/s10714-010-0988-2, arXiv:0911.2187.
[23] H. Falcke, S.B. Markoff, Toward the event horizon – the supermassive black hole in the Galactic Center, Class. Quantum Gravity 30 (24) (2013) 244003, http://dx.doi.org/10.1088/0264-9381/30/24/244003, arXiv:1311.1841.
[24] D. Psaltis, Probes and tests of strong-field gravity with observations in the electromagnetic spectrum, Living Rev. Relativ. 11 (2008) 9, http://dx.doi.org/10.12942/lrr-2008-9, arXiv:0806.1531.
[25] C. Goddi, H. Falcke, M. Kramer, L. Rezzolla, C. Brinkerink, T. Bronzwaer, R. Deane, M. De Laurentis, G. Desvignes, J.R.J. Davelaar, F. Eisenhauer, R. Eatough, R. Fraga-Encinas, C.M. Fromm, S. Gillessen, A. Grenzebach, S. Issaoun, M. Janßen, R. Konoplya, T.P. Krichbaum, R. Laing, K. Liu, R.-S. Lu, Y. Mizuno, M. Moscibrodzka, C. Müller, H. Olivares, O. Porth, O. Pfuhl, E. Ros, F. Roelofs, K. Schuster, R. Tilanus, P. Torne, I. van Bemmel, H.J. van Langevelde, N. Wex, Z. Younsi, A. Zhidenko, BlackHoleCam: fundamental physics of the Galactic center, arXiv:1606.08879.
[26] D. Psaltis, F. Özel, C.-K. Chan, D.P. Marrone, A general relativistic null hypothesis test with event horizon telescope observations of the black hole shadow in Sgr A*, Astrophys. J. 814 (2015) 115, http://dx.doi.org/10.1088/0004-637X/814/2/115, arXiv:1411.1454.

[27] A.E. Broderick, T. Johannsen, A. Loeb, D. Psaltis, Testing the no-hair theorem with event horizon telescope observations of Sagittarius A*, Astrophys. J. 784 (2014) 7, http://dx.doi.org/10.1088/0004-637X/784/1/7, arXiv:1311.5564.

[28] V. Bozza, Gravitational lensing in the strong field limit, Phys. Rev. D 66 (10) (2002) 103001, http://dx.doi.org/10.1103/PhysRevD.66.103001, arXiv:gr-qc/0208075.

[29] Z. Li, C. Bambi, Destroying the event horizon of regular black holes, Phys. Rev. D 87 (12) (2013) 124022, http://dx.doi.org/10.1103/PhysRevD.87.124022, arXiv:1304.6592.

[30] S.W. Hawking, Gravitational radiation from colliding black holes, Phys. Rev. Lett. 26 (1971) 1344–1346, http://dx.doi.org/10.1103/PhysRevLett.26.1344.

[31] B.P. Abbott, R. Abbott, T.D. Abbott, M.R. Abernathy, F. Acernese, K. Ackley, C. Adams, T. Adams, P. Addesso, R.X. Adhikari, et al., Observation of gravitational waves from a binary black hole merger, Phys. Rev. Lett. 116 (6) (2016) 061102, http://dx.doi.org/10.1103/PhysRevLett.116.061102, arXiv:1602.03837.

[32] B.P. Abbott, R. Abbott, T.D. Abbott, M.R. Abernathy, F. Acernese, K. Ackley, C. Adams, T. Adams, P. Addesso, R.X. Adhikari, et al., GW151226: observation of gravitational waves from a 22-solar-mass binary black hole coalescence, Phys. Rev. Lett. 116 (24) (2016) 241103, http://dx.doi.org/10.1103/PhysRevLett.116.241103, arXiv:1606.04855.

[33] V. Bozza, S. Capozziello, G. Iovane, G. Scarpetta, Strong field limit of black hole gravitational lensing, Gen. Relativ. Gravit. 33 (2001) 1535–1548, http://dx.doi.org/10.1023/A:1012292927358, arXiv:gr-qc/0102068.

[34] K.S. Virbhadra, D. Narasimha, S.M. Chitre, Role of the scalar field in gravitational lensing, Astron. Astrophys. 337 (1998) 1–8, arXiv:astro-ph/9801174.

[35] S. Weinberg, Gravitation and Cosmology: Principles and Applications of the General Theory of Relativity, Wiley, New York, 1972.

[36] K.S. Virbhadra, G.F.R. Ellis, Schwarzschild black hole lensing, Phys. Rev. D 62 (8) (2000) 084003, http://dx.doi.org/10.1103/PhysRevD.62.084003, arXiv:astro-ph/9904193.

[37] C.-M. Claudel, K.S. Virbhadra, G.F.R. Ellis, The geometry of photon surfaces, J. Math. Phys. 42 (2001) 818–838, http://dx.doi.org/10.1063/1.1308507, arXiv:gr-qc/0005050.

[38] V. Bozza, L. Mancini, Time delay in black hole gravitational lensing as a distance estimator, Gen. Relativ. Gravit. 36 (2004) 435–450, http://dx.doi.org/10.1023/B:GERG.0000010486.58026.4f, arXiv:gr-qc/0305007.

[39] S. Sahu, M. Patil, D. Narasimha, P.S. Joshi, Time delay between relativistic images as a probe of cosmic censorship, Phys. Rev. D 88 (10) (2013) 103002, http://dx.doi.org/10.1103/PhysRevD.88.103002, arXiv:1310.5350.

[40] S. Gillessen, F. Eisenhauer, T.K. Fritz, H. Bartko, K. Dodds-Eden, O. Pfuhl, T. Ott, R. Genzel, The orbit of the star S2 around SGR A* from very large telescope and Keck data, Astrophys. J. Lett. 707 (2009) L114–L117, http://dx.doi.org/10.1088/0004-637X/707/2/L114, arXiv:0910.3069.

[41] J.M. Bardeen, Timelike and null geodesics in the Kerr metric, in: C. Dewitt, B.S. Dewitt (Eds.), Black Holes (Les Astres Occlus), Gordon and Breach, 1973, pp. 215–239.

[42] J.L. Synge, The escape of photons from gravitationally intense stars, Mon. Not. R. Astron. Soc. 131 (1966) 463, http://dx.doi.org/10.1093/mnras/131.3.463.

[43] J.-P. Luminet, Image of a spherical black hole with thin accretion disk, Astron. Astrophys. 75 (1979) 228–235.

[44] V. Bozza, Quasiequatorial gravitational lensing by spinning black holes in the strong field limit, Phys. Rev. D 67 (10) (2003) 103006, http://dx.doi.org/10.1103/PhysRevD.67.103006, arXiv:gr-qc/0210109.

[45] R.-S. Lu, A.E. Broderick, F. Baron, J.D. Monnier, V.L. Fish, S.S. Doeleman, V. Pankratius, Imaging the supermassive black hole shadow and jet base of m87 with the event horizon telescope, Astrophys. J. 788 (2) (2014) 120.

[46] R.-S. Lu, F. Roelofs, V.L. Fish, H. Shiokawa, S.S. Doeleman, C.F. Gammie, H. Falcke, T.P. Krichbaum, J.A. Zensus, Imaging an event horizon: mitigation of source variability of Sagittarius A*, Astrophys J. 817 (2016) 173, http://dx.doi.org/10.3847/0004-637X/817/2/173, arXiv:1512.08543.

27

The GUP effect on Hawking radiation of the 2 + 1 dimensional black hole

Ganim Gecim, Yusuf Sucu *

Department of Physics, Faculty of Science, Akdeniz University, 07058 Antalya, Turkey

ARTICLE INFO

Editor: S. Dodelson

Keywords:
Quantum gravity
GUP
Tunneling
Hawking temperature

ABSTRACT

We investigate the Generalized Uncertainty Principle (GUP) effect on the Hawking radiation of the 2 + 1 dimensional Martinez–Zanelli black hole by using the Hamilton–Jacobi method. In this connection, we discuss the tunneling probabilities and Hawking temperature of the spin-1/2 and spin-0 particles for the black hole. Therefore, we use the modified Klein–Gordon and Dirac equations based on the GUP. Then, we observe that the Hawking temperature of the scalar and Dirac particles depend on not only the black hole properties, but also the properties of the tunneling particle, such as angular momentum, energy and mass. And, in this situation, we see that the tunneling probability and the Hawking radiation of the Dirac particle is different from that of the scalar particle.

1. Introduction

The discovery of the black hole radiation, known as Hawking radiation in the literature, is one of the milestones to construct a consistent connection between the relativity theory, the statistical mechanics and the quantum mechanics. The nature of a black hole has been started to be investigated in the framework of the thermodynamical and the quantum mechanical concepts since 1970 [1–6]. Hawking investigated the thermodynamical properties of a black hole in the frame of quantum field theory based on the Heisenberg uncertainty principle on a curved spacetime. Since then, the Hawking radiation has been investigated as a quantum tunneling effect of the relativistic particles from a black hole [7–14]. Also, the Hawking radiation as a tunneling process of the particles from various black holes has been studied, extensively, in the literature in both 3 + 1 and 2 + 1 dimensions [13–22].

On the other hand, the suitable candidate quantum gravity theories, such as string theory and loop quantum gravity theory, indicate the presence of a minimal observable length in Planck scale [23–26]. The existence of such a minimal length leads to the generalized Heisenberg uncertainty principle (GUP). The GUP can be expressed as [27,28]

$$\Delta x \Delta p \geq \frac{h}{2}\left[1+\beta(\Delta p)^2\right] \tag{1}$$

where $\beta = \beta_0/M_p^2$, the M_p^2 is the Planck mass and β_0 is the dimensionless parameter. Then, the modified commutation relation becomes

$$\left[x_\mu, p_\nu\right] = i\hbar\delta_{\mu\nu}\left[1+\beta p^2\right], \tag{2}$$

where x_μ and p_μ are the modified position and the momentum operators, respectively, defined by

$$x_\mu = x_{0\mu}$$
$$p_\mu = p_{0\mu}(1+\beta p_{0\mu}^2), \tag{3}$$

where the $x_{0\mu}$ and $p_{0\mu}$ are the standard position and momentum operators, respectively, and they satisfy the usually commutation relation $\left[x_{0\mu}, p_{0\nu}\right] = i\hbar\delta_{\mu\nu}$. These modified relations play an important role in physics. For example, in recent years, using the GUP, the thermodynamics properties of the black holes were investigated via a particle tunneling from the black holes. To include the quantum gravity effect, the Klein–Gordon and Dirac equations are modified by the GUP framework [29]. With these modified relativistic wave equations, the corrected Hawking temperature of various 3 + 1 and higher dimensional black holes computed via a particle tunneling process [30–40]. In this motivation, we will investigate the Hawking radiation of the 2 + 1 dimensional Martinez–Zanelli black hole by the scalar and Dirac particles tunneling process under the effect of the GUP. The metric of the Martinez–Zanelli black hole is given by [41]

$$ds^2 = F(r)dt^2 - \frac{1}{F(r)}dr^2 - r^2 d\theta^2 \tag{4}$$

* Corresponding author.
E-mail addresses: gecimganim@gmail.com (G. Gecim), ysucu@akdeniz.edu.tr (Y. Sucu).

where

$$F(r) = \frac{1}{l^2}\left[r^2 - 3B^2 - \frac{2B^3}{r}\right] = \frac{(r+B)^2(r-2B)}{rl^2},$$

and $l^2 = -1/\Lambda$ is the cosmological constant and B is the mass parameter related to the black hole mass M, as $B = \sqrt{Ml^2/3}$ [42]. Hence, the black hole has a singularity at $r = 0$ surrounded by horizon located at $r_h = 2B$ under the condition $B \neq 0$.

The organization of this work is as follows: In Section 2, we modify the Klein-Gordon equation by using the GUP. Subsequently, from the modified Klein-Gordon equation written in the $2+1$ dimensional Martinez-Zanelli Black hole background, we calculate the tunneling possibility of the scalar particle by using the semi-classical method, and then, we find the Hawking temperature. In Section 3, the modified Dirac equation is written in the $2+1$ dimensional Martinez-Zanelli black hole, and then, the tunneling probability of the Dirac particle from the black hole and its Hawking temperature is also calculated. Finally, in conclusion, we evaluate and summarize the results.

2. The modified Klein-Gordon equation and the scalar particle tunneling

To investigate the quantum gravity effect on the tunneling process of the scalar particles from the black hole and on its Hawking temperature, we will discuss the modified Klein-Gordon equation under the GUP relations. The standard Klein-Gordon equation can be written as [43]

$$p_{0\mu} p_0^{\mu} \phi = m_0^2 \phi, \tag{5}$$

or its explicit form is

$$-(i\hbar)^2 \partial_t \partial^t \phi = \left[(-i\hbar)^2 \partial_i \partial^i - m_0^2\right]\phi, \tag{6}$$

where ϕ is the wave function of the scalar particles. On the other hand, in the context of the GUP, the modified energy relation is given by

$$\widetilde{E} = E\left(1 - \beta E^2\right) = E\left[1 - \beta\left(p^2 + m_0^2\right)\right] \tag{7}$$

where $E^2 = p^2 + m_0^2$. Then, the square of the momentum operator can be obtained by using the Eq. (3) as follows;

$$p^2 = p_\mu p^\mu \simeq -\hbar^2\left[\partial_i \partial^i - 2\beta\left(\partial_j \partial^j\right)\left(\partial_j \partial^j\right)\right] \tag{8}$$

where the higher order terms of the β parameter are neglected. Then, using the Eq. (7) and Eq. (8) in the standard Klein-Gordon equation, the modified Klein-Gordon equation is written as follows;

$$\begin{aligned} &-(i\hbar)^2 \partial_t \partial^t \Phi \\ &= \left[(-i\hbar)^2 \partial_i \partial^i - m_0^2\right]\left[1 - 2\beta\left(-\hbar^2 \partial_i \partial^i + m_0^2\right)\right]\Phi, \end{aligned} \tag{9}$$

where Φ is the generalized wave function of the scalar particles. Hence, the modified Klein-Gordon equation in the Martinez-Zanelli black hole background is

$$\begin{aligned} &\frac{\hbar^2}{F(r)}\frac{\partial^2 \Phi}{\partial t^2} - \hbar^2 F(r)\frac{\partial^2 \Phi}{\partial r^2} - \frac{\hbar^2}{r^2}\frac{\partial^2 \Phi}{\partial \phi^2} + 2\beta F(r)\hbar^4 \frac{\partial^2}{\partial r^2}\left[F(r)\frac{\partial^2 \Phi}{\partial r^2}\right] \\ &+ \frac{2\beta\hbar^4}{r^2}\frac{\partial^2}{\partial \phi^2}\left[\frac{1}{r^2}\frac{\partial^2 \Phi}{\partial \phi^2}\right] + m_0^2\left(1 - 2\beta m_0^2\right)\Phi = 0. \end{aligned} \tag{10}$$

To investigate the tunneling radiation of the Martinez-Zanelli black hole with the Eq. (10), we employ the wave function of the scalar particle as

$$\Phi(t, r, \phi) = A e^{\frac{i}{\hbar} S(t, r, \phi)} \tag{11}$$

where A is a constant and $S(t, r, \phi)$ is the classical action term for the outgoing particle trajectory. Substituting the Eq. (11) into the Eq. (10) and neglecting the higher order terms of $\hbar$, we get the equation of motion of the scalar particle as

$$\begin{aligned} &\left(\frac{\partial S}{\partial t}\right)^2 - F^2(r)\left(\frac{\partial S}{\partial r}\right)^2 - \frac{F(r)}{r^2}\left(\frac{\partial S}{\partial \phi}\right)^2 - 2\beta F^3(r)\left(\frac{\partial S}{\partial r}\right)^4 \\ &- \frac{2\beta F(r)}{r^4}\left(\frac{\partial S}{\partial \phi}\right)^4 - m_0^2\left(1 - 2\beta m_0^2\right)F(r) = 0. \end{aligned} \tag{12}$$

Due to the commuting Killing vectors (∂_t) and (∂_ϕ) we can separate the $S(t, r, \phi)$, in terms of the variables t, r and ϕ, such as $S(t, r, \phi) = -Et + j\phi + K(r)$, where E and j are the energy and angular momentum of the particle, respectively, and $K(r) = K_0(r) + \beta K_1(r)$ [37]. And, from the Eq. (12), the radial integral, $K(r)$, becomes as follows;

$$\begin{aligned} K_\pm(r) = \pm \int &\frac{\sqrt{E^2 - F(r)\left(m_0^2 + \frac{j^2}{r^2}\right)}}{F(r)} \\ &\times \left[1 + \beta\left(\frac{F(r)\left(m_0^2 - \frac{j^4}{r^4}\right)}{E^2 - F(r)\left(m_0^2 + \frac{j^2}{r^2}\right)}\right.\right. \\ &\left.\left. - \frac{E^2 - F(r)\left(m_0^2 + \frac{j^2}{r^2}\right)}{F(r)}\right)\right] dr \end{aligned}$$

and it is computed as

$$K_\pm(r_h) = \pm i\pi \frac{l^2 E}{972 B^3}\left[216 B^2 + \beta\left(324 m_0^2 B^2 + 16 E^2 l^2 + 81 j^2\right)\right]$$

where $K_+(r_h)$ is outgoing and $K_-(r_h)$ is incoming solutions of the radial part. The total imaginary part of the action is $ImS(t, r, \phi) = ImK_\pm(r) = ImK_+(r) - ImK_-(r)$. Hence, the two kind probabilities of the crossing from the outer horizon, from outside to inside and from inside to outside, are given by [13,18,44]

$$P_{out} = \exp\left(-\frac{2}{h} ImK_+(r_h)\right) \tag{13}$$

and

$$P_{in} = \exp\left(-\frac{2}{h} ImK_-(r_h)\right), \tag{14}$$

respectively. Then, the tunneling probability of the scalar particle is written as

$$\begin{aligned} \Gamma &= \frac{P_{out}}{P_{in}} \\ &= \exp\left\{-\frac{\pi l^2 E}{243\hbar B^3}\left[216 B^2 + \beta\left(324 m_0^2 B^2 + 16 E^2 l^2 + 81 j^2\right)\right]\right\}. \end{aligned}$$

Hence, the modified Hawking temperature is obtained from the lowest order in the expansion of the classical action in terms of the particle energy,

$$\Gamma = \exp\left(-\frac{2}{\hbar} ImS\right) = \exp\left(-\frac{E}{T'_H}\right) \tag{15}$$

where T'_H is the modified Hawking temperature of the outer horizon, and it is given by

$$T'_H = \frac{9\hbar B}{8\pi l^2}\left[1+\beta\frac{324m_0^2B^2+16E^2l^2+81j^2}{216B^2}\right]^{-1}.$$

If, at first, we expand the T'_H in terms of the β powers, and, second, neglect the higher order of the β terms, then we get the modified Hawking temperature of the Martinez–Zanelli black hole as follows;

$$T'_H = T_H\left[1-\beta\frac{324m_0^2B^2+16E^2l^2+81j^2}{216B^2}\right] \tag{16}$$

where the $T_H = \frac{9\hbar B}{8\pi l^2}$ is the standard Hawking temperature of the black hole. From the T'_H expression, we see that the modified Hawking temperature is related to not only the mass parameter of the black hole, but also the angular momentum, energy and mass of the emitted scalar particle from the black hole, and it is lower than the standard Hawking temperature.

3. The modified Dirac equation and Fermion tunneling

The Dirac equation in a $(2+1)$ dimensional spacetime is given by the following representation [45],

$$\left\{i\bar{\sigma}^\mu(x)\left[\partial_\mu-\Gamma_\mu(x)\right]\right\}\Psi(x)=\frac{m_0}{\hbar}\Psi(x). \tag{17}$$

In this representation; the Dirac spinor, $\Psi(x)$, has only two components corresponding positive and negative energy eigenstates, which the each one has only one spin polarization. $\bar{\sigma}^\mu(x)$ are the spacetime dependent Dirac matrices and they are written in terms of the constant Dirac matrices, $\bar{\sigma}^i$, by using triads, $e^\mu_{(i)}(x)$, as follows;

$$\bar{\sigma}^\mu(x)=e^\mu_{(i)}(x)\bar{\sigma}^i, \tag{18}$$

where $\bar{\sigma}^i$ are the Dirac matrices in flat spacetime and they are given by

$$\bar{\sigma}^i=(\bar{\sigma}^0,\bar{\sigma}^1,\bar{\sigma}^2) \tag{19}$$

with

$$\bar{\sigma}^0=\sigma^3\ ,\bar{\sigma}^1=i\sigma^1,\ \bar{\sigma}^2=i\sigma^2, \tag{20}$$

where σ^1, σ^2 and σ^3 are Pauli matrices, and $\Gamma_\mu(x)$ are the spin affine connection by the following definition,

$$\Gamma_\mu(x)=\frac{1}{4}g_{\lambda\alpha}(e^i_{\nu,\mu}e^\alpha_i-\Gamma^\alpha_{\nu\mu})s^{\lambda\nu}(x). \tag{21}$$

Here, $\Gamma^\alpha_{\nu\mu}$ is the Christoffell symbol, and $g_{\mu\nu}(x)$ is the metric tensor that is given in terms of the triads as follows,

$$g_{\mu\nu}(x)=e^{(i)}_\mu(x)e^{(j)}_\nu(x)\eta_{(i)(j)}, \tag{22}$$

where μ and ν are a curved spacetime indices running from 0 to 2. i and j are flat spacetime indices running from 0 to 2 and $\eta_{(i)(j)}$ is the metric of the $(2+1)$ dimensional Minkowski spacetime, with signature $(1,-1,-1)$, and $s^{\lambda\nu}(x)$ is a spin operator defined as

$$s^{\lambda\nu}(x)=\frac{1}{2}[\bar{\sigma}^\lambda(x),\bar{\sigma}^\nu(x)]. \tag{23}$$

Using the Eq. (3), Eq. (7) and Eq. (8) in the Dirac equation, the generalized Dirac equation becomes

$$-i\bar{\sigma}^0(x)\partial_0\widetilde{\Psi}=\left(i\bar{\sigma}^i(x)\partial_i-i\bar{\sigma}^\mu(x)\Gamma_\mu-\frac{m_0}{\hbar}\right)\times\left(1+\beta\hbar^2\partial_j\partial^j-\beta m_0^2\right)\widetilde{\Psi}, \tag{24}$$

and it is rewritten as

$$\begin{aligned}&[i\bar{\sigma}^0(x)\partial_0+i\bar{\sigma}^i(x)\left(1-\beta m_0^2\right)\partial_i+i\beta\hbar^2\bar{\sigma}^i(x)\partial_i\left(\partial_j\partial^j\right)\\&-\frac{m_0}{\hbar}\left(1+\beta\hbar^2\partial_j\partial^j-\beta m_0^2\right)\\&-i\bar{\sigma}^\mu(x)\Gamma_\mu\left(1+\beta\hbar^2\partial_j\partial^j-\beta m_0^2\right)]\widetilde{\Psi}=0,\end{aligned} \tag{25}$$

where the $\widetilde{\Psi}$ is the generalized Dirac spinor.

To calculate the tunneling probability of a Dirac particle from the black hole, we use the following ansatz for the wave function;

$$\widetilde{\Psi}(x)=\exp\left(\frac{i}{\hbar}S(t,r,\phi)\right)\begin{pmatrix}A(t,r,\phi)\\B(t,r,\phi)\end{pmatrix} \tag{26}$$

where the $A(t,r,\phi)$ and $B(t,r,\phi)$ are the functions of space–time. Inserting the Eq. (26) in Eq. (25), we get the resulting equations to leading order in $\hbar$ and β as follows;

$$\begin{aligned}&A[\frac{1}{\sqrt{F(r)}}\frac{\partial S}{\partial t}+m_0\left(1-\beta m_0^2\right)+\frac{\beta m_0}{r^2}\left(\frac{\partial S}{\partial\phi}\right)^2\\&+\beta m_0F(r)\left(\frac{\partial S}{\partial r}\right)^2]+B[i\sqrt{F(r)}\left(1-\beta m_0^2\right)\frac{\partial S}{\partial r}\\&+\frac{1-\beta m_0^2}{r}\frac{\partial S}{\partial\phi}+i\beta F(r)\sqrt{F(r)}\left(\frac{\partial S}{\partial r}\right)^3\\&+i\frac{\beta\sqrt{F(r)}}{r^2}\left(\frac{\partial S}{\partial r}\right)\left(\frac{\partial S}{\partial\phi}\right)^2+\frac{\beta F(r)}{r}\left(\frac{\partial S}{\partial\phi}\right)\left(\frac{\partial S}{\partial r}\right)^2\\&+\frac{\beta}{r^3}\left(\frac{\partial S}{\partial\phi}\right)^3]=0\\&A[-i\sqrt{F(r)}\left(1-\beta m_0^2\right)\frac{\partial S}{\partial r}+\frac{1-\beta m_0^2}{r}\frac{\partial S}{\partial\phi}\\&-i\beta F(r)\sqrt{F(r)}\left(\frac{\partial S}{\partial r}\right)^3-i\frac{\beta\sqrt{F(r)}}{r^2}\left(\frac{\partial S}{\partial r}\right)\left(\frac{\partial S}{\partial\phi}\right)^2\\&+\frac{\beta F(r)}{r}\left(\frac{\partial S}{\partial\phi}\right)\left(\frac{\partial S}{\partial r}\right)^2+\frac{\beta}{r^3}\left(\frac{\partial S}{\partial\phi}\right)^3]\\&+B[\frac{1}{\sqrt{F(r)}}\frac{\partial S}{\partial t}-m_0\left(1-\beta m_0^2\right)-\frac{\beta m_0}{r^2}\left(\frac{\partial S}{\partial\phi}\right)^2\\&-\beta m_0F(r)\left(\frac{\partial S}{\partial r}\right)^2]=0.\end{aligned} \tag{27}$$

These two equations have nontrivial solutions for the $A(t,r,\phi)$ and $B(t,r,\phi)$ in case the determinant of the coefficient matrix is vanished. Accordingly, when neglecting the terms containing higher order of the β, then we get

$$\begin{aligned}&\frac{1}{F(r)}\left(\frac{\partial S}{\partial t}\right)^2-F(r)\left(\frac{\partial S}{\partial r}\right)^2-\frac{2\beta}{r^4}\left(\frac{\partial S}{\partial\phi}\right)^4-2\beta F^2(r)\left(\frac{\partial S}{\partial r}\right)^4\\&-\frac{1}{r^2}\left(\frac{\partial S}{\partial\phi}\right)^2-\frac{4F(r)\beta}{r^2}\left(\frac{\partial S}{\partial r}\right)^2\left(\frac{\partial S}{\partial\phi}\right)^2+2\beta m_0^4-m_0^2=0.\end{aligned}$$

Due to the Killing vectors (∂_t) and (∂_ϕ), we can separate the variables for $S(t,r,\phi)$ as $S(t,r,\phi)=-Et+j\phi+K(r)$, where E and j are the energy and angular momentum of the particle, respectively, and $K(r)=K_0(r)+\beta K_1(r)$ [37]. Then, the integral of the radial equation, $K(r)$, becomes as follows;

$$K_{\pm}(r) = \pm \int \frac{\sqrt{E^2 - F(r)\left(m_0^2 + \frac{j^2}{r^2}\right)}}{F(r)} \times \left[1 + \beta\left(\frac{2E^2 m_0^2 F(r) - E^4}{E^2 - F(r)\left(m_0^2 + \frac{j^2}{r^2}\right)}\right)\right] dr$$

and it is computed as

$$K_{\pm}(r_h) = \pm i\pi \frac{l^2 E}{972 B^3}\left[216B^2 + \beta\left(324 m_0^2 B^2 + 16E^2 l^2 - 27 j^2\right)\right]$$

Thus, from the Eq. (13) and Eq. (14), the tunneling probability of the Dirac particle is given by

$$\Gamma = \frac{P_{out}}{P_{in}} = \exp\left\{-\frac{\pi l^2 E}{243\hbar B^3}\left[216B^2 + \beta\left(324 m_0^2 B^2 + 16E^2 l^2 - 27 j^2\right)\right]\right\}.$$

Furthermore, from the Eq. (15), the modified Hawking temperature becomes as follows

$$T'_H = \frac{9\hbar B}{8\pi l^2}\left[1 + \beta\frac{324 m_0^2 B^2 + 16E^2 l^2 - 27 j^2}{216 B^2}\right]^{-1} = T_H\left[1 - \beta\frac{324 m_0^2 B^2 + 16E^2 l^2 - 27 j^2}{216 B^2}\right], \tag{28}$$

where the $T_H = \frac{9\hbar B}{8\pi l^2}$ is the standard Hawking temperature of the Martinez–Zanelli black hole. As in the case of the scalar particle tunneling, the corrected Hawking temperature of the tunneling Dirac particle is related to not only the mass parameter of the Martinez–Zanelli black hole, but also depends on the angular momentum, energy and mass of the emitted Dirac particle, and it is lower than the standard Hawking temperature.

4. Conclusion

In this paper we have studied the issue of the quantum gravity effect on the Hawking radiation of the 2+1 dimensional Martinez–Zanelli black hole by using the particle tunneling method. To take into account the quantum gravity effects, we modified the Dirac and Klein–Gordon equations by the generalized fundamental commutation relations to discuss the tunneling radiation of fermions and scalar particles, respectively. The results showed that the corrected Hawking temperature is not only determined by the mass parameter of the Martinez–Zanelli black hole, but also it is affected by the quantum properties (i.e., the angular momentum, energy and mass) of the emitted fermions and scalar particles. The other important results are given as follows:

- According to Eq. (16), the corrected Hawking temperature of the tunneling scalar particle is lower than the standard temperature.
- In Eq. (28), when $324m_0^2B^2 + 16E^2l^2 > 27j^2$, the corrected Hawking temperature of the tunneling fermions is lower than the standard temperature. However, when $324m_0^2B^2 + 16E^2l^2 < 27j^2$, the corrected temperature is higher than the standard temperature. If $324m_0^2B^2 + 16E^2l^2 = 27j^2$, then the contribution of the GUP effect is canceled, and radiation temperature of the tunneling fermions reduce to the standard temperature.
- By comparing the Eq. (28) with Eq. (16), we can say that the radiation temperature of the tunneling fermions higher than the scalar particles temperature, even if their masses, energies, and angular momentums are same.

Finally, thanks to the GUP effect, we can determine whether the radiated particle from a black hole is the scalar particle or the Dirac particle.

Acknowledgement

This work was supported by Akdeniz University, Scientific Research Projects Unit.

References

[1] J.M. Greif, Junior thesis, Princeton University, 1969 (unpublished).
[2] B. Carter, Nature 238 (1972) 71.
[3] J.D. Bekenstein, Phys. Rev. D 7 (1973) 2333.
[4] S.W. Hawking, Nature 248 (1974) 30.
[5] S.W. Hawking, Commun. Math. Phys. 43 (1975) 199.
[6] S.W. Hawking, Phys. Rev. D 13 (1976) 191.
[7] S. Shankaranarayanan, K. Srinivasan, T. Padmanabhan, Mod. Phys. Lett. A 16 (2001) 571.
[8] K. Srinivasan, T. Padmanabhan, Phys. Rev. D 60 (1999) 024007.
[9] E.C. Vagenas, Nuovo Cimento B 117 (2002) 899.
[10] P. Kraus, F. Wilczek, Nucl. Phys. B 433 (1995) 403.
[11] M.K. Parikh, F. Wilczek, Phys. Rev. Lett. 85 (2010) 5042.
[12] M. Arzano, A.J.M. Medved, E.C. Vagenas, J. High Energy Phys. 0509 (2005) 037.
[13] R. Kerner, R.B. Mann, Phys. Rev. D 73 (2006) 104010.
[14] R. Kerner, R.B. Mann, Class. Quantum Gravity 25 (2008) 095014.
[15] D.Y. Chen, Q.Q. Jian, X.T. Zu, Class. Quantum Gravity 25 (2008) 205022.
[16] J. Zhang, Z. Zhao, Phys. Lett. B 638 (2006) 110.
[17] D.Y. Chen, H. Yang, X.T. Zu, Phys. Lett. B 681 (2009) 463.
[18] R. Li, J.R. Ren, Phys. Lett. B 661 (2008) 370–372.
[19] H.L. Li, S.Z. Yang, Q.Q. Jiang, D.J. Qi, Phys. Lett. B 641 (2006) 139.
[20] G. Gecim, Y. Sucu, J. Cosmol. Astropart. Phys. 02 (2013) 023.
[21] G. Gecim, Y. Sucu, Astropart. Space Sci. 357 (2015) 105.
[22] De-Jiang Qi, Int. J. Theor. Phys. 52 (2013) 345.
[23] P.K. Townsend, Phys. Rev. D 15 (1977) 2795.
[24] D. Amati, M. Ciafaloni, G. Veneziano, Phys. Lett. B 216 (1989) 41.
[25] K. Konishi, G. Paffuti, P. Provero, Phys. Lett. B 234 (1990) 276.
[26] L.J. Garay, Int. J. Mod. Phys. A 10 (1995) 145.
[27] A. Kempf, G. Mangano, R.B. Mann, Phys. Rev. D 52 (1995) 1108.
[28] S. Hossenfelder, et al., Phys. Lett. B 575 (2003) 85–99.
[29] K. Nozari, M. Karami, Mod. Phys. Lett. A 20 (2005) 3095–3103.
[30] D. Chen, H. Wu, H. Yang, Adv. High Energy Phys. 2013 (2013) 432412.
[31] D. Chen, Q.Q. Jiang, P. Wang, H. Yang, J. High Energy Phys. 11 (2013) 176.
[32] D.Y. Chen, H.W. Wu, H. Yang, J. Cosmol. Astropart. Phys. 03 (2014) 036.
[33] D. Chen, H. Wu, H. Yang, S. Yang, Int. J. Mod. Phys. A 29 (2014) 1430054.
[34] X.X. Zeng, Y. Chen, Gen. Relativ. Gravit. 47 (2015) 47.
[35] H.L. Li, Z.W. Feng, X.T. Zu, Gen. Relativ. Gravit. 48 (2016) 18, arXiv:1410.4758v1 [gr-qc].
[36] P. Wang, H. Yang, S. Ying, arXiv:1410.5065v1 [gr-qc].
[37] Z.Y. Liu, J.R. Ren, Commun. Theor. Phys. 62 (2014) 819–823.
[38] B. Mu, P. Wang, H. Yang, Adv. High Energy Phys. 2015 (2015) 898916.
[39] G. Li, X. Zu, Int. J. Adv. Mater. Prod. 03 (2015) 134–139.
[40] M.A. Anacleto, F.A. Brito, E. Passos, Phys. Lett. B 749 (2015) 181–186.
[41] C. Martinez, J. Zanelli, Phys. Rev. D 54 (1996) 3830.
[42] E. Ayon-Beato, A. Garcia, A. Macias, J.M. Perez-Sanchez, Phys. Lett. B 495 (2000) 164–168.
[43] W. Greiner, Relativistic Quantum Mechanics: Wave Equations, 3rd edition, Springer, 2000.
[44] G.E. Volovik, Exotic Properties of Superfluid 3He, World Scientific, Singapore, 1992.
[45] Y. Sucu, N. Unal, J. Math. Phys. 48 (2007) 052503.

28

The H_0 tension in light of vacuum dynamics in the universe

Joan Solà *, Adrià Gómez-Valent, Javier de Cruz Pérez

Departament de Física Quàntica i Astrofísica, and Institute of Cosmos Sciences, Universitat de Barcelona, Av. Diagonal 647, E-08028 Barcelona, Catalonia, Spain

ARTICLE INFO

Editor: M. Trodden

ABSTRACT

Despite the outstanding achievements of modern cosmology, the classical dispute on the precise value of H_0, which is the first ever parameter of modern cosmology and one of the prime parameters in the field, still goes on and on after over half a century of measurements. Recently the dispute came to the spotlight with renewed strength owing to the significant tension (at $> 3\sigma$ c.l.) between the latest Planck determination obtained from the CMB anisotropies and the local (distance ladder) measurement from the Hubble Space Telescope (HST), based on Cepheids. In this work, we investigate the impact of the running vacuum model (RVM) and related models on such a controversy. For the RVM, the vacuum energy density ρ_Λ carries a mild dependence on the cosmic expansion rate, i.e. $\rho_\Lambda(H)$, which allows to ameliorate the fit quality to the overall SNIa+BAO+$H(z)$+LSS+CMB cosmological observations as compared to the concordance ΛCDM model. By letting the RVM to deviate from the vacuum option, the equation of state $w = -1$ continues to be favored by the overall fit. Vacuum dynamics also predicts the following: i) the CMB range of values for H_0 is more favored than the local ones, and ii) smaller values for $\sigma_8(0)$. As a result, a better account for the LSS structure formation data is achieved as compared to the ΛCDM, which is based on a rigid (i.e. non-dynamical) Λ term.

1. Introduction

The most celebrated fact of modern observational cosmology is that the universe is in accelerated expansion [1,2]. At the same time, the most paradoxical reality check is that we do not honestly understand the primary cause for such an acceleration. The simplest picture is to assume that it is caused by a strict cosmological term, Λ, in Einstein's equations, but its fundamental origin is unknown [3]. Together with the assumption of the existence of dark matter (DM) and the spatial flatness of the Friedmann–Lemaître–Robertson–Walker (FLRW) metric (viz. the metric that expresses the homogeneity and isotropy inherent to the cosmological principle), we are led to the "concordance" ΛCDM model, i.e. the standard model of cosmology [4]. The model is consistent with a large body of observations, and in particular with the high precision data from the cosmic microwave background (CMB) anisotropies [5,6]. Many alternative explanations of the cosmic acceleration beyond a Λ-term are possible (including quintessence and the like, see e.g. the review [7]) and are called dark energy (DE) [8].

The current situation with cosmology is reminiscent of the prediction by the famous astronomer A. Sandage in the sixties, who asserted that the main task of future observational cosmology would be the search for two parameters: the Hubble constant H_0 and the deceleration parameter q_0 [9]. The first of them is the most important distance (and time) scale in cosmology prior to any other cosmological quantity. Sandage's last published value with Tammann (in 2010) is 62.3 km/s/Mpc [10] – slightly revised in Ref. [11] as $H_0 = 64.1 \pm 2.4$ km/s/Mpc. There is currently a significant tension between CMB measurements of H_0 [5,12] – not far away from this value – and local determinations emphasizing a higher range above 70 km/s/Mpc [13,14]. As for q_0, its measurement is tantamount to determining Λ in the context of the concordance model. On fundamental grounds, however, understanding the value of Λ is not just a matter of observation; in truth and in fact, it embodies one of the most important and unsolved conundrums of theoretical physics and cosmology: the cosmological constant problem, see e.g. [3,7,15,16]. The problem is connected to the fact that the Λ-term is usually associated with the vacuum energy density, $\rho_\Lambda = \Lambda/(8\pi G)$, with G Newton's coupling. The prediction for ρ_Λ in quantum field theory (QFT) overshoots the measured value $\rho_\Lambda \sim 10^{-47}$ GeV^4 (in natural units $c = \hbar = 1$) by many orders of magnitude [16].

* Corresponding author.
E-mail addresses: sola@fqa.ub.edu (J. Solà), adriagova@fqa.ub.edu (A. Gómez-Valent), decruz@fqa.ub.edu (J. de Cruz Pérez).

Concerning the prime parameter H_0, the tension among the different measurements is inherent to its long and tortuous history. Let us only recall that after Baade's revision (by a factor of one half [17]) of the exceedingly large value ~ 500 km/s/Mpc originally estimated by Hubble (which implied a universe of barely two billion years old only), the Hubble parameter was subsequently lowered to 75 km/s/Mpc and finally to $H_0 = 55 \pm 5$ km/s/Mpc, where it remained for 20 years (until 1995), mainly under the influence of Sandage's devoted observations [18]. Shortly after that period the first measurements of the nonvanishing, positive, value of Λ appeared [1,2] and the typical range for H_0 moved upwards to ~ 65 km/s/Mpc. In the meantime, many different observational values of H_0 have piled up in the literature using different methods (see e.g. the median statistical analysis of > 550 measurements considered in [19,20]). As mentioned above, two kinds of *precision* (few percent level) measurements of H_0 have generated considerable perplexity in the recent literature, specifically between the latest Planck values ($H_0^{\rm Planck}$) obtained from the CMB anisotropies, and the local HST measurement (based on distance ladder estimates from Cepheids). The latter, obtained by Riess et al. [13], is $H_0 = 73.24 \pm 1.74$ km/s/Mpc and will be denoted $H_0^{\rm Riess}$. It can be compared with the CMB value $H_0 = 67.51 \pm 0.64$ km/s/Mpc, as extracted from Planck 2015 TT,TE,EE+lowP+lensing data [5], or with $H_0 = 66.93 \pm 0.62$ km/s/Mpc, based on Planck 2015 TT,TE,EE+SIMlow data [12]. In both cases there is a tension above 3σ c.l. (viz. 3.1σ and 3.4σ, respectively) with respect to the local measurement. This situation, and in general a certain level of tension with some independent observations in intermediate cosmological scales, has stimulated a number of discussions and possible solutions in the literature, see e.g. [21–29].

We wish to reexamine here the $H_0^{\rm Riess} - H_0^{\rm Planck}$ tension, but not as an isolated conflict between two particular sources of observations, but rather in light of the overall fit to the cosmological data SNIa+BAO+$H(z)$+LSS+CMB. Recently, it has been demonstrated that by letting the cosmological vacuum energy density to slowly evolve with the expansion rate, $\rho_\Lambda = \rho_\Lambda(H)$, the global fit can be improved with respect to the ΛCDM at a confidence level of 3–4σ [30–34]. We devote this work to show that the dynamical vacuum models (DVMs) can still give a better fit to the overall data, even if the local HST measurement of the Hubble parameter is taken into account. However we find that our best-fit values of H_0 are much closer to the value extracted from CMB measurements [5,12]. Our analysis also corroborates that the large scale structure formation data (LSS) are crucial in distinguishing the rigid vacuum option from the dynamical one.

2. Dynamical vacuum models and beyond

Let us consider a generic cosmological framework described by the spatially flat FLRW metric, in which matter is exchanging energy with a dynamical DE medium with a phenomenological equation of state (EoS) $p_\Lambda = w\rho_\Lambda$, where $w = -1 + \epsilon$ (with $|\epsilon| \ll 1$). Such medium is therefore of quasi-vacuum type, and for $w = -1$ (i.e. $\epsilon = 0$) we precisely recover the genuine vacuum case. Owing, however, to the exchange of energy with matter, $\rho_\Lambda = \rho_\Lambda(\zeta)$ is in all cases a *dynamical* function that depends on a cosmic variable $\zeta = \zeta(t)$. We will identify the nature of $\zeta(t)$ later on, but its presence clearly indicates that ρ_Λ is no longer associated to a strictly rigid cosmological constant as in the ΛCDM. The Friedmann and acceleration equations read, however, formally identical to the standard case:

$$3H^2 = 8\pi\, G\, (\rho_m + \rho_r + \rho_\Lambda(\zeta)) \tag{1}$$

$$3H^2 + 2\dot{H} = -8\pi\, G\, (p_r + p_\Lambda(\zeta))\,. \tag{2}$$

Here $H = \dot{a}/a$ is the Hubble function, $a(t)$ the scale factor as a function of the cosmic time, ρ_r is the energy density of the radiation component (with pressure $p_r = \rho_r/3$), and $\rho_m = \rho_b + \rho_{dm}$ involves the contributions from baryons and cold DM. The local conservation law associated to the above equations reads:

$$\dot{\rho}_r + 4H\rho_r + \dot{\rho}_m + 3H\rho_m = Q\,, \tag{3}$$

where

$$Q = -\dot{\rho}_\Lambda - 3H(1+w)\rho_\Lambda\,. \tag{4}$$

For $w = -1$ the last equation boils down to just $Q = -\dot{\rho}_\Lambda$, which is nonvanishing on account of $\rho_\Lambda(t) = \rho_\Lambda(\zeta(t))$. The simplest case is, of course, that of the concordance model, in which $\rho_\Lambda = \rho_{\Lambda 0} =$ const and $w = -1$, so that $Q = 0$ trivially. However, for $w \neq -1$ we can also have $Q = 0$ in a nontrivial situation, which follows from solving Eq. (4). It corresponds to the XCDM parametrization [35], in which the DE density is dynamical and self-conserved. It is easily found in terms of the scale factor:

$$\rho_\Lambda^{XCDM}(a) = \rho_{\Lambda 0}\, a^{-3(1+w)} = \rho_{\Lambda 0}\, a^{-3\epsilon}\,, \tag{5}$$

where $\rho_{\Lambda 0}$ is the current value. From (3) it then follows that the total matter component is also conserved. After equality it leads to separate conservation of cold matter and radiation. In general, Q can be a nonvanishing interaction source allowing energy exchange between matter and the quasi-vacuum medium under consideration; Q can either be given by hand (e.g. through an *ad hoc* ansatz), or can be suggested by some specific theoretical framework. In any case the interaction source must satisfy $0 < |Q| \ll \dot{\rho}_m$ since we do not wish to depart too much from the concordance model. Despite matter is exchanging energy with the vacuum or quasi-vacuum medium, we shall assume that radiation and baryons are separately self-conserved, i.e. $\dot{\rho}_r + 4H\rho_r = 0$ and $\dot{\rho}_b + 3H\rho_b = 0$, so that their energy densities evolve in the standard way: $\rho_r(a) = \rho_{r0}\, a^{-4}$ and $\rho_b(a) = \rho_{b0}\, a^{-3}$. The dynamics of ρ_Λ can therefore be associated to the exchange of energy exclusively with the DM (through the nonvanishing source Q) and/or with the possibility that the DE medium is not exactly the vacuum, $w \neq -1$, but close to it $|\epsilon| \ll 1$. Under these conditions, the coupled system of conservation equations (3)–(4) reduces to

$$\dot{\rho}_{dm} + 3H\rho_{dm} = Q \tag{6}$$

$$\dot{\rho}_\Lambda + 3H\epsilon\rho_\Lambda = -Q\,. \tag{7}$$

In the following we shall for definiteness focus our study of the dynamical vacuum (and quasi-vacuum) models to the three interactive sources:

$$\text{Model I } (w\text{RVM}):\ Q = \nu H(3\rho_m + 4\rho_r) \tag{8}$$

$$\text{Model II } (wQ_{dm}):\ Q_{dm} = 3\nu_{dm} H\rho_{dm} \tag{9}$$

$$\text{Model III } (wQ_\Lambda):\ Q_\Lambda = 3\nu_\Lambda H\rho_\Lambda\,. \tag{10}$$

Here $\nu_i = \nu, \nu_{dm}, \nu_\Lambda$ are small dimensionless constants, $|\nu_i| \ll 1$, which are determined from the overall fit to the data, see e.g. Tables 1 and 2. The ordinal number names I, II and III will be used for short, but the three model names are preceded by w to recall that, in the general case, the equation of state (EoS) is near the vacuum one (that is, $w = -1 + \epsilon$). These dynamical quasi-vacuum models are also denoted as wDVMs. In the particular case $w = -1$ (i.e. $\epsilon = 0$) we recover the dynamical vacuum models (DVMs), which were previously studied in detail in [34], and in this case the names of the models will not be preceded by w.

In all of the above (w)DVMs, the cosmic variable ζ can be taken to be the scale factor, $\zeta = a(t)$, since they are all analytically solvable in terms of it, as we shall see in a moment. Model I with

Table 1
Best-fit values for the ΛCDM, XCDM, the three dynamical vacuum models (DVMs) and the three dynamical quasi-vacuum models (wDVMs), including their statistical significance (χ^2-test and Akaike and Bayesian information criteria AIC and BIC). For detailed description of the data and a full list of references, see [31] and [34]. The quoted number of degrees of freedom (dof) is equal to the number of data points minus the number of independent fitting parameters (4 for the ΛCDM, 5 for the XCDM and the DVMs, and 6 for the wDVMs). For the CMB data we have used the marginalized mean values and covariance matrix for the parameters of the compressed likelihood for Planck 2015 TT,TE,EE+lowP+lensing data from [36]. Each best-fit value and the associated uncertainties have been obtained by marginalizing over the remaining parameters.

Model	H_0 (km/s/Mpc)	ω_b	n_s	Ω_m^0	ν_i	w	$\chi^2_{\rm min}/dof$	ΔAIC	ΔBIC
ΛCDM	68.83 ± 0.34	0.02243 ± 0.00013	0.973 ± 0.004	0.298 ± 0.004	-	-1	84.40/85	-	-
XCDM	67.16 ± 0.67	0.02251 ± 0.00013	0.975 ± 0.004	0.311 ± 0.006	-	-0.936 ± 0.023	76.80/84	5.35	3.11
RVM	67.45 ± 0.48	0.02224 ± 0.00014	0.964 ± 0.004	0.304 ± 0.005	0.00158 ± 0.00041	-1	68.67/84	13.48	11.24
Q_{dm}	67.53 ± 0.47	0.02222 ± 0.00014	0.964 ± 0.004	0.304 ± 0.005	0.00218 ± 0.00058	-1	69.13/84	13.02	10.78
Q_Λ	68.84 ± 0.34	0.02220 ± 0.00015	0.964 ± 0.005	0.299 ± 0.004	0.00673 ± 0.00236	-1	76.30/84	5.85	3.61
wRVM	67.08 ± 0.69	0.02228 ± 0.00016	0.966 ± 0.005	0.307 ± 0.007	0.00140 ± 0.00048	-0.979 ± 0.028	68.15/83	11.70	7.27
wQ_{dm}	67.04 ± 0.69	0.02228 ± 0.00016	0.966 ± 0.005	0.308 ± 0.007	0.00189 ± 0.00066	-0.973 ± 0.027	68.22/83	11.63	7.20
wQ_Λ	67.11 ± 0.68	0.02227 ± 0.00016	0.965 ± 0.005	0.313 ± 0.006	0.00708 ± 0.00241	-0.933 ± 0.022	68.24/83	11.61	7.18

Table 2
The same as Table 1 but adding the $H_0^{\rm Riess}$ local measurement from Riess et al. [13].

Model	H_0 (km/s/Mpc)	ω_b	n_s	Ω_m^0	ν_i	w	$\chi^2_{\rm min}/dof$	ΔAIC	ΔBIC
ΛCDM	68.99 ± 0.33	0.02247 ± 0.00013	0.974 ± 0.003	0.296 ± 0.004	-	-1	90.59/86	-	-
XCDM	67.98 ± 0.64	0.02252 ± 0.00013	0.975 ± 0.004	0.304 ± 0.006	-	-0.960 ± 0.023	87.38/85	0.97	−1.29
RVM	67.86 ± 0.47	0.02232 ± 0.00014	0.967 ± 0.004	0.300 ± 0.004	0.00133 ± 0.00040	-1	78.96/85	9.39	7.13
Q_{dm}	67.92 ± 0.46	0.02230 ± 0.00014	0.966 ± 0.004	0.300 ± 0.004	0.00185 ± 0.00057	-1	79.17/85	9.18	6.92
Q_Λ	69.00 ± 0.34	0.02224 ± 0.00016	0.965 ± 0.005	0.297 ± 0.004	0.00669 ± 0.00234	-1	82.48/85	5.87	3.61
wRVM	67.95 ± 0.66	0.02230 ± 0.00015	0.966 ± 0.005	0.300 ± 0.006	0.00138 ± 0.00048	-1.005 ± 0.028	78.93/84	7.11	2.66
wQ_{dm}	67.90 ± 0.66	0.02230 ± 0.00016	0.966 ± 0.005	0.300 ± 0.006	0.00184 ± 0.00066	-0.999 ± 0.028	79.17/84	6.88	2.42
wQ_Λ	67.94 ± 0.65	0.02227 ± 0.00016	0.966 ± 0.005	0.306 ± 0.006	0.00689 ± 0.00237	-0.958 ± 0.022	78.98/84	7.07	2.61

$w=-1$ is the running vacuum model (RVM), see [16,33,34,37]. It is special in that the interaction source indicated in (8) is not *ad hoc* but follows from an expression for the dynamical vacuum energy density, $\rho_\Lambda(\zeta)$, in which ζ is not just the scale factor but the full Hubble rate: $\zeta=H(a)$. The explicit RVM form reads

$$\rho_\Lambda(H)=\frac{3}{8\pi G}\left(c_0+\nu H^2\right). \tag{11}$$

The additive constant $c_0=H_0^2\left(\Omega_\Lambda^0-\nu\right)$ is fixed from the condition $\rho_\Lambda(H_0)=\rho_{\Lambda 0}$, with $\Omega_\Lambda^0=1-\Omega_m^0-\Omega_r^0$. Combining the Friedmann and acceleration equations (1)–(2), we find $\dot{H}=-(4\pi G/3)(3\rho_m+4\rho_r+3\epsilon\rho_\Lambda)$, and upon differentiating (11) with respect to the cosmic time we are led to $\dot{\rho}_\Lambda=-\nu H(3\rho_m+4\rho_r+3\epsilon\rho_\Lambda)$. Thus, for $\epsilon=0$ (vacuum case) we indeed find $\dot{\rho}_\Lambda=-Q$ for Q as in (8). However, for the quasi-vacuum case ($0<|\epsilon|\ll 1$) Eq. (7) does not hold if $\rho_\Lambda(H)$ adopts the form (11). This RVM form is in fact specific to the pure vacuum EoS ($w=-1$), and it can be motivated in QFT in curved spacetime through a renormalization group equation for $\rho_\Lambda(H)$, what explains the RVM name [16]. In it, ν plays the role of the β-function coefficient for the running of ρ_Λ with the Hubble rate. Thus, we naturally expect $|\nu|\ll 1$ in QFT, see [16,38]. Interestingly, the RVM form (11) can actually be extended with higher powers of H^n (typically $n=4$) to provide an effective description of the cosmic evolution from the inflationary universe up to our days [37,39]. Models II and III are purely phenomenological models instead, in which the interaction source Q is introduced by hand, see e.g. Refs. [26,40–42] and references therein.

The energy densities for the wDVMs can be computed straightforwardly. For simplicity, we shall quote here the leading parts only. The exact formulas containing the radiation terms are more cumbersome. In the numerical analysis we have included the full expressions. Details will be shown elsewhere. For the matter densities, we find:

$$\begin{aligned}
\rho_{dm}^{\rm I}(a)&=\rho_{dm0}\,a^{-3(1-\nu)}+\rho_{b0}\left(a^{-3(1-\nu)}-a^{-3}\right)\\
\rho_{dm}^{\rm II}(a)&=\rho_{dm0}\,a^{-3(1-\nu_{dm})}\\
\rho_{dm}^{\rm III}(a)&=\rho_{dm0}\,a^{-3}+\frac{\nu_\Lambda}{\nu_\Lambda+w}\rho_{\Lambda 0}\left(a^{-3}-a^{-3(\epsilon+\nu_\Lambda)}\right),
\end{aligned} \tag{12}$$

and for the quasi-vacuum energy densities:

$$\begin{aligned}
\rho_\Lambda^{\rm I}(a)&=\rho_{\Lambda 0}a^{-3\epsilon}-\frac{\nu\,\rho_{m0}}{\nu+w}\left(a^{-3(1-\nu)}-a^{-3\epsilon}\right)\\
\rho_\Lambda^{\rm II}(a)&=\rho_{\Lambda 0}a^{-3\epsilon}-\frac{\nu_{dm}\,\rho_{dm0}}{\nu_{dm}+w}\left(a^{-3(1-\nu_{dm})}-a^{-3\epsilon}\right)\\
\rho_\Lambda^{\rm III}(a)&=\rho_{\Lambda 0}\,a^{-3(\epsilon+\nu_\Lambda)}.
\end{aligned} \tag{13}$$

Two specific dimensionless parameters enter each formula, $\nu_i=(\nu,\nu_{dm},\nu_\Lambda)$ and $w=-1+\epsilon$. They are part of the fitting vector of free parameters for each model, as explained in detail in the caption of Table 1. For $\nu_i\to 0$ the models become noninteractive and they all reduce to the XCDM model case (5). For $w=-1$ we recover the DVMs results previously studied in [34]. Let us also note that for $\nu_i>0$ the vacuum decays into DM (which is thermodynamically favorable [34]) whereas for $\nu_i<0$ is the other way around. Furthermore, when w enters the fit, the effective behavior of the wDVMs is quintessence-like for $w>-1$ (i.e. $\epsilon>0$) and phantom-like for $w<-1$ ($\epsilon<0$).

Given the energy densities (12) and (13), the Hubble function immediately follows. For example, for Model I:

$$H^2(a)=H_0^2\left[a^{-3\epsilon}+\frac{w}{w+\nu}\Omega_m^0\left(a^{-3(1-\nu)}-a^{-3\epsilon}\right)\right]. \tag{14}$$

Similar formulas can be obtained for Models II and III. For $w=-1$ they all reduce to the DVM forms previously found in [34]. And of course they all ultimately boil down to the ΛCDM form in the limit $(w,\nu_i)\to(-1,0)$.

3. Structure formation: the role of the LSS data

The analysis of structure formation plays a crucial role in comparing the various models. For the ΛCDM and XCDM we use the standard perturbations equation [4]

$$\ddot{\delta}_m+2H\,\dot{\delta}_m-4\pi G\rho_m\,\delta_m=0, \tag{15}$$

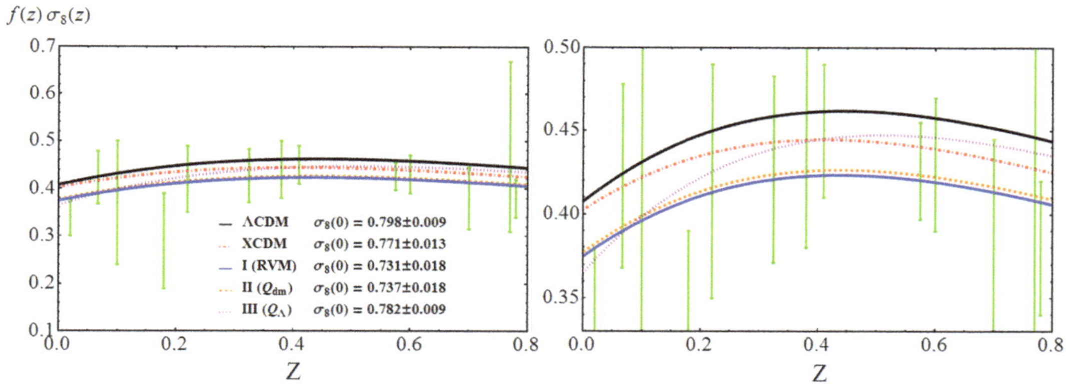

Fig. 1. Left: The LSS structure formation data ($f(z)\sigma_8(z)$) versus the predicted curves by Models I, II and III, see equations (8)–(10) for the case $w = -1$, i.e. the dynamical vacuum models (DVMs), using the best-fit values in Table 1. The XCDM curve is also shown. The values of $\sigma_8(0)$ that we obtain for the models are also indicated. **Right**: Zoomed window of the plot on the left, which allows to better distinguish the various models.

with, however, the Hubble function corresponding to each one of these models. For the wDVMs, a step further is needed: the perturbations equation not only involves the modified Hubble function but the equation itself becomes modified. Trading the cosmic time for the scale factor and extending the analysis of [34,43,44] for the case $w \neq -1$ ($\epsilon \neq 0$), we find

$$\delta_m'' + \frac{A(a)}{a}\delta_m' + \frac{B(a)}{a^2}\delta_m = 0\,, \tag{16}$$

where the prime denotes differentiation with respect to the scale factor, and the functions $A(a)$ and $B(a)$ are found to be as follows:

$$A(a) = 3 + \frac{aH'}{H} + \frac{\Psi}{H} - 3r\epsilon \tag{17}$$

$$B(a) = -\frac{4\pi G\rho_m}{H^2} + 2\frac{\Psi}{H} + \frac{a\Psi'}{H} - 15r\epsilon - 9\epsilon^2 r^2 + 3\epsilon(1+r)\frac{\Psi}{H} - 3r\epsilon\frac{aH'}{H}\,. \tag{18}$$

Here $r \equiv \rho_\Lambda/\rho_m$ and $\Psi \equiv -\dot{\rho}_\Lambda/\rho_m$. For $\nu_i = 0$ we have $\Psi = 3Hr\epsilon$, and after a straightforward calculation one can show that (16) can be brought to the standard form Eq. (15).

To solve the above perturbations equations we have to fix the initial conditions on δ_m and δ_m' for each model at high redshift, namely when non-relativistic matter dominates over radiation and DE, see [34]. Functions (17) and (18) are then approximately constant and Eq. (16) admits power-law solutions $\delta_m(a) = a^s$. From explicit calculation we find that the values of s for each model turn out to be:

$$s^{\rm I} = 1 + \frac{3}{5}\nu\left(\frac{1}{w} - 4\right) + \mathcal{O}(\nu^2)$$

$$s^{\rm II} = 1 - \frac{3}{5}\nu_{dm}\left(1 + 3\frac{\Omega_{dm}^0}{\Omega_m^0} - \frac{1}{w}\right) + \mathcal{O}(\nu_{dm}{}^2) \tag{19}$$

$$s^{\rm III} = 1\,.$$

We can check that for $w = -1$ all of the above equations (16)–(19) render the DVM results previously found in [34]. The generalization that we have made to $w \neq -1$ ($\epsilon \neq 0$) has introduced several nontrivial extra terms in equations (17)–(19).

The analysis of the linear LSS regime is usually implemented with the help of the weighted linear growth $f(z)\sigma_8(z)$, where $f(z) = d\ln\delta_m/d\ln a$ is the growth rate and $\sigma_8(z)$ is the rms mass fluctuation on $R_8 = 8\,h^{-1}$ Mpc scales. It is computed as follows (see e.g. [31,34]):

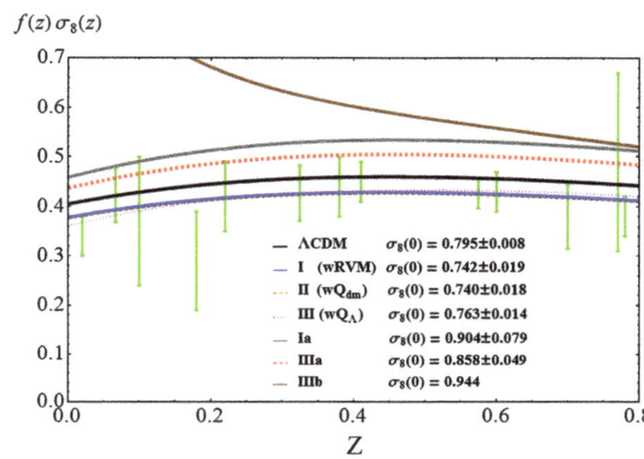

Fig. 2. The LSS structure formation data ($f(z)\sigma_8(z)$) and the theoretical predictions for models (8)–(10), using the best-fit values in Tables 2 and 3. The curves for the cases Ia, IIIa correspond to special scenarios for Models I and III where the agreement with the Riess et al. local value $H_0^{\rm Riess}$ [13] is better (cf. Table 3). The price, however, is that the concordance with the LSS data is now spoiled. Case IIIb is our theoretical prediction for the scenario proposed in [26], aimed at optimally relaxing the tension with $H_0^{\rm Riess}$. Unfortunately, the last three scenarios lead to phantom-like DE and are in serious disagreement with the LSS data.

$$\sigma_8(z) = \sigma_{8,\Lambda}\frac{\delta_m(z)}{\delta_m^\Lambda(0)}\sqrt{\frac{\int_0^\infty k^{n_s+2}T^2(\mathbf{p},k)W^2(kR_8)dk}{\int_0^\infty k^{n_{s,\Lambda}+2}T^2(\mathbf{p}_\Lambda,k)W^2(kR_{8,\Lambda})dk}}\,, \tag{20}$$

where W is a top-hat smoothing function and $T(\mathbf{p},k)$ the transfer function. The fitting parameters for each model are contained in $\mathbf{p}$. Following the mentioned references, we have defined as fiducial model the ΛCDM at fixed parameter values from the Planck 2015 TT,TE,EE+lowP+lensing data [5]. These fiducial values are collected in $\mathbf{p}_\Lambda$. In Figs. 1–2 we display $f(z)\sigma_8(z)$ for the various models using the fitted values of Tables 1–3. We remark that our BAO and LSS data include the bispectrum data points from Ref. [45] – see [34] for a full-fledged explanation of our data sets. In the next section, we discuss our results for the various models and assess their ability to improve the ΛCDM fit as well as their impact on the H_0 tension.

4. Discussion

Following [34] the statistical analysis of the various models is performed in terms of a joint likelihood function, which is the product of the likelihoods for each data source and includes the corresponding covariance matrices. As indicated in the caption of Table 1, the ΛCDM has 4 parameters, whereas the XCDM and the

Table 3
Best-fit values for the ΛCDM and models RVM, Q_Λ, wRVM and wQ_Λ by making use of the CMB+BAO data only. In contrast to Tables 1–2, we now fully dispense with the LSS data (see [31,34]) to test its effect. The starred/non-starred cases correspond respectively to adding or not the local value $H_0^{\rm Riess}$ from [13] as data point in the fit. The AIC and BIC differences of the starred models are computed with respect to the ΛCDM*. We can see that under these conditions models tend to have ΔAIC, ΔBIC < 0, including the last two starred scenarios, which are capable of significantly approaching $H_0^{\rm Riess}$.

Model	H_0 (km/s/Mpc)	ω_b	n_s	Ω_m^0	ν_i	w	$\chi^2_{\rm min}/dof$	ΔAIC	ΔBIC
ΛCDM	68.23 ± 0.38	0.02234 ± 0.00013	0.968 ± 0.004	0.306 ± 0.005	-	-1	13.85/11	-	-
RVM	67.70 ± 0.69	0.02227 ± 0.00016	0.965 ± 0.005	0.306 ± 0.005	0.0010 ± 0.0010	-1	13.02/10	-3.84	-1.88
Q_Λ	68.34 ± 0.40	0.02226 ± 0.00016	0.965 ± 0.005	0.305 ± 0.005	0.0030 ± 0.0030	-1	12.91/10	-3.73	-1.77
wRVM	66.34 ± 2.30	0.02228 ± 0.00016	0.966 ± 0.005	0.313 ± 0.012	0.0017 ± 0.0016	-0.956 ± 0.071	12.65/9	-9.30	-4.22
wQ_Λ	66.71 ± 1.77	0.02226 ± 0.00016	0.965 ± 0.005	0.317 ± 0.014	0.0070 ± 0.0054	-0.921 ± 0.082	12.06/9	-8.71	-3.63
ΛCDM*	68.46 ± 0.37	0.02239 ± 0.00013	0.969 ± 0.004	0.303 ± 0.005	-	-1	21.76/12	-	-
RVM*	68.48 ± 0.67	0.02240 ± 0.00015	0.969 ± 0.005	0.303 ± 0.005	0.0000 ± 0.0010	-1	21.76/11	-4.36	-2.77
Q_Λ*	68.34 ± 0.39	0.02224 ± 0.00016	0.966 ± 0.005	0.302 ± 0.005	0.0034 ± 0.0030	-1	20.45/11	-3.05	-1.46
Ia (wRVM*)	70.95 ± 1.46	0.02231 ± 0.00016	0.967 ± 0.005	0.290 ± 0.008	-0.0008 ± 0.0010	-1.094 ± 0.050	18.03/10	-5.97	-1.82
IIIa (wQ_Λ*)	70.27 ± 1.33	0.02228 ± 0.00016	0.966 ± 0.005	0.291 ± 0.010	-0.0006 ± 0.0042	-1.086 ± 0.065	18.64/10	-6.58	-2.43

DVMs have 5, and finally any of the wDVMs has 6. Thus, for a fairer comparison of the various nonstandard models with the concordance ΛCDM we have to invoke efficient criteria in which the presence of extra parameters in a given model is conveniently penalized so as to achieve a balanced comparison with the model having less parameters. The Akaike information criterion (AIC) and the Bayesian information criterion (BIC) are known to be extremely valuable tools for a fair statistical analysis of this kind. They can be thought of as a modern quantitative formulation of Occam's razor. They read [46–48]:

$$\mathrm{AIC} = \chi^2_{\rm min} + \frac{2nN}{N-n-1}\,, \quad \mathrm{BIC} = \chi^2_{\rm min} + n\ln N\,, \tag{21}$$

where n is the number of independent fitting parameters and N the number of data points. The bigger are the (positive) differences ΔAIC and ΔBIC with respect to the model having smaller values of AIC and BIC the higher is the evidence against the model with larger AIC and BIC. Take, for instance, Tables 1 and 2, where in all cases the less favored model is the ΛCDM (thus with larger AIC and BIC). For ΔAIC and ΔBIC in the range 6–10 one speaks of "strong evidence" against the ΛCDM, and hence in favor of the nonstandard models being considered. This is typically the situation for the RVM and Q_{dm} vacuum models in Table 2 and for the three wDVMs in Table 1. Neither the XCDM nor the Q_Λ vacuum model attain the "strong evidence" threshold in Tables 1 or 2. The XCDM parametrization, which is used as a baseline for comparison of the dynamical DE models, is nevertheless capable of detecting significant signs of dynamical DE, mainly in Table 1 (in which $H_0^{\rm Riess}$ is excluded), but not so in Table 2 (where $H_0^{\rm Riess}$ is included). In contrast, model Q_Λ does not change much from Table 1 to Table 2.

In actual fact, the vacuum model III (Q_Λ) tends to remain always fairly close to the ΛCDM. Its dynamics is weaker than that of the main DVMs (RVM and Q_{dm}). Being $|\nu_i| \ll 1$ for all the DVMs, the evolution of its vacuum energy density is approximately logarithmic: $\rho_\Lambda^{\rm III} \sim \rho_{\Lambda 0}(1-3\nu_\Lambda \ln a)$, as it follows from (13) with $\epsilon = 0$. Thus, it is significantly milder in comparison to that of the main DVMs, for which $\rho_\Lambda^{\rm I,II} \sim \rho_{\Lambda 0}\left[1+(\Omega_m^0/\Omega_\Lambda^0)\nu_i(a^{-3}-1)\right]$. The performance of Q_Λ can only be slightly better than that of ΛCDM, a fact that may have not been noted in previous studies – see [21, 26,40–42] and references therein.

According to the same jargon, when the differences ΔAIC and ΔBIC are both above 10 one speaks of "very strong evidence" against the unfavored model (the ΛCDM, in this case), wherefore in favor of the dynamical vacuum and quasi-vacuum models. It is certainly the case of the RVM and Q_{dm} models in Table 1, which are singled out as being much better than the ΛCDM in their ability to describe the overall observations. From Table 1 we can see that the best-fit values of ν_i for these models are secured at a confidence level of $\sim 3.8\sigma$. These two models are indeed the most conspicuous ones in our entire analysis, and remain strongly favored even if $H_0^{\rm Riess}$ [13] is included (cf. Table 2). In the last case, the best-fit values of ν_i for the two models are still supported at a fairly large c.l. ($\sim 3.2\sigma$). This shows that the overall fit to the data in terms of dynamical vacuum is a real option since the fit quality is not exceedingly perturbed in the presence of the data point $H_0^{\rm Riess}$. However, the optimal situation is really attained in the absence of that point, not only because the fit quality is then higher but also because that point remains out of the fit range whenever the large scale structure formation data (LSS) are included. For this reason we tend to treat that input as an outlier – see also [49] for an alternative support to this possibility, which we comment later on. In the following, we will argue that a truly consistent picture with all the data is only possible for H_0 in the vicinity of $H_0^{\rm Planck}$ rather than in that of $H_0^{\rm Riess}$.

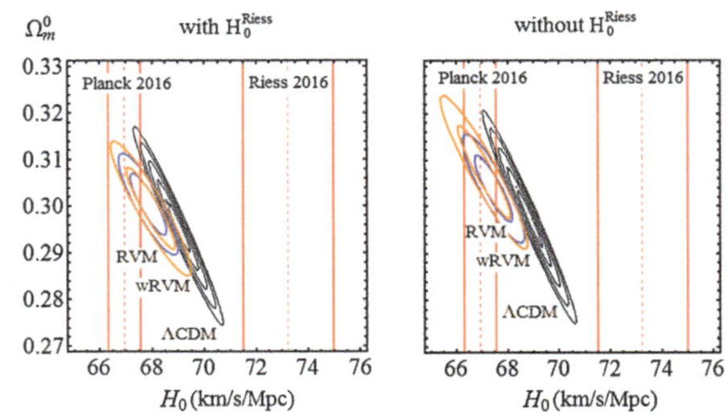

Fig. 3. Contour plots for the RVM (blue) and wRVM (orange) up to 2σ, and for the ΛCDM (black) up to 5σ in the (H_0, Ω_m^0)-plane. Shown are the two relevant cases under study: the plot on the left is for when the local H_0 value of Riess et al. [13] is included in the fit (cf. Table 2), and the plot on the right is for when that local value is *not* included (cf. Table 1). Any attempt at reaching the $H_0^{\rm Riess}$ neighborhood enforces to pick too small values $\Omega_m^0 < 0.27$ through extended contours that go beyond 5σ c.l. We also observe that the two (w)RVMs are much more compatible (already at 1σ) with the $H_0^{\rm Planck}$ range than the ΛCDM. The latter, instead, requires some of the most external contours to reach the $H_0^{\rm Planck}$ 1σ region whether $H_0^{\rm Riess}$ is included or not in the fit. Thus, remarkably, in both cases when the full data string SNIa+BAO+$H(z)$+LSS+CMB enters the fit the ΛCDM has difficulties to overlap also with the $H_0^{\rm Planck}$ range at 1σ, in contrast to the RVM and wRVM. (For interpretation of the references to color in this figure legend, the reader is referred to the web version of this article.)

The conclusion is that the $H_0^{\rm Riess}$–$H_0^{\rm Planck}$ tension cannot be relaxed without unduly forcing the overall fit, which is highly sensitive to the LSS data. It goes without saying that one cannot have a prediction that matches both H_0 regions at the same time, so at some point new observations (or the discovery of some systematic in one of the experiments) will help to consolidate one of the two ranges of values and exclude definitely the other. At present no fa-

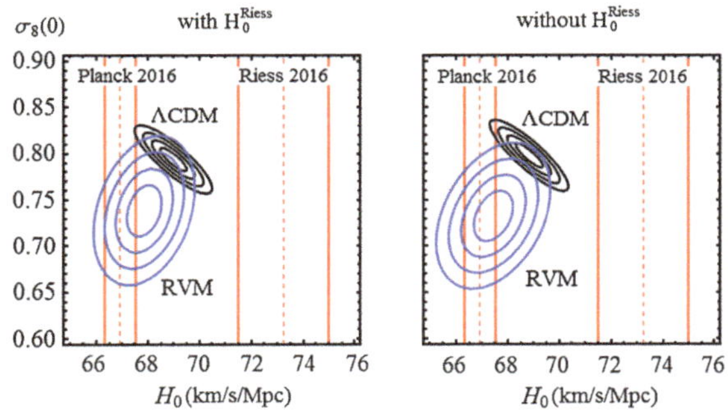

Fig. 4. Contour lines for the ΛCDM (black) and RVM (blue) up to 4σ in the $(H_0, \sigma_8(0))$-plane. As in Fig. 3, we present in the *left plot* the case when the local H_0 value of Riess et al. [13] is included in the fit (cf. Table 2), whereas in the *right plot* the case when that local value is *not* included (cf. Table 1). Again, any attempt at reaching the $H_0^{\rm Riess}$ neighborhood enforces to extend the contours beyond the 5σ c.l., which would lead to a too low value of Ω_m^0 in both cases (cf. Fig. 3) and, in addition, would result in a too large value of $\sigma_8(0)$ for the RVM. Notice that H_0 and $\sigma_8(0)$ are positively correlated in the RVM (i.e. H_0 decreases when $\sigma_8(0)$ decreases), whilst they are anticorrelated in the ΛCDM (H_0 increases when $\sigma_8(0)$ decreases, and vice versa). It is precisely this opposite correlation feature with respect to the ΛCDM what allows the RVM to improve the LSS fit in the region where both H_0 and $\sigma_8(0)$ are smaller than the respective ΛCDM values (cf. Fig. 1). This explains why the Planck range for H_0 is clearly preferred by the RVM, as it allows a much better description of the LSS data. (For interpretation of the references to color in this figure legend, the reader is referred to the web version of this article.)

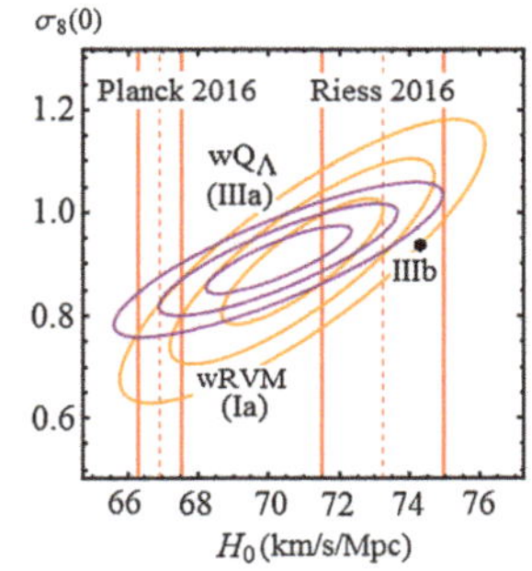

Fig. 5. Contour lines for the models wRVM (Ia) and wQ$_\Lambda$ (IIIa) up to 3σ in the $(H_0, \sigma_8(0))$-plane, depicted in orange and purple, respectively, together with the isolated point (in black) extracted from the analysis of Ref. [26], which we call IIIb. The cases Ia, IIIa and IIIb correspond to special scenarios with $w \neq -1$ for Models I and III in which the value $H_0^{\rm Riess}$ is included as a data point and then a suitable strategy is followed to optimize the fit agreement with such value. The strategy consists to exploit the freedom in w and remove the LSS data from the fit analysis. The plot clearly shows that some agreement is indeed possible, but only if w takes on values in the phantom region ($w < -1$) (see text) and at the expense of an anomalous (too large) value of the parameter $\sigma_8(0)$, what seriously spoils the concordance with the LSS data, as can be seen in Fig. 2. (For interpretation of the references to color in this figure legend, the reader is referred to the web version of this article.)

vorable fit can be obtained from the ΛCDM that is compatible with any of the two H_0 ranges. This is transparent from Figs. 3 and 4, in which the ΛCDM remains always in between the two regions. However, our work shows that a solution (with minimum cost) is possible in terms of vacuum dynamics. Such solution, which inevitably puts aside the $H_0^{\rm Riess}$ range, is however compatible with all the remaining data and tends to favor the Planck range of H_0 values. The DVMs can indeed provide an excellent fit to the overall cosmological observations and be fully compatible with both the $H_0^{\rm Planck}$ value and at the same time with the needed low values of the $\sigma_8(0)$ observable, these low values of $\sigma_8(0)$ being crucial to fit the structure formation data. Such strategy is only possible in the presence of vacuum dynamics, whilst it is impossible with a rigid Λ-term, i.e. is not available to the ΛCDM.

In Fig. 1 we confront the various models with the LSS data when the local measurement $H_0^{\rm Riess}$ is not included in our fit. The differences can be better appraised in the plot on the right, where we observe that the RVM and Q_{dm} curves stay significantly lower than the ΛCDM one (hence matching better the data than the ΛCDM), whereas those of XCDM and Q_Λ remain in between.

Concerning the wDVMs, namely the quasi-vacuum models in which an extra parameter is at play (the EoS parameter w), we observe a significant difference as compared to the DVMs (with vacuum EoS $w = -1$): they *all* provide a similarly good fit quality, clearly superior to that of the ΛCDM (cf. Tables 1 and 2) but in all cases below that of the main DVMs (RVM and Q_{dm}), whose performance is outstanding.

In Table 3, in an attempt to draw our fit nearer and nearer to $H_0^{\rm Riess}$ [13], we test the effect of ignoring the LSS structure formation data, thus granting more freedom to the fit parameter space. We perform this test using the ΛCDM and models (w)RVM and (w)Q_Λ (i.e. models I and III and testing both the vacuum and quasi-vacuum options), and we fit them to the CMB+BAO data alone. We can see that the fit values for H_0 increase in all starred scenarios (i.e. those involving the $H_0^{\rm Riess}$ data point in the fit), and specially for the cases Ia and IIIa in Table 3. Nonetheless, these lead to $\nu_i < 0$ and $w < -1$ (and hence imply phantom-like DE); and, what is worse, the agreement with the LSS data is ruined (cf. Fig. 2) since the corresponding curves are shifted too high (beyond the ΛCDM one). In the same figure we superimpose one more scenario, called IIIb, corresponding to a rather acute phantom behavior ($w = -1.184 \pm 0.064$). The latter was recently explored in [26] so as to maximally relax the H_0 tension – see also [21]. Unfortunately, we find (see Fig. 2) that the associated LSS curve is completely strayed since it fails to minimally describe the $f\sigma_8$ data (LSS).

In Fig. 3 we demonstrate in a very visual way that, in the context of the overall observations (i.e. SNIa+BAO+$H(z)$+LSS+CMB), whether including or not including the data point $H_0^{\rm Riess}$ (cf. Tables 1 and 2), it becomes impossible to getting closer to the local measurement $H_0^{\rm Riess}$ unless we go beyond the 5σ contours and end up with a too low value $\Omega_m^0 < 0.27$. These results are aligned with those of [50], in which the authors are also unable to accommodate the $H_0^{\rm Riess}$ value when a string of SNIa+BAO+$H(z)$+LSS+CMB data (similar but *not* equal to the one used by us) is taken into account. Moreover, we observe in Fig. 3 not only that both the RVM and wRVM remain much closer to $H_0^{\rm Planck}$ than to $H_0^{\rm Riess}$, but also that they are overlapping with the $H_0^{\rm Planck}$ range much better than the ΛCDM does. The latter is seen to have serious difficulties in reaching the Planck range unless we use the most external regions of the elongated contours shown in Fig. 3.

Many other works in the literature have studied the existing H_0 tension. For instance, in [28] the authors find $H_0 = 69.13 \pm 2.34$ km/s/Mpc assuming the ΛCDM model. Such result almost coincides with the central values of H_0 that we obtain in Tables 1 and 2 for the ΛCDM. This fact, added to the larger uncertainties of the result, seems to relax the tension. Let us, however, notice that the value of [28] has been obtained using BAO data only, what explains the larger uncertainty that they find. In our case, we have considered a much more complete data set, which includes CMB and LSS data as well. This is what has allowed us to better constrain H_0 with smaller errors and conclude that when a larger data set (SNIa+BAO+$H(z)$+LSS+CMB) is used, the fitted value of the Hubble parameter for the ΛCDM is incompatible with the Planck best-fit value at about 4σ c.l. Thus, the ΛCDM model seems to be in conflict not only with the local HST estimation of H_0, but also with the Planck one!

Finally, in Figs. 4 and 5 we consider the contour plots (up to 4σ and 3σ, respectively) in the $(H_0, \sigma_8(0))$-plane for different situations. Specifically, in the case of Fig. 4 the plots on the left and on the right are in exact correspondence with the situations previously presented in the left and right plots of Fig. 3, respectively.[1] As expected, the contours in the left plot of Fig. 4 are slightly shifted ("attracted") to the right (i.e. towards the $H_0^{\rm Riess}$ region) as compared to those in the right plot because in the former $H_0^{\rm Riess}$ was included as a data point in the fit, whereas $H_0^{\rm Riess}$ was not included in the latter. Therefore, in the last case the contours for the RVM are more centered in the $H_0^{\rm Planck}$ region and at the same time centered at relatively low values of $\sigma_8(0) \simeq 0.73$–0.74, which are precisely those needed for a perfect matching with the experimental data points on structure formation (cf. Fig. 1). On the other hand, in the case of Fig. 5 the contour lines correspond to the fitting sets Ia, IIIa of Table 3 (in which BAO and CMB data, but *no* LSS formation data, are involved). As can be seen, the contour lines in Fig. 5 can attain the Riess 2016 region for H_0, but they are centered at rather high values (~ 0.9) of the parameter $\sigma_8(0)$. These are clearly higher than the needed values $\sigma_8(0) \simeq 0.73$–0.74. This fact demonstrates once more that such option leads to a bad description of the structure formation data. The isolated point in Fig. 5 is even worst: it corresponds to the aforementioned theoretical prediction for the scenario IIIb proposed in [26], in which the $H_0^{\rm Riess}$ region can be clearly attained but at the price of a serious disagreement with the LSS data. Here we can see, with pristine clarity, that such isolated point, despite it comfortably reaches the $H_0^{\rm Riess}$ region, it attains a value of $\sigma_8(0)$ near 1, thence completely strayed from the observations. This is, of course, the reason why the upper curve in Fig. 2 fails to describe essentially all points of the $f(z)\sigma_8(z)$ observable. So, as it turns, it is impossible to reach the $H_0^{\rm Riess}$ region without paying a high price, no matter what strategy is concocted to approach it in parameter space.

As indicated, we must still remain open to the possibility that the $H_0^{\rm Planck}$ and/or $H_0^{\rm Riess}$ measurements are affected by some kind of (unknown) systematic errors, although some of these possibilities may be on the way of being ruled out by recent works. For instance, in [51] the authors study the systematic errors in Planck's data by comparing them with the South Pole Telescope data. Their conclusion is that there is no evidence of systematic errors in Planck's results. If confirmed, the class of the (w)RVMs studied here would offer a viable solution to both the H_0 and $\sigma_8(0)$ existing tensions in the data, which are both unaccountable within the ΛCDM. Another interesting result is the "blinded" determination of H_0 from [27], based on a reanalysis of the SNIa and Cepheid variables data from the older work by Riess et al. [14]. These authors find $H_0 = 72.5 \pm 3.2$ km/s/Mpc, which should be compared with $H_0 = 73.8 \pm 2.4$ km/s/Mpc [14]. Obviously, the tension with $H_0^{\rm Planck}$ diminished since the central value decreased and in addition the uncertainty has grown by $\sim 33\%$. We should now wait for a similar reanalysis to be made on the original sample used in [13], i.e. the one supporting the value $H_0^{\rm Riess}$, as planned in [27]. In [52] they show that by combining the latest BAO results with WMAP, Atacama Cosmology Telescope (ACT), or South Pole Telescope (SPT) CMB data produces values of H_0 that are 2.4–3.1σ lower than the distance ladder, independent of Planck. These authors conclude from their analysis that it is not possible to explain the H_0 disagreement solely with a systematic error specific to the Planck data. Let us mention other works, see e.g. [24,29], in which a value closer to $H_0^{\rm Riess}$ is found and the tension is not so severely loosened; or the work [53], which excludes systematic bias or uncertainty in the Cepheid calibration step of the distance ladder measurement by [13]. Finally, we recall the aforementioned recent study [49], where the authors run a new (dis)cordance test to compare the constraints on H_0 from different methods and conclude that the local measurement is an outlier compared to the others, what would favor a systematics-based explanation. Quite obviously, the search for a final solution to the H_0 tension is still work in progress.

5. Conclusions

The present updated analysis of the cosmological data SNIa + BAO + $H(z)$ + LSS + CMB disfavors the hypothesis $\Lambda =$ const. as compared to the dynamical vacuum models (DVMs). This is consistent with our most recent studies [30–34]. Our results suggest a dynamical DE effect near 3σ within the standard XCDM parametrization and near 4σ for the best DVMs. Here we have extended these studies in order to encompass the class of quasi-vacuum models (wDVMs), where the equation of state parameter w is near (but not exactly equal) to -1. The new degree of freedom w can then be used to try to further improve the overall fit to the data. But it can also be used to check if values of w different from -1 can relax the existing tension between the two sets of measurements of the H_0 parameter, namely those based: i) on the CMB measurements by the Planck collaboration [5,12], and ii) on the local measurement (distance ladder method) using Cepheid variables [13].

Our study shows that the RVM with $w = -1$ remains as the preferred DVM for the optimal fit of the data. At the same time it favors the CMB measurements of H_0 over the local measurement. Remarkably, we find that not only the CMB and BAO data, but also the LSS formation data (i.e. the known data on $f(z)\sigma_8(z)$ at different redshifts), are essential to support the CMB measurements of H_0 over the local one. We have checked that if the LSS data are not considered (while the BAO and CMB are kept), then there is a unique chance to try to accommodate the local measurement of H_0, but only at the expense of a phantom-like behavior (i.e. for $w < -1$). In this region of the parameter space, however, we find that the agreement with the LSS formation data is manifestly lost, what suggests that the $w < -1$ option is ruled out. There is no other window in the parameter space where to accommodate the local H_0 value in our fit. In contrast, when the LSS formation data are restored, the fit quality to the overall SNIa+BAO+$H(z)$+LSS+CMB observations improves dramatically and definitely favors the Planck range for H_0 as well as smaller values for $\sigma_8(0)$ as compared to the ΛCDM.

In short, our work suggests that signs of dynamical vacuum energy are encoded in the current cosmological observations. They appear to be more in accordance with the lower values of H_0 obtained from the Planck (CMB) measurements than with the higher range of H_0 values obtained from the present local (distance ladder) measurements, and provide smaller values of $\sigma_8(0)$ that are in better agreement with structure formation data as compared to the ΛCDM. We hope that with new and more accurate observations, as well as with more detailed analyses, it will be possible to assess the final impact of vacuum dynamics on the possible solution of the current tensions in the ΛCDM.

6. Acknowledgments

We are partially supported by MINECO FPA2016-76005-C2-1-P, Consolider CSD2007-00042, 2014-SGR-104 (Generalitat de Catalunya) and MDM-2014-0369 (ICCUB).

7. Note added in proof

Since the first version of this work appeared in preprint form, arXiv:1705.06723, new analyses of the cosmological data have appeared, in particular the one-year results by the DES collaboration

[1] The $H_0^{\rm Planck}$ band indicated in Figs. 3–5 is that of [12], which has no significant differences with that of [5].

(DES Y1 for short) [54]. They do not find evidence for dynamical DE, and the Bayes factor indicates that the DES Y1 and Planck data sets are consistent with each other in the context of ΛCDM. However, in our previous works – see in particular [31,34] – we explained why the Planck results did not report evidence on dynamical DE. For instance, in [5] they did not use LSS (RSD) data, and in [6] they only used a limited set of BAO and LSS points. In the mentioned works [31,34] we have shown that under the same conditions we recover their results, but when we use the full data string, which involves not only CMB but also the rich BAO+LSS data set, we do obtain instead positive indications of dynamical DE. A similar situation occurs with DES Y1; they do not use direct data on LSS structure formation despite they recognize that smaller values of $\sigma_8(0)$ than those predicted by the ΛCDM are necessary to solve the tension existing between the concordance model and the LSS observations. In contrast, let us finally mention that our positive result on dynamical DE is consistent with the recent analysis by Gong-Bo Zhao et al. [55], who report on a signal of dynamical DE at 3.5σ c.l. using similar data ingredients as in our analysis.

References

[1] A.G. Riess, et al., Astron. J. 116 (1998) 1009.
[2] S. Perlmutter, et al., Astrophys. J. 517 (1999) 565.
[3] S. Weinberg, Rev. Mod. Phys. 61 (1989) 1.
[4] P.J.E. Peebles, Principles of Physical Cosmology, Princeton Univ. Press, Princeton, 1993.
[5] P.A.R. Ade, et al., Planck 2015 results. XIII, Astron. Astrophys. 594 (2016) A13.
[6] P.A.R. Ade, et al., Planck 2015 results. XIV, Astron. Astrophys. 594 (2016) A14.
[7] P.J.E. Peebles, B. Ratra, Rev. Mod. Phys. 75 (2003) 559.
[8] L. Amendola, S. Tsujikawa, Dark Energy, Cambridge Univ. Press, Cambridge, 2010, 2015.
[9] A. Sandage, Astrophys. J. 133 (1961) 355.
[10] G.A. Tammann, A. Sandage, The Hubble Constant and HST, in: F. Macchetto (Ed.), The Impact of HST on European Astronomy, in: Astrophysics and Space Science Proceedings, Springer, Dordrecht, 2010, p. 289.
[11] G.A. Tammann, B. Reindl, in: IAU Symp., vol. 289, 2013, p. 13;
G.A. Tammann, B. Reindl, Astron. Astrophys. 549 (2013) A136.
[12] N. Aghanim, et al., Planck intermediate results. XLVI, Astron. Astrophys. 596 (2016) A107.
[13] A.G. Riess, et al., Astrophys. J. 826 (2016) 56.
[14] A.G. Riess, et al., Astrophys. J. 730 (2011) 119, Erratum: Astrophys. J. 732 (2011) 129.
[15] T. Padmanabhan, Phys. Rep. 380 (2003) 235.
[16] J. Solà, J. Phys. Conf. Ser. 453 (2013) 012015.
[17] W. Baade, Astrophys. J. 100 (1944) 137.
[18] G.A. Tammann, Publ. Astron. Soc. Pac. 108 (1996) 1083.
[19] G. Chen, B. Ratra, Publ. Astron. Soc. Pac. 123 (2011) 1127.
[20] S. Bethapudi, S. Desai, Eur. Phys. J. Plus 132 (2017) 78.
[21] E.D. Valentino, A. Melchiorri, J. Silk, Phys. Lett. B 761 (2016) 242.
[22] J.L. Bernal, L. Verde, A.G. Riess, J. Cosmol. Astropart. Phys. 1610 (2016) 019.
[23] A. Shafieloo, D.K. Hazra, J. Cosmol. Astropart. Phys. 1704 (2017) 012.
[24] W. Cardona, M. Kunz, V. Pettorino, J. Cosmol. Astropart. Phys. 1703 (2017) 056.
[25] E.D. Valentino, A. Melchiorri, E.V. Linder, J. Silk, Phys. Rev. D 96 (2017) 023523.
[26] E.D. Valentino, A. Melchiorri, O. Mena, Phys. Rev. D 96 (2017) 043503.
[27] B.R. Zhang, et al., Mon. Not. R. Astron. Soc. 471 (2017) 2254.
[28] Y. Wang, X. Lixin, G.B. Zhao, A measurement of the Hubble constant using galaxy redshift surveys, arXiv:1706.09149.
[29] S.M. Feeney, D.J. Mortlock, N. Dalmasso, Clarifying the Hubble constant tension with a Bayesian hierarchical model of the local distance ladder, arXiv: 1707.00007.
[30] J. Solà, A. Gómez-Valent, J. de Cruz Pérez, Astrophys. J. 811 (2015) L14.
[31] J. Solà, A. Gómez-Valent, J. de Cruz Pérez, Astrophys. J. 836 (2017) 43.
[32] J. Solà, A. Gómez-Valent, J. de Cruz Pérez, Mod. Phys. Lett. A 32 (2017) 1750054.
[33] J. Solà, Int. J. Mod. Phys. A 31 (2016) 1630035.
[34] J. Solà, J. de Cruz Pérez, A. Gómez-Valent, Towards the firsts compelling signs of vacuum dynamics in modern cosmological observations, arXiv:1703.08218;
J. Solà, J. de Cruz Pérez, A. Gómez-Valent, R.C. Nunes, Dynamical Vacuum against a rigid Cosmological Constant, arXiv:1606.00450.
[35] S.M. Turner, M. White, Phys. Rev. D 56 (1997) R4439.
[36] Y. Wang, M. Dai, Phys. Rev. D 94 (2016) 083521.
[37] J. Solà, A. Gómez-Valent, Int. J. Mod. Phys. D 24 (2015) 1541003.
[38] J. Solà, J. Phys. A 41 (2008) 164066.
[39] J.A.S. Lima, S. Basilakos, J. Solà, Mon. Not. R. Astron. Soc. 431 (2013) 923.
[40] V. Salvatelli, et al., Phys. Rev. Lett. 113 (2014) 181301.
[41] R. Murgia, S. Gariazzo, N. Fornengo, J. Cosmol. Astropart. Phys. 1604 (2016) 014.
[42] Y.H. Li, J.F. Zhang, X. Zhang, Phys. Rev. D 93 (2016) 023002.
[43] A. Gómez-Valent, J. Solà, S. Basilakos, J. Cosmol. Astropart. Phys. 1501 (2015) 004.
[44] A. Gómez-Valent, J. Solà, Mon. Not. R. Astron. Soc. 448 (2015) 2810.
[45] H. Gil-Marín, et al., Mon. Not. R. Astron. Soc. 465 (2017) 1757.
[46] H. Akaike, IEEE Trans. Autom. Control 19 (1974) 716.
[47] G. Schwarz, Ann. Stat. 6 (1978) 461.
[48] R.E. Kass, A. Raftery, J. Am. Stat. Assoc. 90 (1995) 773.
[49] W. Lin, M. Ishak, Cosmological discordances II: Hubble constant, Planck and large-scale-structure data sets, arXiv:1708.09813.
[50] Z. Zhai, M. Blanton, A. Slosar, J. Tinker, An evaluation of cosmological models from expansion and growth of structure measurements, arXiv:1705.10031.
[51] A. Aylor, et al., A comparison of cosmological parameters determined from CMB temperature power spectra from the South Pole telescope and the Planck Satellite, arXiv:1706.10286.
[52] G.E. Addison, et al., Elucidating ΛCDM: impact of baryon acoustic oscillation measurements on the Hubble constant discrepancy, arXiv:1707.06547.
[53] B. Follin, L. Knox, Insensitivity of the distance ladder Hubble constant determination to Cepheid calibration modeling choices, arXiv:1707.01175.
[54] DES Collaboration, T.M.C. Abbott, et al., Dark energy survey year 1 results: cosmological constraints from galaxy clustering and weak lensing, e-Print: arXiv:1708.01530.
[55] Gong-Bo Zhao, et al., Dynamical dark energy in light of the latest observations, Nat. Astron. 1 (2017) 627, arXiv:1701.08165.

29

Ultra slow-roll inflation demystified

Konstantinos Dimopoulos

Consortium for Fundamental Physics, Physics Department, Lancaster University, Lancaster LA1 4YB, UK

A R T I C L E   I N F O

Editor: M. Trodden

A B S T R A C T

Ultra-slow-roll (USR) inflation is a new mode of inflation which corresponds to the occasions when the inflaton field must traverse an extremely flat part of the scalar potential, when the usual slow-roll (SR) fails. We investigate USR and obtain an estimate for how long it lasts, given the initial kinetic density of the inflaton. We also find that, if the initial kinetic density is small enough, USR can be avoided and the usual SR treatment is valid. This has important implications for inflection-point inflation.

1. Introduction

Cosmic inflation is an organic component of the concordance model of cosmology. It is a period of exponential expansion in the early Universe, which determines the initial conditions for the subsequent Hot Big Bang cosmology. In particular, it makes the Universe spatially flat, large and uniform but also provides the necessary deviations from perfect uniformity in the form of the primordial curvature perturbation, which accounts for the eventual formation of the large scale structure. Typically, inflation is modelled through the inflationary paradigm, which suggests that the Universe undergoes inflation when dominated by the potential density of a scalar field (inflaton). This potential density remains roughly constant during inflation. As a result, the generated curvature perturbation is almost scale-invariant, as suggested by observations. In order to keep the potential density roughly constant, the variation of the field must be very small throughout inflation. Because the inflaton's equation of motion is the same as a body rolling down a potential slope subject to friction, we need this roll to be slow for the inflaton, in field space, so as to keep the potential density roughly unchanged. Thus, in the inflationary paradigm, the inflaton undergoes slow-roll (SR) during inflation. Indeed, the latest CMB data favours single-field slow-roll inflation [1].

The SR solution is an attractor [2] as long as the potential is flat enough to support it. However, it was recently realised that SR may end not only when the potential becomes steep and curved, as is for the end of inflation, but also when it suddenly becomes extremely flat, too flat for the regular SR assumptions to apply. In this case, the system engages in so-called ultra slow-roll (USR) inflation. This new mode of inflation has been hitherto unknown. It can have a profound impact on inflationary observables, so it must be taken into account. However, even though diagnosed, USR has not been fully understood, with most of its dynamics traced numerically. In this letter, we attempt to demystify USR and provide a conceptual understanding of its dynamics. Ignoring USR can lead to important miscalculations of inflationary observables.

USR arises when the potential becomes extremely flat, so much so that, SR would force the kinetic density of the field to reduce faster than it would if the field were in free-fall $\rho_{\rm kin} \equiv \frac{1}{2}\dot{\phi}^2 \propto a^{-6}$, which of course cannot happen. Thus, the system departs from SR and the field engages in USR, during which the kinetic density decreases as in free-fall, until the system can get back to SR, when the decreasing $|\ddot{\phi}|$ catches up with the slope of the potential $|V'|$, or until inflation ends, e.g. by a phase transition. Note that, even though the slope is very small, we still have potential domination $V > \rho_{\rm kin}$ so inflation continues. USR was first investigated in Ref. [3], which was followed by Refs. [4,5] and recently by Ref. [6]. In Refs. [3] and [5] a constant potential is assumed, which cannot exhibit SR. In Ref. [4] it was shown that USR is not an attractor solution and the system departs from it as soon as the conditions which enforce USR allow it. But which conditions are these?

In this letter we explore this question. To obtain an insight of the dynamics of USR, we study USR in linear inflation and then generalise our findings for an arbitrary inflation model. We particularly consider inflection-point inflation because it can lead to USR. It is fair to say that the community seems little aware of USR, so the hope is that our treatment may be revealing of USR's nature. This is a particularly acute problem in models of inflection-point inflation, where a region of USR exists around the inflection point. In USR this region is traversed in a moderate number of e-folds. However, were SR assumed, this number would grow substantially. As inflationary observables are determined by the correct number

E-mail address: k.dimopoulos1@lancaster.ac.uk.

of e-folds, this can have profound implications on inflationary predictions and on the viability of inflection-point models.

We use natural units, where $c=\hbar=1$ and $8\pi G=m_P^{-2}$, with $m_P=2.43\times10^{18}$ GeV being the reduced Planck mass.

2. Ultra-slow roll inflation

To explore USR inflation, we will look closely at the Klein–Gordon equation of motion of the canonical homogeneous inflaton field ϕ:

$$\ddot{\phi}+3H\dot{\phi}+V'=0\,, \tag{1}$$

where $H\equiv\dot{a}/a$ is the Hubble parameter (with a being the scale factor) V is the scalar potential and the dot {prime} denotes derivative with respect to the cosmic time {the inflaton field}. We name each term of the above as the acceleration, the friction and the slope term respectively. We also employ the flat Friedman equation during inflation, when the Universe is dominated by the inflaton field:

$$3H^2m_P^2=\frac{1}{2}\dot{\phi}^2+V\,. \tag{2}$$

We define two slow-roll parameters

$$\epsilon\equiv-\dot{H}/H^2 \tag{3}$$

and

$$\epsilon_2\equiv\frac{\dot{\epsilon}}{\epsilon H}=-6-\frac{2V'}{H\dot{\phi}}+2\epsilon=\frac{2\ddot{\phi}}{H\dot{\phi}}+2\epsilon\,, \tag{4}$$

where we have employed Eqs. (1) and (2). It is easy to show that

$$\epsilon=\frac{3}{2}(1+w)\,, \tag{5}$$

where w is the barotropic parameter of the homogeneous inflaton field, given by

$$w=\frac{\rho_{\rm kin}-V}{\rho_{\rm kin}+V}\,, \tag{6}$$

where $\rho_{\rm kin}\equiv\frac{1}{2}\dot{\phi}^2$. For inflation we need $w<-\frac{1}{3}$, which means $V>2\rho_{\rm kin}$. From Eq. (5), we see that inflation (accelerated expansion) occurs when $\epsilon<1$.

Now, in the usual SR, the acceleration term in Eq. (1) is negligible, so the latter becomes

$$3H\dot{\phi}\simeq-V\,, \tag{7}$$

which shows that the friction term is locked to the slope term. In this case, Eq. (4) becomes

$$\epsilon_2=-2\eta+4\epsilon\,, \tag{8}$$

where the usual SR parameters are

$$\epsilon\simeq\epsilon_{\rm SR}\equiv\frac{1}{2}m_P^2\left(\frac{V'}{V}\right)^2\quad\text{and}\quad\eta\equiv m_P^2\frac{V''}{V}\,. \tag{9}$$

During SR, $\epsilon,|\eta|\ll1$, which means that $|\epsilon_2|\ll1$.

However, if the potential suddenly becomes extremely flat, the slope term in Eq. (1) may reduce drastically, which means that it virtually disappears. The equation is then rendered

$$\ddot{\phi}+3H\dot{\phi}\simeq0\,, \tag{10}$$

which shows that the friction term is now locked with the acceleration term. In this case, Eq. (4) becomes

$$\epsilon_2=-6+2\epsilon\,. \tag{11}$$

During inflation $\epsilon<1$, which means $|\epsilon_2|\approx6$. Thus, if during inflation, the potential becomes suddenly very flat, $|\epsilon_2|$, which is initially small grows to larger than unity, SR is applicable no-more and a period of USR begins.

Intuitively, one can understand this as follows. If we are in SR but the slope $|V'|$ reduces drastically, it initially drags with it the friction term, by virtue of Eq. (7). This decreases the value of $|\dot{\phi}|$, i.e. the kinetic density $\rho_{\rm kin}=\frac{1}{2}\dot{\phi}^2$, but this value cannot decrease arbitrarily quickly. The fastest it can decrease is $\rho_{\rm kin}\propto a^{-6}$, which we call free-fall because it corresponds to a field with no potential density $V=0$, such that its equation of motion is Eq. (10). Therefore, if the kinetic density of SR is forced (by the decreasing slope) to reduce faster than free-fall then the system breaks away from SR. In SR the acceleration term is negligible, because it is very small, compared to the friction and slope terms, which are locked together as shown in Eq. (7). However, if the slope reduces drastically and drags the friction term with it, they both become small too and eventually comparable to the acceleration term. So all three terms in Eq. (1) are comparable. When this happens, the friction term changes allegiances and becomes locked with the acceleration term, resulting in USR.

Now, once in USR, the field becomes oblivious of the potential, as demonstrated by Eq. (10). This is similar to the kination period of quintessential inflation models [7,8] but there is a crucial difference. In kination, the Universe is dominated by $\rho_{\rm kin}$, while in USR inflation, we still have potential domination and $V>\rho_{\rm kin}$. Being oblivious to the potential, the inflaton field can even climb up an ultra-shallow V [4]. Indeed, when the system enters the USR regime, it "flies over" the flat patch of the potential, sliding on its decreasing kinetic density. In that sense, the term ultra-SR is actually a misnomer, because the field rolls faster than it would have done if SR were still applicable over the extremely flat region.

Indeed, if $|V'|$ decreases to almost zero, so does $\epsilon_{\rm SR}$. In SR the number of elapsing e-folds is

$$\Delta N=\frac{1}{m_P}\int_{\phi_1}^{\phi_2}\frac{{\rm d}\phi}{\sqrt{2\epsilon_{\rm SR}}}\,, \tag{12}$$

which increases substantially if $\epsilon_{\rm SR}$ becomes extremely small. In contrast, in USR ϵ does not decrease too much, so we have $\epsilon_{\rm SR}\ll\epsilon<1$. The number of elapsing e-folds is given in general by

$$\Delta N=-\int\frac{{\rm d}H}{\epsilon H} \tag{13}$$

and in USR it can be much smaller compared to SR if $\epsilon_{\rm SR}\ll\epsilon$. Thus, when considering an inflation model that results in periods of USR, but only SR is assumed, there is a danger of overestimating the number of e-folds it takes for the field to roll down.

It is evident that USR depends on having substantial kinetic density, which cannot decrease faster than free-fall. However, if one begins inflation at the extremely flat region with very small kinetic density, then SR may be attained, quickly, even immediately. Now, the initial conditions for inflation are shrouded by the no-hair theorem, which renders them academic, because all memory is lost once the inflationary attractor is reached. Thus, provided inflation begins comfortably before the cosmological scales exit the horizon, the initial conditions of the inflaton field can be taken to correspond to kinetic density small enough to avoid USR despite an extremely flat scalar potential. This can rescue inflation models such as inflection-point inflation, which may have problems with USR. To quantify how small the initial kinetic density needs to be, we first investigate linear inflation.

3. Ultra-slow-roll in linear inflation and beyond

We consider the inflation model:

$$V = V_0 + M^3\phi\,, \tag{14}$$

where V_0 is a constant density scale and M is a mass scale. Then the Klein–Gordon Eq. (1) becomes:

$$\ddot{\phi} + 3H_0\dot{\phi} + M^3 = 0\,, \tag{15}$$

where $H_0^2 \equiv V_0/3m_P^2$ and we assumed $V_0 \gg M^3\phi$. The above has the general solution

$$\dot{\phi} = Ce^{-3\Delta N} - \frac{M^3}{3H_0}\,, \tag{16}$$

where $\Delta N = H_0\Delta t$ is the elapsing e-folds and C is a constant. We also find

$$\ddot{\phi} = -3H_0Ce^{-3\Delta N}\,. \tag{17}$$

If initially ($\Delta N = 0$) the velocity of the field is $\dot{\phi}_0 = 0$, then $C = M^3/3H_0$ and the Klein–Gordon suggests that $\ddot{\phi}_0 = -M^3$. Then, as time continues, the above suggest that the Klein–Gordon becomes

$$-M^3e^{-3\Delta N} + M^3(e^{-3\Delta N} - 1) + M^3 = 0\,. \tag{18}$$

Notice that, even though the friction term begins as zero it soon (in a single e-fold) dominates over the acceleration term and the slow-roll (SR) condition is recovered, where $\ddot{\phi}$ is negligible and V' is balanced by $3H\dot{\phi}$. Thus if we start with zero velocity, we have SR immediately afterwards.

Now suppose that, originally $\dot{\phi} \neq 0$. If $|C| \ll M^3/3H_0$ then $\dot{\phi} \simeq \dot{\phi}_0 \simeq -M^3/3H_0$ (cf. Eq. (16)), which means that the friction term is $3H\dot{\phi} \simeq -M^3 = V'$ and we have SR. Thus we always obtain immediately SR if $|C| \leq M^3/3H_0$. If $|C| \gg M^3/3H_0$ then $3H_0|\dot{\phi}_0| \simeq 3H_0|C| > M^3$, which means that the friction term initially dominates over the slope term and is balanced by the acceleration term, $|\ddot{\phi}_0| = 3H_0|C|$ according to Eq. (17). Thus, the Klein–Gordon is $\ddot{\phi} + 3H\dot{\phi} \simeq 0$ (cf. Eq. (10)), which gives rise to USR. USR continues until $3H_0|C|e^{-3\Delta N} = M^3$, when all three terms in the Klein–Gordon become comparable. Afterwards, the friction term becomes $3H\dot{\phi} \simeq -M^3$, which counterbalances the slope term, while the acceleration term becomes negligible. Thus, we recover SR.

Therefore, USR lasts

$$\Delta N_{\rm USR} = \frac{1}{3}\ln\left(\frac{3H_0|C|}{M^3}\right) = \frac{1}{3}\ln\left(\frac{3H_0\sqrt{2\rho_{\rm kin}^0}}{M^3}\right), \tag{19}$$

where $\rho_{\rm kin}^0 \equiv \frac{1}{2}\dot{\phi}_0^2$ is the initial kinetic density, which is $\rho_{\rm kin}^0 \simeq \frac{1}{2}C^2$ for large $|C|$.

All in all, we find that, to obtain a sizeable period of USR, we need

$$|C| \gg M^3/3H_0 \quad \Leftrightarrow \quad \rho_{\rm kin}^0 > \frac{M^6}{18H_0^2}. \tag{20}$$

Otherwise, we have SR only. Note that, if $M = 0$ and the potential is exactly flat, SR is never recovered [3].

We may generalise the above for an arbitrary potential, as follows. At extremely flat region of the potential we set $M^3 \equiv V'(\phi_f)$ and $H_0^2 = V(\phi_f)/3m_P^2$ and enforce the bound in Eq. (20), where ϕ_f corresponds to the flattest part of the potential. Thus, to avoid USR we need

$$\rho_{\rm kin}(\phi_f) \equiv \frac{1}{2}\dot{\phi}_f^2 \leq \left.\frac{(V')^2m_P^2}{6V}\right|_{\phi_f} = \frac{1}{3}\epsilon_{\rm SR}(\phi_f)V(\phi_f)\,. \tag{21}$$

The above makes sense, because the kinetic density in SR is

$$\rho_{\rm kin}^{\rm SR} = \frac{1}{2}\left(\frac{V'}{3H}\right)^2 = \frac{1}{3}\epsilon_{\rm SR}V\,, \tag{22}$$

where we used Eqs. (7) and (9). Thus, the bound in Eq. (21) really requests that the kinetic density in the flat patch be at most the one corresponding to SR: $\rho_{\rm kin}(\phi_f) \leq \rho_{\rm kin}^{\rm SR}$. This makes sense because if one has kinetic density in excess of $\rho_{\rm kin}^{\rm SR}$, the friction term in Eq. (1) cannot be balanced by the slope term and we have USR.

In view of the above, we can also recast Eq. (19) as

$$\Delta N_{\rm USR} = \frac{1}{6}\ln\left(\frac{3\rho_{\rm kin}^0}{\epsilon_{\rm SR}V}\right), \tag{23}$$

where we used the potentially dominated Friedman equation.

4. Ultra-slow-roll in inflection-point inflation

We now focus on inflection-point inflation, which may feature USR. Inflection point inflation corresponds to the case of a flat step on the otherwise steep potential wall. This step is formed because of opposing terms in the potential which almost cancel each other. There are many model realisations, most notably A-term inflation [9], MSSM inflation [10] and many others [11]. However, in the vast majority of these works the USR phase has not been considered, which may cast doubt on some of their findings.

To avoid the USR period, one only needs to assume that the initial kinetic density is small enough according to the bound in Eq. (21), where ϕ_f now corresponds to the inflection point, which is the flattest part of the potential plateau. This can be understood as follows. The potential for inflection-point inflation can be crudely approximated by three consecutive segments of linear potential. Inflation only takes place along the flattest segment, and it is similar to linear inflation.

While rolling from large values of ϕ to small, when the field reaches the flat segment then there is an abrupt reduction in $|V'|$. Because the friction term in the Klein–Gordon was at least as large as the slope term before reaching the flat segment (i.e. we had SR or free-fall), afterwards, the friction term cannot be balanced by the (substantially reduced) slope term. Thus, the acceleration term rushes to balance it and we have USR.

Now, during USR, we have $\rho_{\rm kin} \propto a^{-6}$ so that

$$\dot{\phi}\ddot{\phi} = \dot{\rho}_{\rm kin} = -6H\rho_{\rm kin} \propto a^{-6}, \tag{24}$$

where we took $H \simeq$constant. Because $|\dot{\phi}| = \sqrt{2\rho_{\rm kin}} \propto a^{-3}$, the above suggests that $|\ddot{\phi}| \propto a^{-3}$. After crossing the inflection point, though, the slope of the potential begins to increase, while the acceleration decreases, as we have seen. At some point, they meet each other and then the friction term changes allegiances and becomes locked with the slope term, so that SR is recovered.

But what if the evolution of the field had already begun at the flat patch? Then, provided the kinetic density is small enough, one can immediately have SR inflation [12]. The bound in Eq. (21) is a conservative estimate on the maximum kinetic density because the slope at the inflection point is smaller than at the rest of the plateau.

5. Quantum diffusion

Now, we briefly discuss quantum diffusion. If the potential is extremely flat, quantum fluctuations of the field may dominate its variation. The quantum variation of the field per Hubble time $\delta t = H^{-1}$ is typically given by the Hawking temperature in de Sitter space $\delta\phi = H/2\pi$. Thus, the kinetic density of quantum fluctuations is

$$\rho_{\text{kin}}^{\text{diff}} = \frac{1}{2}\left(\frac{\delta\phi}{\delta t}\right)^2 = \frac{H^4}{8\pi^4}. \tag{25}$$

This should be interpreted as the lowest value the kinetic density can have. If $|V'| < \frac{3}{2\pi}H^3$ then quantum diffusion overwhelms SR,[1] so that the number of USR e-folds is

$$\Delta N_{\text{USR}} = \frac{1}{6}\ln\left(\frac{\rho_{\text{kin}}^0}{\rho_{\text{kin}}^{\text{diff}}}\right) = \frac{1}{3}\ln\left(6\pi\,\frac{\sqrt{2\rho_{\text{kin}}^0}\,m_P^2}{V}\right), \tag{26}$$

where we used $\rho_{\text{kin}} \propto a^{-6}$ in USR and the potential dominated Friedman equation.

6. Perturbations

We now comment briefly on the curvature perturbation during USR inflation. This has been studied extensively in Ref. [5]. Here we note that, during USR there is a spike in the curvature perturbation, which may potentially lead to the copious production of primordial black holes, that can substantially contribute to the dark matter in the Universe [6,13]. This can be understood as follows.

For the spectrum of the curvature perturbation we have

$$\sqrt{\mathcal{P}} = \frac{H^2}{2\pi\dot{\phi}} \Rightarrow \mathcal{P} = \frac{H^2}{8\pi^2 m_P^2\epsilon}, \tag{27}$$

where we used that $2m_P^2\dot{H} = -\dot{\phi}^2$ and Eq. (3).

In SR inflation the variation of $\epsilon = \epsilon_{\text{SR}}$ is very small, so $\mathcal{P}$ remains roughly constant, which corresponds to an almost scale-invariant spectrum of perturbations. Indeed, the variation of ϵ is traced by $\epsilon_2 \equiv \dot{\epsilon}/\epsilon H$ (cf. Eq. (4)). In SR, Eq. (8) suggests that $|\epsilon_2| = |4\epsilon - 2\eta| \ll 1$.

Things are different during USR, though. Because $\epsilon = \frac{3}{2}\dot{\phi}^2/V$, where $V \simeq 3m_P^2H^2$ and $\dot{\phi}^2 = 2\rho_{\text{kin}} \propto a^{-6}$ during USR inflation, we find $\epsilon \propto a^{-6} \propto e^{-6\Delta N}$, where ΔN is the elapsing USR e-folds. Thus, we obtain that $\mathcal{P} \propto e^{6\Delta N}$ and the curvature perturbation grows exponentially during USR inflation.[2]

Note, though, that, were USR not considered, according to Eq. (27), the usual SR would lead to an even more dramatic increase in $\mathcal{P}$ because $\epsilon_{\text{SR}} \ll \epsilon$ when traversing an extremely flat patch of the potential. This, however, does not happen as the field "overshoots" the flat patch [6] "surfing" on its decreasing kinetic density.

7. Conclusions

In conclusion, we have investigated ultra-slow-roll (USR) inflation, which may take place when the inflationary potential becomes extremely flat. We have showed that this is a temporary phase of inflation, not an attractor, and obtained an estimate of how many e-folds it lasts, depending on the initial kinetic density of the inflaton field. We have discussed how the field can depart from the usual slow-roll (SR) when crossing an extremely flat patch in the scalar potential. SR would force the field to spend a lot of time traversing the flat patch. Instead, the field "glosses over" the flat patch in a moderate number of e-folds. Because the number of e-folds is of paramount importance when calculating inflationary observables, we argued that USR has to be taken into account, when necessary. In particular, we looked into inflection-point inflation, which exhibits a flat patch near the inflection point in the potential, that may give rise to USR. Models which do not take this into account are in danger of miscalculating the values of inflationary observables. However, this danger can be averted if one assumes that the field begins its evolution already on the flat patch (e.g. near the inflection point) with small initial kinetic density. We obtained a conservative bound on the initial kinetic density of the field, which manages to avoid USR inflation and render the SR treatment valid.

[1] Because $|\dot{\phi}| = |V'|/3H < H^2/2\pi = \frac{\delta\phi}{\delta t}$.

[2] Note that $\mathcal{P}$ must be evaluated at the end of USR and not at horizon exit [3,5].

Acknowledgements

My research is supported (in part) by the Lancaster–Manchester–Sheffield Consortium for Fundamental Physics under STFC grant: ST/L000520/1. I would like to thank C. Germani and C. Owen for discussions.

References

[1] P.A.R. Ade, et al., Planck Collaboration, Astron. Astrophys. 594 (2016) A20.
[2] D.H. Lyth, A.R. Liddle, Cambridge Univ. Pr., Cambridge, UK, 2009, 497 pp.
[3] W.H. Kinney, Phys. Rev. D 72 (2005) 023515.
[4] J. Martin, H. Motohashi, T. Suyama, Phys. Rev. D 87 (2) (2013) 023514.
[5] M.H. Namjoo, H. Firouzjahi, M. Sasaki, Europhys. Lett. 101 (2013) 39001;
S. Mooij, G.A. Palma, J. Cosmol. Astropart. Phys. 1511 (11) (2015) 025;
A.E. Romano, S. Mooij, M. Sasaki, Phys. Lett. B 761 (2016) 119.
[6] C. Germani, T. Prokopec, arXiv:1706.04226 [astro-ph.CO].
[7] B. Spokoiny, Phys. Lett. B 315 (1993) 40;
M. Joyce, T. Prokopec, Phys. Rev. D 57 (1998) 6022;
C. Pallis, J. Cosmol. Astropart. Phys. 0510 (2005) 015;
C. Pallis, Nucl. Phys. B 751 (2006) 129;
M.E. Gomez, S. Lola, C. Pallis, J. Rodriguez-Quintero, J. Cosmol. Astropart. Phys. 0901 (2009) 027.
[8] K. Dimopoulos, Nucl. Phys. B, Proc. Suppl. 95 (2001) 70;
K. Dimopoulos, Phys. Rev. D 68 (2003) 123506;
K. Dimopoulos, C. Owen, J. Cosmol. Astropart. Phys. 1706 (06) (2017) 027.
[9] J.C. Bueno Sanchez, K. Dimopoulos, D.H. Lyth, J. Cosmol. Astropart. Phys. 0701 (2007) 015;
R. Allahverdi, A. Kusenko, A. Mazumdar, J. Cosmol. Astropart. Phys. 0707 (2007) 018;
J. Garcia-Bellido, AIP Conf. Proc. 878 (2006) 277;
C.M. Lin, K. Cheung, Mod. Phys. Lett. A 25 (2010) 1425.
[10] R. Allahverdi, K. Enqvist, J. Garcia-Bellido, A. Mazumdar, Phys. Rev. Lett. 97 (2006) 191304;
D.H. Lyth, J. Cosmol. Astropart. Phys. 0704 (2007) 006;
R. Allahverdi, K. Enqvist, J. Garcia-Bellido, A. Jokinen, A. Mazumdar, J. Cosmol. Astropart. Phys. 0706 (2007) 019;
T. Matsuda, J. Cosmol. Astropart. Phys. 0706 (2007) 029;
K. Enqvist, L. Mether, S. Nurmi, J. Cosmol. Astropart. Phys. 0711 (2007) 014;
Z. Lalak, K. Turzynski, Phys. Lett. B 659 (2008) 669;
S. Nurmi, J. Cosmol. Astropart. Phys. 0801 (2008) 016;
T. Matsuda, Nucl. Phys. B 822 (2009) 88;
K. Kamada, J. Yokoyama, Prog. Theor. Phys. 122 (2010) 969;
R. Allahverdi, A. Ferrantelli, J. Garcia-Bellido, A. Mazumdar, Phys. Rev. D 83 (2011) 123507;
S. Choudhury, A. Mazumdar, S. Pal, J. Cosmol. Astropart. Phys. 1307 (2013) 041.
[11] N. Itzhaki, E.D. Kovetz, J. High Energy Phys. 0710 (2007) 054;
R. Allahverdi, B. Dutta, A. Mazumdar, Phys. Rev. D 78 (2008) 063507;
M. Badziak, M. Olechowski, J. Cosmol. Astropart. Phys. 0902 (2009) 010;
K. Enqvist, A. Mazumdar, P. Stephens, J. Cosmol. Astropart. Phys. 1006 (2010) 020;
S.M. Choi, H.M. Lee, Eur. Phys. J. C 76 (6) (2016) 303;
N. Okada, D. Raut, Phys. Rev. D 95 (3) (2017) 035035;
N. Okada, S. Okada, D. Raut, Phys. Rev. D 95 (5) (2017) 055030.
[12] K. Dimopoulos, C. Owen, A. Racioppi, arXiv:1706.09735 [hep-ph].
[13] G. Ballesteros, M. Taoso, arXiv:1709.05565 [hep-ph];
Y. Gong, arXiv:1707.09578 [astro-ph.CO];
K. Kannike, L. Marzola, M. Raidal, H. Veermäe, J. Cosmol. Astropart. Phys. 1709 (09) (2017) 020;
J.M. Ezquiaga, J. Garcia-Bellido, E. Ruiz Morales, arXiv:1705.04861 [astro-ph.CO];
J. Garcia-Bellido, E. Ruiz Morales, Phys. Dark Universe 18 (2017) 47;
H. Motohashi, W. Hu, Phys. Rev. D 96 (6) (2017) 063503.

30

Universal features of quantum bounce in loop quantum cosmology

Tao Zhu [a,b], Anzhong Wang [a,b,*], Klaus Kirsten [c], Gerald Cleaver [d], Qin Sheng [c]

[a] Institute for Advanced Physics & Mathematics, Zhejiang University of Technology, Hangzhou, 310032, China
[b] GCAP-CASPER, Physics Department, Baylor University, Waco, TX 76798-7316, USA
[c] GCAP-CASPER, Mathematics Department, Baylor University, Waco, TX 76798-7328, USA
[d] EUCOS-CASPER, Physics Department, Baylor University, Waco, TX 76798-7316, USA

ARTICLE INFO

Editor: S. Dodelson

ABSTRACT

In this Letter, we study analytically the evolutions of the flat Friedmann–Lemaitre–Robertson–Walker (FLRW) universe and its linear perturbations in the framework of *the dressed metric approach* in loop quantum cosmology (LQC). Assuming that the evolution of the background is dominated by the kinetic energy of the inflaton at the quantum bounce, we find that both evolutions of the background and its perturbations are independent of the inflationary potentials during the pre-inflationary phase. During this period the effective potentials of the perturbations can be well approximated by a Pöschl–Teller (PT) potential, from which we find analytically the mode functions and then calculate the corresponding Bogoliubov coefficients at the onset of the slow-roll inflation, valid for any inflationary model with a single scalar field. Imposing the Bunch–Davies (BD) vacuum in the contracting phase prior to the bounce when the modes are all inside the Hubble horizon, we show that particles are generically created due to the pre-inflation dynamics. Matching them to those obtained in the slow-roll inflationary phase, we investigate the effects of the pre-inflation dynamics on the scalar and tensor power spectra and find features that can be tested by current and forthcoming observations. In particular, to be consistent with the Planck 2015 data, we find that the universe must have expanded at least 141 e-folds since the bounce.

1. Introduction

The paradigm of cosmic inflation has achieved remarkable successes in solving several problems of the standard big bang cosmology and predicting the primordial perturbation spectra whose evolutions explain both the formation of the large scale structure of the universe and the small inhomogeneities in the cosmic microwave background (CMB) [1]. Now they are matched to observations with unprecedented precisions [2–4]. However, such successes are contingent on the understanding of physics in much earlier epochs when energies were about the Planck scale. This leads to several conceptual issues. For example, to be consistent with observations, the universe must have expanded at least 60 e-folds during its inflationary phase. However, if the universe had expanded a little bit more than 70 e-folds during inflation (as it is the case in a large class of inflationary models [5]), then one can show that the wavelengths of all fluctuation modes which are currently inside the Hubble radius were smaller than the Planck length at the beginning of the period of inflation. This was referred to as the trans-Planckian issue in [6], and leads to the question about the validity of the assumption: *the matter fields are quantum in nature but the spacetime is still classical*, which are used at the beginning of inflation in order to make predictions [1]. In addition, insisting on the use of general relativity (GR) to describe the inflationary process will inevitably lead to an initial singularity [7]. Moreover, the inflation paradigm usually sets the BD vacuum state at the time when the wavelength of fluctuations were well within the Hubble horizon during the inflationary process. However, such treatment ignores the pre-inflationary dynamics which could lead to non-BD states at the onset of inflation, even when these modes were well inside the Hubble horizon during inflation. For more detail about the sensibility of the inflationary paradigm to Planckian physics, we refer the readers to [6,8].

All the issues mentioned above are closely related to the fact that we are working in the regime where GR is known to break down. One believes that new physics in this regime – a quantum theory of gravity, will provide a complete description of inflation as

* Corresponding author.
E-mail addresses: Tao_Zhu@baylor.edu (T. Zhu), anzhong_wang@baylor.edu (A. Wang), klaus_kirsten@baylor.edu (K. Kirsten), gerald_cleaver@baylor.edu (G. Cleaver), qin_sheng@baylor.edu (Q. Sheng).

well as its pre-inflationary dynamics. LQC is one of such theories that offers a framework to address these issues, in which the inflationary scenarios can be extended from the onset of the slow-roll inflation back to the Planck scale in a self-consistent way [9–11]. Remarkably, the quantum geometry effects of LQC at the Planck scale provide a natural resolution of the big bang singularity (see [12–15] and references therein). In such a picture, the singularity is replaced by a quantum bounce, and the universe that starts at the bounce can eventually evolve to the desired slow-roll inflation [16–23]. An important question now is whether the quantum bounce can leave any observational signatures to current/forthcoming observations, so LQC can be placed directly under experimental tests. The answer to this question is affirmative. In fact, with some (reasonable) assumptions and choice of the initial conditions, the *deformed algebra approach* already leads to inconsistence with current observations [21]. Note that in general there are two main approaches to implement cosmological perturbations in the framework of LQC, the *dressed metric* and *deformed algebra approaches* [12–14]. In both, the primordial perturbations have been intensively studied *numerically* [10,11,19–23].

One of our purposes of this Letter, in contrast to the previous numerical studies, is to present an *analytical* analysis of the effects of the quantum bounce and pre-inflation dynamics on the evolutions of both background and spectra of the scalar and tensor perturbations, in the framework of the dressed metric approach [9–11]. It is expected that such an analysis will provide a more complete understanding of the problem and deeper insights. In the following, we will focus on the case that the kinetic energy of the inflaton dominates the evolutions at the bounce, because a potential dominated bounce is either not able to produce the desired slow-roll inflation [22], or leads to a large amount of e-folds of expansion. This will wash out all the observational information about the pre-inflation dynamics and the resulting perturbations are the same as those given in GR [12–14]. Assuming that the influence of the potential at the bounce is negligible, our studies show that:

- During the pre-inflationary phase, the evolutions of the background and the scalar and tensor perturbations are independent of the inflationary potentials. Thus, the evolution of the background is the same for any chosen potential, and in this sense we say that it is *universal.*
- During this phase the potentials of the scalar and tensor perturbations can be well approximated by an effective PT potential, for which analytic solutions of the mode functions can be found. The Bogoliubov coefficients at the onset of the slow-roll inflation can thereby be calculated [cf. (13)], which are valid for any slow-roll inflationary model with a single scalar field. Assuming that the universe is in the BD vacuum in the contracting phase (the moments where $t \lesssim -t_s$ as shown in Fig. 2) we find that particle creations occur generically during the pre-inflation phase.
- Oscillations always happen in the power spectra, and their phases for both scalar and tensor perturbations are the same, in contrast to other theories of quantum gravity [6,24].
- Fitting the power spectra to the Planck 2015 data [4], we find the lower bound for $N_{\text{tot}} \equiv \ln(a_0/a_B) > 141$ (95% C.L.), where a_B and a_0 denote the expansion factor at the bounce and current time, respectively. Details of the calculations will be reported elsewhere [25].

2. Quantum bounce

In LQC, the semi-classical dynamics of a flat FLRW universe with a single scalar field ϕ and potential $V(\phi)$ is described by [9–11],

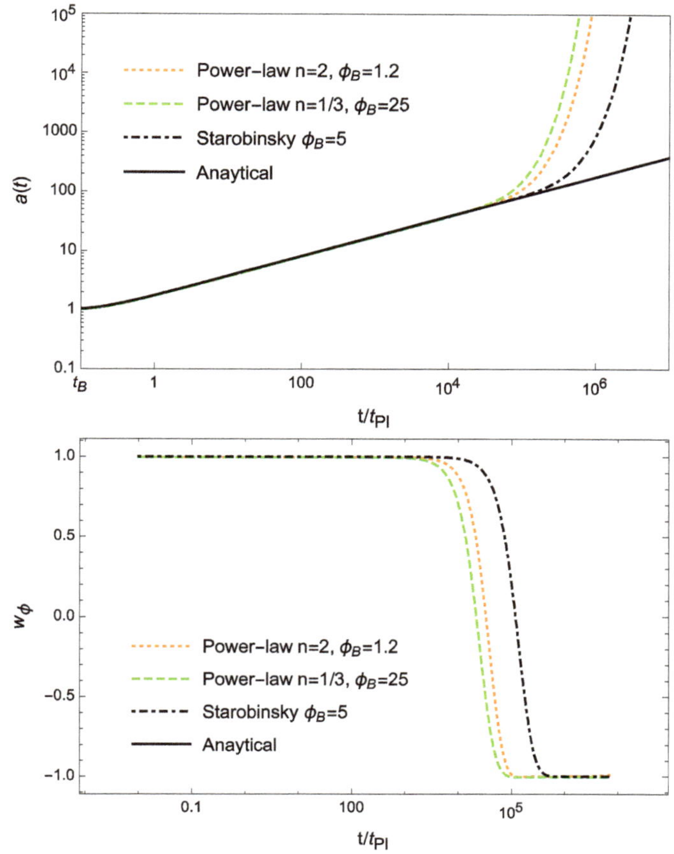

Fig. 1. Evolutions of $a(t)$ and w_ϕ for the power-law $V(\phi) = \frac{1}{2}m^{4-n}\phi^n$ and Starobinsky $V(\phi) = \frac{3}{4}M^2M_{\text{Pl}}^2(1 - e^{-\sqrt{2/3}\phi/M_{\text{Pl}}})^2$ potentials. Solution (3) is also shown. We choose $m = 1.3 \times 10^{-6}$ for $n = 2$, $m = 1.1 \times 10^{-3}$ for $n = 1/3$, and $M = 2.5 \times 10^{-6}$ for the Starobinsky potential. In all the cases we set $m_{\text{Pl}} = 1$.

$$H^2 = \frac{8\pi}{3m_{\text{Pl}}^2}\rho\left(1 - \frac{\rho}{\rho_c}\right), \tag{1}$$

$$\ddot{\phi} + 3H\dot{\phi} + V_{,\phi} = 0, \tag{2}$$

where $H \equiv \dot{a}/a$ is the Hubble parameter, a dot denotes the derivative with respect to the cosmic time t, and ρ_c is the maximum energy density, with $\rho \equiv \dot{\phi}^2/2 + V(\phi) \leq \rho_c$. Eq. (1) shows that the big bang singularity now is replaced by a non-singular quantum bounce at $\rho = \rho_c$ [cf. Fig. 1]. The background evolution has been extensively studied, and one of the main results is that, following the bounce, a desired slow-roll inflation phase is almost inevitable, provided that the evolution is dominated initially by the kinetic energy of the scalar field at the quantum bounce [12,17,18,22]. In this Letter, we will focus on this case. Then, ignoring the potential term $V(\phi)$, from Eqs. (1) and (2) we find

$$a(t) = a_{\text{B}}\left(1 + \gamma_B \frac{t^2}{t_{\text{Pl}}^2}\right)^{1/6}, \tag{3}$$

where $a_{\text{B}} \equiv a(t_B), \gamma_B \equiv 24\pi\rho_c/m_{\text{Pl}}^4$, and t_{Pl} denotes the Planck time. In writing the above expression we also set $t_{\text{B}} = 0$. In Fig. 1 we display the above analytical solution and the equation of state

$$w_\phi \equiv \frac{\dot{\phi}^2 - 2V(\phi)}{\dot{\phi}^2 + 2V(\phi)}, \tag{4}$$

together with several numerical solutions of $a(t)$ for different potentials. From this figure, specially the curves of w_ϕ, we can see that the universe experiences three different phases: *bouncing, transition, and slow-roll inflation.* During the bouncing phase, w_ϕ remains almost one until $t/t_{Pl} \simeq 10^4$. Then, it suddenly drops from 1 to -1 at $t/t_{Pl} \simeq 10^5$. This transition phase is very short in comparison to the other two, and the kinetic energy of the scalar field drops almost 12 orders from the beginning of this phase to the end of it. Afterwards, the potential energy $V(\phi)$ dominates the evolution, and w_ϕ remains practically -1 during the whole slow-roll inflation phase. The end of this transition phase can be well defined as the moment where $\ddot{a}(t=t_i)=0$, as shown in Fig. 1. Afterward, the expansion of the universe will be accelerating $\ddot{a}(t>t_i)>0$. However, unlike t_i, the starting point of the transition phase is not abrupt, even though the division is very clean in concept, as one can see from Fig. 1. Fortunately, the results are not sensitive to such a choice at all, as argued below and shown in detail in [25]. In particular, we find that the choices of $w_\phi=0.95$ and $w_\phi=2/3$ make no (observational) difference in the power spectra and the total e-folds of the expansion of the universe.

During the bouncing phase, *the evolution of $a(t)$ is independent of the choice of ϕ_B and the choice of the potential $V(\phi)$ of the scalar field.* This is because $V(\phi)$ remains very small and the kinetic energy is completely dominant during this whole phase. For example, for the potential $V(\phi)=V_0\phi^n$ with $n=2$, we find that $V(\phi)/m_{\rm Pl}^4 \in (2\times 10^{-11}, 4.5\times 10^{-11})$; for $n=1/3$, $V(\phi)/m_{\rm Pl}^4 \in (9\times 10^{-12}, 1.2\times 10^{-11})$; and for the Starobinsky potential, we have $V(\phi)/m_{\rm Pl}^4 \in (7\times 10^{-13}, 7.3\times 10^{-13})$. This explains why the evolution of $a(t)$ is universal during this period.

3. Primordial power spectra

The linear perturbations in the dressed metric approach [9,10] were studied numerically in detail with the inflationary potential $V(\phi)=m^2\phi^2/2$ [11]. In this Letter, our goals are two-fold: First, we study these perturbations *analytically*, and provide their explicit expressions. Second, we show that they are independent of the choices of the slow-roll inflationary potentials, so they are *universal*. In fact, this follows directly from the universality of the evolution of $a(t)$ during this phase. To show this, let us start with the scalar and tensor perturbations [9–11],

$$\mu_k^{(s,t)}(\eta)'' + \left(k^2 - \frac{a''}{a} + U^{(s,t)}(\eta)\right)\mu_k^{(s,t)}(\eta) = 0, \tag{5}$$

where $U^{(s)}(\eta) \equiv a^2\left(\mathfrak{f}^2 V(\phi) + 2\mathfrak{f}V_{,\phi}(\phi) + V_{,\phi\phi}(\phi)\right)$, $U^{(t)}(\eta)=0$, with $\mathfrak{f} \equiv \sqrt{24\pi G}\dot{\phi}/\sqrt{\rho}$. $\mu_k^{(s,t)}(\eta)$ denote the Mukhanov–Sasaki variables with $\mu_k^{(s)}(\eta) = z_s\mathcal{R}$ and $\mu_k^{(t)}(\eta) = ah_k/2$, where $\mathcal{R}$ denotes the comoving curvature perturbations, h_k the tensor perturbations, and $z_s \equiv a\dot{\phi}/H$. A prime denotes the derivative with respect to the conformal time $\eta(t) = \int_{t_{\rm end}}^t dt'/a(t')$, where $t_{\rm end}$ is the time when the inflation ends. Near the bounce, $U^{(s)}(\eta)$ is negligible [10,11,25]. During the transition phase, ρ drops down to about $10^{-12}\rho_c$, and $(a''/a - U^{(s)}(\eta)) \to z_s''/z_s$, so thereafter the perturbations reduce precisely to those of GR [10,13].

The evolutions of the perturbations depend on both background and wavenumber k. As we consider only the case in which the kinetic energy dominates the evolution of the background at the bounce, both scalar and tensor perturbations follow the same equation of motion during the bouncing phase ($t/t_{Pl} \lesssim 10^4$). In this case, the term a''/a in Eq. (5) defines a typical radius $\lambda = \sqrt{a/a''}$ for $a''>0$, which plays the same role as that of the comoving Hubble radius $L_H = (aH)^{-1}$ often used in GR. However, for a better understanding, we find that here it is more proper to use a/a'', as shown schematically in Fig. 2. For example, when the modes are inside the radius ($1/k^2 < \lambda^2$), the solution of Eq. (5) is of the form, $e^{\pm i\int\sqrt{k^2-a''/a}d\eta}$. When the modes are outside of the "horizon" (radius) ($1/k^2 > \lambda^2$), it is of the form, $e^{\pm\int\sqrt{a''/a-k^2}d\eta}$. The term a''/a has its maximum at the bounce, $a''/a|_{t=0} = a_B^2\gamma_B m_{\rm Pl}^2/3$, which defines a typical scale $k_B = \sqrt{\gamma_B/3}a_B m_{\rm Pl}$ (the blue solid curve in Fig. 2), so we can use it to classify different modes. Some modes with large values of $k^2 \gg k_B^2$ (the region below the low (orange) dashed line in Fig. 2) are inside the horizon all the time until they exit the Hubble horizon during the slow-roll inflation. Some of the modes with smaller $k^2 \ll k_B^2$ (the region above the upper (green) dashed line in Fig. 2) exit and re-enter the horizon during the bouncing process, and will finally re-exit the Hubble horizon during the slow-roll inflation. Since the modes with $k \gg k_B$ are inside the horizon during the whole pre-inflationary phase, they will have the same power-law spectra as those given in GR [1]. We are interested in the modes with $k \simeq k_B$ (the shaded region in Fig. 2). However, the perturbations for these modes have different behaviors when they are inside or outside the horizon, which makes Eq. (5) extremely difficult to be solved analytically.

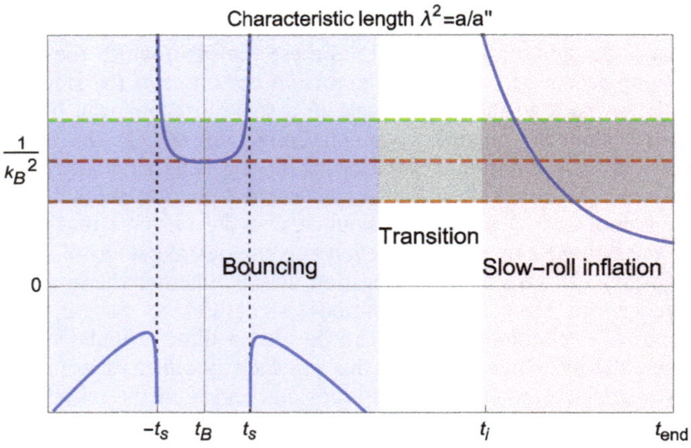

Fig. 2. Schematic plot of $\lambda^2[\equiv a/a'']$, where $a''/a|_{t=t_s}=0$ with $t_s \simeq 0.2t_{\rm Pl}$, and $\ddot{a}(t_i)=0$ with t_i being the starting time of the inflationary phase. During the slow-roll inflation, $a/a''=L_H/2$. The expansion factor $a(t)$ can be analytically extended to a contracting phase $t<t_{\rm B}$. (For interpretation of the references to color in this figure, the reader is referred to the web version of this article.)

In this Letter, we first present an analytical solution of Eq. (5) by using an effective Pöschl-Teller (PT) potential. To this goal, let us first consider the quantity,

$$\mathscr{V}(\eta) \equiv \frac{a''}{a} = a_{\rm B}^2\frac{\gamma_B m_{\rm Pl}^2(3-\gamma_B t^2/t_{\rm Pl}^2)}{9(1+\gamma_B t^2/t_{\rm Pl}^2)^{5/3}}. \tag{6}$$

If we consider Eq. (5) as the Schrödinger equation, then $\mathscr{V}(\eta)$ serves as an effective barrier during the bouncing phase. Such a potential can be approximated by a PT potential for which we know the analytical solution,

$$\mathscr{V}_{\rm PT}(\eta) = \mathscr{V}_0\cosh^{-2}\alpha(\eta-\eta_{\rm B}), \tag{7}$$

where $\mathscr{V}_0 = a_{\rm B}^2\gamma_B m_{\rm Pl}^2/3$ and $\alpha^2 \equiv 2a_{\rm B}^2\gamma_B m_{\rm Pl}^2 = 6k_{\rm B}^2$. From Fig. 3 we can see that $\mathscr{V}_{\rm PT}(\eta)$ mimics $\mathscr{V}(\eta)$ very well. Introducing x and $\mathcal{Y}(x)$ via $x(\eta) = 1/(1+e^{-2\alpha(\eta-\eta_{\rm B})})$, $\mathcal{Y}(x) = [x(1-x)]^{ik/(2\alpha)}\mu_k(\eta)$, we find that Eq. (5) reduces to,

$$x(1-x)\mathcal{Y}'' + [c_3 - (c_1+c_2+1)x]\mathcal{Y}' - c_1c_2\mathcal{Y} = 0, \tag{8}$$

where $\mathcal{Y}' \equiv d\mathcal{Y}/dx$ and

$$c_1 \equiv \frac{1}{2} + \frac{1}{2\alpha}\sqrt{\alpha^2 - 4\mathscr{V}_0} - \frac{ik}{\alpha},$$
$$c_2 \equiv \frac{1}{2} - \frac{1}{2\alpha}\sqrt{\alpha^2 - 4\mathscr{V}_0} - \frac{ik}{\alpha},$$
$$c_3 \equiv 1 - \frac{ik}{\alpha}. \tag{9}$$

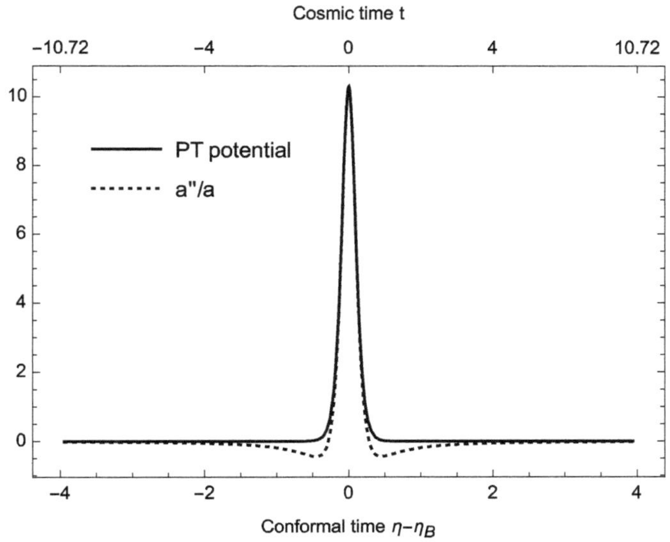

Fig. 3. Comparison between the effective potential given by Eq. (6) and the PT potential in Eq. (7). In this plot, we have set $a_B = 1$, $m_{Pl} = 1$.

This equation is the standard hypergeometric equation, and its general solution is given by,

$$\mu_k^{(\mathrm{PT})}(\eta) = a_k x^{ik/(2\alpha)}(1-x)^{-ik/(2\alpha)} \times {}_2F_1(c_1 - c_3 + 1, c_2 - c_3 + 1, 2 - c_3, x) + b_k[x(1-x)]^{-ik/(2\alpha)}\, {}_2F_1(c_1, c_2, c_3, x). \tag{10}$$

Here a_k and b_k are two integration constants to be determined by the initial conditions.

To impose them, let us first specify the initial time. A natural choice is right at the bounce, at which the initial state can be constructed as the *fourth-order adiabatic vacuum* [9,10]. While such constructions work well for large k, however, ambiguity remains for modes with $k < k_B$ [10]. Another choice that has been frequently used is a time during the contracting phase when the modes are well within the characteristic length λ, which is $t \lesssim -t_s$ as shown in Fig. 2 [11,19–22,27]. In this Letter, we also shall make that choice, as the main conclusions will not sensitively depend on these choices, as shown in [11,25,26], and we require that at this initial time the state should be the BD vacuum. Then, we find

$$a_k = 0,\ b_k = \frac{e^{ik\eta_B}}{\sqrt{2k}}. \tag{11}$$

It should be noted that $\mu_k^{(\mathrm{PT})}(\eta)$ of Eq. (10) and the above initial conditions are valid for any value of k. In particular, at the bounce it reduces to the one obtained in [10] with the *fourth-order adiabatic vacuum* for large $k > k_B$. This further confirms our above arguments. In Fig. 4 we compare our analytical approximate solution with the numerical (exact) one, which shows that they match extremely well during the bouncing phase. After this period, the universe soon sets to the slow-roll inflation phase, and the mode functions of tensor and scalar perturbations are the well-known solutions given in GR [1]. When all the relevant modes are inside the Hubble horizon ($t < t_i$ as shown in Fig. 2), they take the asymptotic form [1],

$$\mu_k^{(s,t)}(\eta) \simeq \frac{1}{\sqrt{2k}}\left(\alpha_k e^{-ik\eta} + \beta_k e^{ik\eta}\right),\ (t < t_i). \tag{12}$$

In GR, one usually imposes the BD vacuum at the beginning of inflation, at which all the (physical) modes are inside the Hubble horizon, so that $\alpha_k^{GR} = 1$, $\beta_k^{GR} = 0$. This in turn leads to the

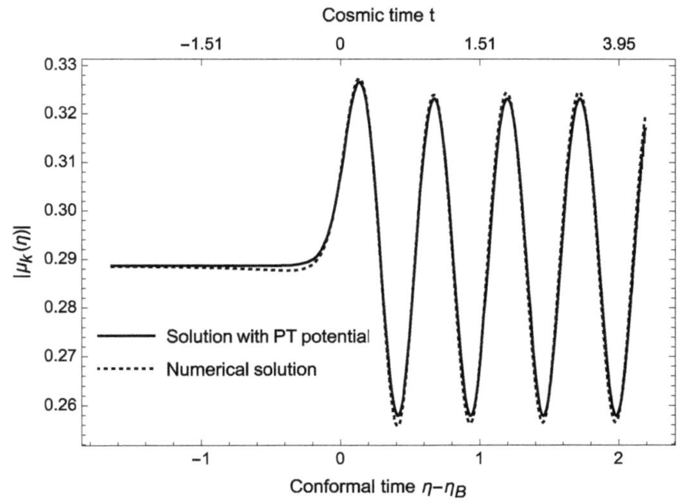

Fig. 4. Comparison between the analytical solution and numerical one with $a_B = 1$, $m_{Pl} = 1$, and $k = 6$.

standard power-law spectra. However, due to the quantum gravitational effects, β_k now does not vanish generically. To see this, we need to match the GR solution to Eq. (10). Taking its limit $t/t_{Pl} \gg 1$ and then comparing it with the GR solution we find

$$\alpha_k = \frac{\Gamma(c_3)\Gamma(c_3 - c_1 - c_2)}{\Gamma(c_3 - c_1)\Gamma(c_3 - c_2)} e^{2ik\eta_B},\quad \beta_k = \frac{\Gamma(c_3)\Gamma(c_1 + c_2 - c_3)}{\Gamma(c_1)\Gamma(c_2)}, \tag{13}$$

where c_n are the constants given by Eq. (9). This represents one of our main results. When $k \simeq k_B$ we find that $|\beta_k|^2 \simeq 10$. That is, particles of such modes were created during the bouncing phase. However, such creation will not alter significantly the evolution of the background, nor the perturbations during the slow-roll inflation period, as shown explicitly in [10]. Then, from Eq. (10) we obtain $\mathcal{P}_{\mathrm{LQC}}^{(s,t)}(k) = |\alpha_k + \beta_k|^2 \mathcal{P}_{\mathrm{GR}}^{(s,t)}(k) \equiv (1 + \delta_{\mathcal{P}})\mathcal{P}_{\mathrm{GR}}^{(s,t)}(k)$, where

$$\delta_{\mathcal{P}} \equiv \left[1 + \cos\left(\frac{\pi}{\sqrt{3}}\right)\right]\mathrm{csch}^2\left(\frac{\pi k}{\sqrt{6}k_B}\right) + \sqrt{2}\cos\left(\frac{\pi}{2\sqrt{3}}\right)\sqrt{\cos\left(\frac{\pi}{\sqrt{3}}\right) + \cosh\left(\frac{2\pi k}{\sqrt{6}k_B}\right)} \times \mathrm{csch}^2\left(\frac{\pi k}{\sqrt{6}k_B}\right)\cos(2k\eta_B + \varphi_k), \tag{14}$$

where

$$\varphi_k \equiv \arctan\left\{\frac{\mathrm{Im}[\Gamma(c_1)\Gamma(c_2)\Gamma^2(c_3 - c_1 - c_2)]}{\mathrm{Re}[\Gamma(c_1)\Gamma(c_2)\Gamma^2(c_3 - c_1 - c_2)]}\right\}. \tag{15}$$

In Fig. 5, we display the ratio between the power spectrum with the bounce effects and the standard power-law spectrum in GR, i.e., $1 + \delta_P$ with δ_P being given by the above equation, as a function of wavenumber. We would like to note that Fig. 5 is consistent with that given in [9,10] (cf. Fig. 1 in the first paper of [9] and Fig. 5 in [10]). While the results obtained in [9,10] are purely numerical, here ours are derived directly from the analytical expression of Eq. (14).

It is remarkable to note that, although it is well-known that quantum gravitational effects often lead to oscillations [6], *in LQC the oscillating phases for both scalar and tensor perturbations are the same.* In Eq. (14), the second term is oscillating very fast and

Table I
The best fitting values of the six cosmological parameters and the constraints on k_B/a_0 and r at 95% C.L for different cosmological models from different data combinations.

Parameter	Planck TT + lowP	Planck TT, TE, EE + lowP	Planck TT + lowP + r	Planck TT, TE, EE + lowP + r
$\Omega_B h^2$	0.022355	0.022193	0.022322	0.022064
$\Omega_c h^2$	0.11893	0.12000	0.11908	0.12071
$100\theta_{MC}$	1.04115	1.04065	1.04080	1.04057
τ	0.077835	0.089272	0.081955	0.085259
$\ln(10^{10}A_s)$	3.088	3.112	3.101	3.104
n_s	0.9662	0.9647	0.9658	0.9607
k_B/a_0	$< 3.12 \times 10^{-4}$	$< 3.05 \times 10^{-4}$	$< 3.14 \times 10^{-4}$	$< 3.14 \times 10^{-4}$
r	-	-	< 0.113	< 0.107

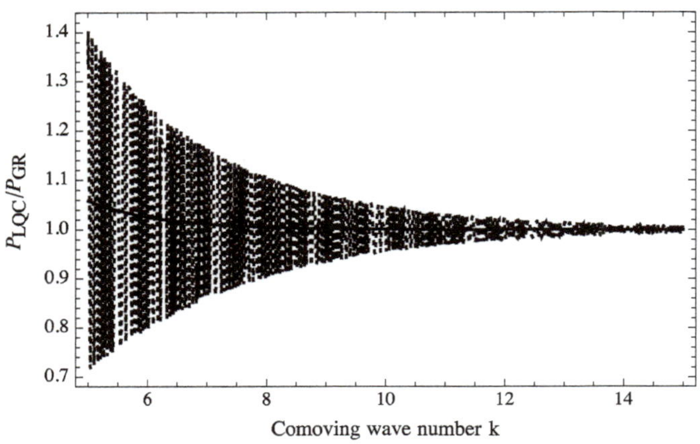

Fig. 5. The ratio $\mathcal{P}^{(s,t)}_{LQC}(k)/\mathcal{P}^{(s,t)}_{GR}(k)$ between the power spectrum with the bounce effects and the standard power-law one obtained in GR. The dotted blue curve denotes the analytical power spectrum, which obviously oscillates rapidly with k. The solid red curve shows the average of the oscillating spectrum. (For interpretation of the references to color in this figure legend, the reader is referred to the web version of this article.)

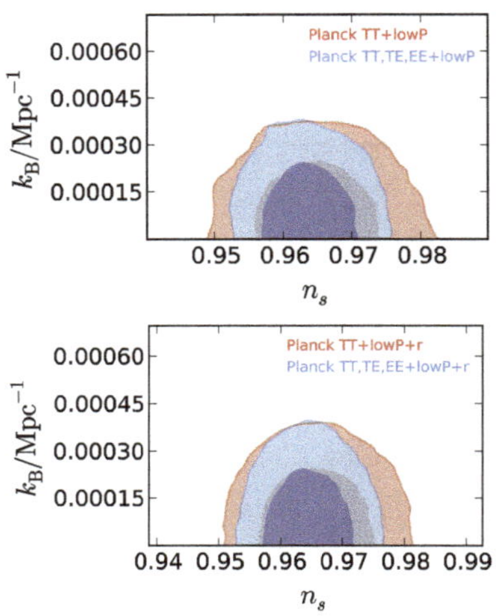

Fig. 6. Observational constraints for $(n_s, k_B/\text{Mpc}^{-1})$ at 68% and 95% C.L. by using the Planck 2015 TT + lowP and TT, TE, EE + lowP data with $a_0 = 1$. The upper panel only considers the scalar spectrum, while the bottom includes the tensor.

can be ignored observationally [9–11]. On the other hand, the first term, proportional to $\text{csch}^2[\pi k/(\sqrt{6}k_B)]$, decreases exponentially as k increases, and the power spectra get enhanced (reduced) for small (large) k. The modes $k \simeq k_B$, of the Planck scale at the bounce, are initially inside the radius defined by $\lambda = \sqrt{|a/a''|}$, and then leave and re-enter it during the bouncing phase. The modes with $k \gg k_B$ are always inside the radius before they leave the Hubble horizon during the slow-roll inflation, thus they finally lead to a standard power spectrum.

It should also be noted that the solution with the PT potential is not valid for the modes with a very small k (i.e., $k \ll |a''/a|$ holds all the time during the bouncing phase). For these modes, if we ignore the k^2 term in Eq. (5), the solution can be approximated by [19],

$$\mu_k(\eta) \simeq a_k a(\eta) + \frac{b_k}{a(\eta)}. \tag{16}$$

However, we are not interested in these modes, as they currently are still outside of the observable universe.

4. Observational constraints

The quantum corrections (14) are k-dependent and expected to be constrained by observations. In the following, we perform the CMB likelihood analysis by using the Planck 2015 data [4], with the MCMC code developed in [28]. We assume the flat cold dark matter model with the effective number of neutrinos $N_{\text{eff}} = 3.046$ and choose the total neutrino mass as $\Sigma m_\nu = 0.06$ eV. We also write

$$\mathcal{P}^{(s,t)}_{GR}(k) = A_{(s,t)}\left(\frac{k}{k_*}\right)^{n^{(s,t)}_{\text{inf}}}, \tag{17}$$

where $k_*(= 0.05\ \text{Mpc}^{-1})$ denotes the pivot scale, $n^{(s)}_{\text{inf}} = n_s - 1$ and $n^{(t)}_{\text{inf}} = n_t$. We vary the seven parameters, $\Omega_B h^2, \Omega_c h^2, \tau, \Theta_s, n_s, A_s, k_B/a_0$ [30]. For the six cosmological parameters except k_B/a_0 $(\Omega_B h^2, \Omega_c h^2, \tau, \Theta_s, n_s, A_s)$, we use the same prior ranges as those adopted in [29], while for the parameter k_B which is related to the bouncing effects, we set the prior range to $k_B \in [10^{-8}, 0.002]$.

In particular, we use the high-l CMB temperature power spectrum (TT) and polarization data (TT, TE, EE) respectively with the low-l polarization data (lowP) from Planck 2015. In Table I, we list the best fit values of the six cosmological parameters and constraints on k_B/a_0 and r at 95% C.L. for different cosmological models from different data combinations.

Marginalizing other parameters, we find that k_B/a_0 is constrained by the Planck TT + lowP (Planck TT, TE, EE + lowP) to

$$\frac{k_B}{a_0} < 3.12 \times 10^{-4}\ \text{Mpc}^{-1}(3.05 \times 10^{-4}), \tag{18}$$

at 95% C.L [cf. Fig. 6]. When we consider the ratio $r = A_{(t)}/A_{(s)}$, the Planck TT + lowP (Planck TT, TE, EE + lowP) data yields

$$\frac{k_B}{a_0} < 3.14 \times 10^{-4}\ \text{Mpc}^{-1}(3.14 \times 10^{-4}), \tag{19}$$

at 95% C.L. These upper bounds show that the observational constraints on the bouncing effects are robust with respect to different data sets (without/with the polarization data included) and whether the tensor spectrum is included or not. In Fig. 7 we show constraints on a couple of cosmological parameters and their respective probability distributions for the CosmoMC runs described

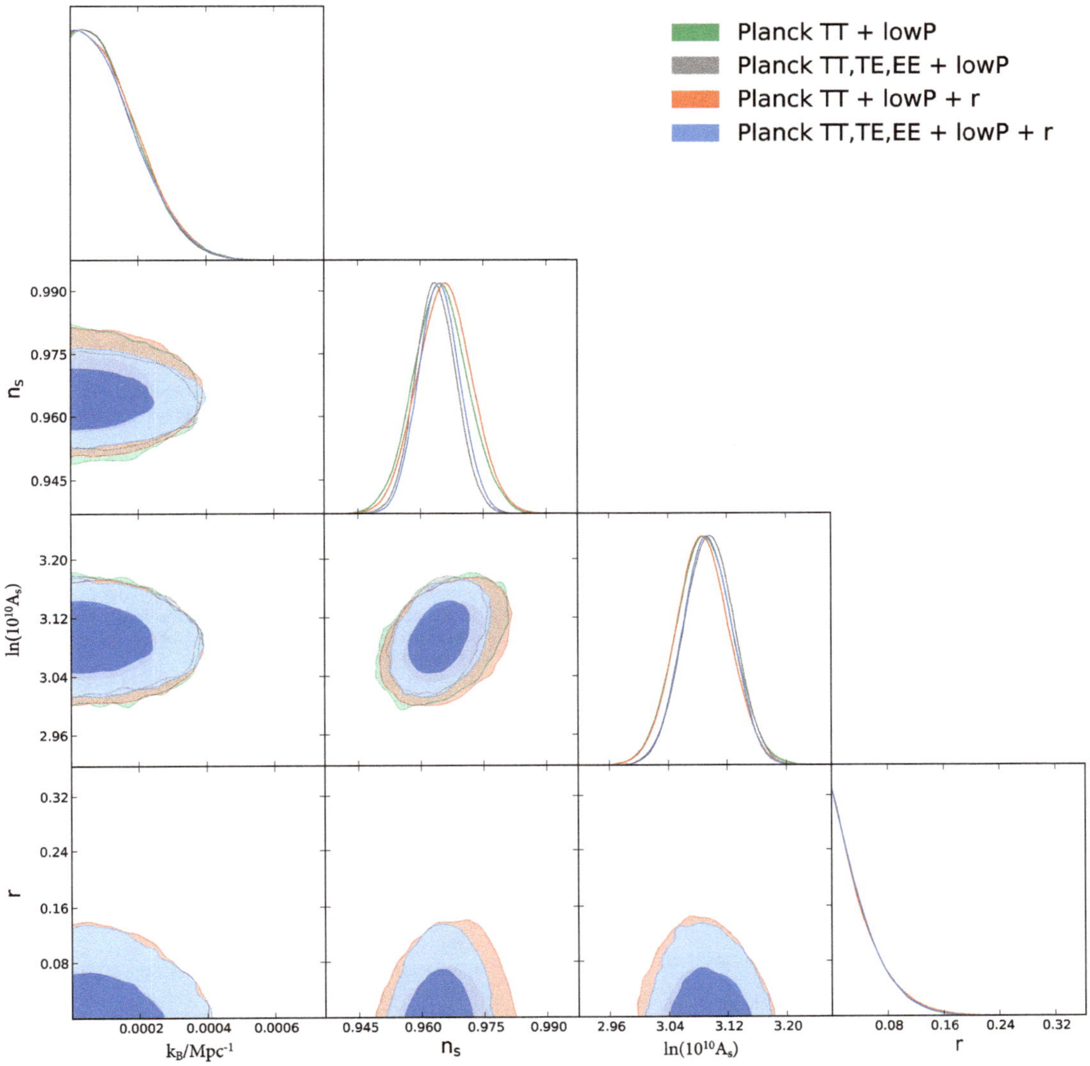

Fig. 7. Observational constraints on a couple of parameters (68% and 95% contour lines) and the probability distributions for $\ln(10^{10}A_s)$, n_s, k_B/a_0, and r by using the Planck 2015 data. Note that in the numerical simulations we set $a_0 = 1$.

above and for the results from the Planck 2015 data. We notice that the colored curves which represent the probability distributions of k_B/a_0 are almost perfectly superposed, which strongly indicates again that the constraints on k_B derived in this paper are robust.

Using the relation

$$\frac{k_B}{a_0} = \sqrt{\frac{\gamma_B}{3}}\frac{a_B}{a_0}m_{Pl} = \sqrt{\frac{\gamma_B}{3}}m_{Pl}e^{-N_{tot}}, \tag{20}$$

where $N_{tot} \equiv \ln(a_0/a_B)$ denotes the total e-folds from the quantum bounce until today, then the above upper bounds on k_B/a_0 can be translated into the constraint on the total *e*-folds N_{tot} as

$$N_{tot} > 141 \quad (95\% \text{ C.L.}), \tag{21}$$

where we have taken $\rho_c = 0.41m_{Pl}^4$ [9,10]. This in turn leads to a lower bound of δN_*,

$$\delta N_* > N_{tot} - N_* - N_{after}, \tag{22}$$

where $\delta N_* \equiv \ln(a_*/a_B)$, $N_* \equiv \ln(a_{end}/a_*)$, and $N_{after} \equiv \ln(a_0/a_{end})$, where a_* denotes the expansion factor at the moment that the current Horizon exited the Hubble horizon during the slow-roll inflation, and a_{end} is that of the end of inflation. Taking $N_* \simeq 60 \simeq N_{after}$, we find

$$\delta N_* \gtrsim 21. \tag{23}$$

Note that our results given by Eqs. (21) and (23) are based on three assumptions: (1) the Universe is filled with a scalar field with its potential $V(\phi)$; (2) the background evolution initially is dominated completely by the kinetic energy of the scalar field; and (3) the Universe is in the BD vacuum state in the contracting phase ($t \lesssim -t_s$, as shown in Fig. 2).

5. Conclusions

In this Letter, we *analytically* studied the evolutions of the background and the linear scalar and tensor perturbations of the FLRW universe in LQC within the framework of *the dressed metric approach* [9–11], and showed that, if the pre-inflationary phase is dominated by the kinematic energy of the inflaton, the evolutions will be *independent of the slow-roll inflationary models* during this phase [cf. Fig. 1 and Eqs. (3) and (14)]. Imposing the BD vacuum in the contracting phase ($t \lesssim -t_s$ as shown in Fig. 2), we obtained the Bogoliubov coefficients (13) at the onset of the slow-roll inflation, which shows clearly that during the pre-inflationary phase, particles are generically created ($\beta_k|_{t=t_i} \neq 0$), and the resulting power spectra are k-dependent. This is in contrast to GR (where the BD vacuum ($\beta_k|_{t=t_i} = 0$) is usually imposed at the onset of the slow-roll inflation [1]. This provides a potential window to test LQC directly by the measurements of CMB and galaxy surveys [31]. In particular, fitting the power spectra to the Planck 2015 temperature (TT+lowP) and polarization (TT, TE, EE+lowP) data, we found the lower bound for $N_{tot} \equiv \ln(a_0/a_B) > 141$ (95% C.L.). That is, to be consistent with current observations of CMB, the universe must have expanded at least 141 e-folds since the bounce.

Acknowledgements

We would like to thank M. Sasaki and W. Zhao for valuable comments and suggestions. This work is supported in part by Ciencia Sem Fronteiras, Grant No. 004/2013 – DRI/CAPES, Brazil (A.W.); Chinese NSF Grants, Nos. 11375153, 11675145 (A.W.), 11675143, 11105120, and 11205133 (T.Z.).

References

[1] A.H. Guth, Inflationary universe: a possible solution to the horizon and flatness problems, Phys. Rev. D 23 (1981) 347;
A. Starobinsky, A new type of isotropic cosmological models without singularity, Phys. Lett. B 91 (1980) 99;
K. Sato, First-order phase transition of a vacuum and the expansion of the universe, Mon. Not. R. Astron. Soc. 195 (1981) 467.

[2] E. Komatsu, et al., WMAP Collaboration, Seven-year Wilkinson Microwave Anisotropy Probe (WMAP) observations: cosmological interpretation, Astrophys. J. Suppl. Ser. 192 (2011) 18;
D. Larson, et al., WMAP Collaboration, Seven-year Wilkinson Microwave Anisotropy Probe (WMAP) observations: power spectra and Wmap-derived parameters, Astrophys. J. Suppl. Ser. 192 (2011) 16.

[3] P. Ade, et al., PLANCK Collaboration, Planck 2013 results. XXII. Constraints on inflation, Astron. Astrophys. 571 (2014) A22, arXiv:1303.5082.

[4] P.A.R. Ade, et al., PLANCK Collaboration, Planck 2015 results. XX. Constraints on inflation, arXiv:1502.02114.

[5] J. Martin, C. Ringeval, V. Vennin, Encyclopaedia Inflationaris, Phys. Dark Universe 5 (2014) 75, arXiv:1303.3787.

[6] R.H. Brandenberger, Inflationary cosmology: progress and problems, arXiv:hep-th/9910410;
J. Martin, R.H. Brandenberger, The trans-Planckian problem of inflationary cosmology, Phys. Rev. D 63 (2001) 123501;
R.H. Brandenberger, J. Martin, Trans-Planckian issues for inflationary cosmology, Class. Quantum Gravity 30 (2013) 113001.

[7] A. Borde, A. Vilenkin, Eternal inflation and the initial singularity, Phys. Rev. Lett. 72 (1994) 3305;
A. Borde, A.H. Guth, A. Vilenkin, Inflationary spacetimes are incomplete in past directions, Phys. Rev. Lett. 90 (2003) 151301.

[8] D. Baumann, TASI lectures on inflation, arXiv:0907.5424;
C.P. Burgess, M. Cicoli, F. Quevedo, String inflation after planck 2013, J. Cosmol. Astropart. Phys. 11 (2013) 003;
D. Baumann, L. McAllister, Inflation and String Theory, Cambridge Monogr. Math. Phys., Cambridge University Press, 2015;
E. Silverstein, TASI lectures on cosmological observables and string theory, arXiv:1606.03640.

[9] I. Agullo, A. Ashtekar, W. Nelson, Quantum gravity extension of the inflationary scenario, Phys. Rev. Lett. 109 (2012) 251301;
I. Agullo, A. Ashtekar, W. Nelson, Phys. Rev. D 87 (2013) 043507.

[10] I. Agullo, A. Ashtekar, W. Nelson, The pre-inflationary dynamics of loop quantum cosmology: confronting quantum gravity with observations, Class. Quantum Gravity 30 (2013) 085014.

[11] I. Agullo, N.A. Morris, Detailed analysis of the predictions of loop quantum cosmology for the primordial power spectra, Phys. Rev. D 92 (2015) 124040.

[12] A. Ashtekar, P. Singh, Loop quantum cosmology: a status report, Class. Quantum Gravity 28 (2011) 213001.

[13] A. Ashtekar, A. Barrau, Loop quantum cosmology: from pre-inflationary dynamics to observations, Class. Quantum Gravity 32 (2015) 234001.

[14] A. Barrau, B. Bolliet, Some conceptual issues in loop quantum cosmology, arXiv:1602.04452.

[15] J. Yang, Y. Ding, Y. Ma, Alternative quantization of the Hamiltonian in loop quantum cosmology, Phys. Lett. B 682 (2009) 1.

[16] A. Ashtekar, D. Sloan, Loop quantum cosmology and slow roll inflation, Phys. Lett. B 694 (2010) 108;
A. Ashtekar, D. Sloan, Probability of inflation in loop quantum cosmology, Gen. Relativ. Gravit. 43 (2011) 3619.

[17] P. Singh, K. Vandersloot, G.V. Vereshchagin, Nonsingular bouncing universes in loop quantum cosmology, Phys. Rev. D 74 (2006) 043510;
J. Mielczarek, T. Cailleteau, J. Grain, A. Barrau, Inflation in loop quantum cosmology: dynamics and spectrum of gravitational waves, Phys. Rev. D 81 (2010) 104049.

[18] X. Zhang, Y. Ling, Inflationary universe in loop quantum cosmology, J. Cosmol. Astropart. Phys. 08 (2007) 012;
L. Chen, J.-Y. Zhu, Loop quantum cosmology: the horizon problem and the probability of inflation, Phys. Rev. D 92 (2015) 084063.

[19] B. Bolliet, J. Grain, C. Stahl, L. Linsefors, A. Barrau, Comparison of primordial tensor power spectra from the deformed algebra and dressed metric approaches in loop quantum cosmology, Phys. Rev. D 91 (2015) 084035.

[20] S. Schander, A. Barrau, B. Bolliet, L. Linsefors, J. Grain, Primordial scalar power spectrum from the Euclidean bounce of loop quantum cosmology, Phys. Rev. D 93 (2016) 023531.

[21] B. Bolliet, A. Barrau, J. Grain, S. Schander, Observational exclusion of a consistent quantum cosmology scenario, Phys. Rev. D 93 (2016) 124011;
J. Grain, The perturbed universe in the deformed algebra approach of loop quantum cosmology, Int. J. Mod. Phys. D 25 (2016) 1642003, arXiv:1606.03271.

[22] B. Bonga, B. Gupt, Inflation with the Starobinsky potential in loop quantum cosmology, Gen. Relativ. Gravit. 48 (2016) 1;
B. Bonga, B. Gupt, Phenomenological investigation of a quantum gravity extension of inflation with the Starobinsky potential, Phys. Rev. D 93 (2016) 063513.

[23] J. Mielczarek, Possible observational effects of loop quantum cosmology, Phys. Rev. D 81 (2010) 063503;
L. Linsefors, T. Cailleteau, A. Barrau, J. Grain, Primordial tensor power spectrum in holonomy corrected Ω loop quantum cosmology, Phys. Rev. D 87 (2013) 107503;
J. Mielczarek, Gravitational waves from the big bounce, J. Cosmol. Astropart. Phys. 11 (2008) 011.

[24] T. Zhu, A. Wang, K. Kirsten, G. Cleaver, Q. Sheng, High-order primordial perturbations with quantum gravitational effects, Phys. Rev. D 93 (2016) 123525.

[25] T. Zhu, A. Wang, K. Kirsten, G. Cleaver, Q. Sheng, Pre-inflationary universe in loop quantum cosmology, arXiv:1705.07544.

[26] I. Agullo, W. Nelson, A. Ashtekar, Preferred instantaneous vacuum for linear scalar fields in cosmological space-times, Phys. Rev. D 91 (2015) 064051.

[27] E. Wilson-Ewing, Lattice loop quantum cosmology: scalar perturbations, Class. Quantum Gravity 29 (2012) 215013;
E. Wilson-Ewing, The matter bounce scenario in loop quantum cosmology, J. Cosmol. Astropart. Phys. 03 (2013) 026.

[28] A. Lewis, S. Bridle, Cosmological parameters from CMB and other data: a Monte Carlo approach, Phys. Rev. D 66 (2002) 103511;
Y.-G. Gong, Q. Wu, A. Wang, Dark energy and cosmic curvature: Monte Carlo Markov chain approach, Astrophys. J. 681 (2008) 27.

[29] P. Ade, et al., PLANCK Collaboration, Planck 2013 results. XVI. Cosmological parameters, Astron. Astrophys. 571 (2014) A16, arXiv:1303.5076.

[30] T. Zhu, A. Wang, G. Cleaver, K. Kirsten, Q. Sheng, Q. Wu, Detecting quantum gravitational effects of loop quantum cosmology in the early universe?, Astrophys. J. Lett. 807 (2015) L17.

[31] K.N. Abazajian, K. Arnold, J. Austermann, B.A. Benson, et al., Inflation physics from the cosmic microwave background and large scale structure, Astropart. Phys. 63 (2015) 55, arXiv:1610.02743.

Vaidya spacetime in massive gravity's rainbow

Yaghoub Heydarzade [a], Prabir Rudra [b], Farhad Darabi [a,*], Ahmed Farag Ali [c,d], Mir Faizal [e,f]

[a] Department of Physics, Azarbaijan Shahid Madani University, Tabriz, 53714-161, Iran
[b] Department of Mathematics, Asutosh College, Kolkata 700 026, India
[c] Netherlands Institute for Advanced Study, Korte Spinhuissteeg 3, 1012 CG Amsterdam, Netherlands
[d] Department of Physics, Faculty of Science, Benha University, Benha, 13518, Egypt
[e] Irving K. Barber School of Arts and Sciences, University of British Columbia-Okanagan, Kelowna, BC V1V 1V7, Canada
[f] Department of Physics and Astronomy, University of Lethbridge, Lethbridge, AB T1K 3M4, Canada

ARTICLE INFO

Editor: N. Lambert

ABSTRACT

In this paper, we will analyze the energy dependent deformation of massive gravity using the formalism of massive gravity's rainbow. So, we will use the Vainshtein mechanism and the dRGT mechanism for the energy dependent massive gravity, and thus analyze a ghost free theory of massive gravity's rainbow. We study the energy dependence of a time-dependent geometry, by analyzing the radiating Vaidya solution in this theory of massive gravity's rainbow. The energy dependent deformation of this Vaidya metric will be performed using suitable rainbow functions.

1. Introduction

It is expected that the usual energy-momentum dispersion relation will get deformed in the UV limit due to quantum gravitational effects. In fact, such a deformation of the usual energy-momentum dispersion relation has been observed to occur in loop quantum gravity [1,2], discrete spacetime [3], string field theory [4], spacetime foam [5], spin-networks [6], and non-commutative geometry [7]. As the usual energy-momentum dispersion relation is fixed by Lorentz symmetry, the deformation of the usual energy-momentum dispersion relation in the UV limit seems to indicate a breaking of Lorentz symmetry in the UV limit. In fact, such a violation of Lorentz symmetry can be used to explain anomalies in ultra-high energy cosmic rays and TeV photons [8,9]. It may be noted that the threshold anomalies are only predicted by deformations where the usual energy-momentum dispersion relation is deformed by a preferred reference frame, and they do not occur in deformation where no such preferred reference frame exists [10]. The deformation of the usual energy-momentum dispersion relation can be explained using the doubly special relativity (DSR) [11,12]. The DSR is an extension of the special theory of relativity, in which the Planck energy and the velocity of light are universal constants. So, just as in special relativity, no object can attain a velocity greater than the velocity of light, in DSR no object can have an energy greater than the Planck energy. The DSR can be generalized to curved spacetime, and the resulting theory is called gravity's rainbow [13]. In this formalism, the spacetime geometry is described by a rainbow of energy dependent metrics, as the geometry of spacetime depends on the energy of the probe. The gravitational dynamics in gravity's rainbow can be studied using rainbow functions [14–22]. The gravity's rainbow has been used to study inflation [23,24], and a resolving of the Big Bang singularity [25–27]. It may be noted that gravity's rainbow is related to Horava–Lifshitz gravity [28,29], and for a specific choice of rainbow functions, it produces the same results as produced by Horava–Lifshitz gravity [30].

The main motivation for gravity's rainbow comes from the observation that the supergravity is a low energy approximation to the string theory [31–36]. This is because according to the renormalization group flow, constants depend on the scale at which a theory [37,38]. Furthermore, the scale at which a theory is measured will depend on the energy of the probe used to measure such a theory. Thus, as the constants in a theory depend explicitly on the scale at which a theory is measured, they also depend implicitly on the energy of the probe used to measure such constants. Now string theory can be viewed as a two dimensional theory, and the target space metric can be regarded as a matrix of coupling constants of this two dimensional theory. As these coupling constants would flow and depend explicitly on the scale at which the theory is measured, they would implicitly depend on the energy of the probe used to perform such a measurement. This would make

* Corresponding author.
E-mail addresses: heydarzade@azaruniv.edu (Y. Heydarzade), prudra.math@gmail.com (P. Rudra), f.darabi@azaruniv.edu (F. Darabi), ahmed.ali@fsc.bu.edu.eg (A.F. Ali), mir.faizal@ubc.ca, mir.faizal@uleth.ca (M. Faizal).

the metric of spacetime depend on the energy of the probe, and thus we would obtain gravity's rainbow [13].

So, the gravity's rainbow can be motivated from string theory, as the energy dependence of the spacetime metric can be motivated from the flowing of target space geometry in string theory. It may be also noted that various solution obtained in string theory have been generalized to massive gravity, which is a theory with massive gravitons. In fact, massive Type IIA supergravity has been studied [39,40]. The Fermionic T-duality has also been studied for massive type IIA supergravity [41]. The relation between the massive IIA supergravity and M-theory has also been investigated [42]. A relation between massive IIA/IIB supergravities has also been analyzed, and it has been demonstrated that a duality exists between such massive supergravities [43,44]. Thus, it is possible to study massive supergravity in string theory, and so massive gravity is also important in string theory. It may be noted that other solutions motivated by string theory has been also studied in massive gravity. In fact, a brane in warped AdS spacetime has been constructed in massive gravity [45]. This was done by analyzing the effect of the mass term for the graviton on a infrared brane. A nonextremal brane has also been analyzed in massive gravity [46]. As there is a good motivation to both study the massive gravity and gravity's rainbow from string theory, it is both interesting and important to study the rainbow deformation of massive gravity. So, the target space metric flows and becomes energy dependent, even in massive supergravities, and the bosonic part of this theory will be described by massive gravity's rainbow. Thus, the rainbow deformation of interesting solution massive gravity has been recently studied [47–50].

It may be noted that massive gravity can also be phenomenologically motivated from accelerated cosmic expansion [51–56]. Even though there are problems with the massive gravity, these problems can be resolved using the Vainshtein mechanism [57, 58]. However, the Vainshtein mechanism produces the Boulware–Deser ghosts [59]. It is possible to resolve this with ghosts fields by using the dRGT mechanism [60–67]. It is possible to have a well defined initial value formulation for massive gravity. In fact, initial value constraints for spherically symmetric deformations of flat space, in such a massive theory of gravity have been studied [68]. It has been demonstrated that even though the energy can be negative and even unbounded from below in certain sector of the theory, there is a physical sector of the theory, in which the energy is positive and the ghosts are suppressed, and that the theory is stable [68]. The negative energy sector remains disjointed, and does not have any effect on the physical sector of this theory. The initial values for cosmological solutions have also been studied in massive gravity [69]. So, the theory has well defined posed initial value formulation, and can be used to analyze the effects of graviton mass on various physical phenomena. The cosmological solutions in massive gravity have also been used to obtain an upper bound on the graviton mass [70]. The open FRW universes have been also studied in massive gravity, and it has been possible to obtain universes with standard curvature and an effective cosmological constant, such a theory of massive gravity [71]. In fact, various different solutions in massive gravity have been studied, and the effect of such a mass on the physics of various systems has been discussed [72–77]. So, massive gravity is a very important theory of modified gravity, and it is important to study different solutions in massive gravity.

In fact, as both rainbow gravity, and massive gravity are motivated from string theory and phenomenology, we will analyze a solution in the rainbow deformation of the massive gravity [47–50]. We will study Vaidya solutions in this theory of massive gravity's rainbow because Vaidya spacetime has used to study interesting physical system [78–81]. The Vaidya spacetime in massive gravity has been constructed [82], and the AdS/CFT has been used to interstage field theory dual to a Vaidya-AdS solutions in massive gravity [83]. The Vaidya spacetime is also important in string theory [85,86]. As Vaidya solution is important in string theory, and string theory can also be used to motivate a rainbow deformation of massive gravity, we will study the Vaidya spacetime, massive gravity's rainbow. It may be noted even though Vaidya solution has been studied in gravity's rainbow [84] it has not been studied in massive gravity's rainbow, and so such it is interesting to analyze the Vaidya solution in massive gravity's rainbow.

2. The massive gravity's rainbow

In this section, we study the time-dependent black hole solution using Vaidya metric. This metric will be made energy dependent using the framework of massive gravity's rainbow [47–50]. The four dimensional action for such a massive theory of gravity, can be written as

$$\mathcal{I}=\int d^4x\sqrt{-g}\left[\mathcal{R}+\mathcal{M}^2\sum_i^4 c_i\mathcal{U}_i(g,f)+\mathcal{L}_m\right], \tag{1}$$

where $\mathcal{M}$ is the mass parameter in the massive gravity. Here f is the reference metric, c_i are constants, and $\mathcal{U}_i$ are symmetric polynomials of the eigenvalues of the $d\times d$ matrix $\mathcal{K}^\mu{}_\nu=\sqrt{g^{\mu\alpha}f_{\alpha\nu}}$. These symmetric polynomials can be written as

$$\begin{aligned}
\mathcal{U}_1&=[\mathcal{K}],\\
\mathcal{U}_2&=[\mathcal{K}]^2-[\mathcal{K}^2],\\
\mathcal{U}_3&=[\mathcal{K}]^3-3[\mathcal{K}][\mathcal{K}^2]+2[\mathcal{K}^3],\\
\mathcal{U}_4&=[\mathcal{K}]^4-6[\mathcal{K}^2][\mathcal{K}]^2+8[\mathcal{K}^3][\mathcal{K}]+3[\mathcal{K}^2]^2-6[\mathcal{K}^4].
\end{aligned} \tag{2}$$

The square root in $\mathcal{K}$ can be defined using $(\sqrt{A})^\mu{}_\nu(\sqrt{A})^\nu{}_\lambda=A^\mu{}_\lambda$ and $\mathcal{K}=\mathcal{K}^\mu{}_\mu$. Now the equation of motion from this action, can be written as

$$G_{\mu\nu}+\mathcal{M}^2\chi_{\mu\nu}=T_{\mu\nu}, \tag{3}$$

where $G_{\mu\nu}$ is the Einstein tensor, and $\chi_{\mu\nu}$ is given by

$$\begin{aligned}
\chi_{\mu\nu}=&-\frac{c_1}{2}\left(\mathcal{U}_1g_{\mu\nu}-\mathcal{K}_{\mu\nu}\right)-\frac{c_2}{2}\left(\mathcal{U}_2g_{\mu\nu}-2\mathcal{U}_1\mathcal{K}_{\mu\nu}+2\mathcal{K}^2_{\mu\nu}\right)\\
&-\frac{c_3}{2}(\mathcal{U}_3g_{\mu\nu}-3\mathcal{U}_2\mathcal{K}_{\mu\nu}+6\mathcal{U}_1\mathcal{K}^2_{\mu\nu}-6\mathcal{K}^3_{\mu\nu})\\
&-\frac{c_4}{2}(\mathcal{U}_4g_{\mu\nu}-4\mathcal{U}_3\mathcal{K}_{\mu\nu}+12\mathcal{U}_2\mathcal{K}^2_{\mu\nu}-24\mathcal{U}_1\mathcal{K}^3_{\mu\nu}\\
&+24\mathcal{K}^4_{\mu\nu}).
\end{aligned} \tag{4}$$

We will analyze a spatial reference metric, in the basis (t,r,θ,ϕ) [88], for this theory of massive gravity. Thus, we can write

$$f_{\mu\nu}=diag(0,0,c^2h_{ij}), \tag{5}$$

where h_{ij} is two dimensional Euclidean metric and c is a positive constant. We will now write the Vaidya metric for this massive theory, deformed by gravity's rainbow. So, we will analyze the rainbow deformation of the Vaidya metric, in the case of advanced time coordinate. These rainbow deformations of this metric can be expressed as [47–50]

$$\begin{aligned}
ds^2=&-\frac{1}{\mathcal{F}^2(E)}\left(1-\frac{m(t,r)}{r}\right)dt^2+\frac{2}{\mathcal{F}(E)\mathcal{G}(E)}dtdr\\
&+\frac{1}{\mathcal{G}^2(E)}r^2d\Omega_2^2,
\end{aligned} \tag{6}$$

where $\mathcal{F}(E)$ and $\mathcal{G}(E)$ are known as the gravity's rainbow functions. It may be noted that here $E = E_s/E_P$, where E_s is the maximum energy that a probe in that system can take, and E_p is the Planck energy. So, as $E_s/E_p \to 0$, $\mathcal{F}(E) = \mathcal{G}(E) = 1$, and the general relativity is recovered in the IR limit of the theory [14–22]. These rainbow functions are motivated from various theoretical and phenomenology considerations. The results from loop quantum gravity and κ-Minkowski noncommutative spacetime, have been used to motivate the following rainbow functions [1,2]

$$\mathcal{F}(E/E_p) = 1 \quad \text{and} \quad \mathcal{G}(E/E_p) = \sqrt{1 - a\left(\frac{E}{E_p}\right)^q}. \tag{7}$$

The modified dispersion relation with constant velocity of light, has been used to motivate the following rainbow functions [87]

$$\mathcal{F}(E/E_p) = \mathcal{G}(E/E_p) = \frac{1}{1 - aE/E_p}. \tag{8}$$

The hard spectra from gamma-ray burster's, has been used to motivate the following rainbow functions [5]

$$\mathcal{F}(E/E_p) = \frac{e^{aE/E_p} - 1}{aE/E_p} \quad \text{and} \quad \mathcal{G}(E/E_p) = 1. \tag{9}$$

The maximum energy of the system depends on the physical systems being analyzed, and for black holes, this energy is equal to the energy of a quantum particle near the horizon. This is because such a particle can be viewed as a probe for the geometry of the black hole. In fact, we can use the uncertainty principle, $\Delta p \geq 1/\Delta x$, to obtain a bound on the energy of such a particle. So, we can write $E_s \geq 1/\Delta x$, where Δx is the uncertainty in position of the particle near the horizon, and it is equal to the radius of the event horizon. Thus, the bound on the energy for a black hole can be written as

$$E_s \geq 1/\Delta x \approx 1/r_+. \tag{10}$$

It may be noted as the black hole evaporates due to the Hawking radiation, its radius reduces, and this changes the bound on this maximum energy. So, this energy is a dynamical function, and thus rainbow functions are also dynamical. Even though, we do not need the explicit dynamical behavior of rainbow functions, it is important to know that they are dynamical, and so they cannot be gauged away by rescaling of the metric.

Now, we assume the total energy-momentum tensor of the field equation (3), can be expressed in the following form

$$T_{\mu\nu} = T^{(n)}_{\mu\nu} + T^{(m)}_{\mu\nu}, \tag{11}$$

where $T^{(n)}_{\mu\nu}$ and $T^{(m)}_{\mu\nu}$ are the energy-momentum tensor for the Vaidya null radiation and the energy-momentum tensor of the perfect fluid, respectively. They can be defined as

$$T^{(n)}_{\mu\nu} = \sigma l_\mu l_\nu,$$
$$T^{(m)}_{\mu\nu} = (\rho + p)(l_\mu n_\nu + l_\nu n_\mu) + p g_{\mu\nu}, \tag{12}$$

where σ, ρ and p are null radiation density, energy density and pressure of the perfect fluid, respectively. In this regard, l_μ and n_μ are linearly independent future pointing null vectors,

$$l_\mu = \left(\frac{1}{\mathcal{F}(E)}, 0, 0, 0\right) \quad \&$$
$$n_\mu = \left(\frac{1}{2\mathcal{F}(E)}\left(1 - \frac{m(t,r)}{r}\right), -\frac{1}{\mathcal{G}(E)}, 0, 0\right), \tag{13}$$

satisfying the following conditions

$$l_\mu l^\mu = n_\mu n^\mu = 0 \quad \& \quad l_\mu n^\mu = -1. \tag{14}$$

Therefore, the non-vanishing components of the total energy-momentum tensor can be written as

$$T_{00} = \frac{\sigma}{\mathcal{F}^2(E)} + \frac{\rho}{\mathcal{F}^2(E)}\left(1 - \frac{m(t,r)}{r}\right), \quad T_{01} = -\frac{\rho}{\mathcal{F}(E)\mathcal{G}(E)},$$
$$T_{22} = \frac{pr^2}{\mathcal{G}^2(E)}, \quad T_{33} = \frac{pr^2 \sin^2\theta}{\mathcal{G}^2(E)}. \tag{15}$$

Using the metric ansatz (5), we obtain

$$\mathcal{K}^\mu{}_\nu = diag\left(0, 0, \frac{c\mathcal{G}(E)}{r}, \frac{c\mathcal{G}(E)}{r}\right). \tag{16}$$

Therefore, we find that

$$(\mathcal{K}^2)^\mu{}_\nu = \mathcal{K}^\mu{}_\alpha \mathcal{K}^\alpha{}_\nu = diag\left(0, 0, \frac{c^2\mathcal{G}^2(E)}{r^2}, \frac{c^2\mathcal{G}^2(E)}{r^2}\right),$$
$$(\mathcal{K}^3)^\mu{}_\nu = \mathcal{K}^\mu{}_\alpha \mathcal{K}^\alpha{}_\beta \mathcal{K}^\beta{}_\nu = diag\left(0, 0, \frac{c^3\mathcal{G}^3(E)}{r^3}, \frac{c^3\mathcal{G}^3(E)}{r^3}\right),$$
$$(\mathcal{K}^4)^\mu{}_\nu = \mathcal{K}^\mu{}_\alpha \mathcal{K}^\alpha{}_\beta \mathcal{K}^\beta{}_\lambda \mathcal{K}^\lambda{}_\nu = diag\left(0, 0, \frac{c^4\mathcal{G}^4(E)}{r^4}, \frac{c^4\mathcal{G}^4(E)}{r^4}\right). \tag{17}$$

We also obtain the following quantities

$$[\mathcal{K}] = \mathcal{K}^\mu{}_\mu = \frac{2c\mathcal{G}(E)}{r}, \quad [\mathcal{K}^2] = (\mathcal{K}^2)^\mu{}_\mu = \frac{2c^2\mathcal{G}^2(E)}{r^2},$$
$$[\mathcal{K}^3] = (\mathcal{K}^3)^\mu{}_\mu = \frac{2c^3\mathcal{G}^3(E)}{r^3}, \quad [\mathcal{K}^4] = (\mathcal{K}^4)^\mu{}_\mu = \frac{2c^4\mathcal{G}^4(E)}{r^4}. \tag{18}$$

Now, using the Eqs. (17), (18), and Eq. (2), we obtain

$$\mathcal{U}_1 = \frac{2c\mathcal{G}(E)}{r}, \quad \mathcal{U}_2 = \frac{2c^2\mathcal{G}^2(E)}{r^2},$$
$$\mathcal{U}_3 = 0, \quad \mathcal{U}_4 = 0. \tag{19}$$

Using the Eqs. (16), (17), (18) and (19), we can obtain the non-vanishing components of the massive gravity term $\chi_{\mu\nu}$ in the field equation (3) as

$$\chi_{00} = \left[\frac{c_1 c\,\mathcal{G}(E)}{r\mathcal{F}^2(E)} + \frac{c_2 c^2\,\mathcal{G}^2(E)}{r^2\mathcal{F}^2(E)}\right]\left(1 - \frac{m}{r}\right),$$
$$\chi_{01} = \chi_{10} = -\frac{1}{r\mathcal{F}(E)}\left(c_1 c + \frac{c_2 c^2\,\mathcal{G}(E)}{r}\right),$$
$$\chi_{22} = -\frac{c_1 c\, r}{2\mathcal{G}(E)},$$
$$\chi_{33} = -\frac{c_1\, c\, r \sin^2\theta}{2\mathcal{G}(E)}. \tag{20}$$

Then, for the 00 component of the field equation (3), we have

$$\frac{\mathcal{G}(E)}{r^3}\left[r\dot{m}\mathcal{F}(E) + r\mathcal{G}(E)m' - \mathcal{G}(E)mm'\right]$$
$$= \sigma + \rho\left(1 - \frac{m}{r}\right) - \mathcal{M}^2\left[\frac{c_1 c\mathcal{G}(E)}{r} + \frac{c_2 c^2\mathcal{G}^2(E)}{r^2}\right]\left(1 - \frac{m}{r}\right), \tag{21}$$

where dot and prime signs denote the derivative with respect to time and radial coordinates, respectively. For the 01 and 10 component, we have

$$-\frac{\mathcal{G}(E)m'}{r^2} = -\frac{\rho}{\mathcal{G}(E)} + \frac{\mathcal{M}^2}{r}\left(c_1 c + \frac{c_2 c^2\,\mathcal{G}(E)}{r}\right). \tag{22}$$

Finally, for the 22 and 33 component, we obtain

$$-\frac{1}{2}rm'' = \frac{pr^2}{\mathcal{G}^2(E)} + \frac{\mathcal{M}^2 c_1 c\, r}{2\mathcal{G}(E)}. \tag{23}$$

Thus, we have been able to analyze the Einstein equation in gravity's rainbow. In the next section, we will analyze Vaidya spacetime in this massive gravity's rainbow.

3. Dynamics of the collapsing system

In this section, we will first find a solution for the field equations describing this model. Then, we will analyze the dynamics of a collapsing system. The matter field will be assumed to follow a barotropic equation of state, which is given by

$$p = k\rho, \tag{24}$$

where k is the barotropic parameter. Now we can use the Eqs. (18), (19) and (20), and obtain an equation describing the behavior of $m(t, r)$ for this system,

$$r^2 m'' + 6km + \frac{(1+3k)\,\mathcal{M}^2 c_1 c\, r}{\mathcal{G}(E)} + 6kc_2 c^2 \mathcal{M}^2 r - 6kf_1(t) = 0, \tag{25}$$

where $f_1(t)$ is an arbitrary function of time. This differential equation can be solved to obtain a solution for $m(t, r)$,

$$\begin{aligned} m(t, r) = {} & f_2(t) r^{\omega_1} + f_3(t) r^{\omega_2} - \frac{\mathcal{M}^2 c_1 c\,(1+3k)\, r}{(2-\omega_1)(2-\omega_2)\,\mathcal{G}(E)} \\ & - c_2 c^2 \mathcal{M}^2 r + f_1(t), \end{aligned} \tag{26}$$

where $\omega_1 = \frac{1}{2}\left(1+\sqrt{1-24k}\right)$, $\omega_2 = \frac{1}{2}\left(1-\sqrt{1-24k}\right)$. Here, $f_2(t)$ and $f_3(t)$ are arbitrary functions of time t. So, from these equations, we obtain the admissible range of k, which is $(-\infty, 1/24]$. Thus, the metric given in Eq. (6), can be expressed as

$$\begin{aligned} ds^2 = {} & \frac{1}{\mathcal{F}^2(E)}\Bigg(-1 + f_2(t) r^{\omega_1 - 1} + f_3(t) r^{\omega_2 - 1} \\ & - \frac{\mathcal{M}^2 c_1 c\,(1+3k)}{(2-\omega_1)(2-\omega_2)\,\mathcal{G}(E)} - c_2 c^2 \mathcal{M}^2 + \frac{f_1(t)}{r}\Bigg) dt^2 \\ & + \frac{2dtdr}{\mathcal{F}(E)\mathcal{G}(E)} + \frac{1}{\mathcal{G}^2(E)} r^2 d\Omega_2^2 . \end{aligned} \tag{27}$$

This metric is the generalized Vaidya metric in Massive gravity's rainbow.

In this generalized Vaidya spacetime, the singularity can be either a naked singularity or a black hole. The nature of this singularity is determined by the existence of outgoing radial null geodesics, which end in the past central singularity at $r = 0$. Such geodesics exist for a locally naked singularity, and do not exist for a black hole. So, in massive gravity's rainbow, the singularity formed from gravitational collapse can be either a naked singularity or a black hole. In general relativity, the cosmic censorship hypothesis states that the gravitational singularity must necessarily be covered by an event horizon. So, according to cosmic censorship hypothesis only black hole can form from a collapsing system. However, it has been demonstrated that inhomogeneous dust cloud may form a naked singularity [89]. Interesting results have also been obtained by studying fluid whose equation of state is different from the equation of state of dust [90]. So, it is possible to generalize the cosmic censorship hypothesis [91]. As we have to investigate the nature of singularities in massive gravity's rainbow, we can use such a generalization of the cosmic censorship hypothesis.

As this system is described by a time-dependent geometry, the radius of shell at r, will also be a function of time t. We will describe such a radius by $R(t,\ r)$. This system starts from an initial time $t = 0$, and at that time, we have $R(0,\ r) = r$. It may be noted that for a inhomogeneous system, different shells may become singular at different times. Now, for this system, we can have future directed radial null geodesics coming out of the singularity. These will have a well defined tangent at the singularity. So, for this system, $\frac{dR}{dr}$ must tend to a finite limit, as the system approach the past singularity. It is possible for the system to reach the points $(t_0,\ r) = 0$. At this point, the singularity $R(t_0, 0) = 0$ occurs and the matter shells are crushed to a zero radius. This singularity at $r = 0$, is called a central singularity.

Now a naked singularity will form in this system, if future directed curves end in the past singularity. So, for such a system, the outgoing null geodesics will end in the past central singularity, which is at $r = 0$ and $t = t_0$. At such a point, $R(t_0, 0) = 0$, and so for these geodesics, we have $R \to 0$ as $r \to 0$ [92]. The equation for these outgoing radial null geodesics can be obtained from the Eq. (6). Thus, by putting $ds^2 = 0$ and $d\Omega_2^2 = 0$, we obtain

$$\frac{dt}{dr} = \frac{2\mathcal{F}(E)}{\mathcal{G}(E)\left(1 - \frac{m(t,r)}{r}\right)}. \tag{28}$$

Here $r = 0,\ t = 0$ corresponds to a singularity in this equation. Now if $X = \frac{t}{r}$, then we can analyze the limiting behavior of X, as the system approaches $r = 0,\ t = 0$. So, if this limiting value of X is denoted by X_0, then we can write

$$\begin{aligned} X_0 = \lim_{t\to 0}\lim_{r\to 0} X = \lim_{t\to 0}\lim_{r\to 0}\frac{t}{r} = \lim_{t\to 0}\lim_{r\to 0}\frac{dt}{dr} \\ = \lim_{t\to 0}\lim_{r\to 0}\frac{2\mathcal{F}(E)}{\mathcal{G}(E)\left(1 - \frac{m(t,r)}{r}\right)}. \end{aligned} \tag{29}$$

We also use Eqs. (25) and (28), and obtain

$$\begin{aligned} \frac{2}{X_0} = \lim_{t\to 0}\lim_{r\to 0}\frac{\mathcal{G}(E)}{\mathcal{F}(E)}\Bigg[1 - f_2(t) r^{\omega_1 - 1} - f_3(t) r^{\omega_2 - 1} \\ + \frac{\mathcal{M}^2 c_1 c\,(1+3k)}{(1+\omega_1)(1+\omega_2)\,\mathcal{G}(E)} + c_2 c^2 \mathcal{M}^2 - \frac{f_1(t)}{r}\Bigg]. \end{aligned}$$

Now, choosing $f_1(t) = \gamma t$, $f_2(t) = \alpha t^{1-\omega_1}$ and $f_3(t) = \beta t^{1-\omega_2}$, we obtain the algebraic equation for X_0, which can be written as

$$\begin{aligned} & \alpha X_0^{1+\omega_2} + \beta X_0^{1+\omega_1} + \gamma X_0^2 - \left(1 + c_2 c^2 \mathcal{M}^2\right) X_0 \\ & - \frac{(1+3k)\,\mathcal{M}^2 c_1 c}{(2-\omega_1)(2-\omega_2)\,\mathcal{G}(E)} + \frac{2\mathcal{F}(E)}{\mathcal{G}(E)} = 0, \end{aligned} \tag{30}$$

where α, β and γ are constants. If we only obtain the non-positive solution of the equation, then a black hole will form in this system. However, a naked singularity can form for positive roots of this equation. Since this equation is highly complicated, it is extremely difficult to find out an analytic solution of X_0. So, we will use numerical methods to find a numerical solutions of X_0. This will be done by assigning particular numerical values to the associated variables. In fact, as a specific rainbow function has been well motivated [5,9], we will use this rainbow functions for analyzing this system,

$$\mathcal{F}(E) = 1, \quad \mathcal{G}(E) = \sqrt{1 - \eta\left(\frac{E_1}{E_p}\right)} \tag{31}$$

In the above expressions, E_p is the Planck energy, given by $E_p = 1/\sqrt{G} = 1.221 \times 10^{19}$ GeV, where G is the gravitational constant and $E_1 = 1.42 \times 10^{-13}$ [5,9]. The value of η has been estimated to be $\eta \approx 1$ [5], and so in our study, we will use $\eta = 1$.

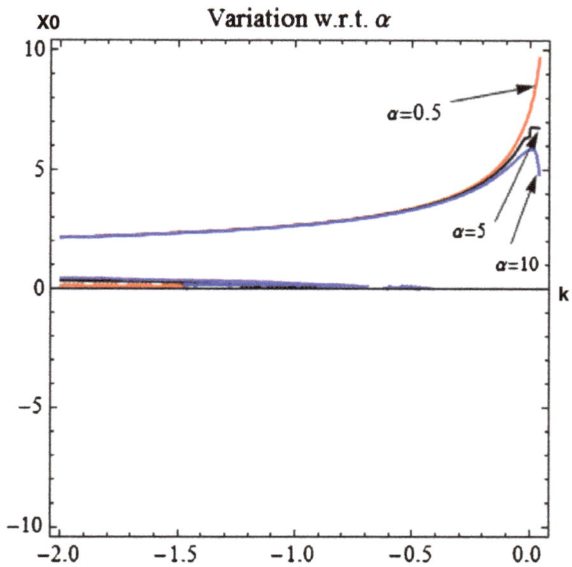

Fig. 1. This figure shows the variation of X_0 with k for different values of α in massive gravity's rainbow. The other parameters are fixed at $\beta = 2$, $\gamma = 3$, $c = 0.8$, $c_1 = 4$, $c_2 = 2$, $\mathcal{M} = 5$, $\eta = 1$, $E_1 = 1.42 \times 10^{-13}$, $E_p = 1.221 \times 10^{19}$. (For interpretation of the references to color in this figure legend, the reader is referred to the web version of this article.)

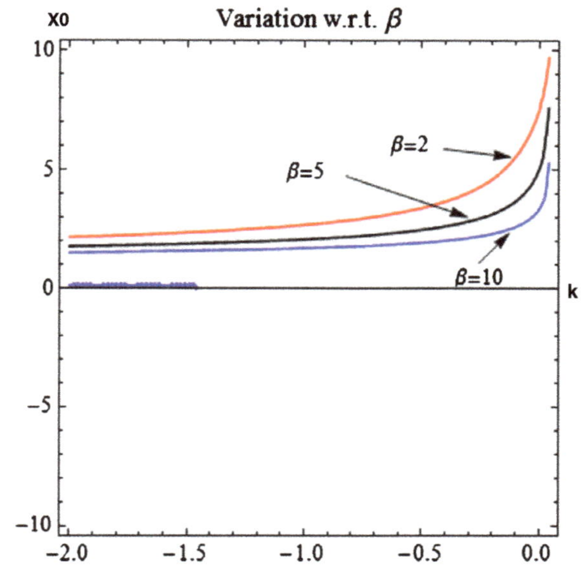

Fig. 2. This figure shows the variation of X_0 with k for different values of β in massive gravity's rainbow. The other parameters are taken as $\alpha = 0.5$, $\gamma = 3$, $c = 0.8$, $c_1 = 4$, $c_2 = 2$, $\mathcal{M} = 5$, $\eta = 1$, $E_1 = 1.42 \times 10^{-13}$, $E_p = 1.221 \times 10^{19}$.

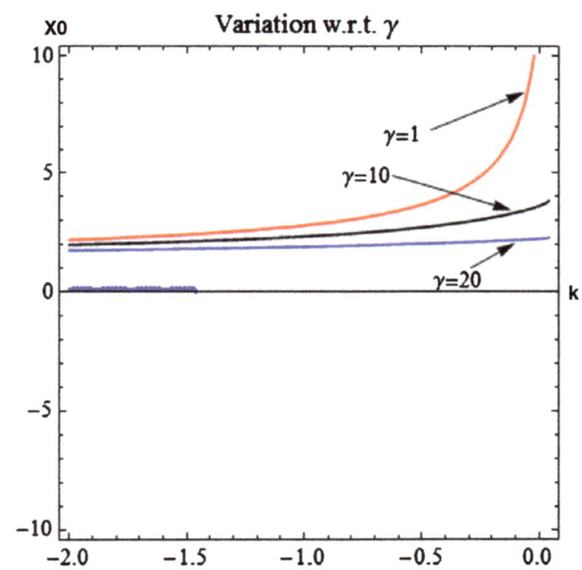

Fig. 3. This figure shows the variation of X_0 with k for different values of γ in massive gravity's rainbow. The other parameters are fixed at $\alpha = 0.5$, $\beta = 2$, $c = 0.8$, $c_1 = 4$, $c_2 = 2$, $\mathcal{M} = 5$, $\eta = 1$, $E_1 = 1.42 \times 10^{-13}$, $E_p = 1.221 \times 10^{19}$.

4. Conclusion and discussion

Now, we will comment on the numerical results obtained in this paper. In Fig. 1, the contours $k - X_0$ were obtained for different numerical values of α, where other parameters were fixed, in the massive gravity's rainbow. The admissible plot range for the equation of state parameter, k is $(-\infty, 1/24]$. In the figure, the plot range for k has been taken as $-2 < k < 1/24$ (from late to early universe). We observe that the trajectories for different values of α almost coincide with each other from $k = -2$ till around $k = -1/3$, i.e., the quintessence and phantom regime (dark energy). But for $k > -1/3$, we observe that this coincidence gradually disappears, and the red line ($\alpha = 0.5$) diverges. However, this does not change the physics of the system much, as all the trajectories remain in the positive level of X_0. So, the singularity formed is a naked singularity. At around $k = 0$, the separation of the trajectories becomes more pronounced. For greater values of α (blue line), we see that there is a decreased tendency of formation of naked singularity compared to lower values of α. Even we see that the blue line starts to take a dip around $k = 0$. The increased significance of α directly reflects on the function $f_2(t)$. Physically $k = 0$ represents the dust regime and $k > 0$ corresponds to early universe. So, the α dependence of the system will be more significant in the earlier than in the later stages of the evolution of the universe. This is because in massive gravity's rainbow, the spacetime is energy dependent, and the energy in the earlier stages of the evolution of the universe is more than the energy at the later stages of the evolution of the universe. So, the rainbow functions are more important in the physics of the early stages of the universe. This is the reasons that the significance of α decreases at the later stages of the evolution of the universe.

In Fig. 2, similar figures are obtained for different values of β, where the other parameters are fixed. We observe that as the values of β increase, the trajectories push downwards towards the k-axis. This indicate an increase in the tendency to form black holes. However, for both Figs. 1 and 2, it is clear that the trajectories remain in the positive X_0 region, and a naked singularity forms from the collapse of this system. In Figs. 3 and 4, the $k - X_0$ plots are obtained for different values of γ and $\mathcal{M}$, respectively. These plots also indicate that a naked singularity is formed from the collapse of this system. We can observe from Fig. 3, the tendency to form a black hole increases with increase in the value of γ. We can also observe from Fig. 4, the tendency to form a black hole decreases with the increase in the value of $\mathcal{M}$. So, the system can form a naked singularity by increasing the value of $\mathcal{M}$, and decreasing the value of γ.

In Figs. 5 and 6, we compare the $k - X_0$ contours of both massive gravity and massive gravity's rainbow. In Fig. 5, the trajectories are for different values of α. From the plot, we can observe that for massive gravity's rainbow, α does not play an important role in the collapsing system, when $k < -1/3$. However, for pure massive gravity α, does not play an important role throughout the domain. Besides, in gravity's rainbow, the tendency to form black holes is greater than that in pure massive gravity. In Fig. 6, similar plots are

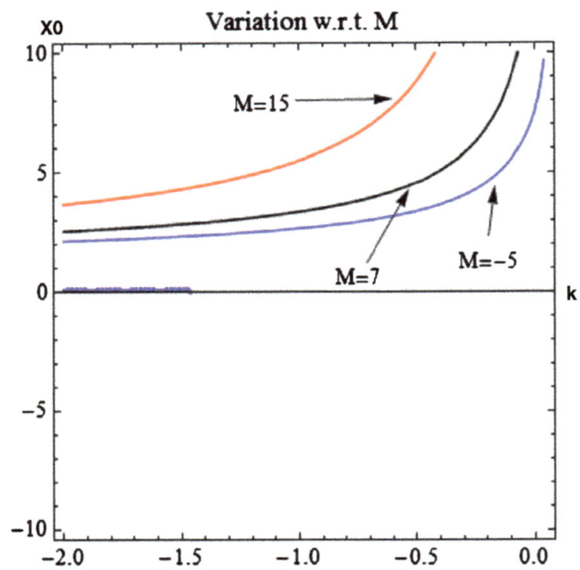

Fig. 4. This figure shows the variation of X_0 with k for different values of $\mathcal{M}$ in massive gravity's rainbow. The other parameters are taken as $\alpha = 0.5$, $\beta = 2$, $\gamma = 3$, $c = 0.8$, $c_1 = 4$, $c_2 = 2$, $\eta = 1$, $E_1 = 1.42 \times 10^{-13}$, $E_p = 1.221 \times 10^{19}$.

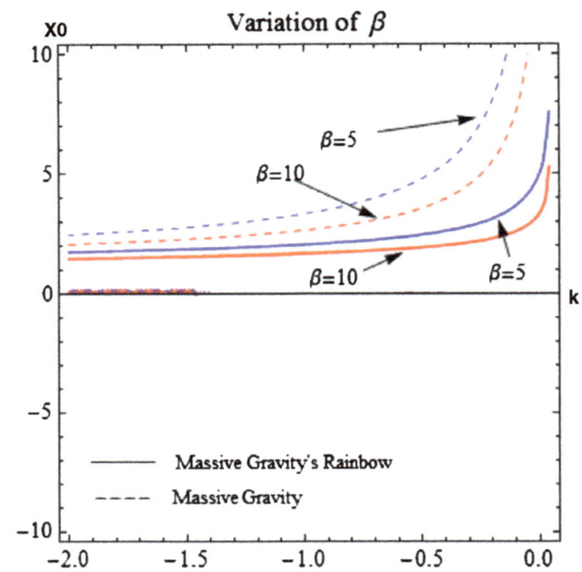

Fig. 6. This figure shows the variation of X_0 with k for different values of β in a comparative scenario between Massive gravity and Massive gravity's rainbow. The other parameters are taken as $\alpha = 0.5$, $\gamma = 3$, $c = 0.8$, $c_1 = 4$, $c_2 = 2$, $\mathcal{M} = 5$, $\eta = 1$, $E_1 = 1.42 \times 10^{-13}$, $E_p = 1.221 \times 10^{19}$.

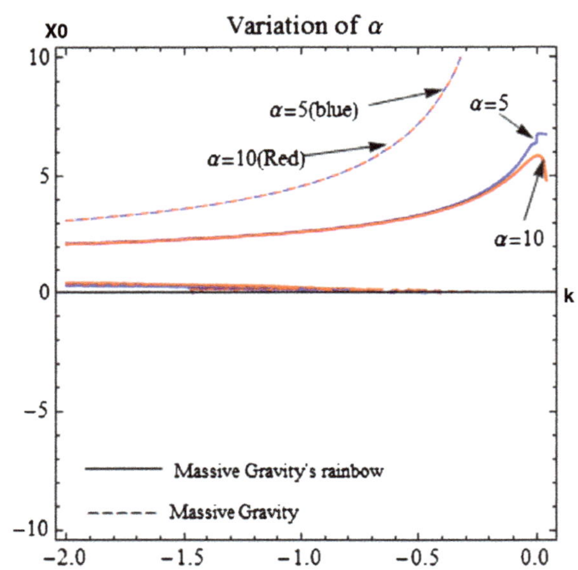

Fig. 5. This figure shows the variation of X_0 with k for different values of α in a comparative scenario between Massive gravity and Massive gravity's rainbow. The other parameters are fixed at $\beta = 2$, $\gamma = 3$, $c = 0.8$, $c_1 = 4$, $c_2 = 2$, $\mathcal{M} = 5$, $\eta = 1$, $E_1 = 1.42 \times 10^{-13}$, $E_p = 1.221 \times 10^{19}$. (For interpretation of the references to color in this figure legend, the reader is referred to the web version of this article.)

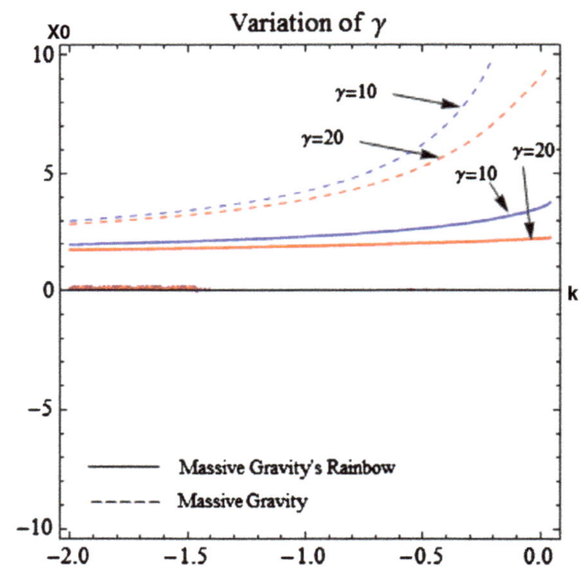

Fig. 7. This figure shows the variation of X_0 with k for different values of γ in a comparative scenario between Massive gravity and Massive gravity's rainbow. The other parameters are fixed at $\alpha = 0.5$, $\beta = 2$, $c = 0.8$, $c_1 = 4$, $c_2 = 2$, $\mathcal{M} = 5$, $\eta = 1$, $E_1 = 1.42 \times 10^{-13}$, $E_p = 1.221 \times 10^{19}$.

obtained for different values of β. Here, it is also confirmed that in gravity's rainbow, there is a greater tendency to form a black hole. The above observation is again established in Figs. 7 and 8, where similar plots are generated by varying γ and $\mathcal{M}$, respectively. Finally we observe that in all the figures, there are small portions of lines which nearly vanish around the k-axis, for small values of k. As this system was very complicated, we could not find an analytical solution for Eq. (30). So, we obtained numerical solution for this equation, using particular values for the parameters. The vanishing lines in the $k - X_0$ plane, are produced from the noise in the numerical solution, and do not have physical significance.

In this paper, we have constructed a theory of massive gravity's rainbow. This was done by analyzing the energy dependent deformation of massive gravity. In the construction of massive gravity, we have used the Vainshtein mechanism and the dRGT mechanism. Then, this theory has been deformed by rainbow functions. We have analyzed radiating Vaidya black hole solution in this theory of massive gravity's rainbow. The effects of both the graviton mass and rainbow deformation have been studied for a time-dependent system. It may be noted that the AdS solution in massive gravity, and the AdS/CFT correspondence corresponding to this AdS solution have been studied [93,94]. In fact, the holographic entanglement entropy for massive gravity has also been studied [95], and it has been demonstrated that for such systems both first order and second order phase transitions can occur. The holographic complexity for massive gravity has also been studied

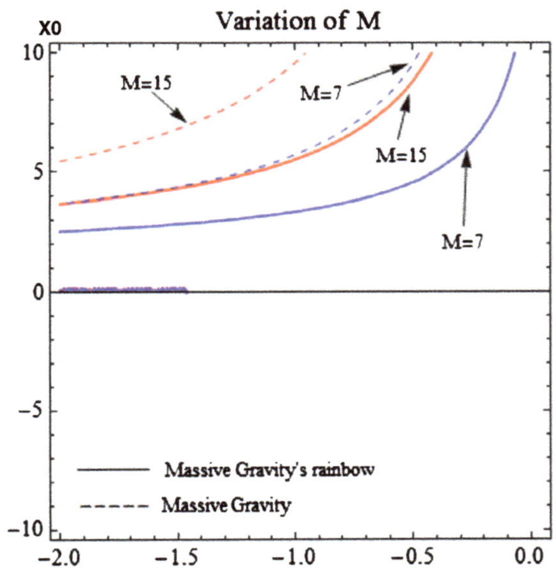

Fig. 8. This figure shows the variation of X_0 with k for different values of $\mathcal{M}$ in a comparative scenario between Massive gravity and Massive gravity's rainbow. The other parameters are taken as $\alpha = 0.5$, $\beta = 2$, $\gamma = 3$, $c = 0.8$, $c_1 = 4$, $c_2 = 2$, $\eta = 1$, $E_1 = 1.42 \times 10^{-13}$, $E_p = 1.221 \times 10^{19}$.

[96]. This holographic complexity of a boundary theory is dual to a volume in the bulk, just as the holographic entanglement entropy is dual to an area in the bulk. It would be interesting to study the rainbow deformation of such solutions. This can be done by making the bulk metric to depend on the energy of the probe. Then, deformation of the bulk metric can be done using suitable rainbow functions. It would be interesting to investigate the holographic entanglement entropy and holographic complexity of massive gravity deformed by suitable rainbow functions.

Acknowledgements

P. Rudra acknowledges University Grants Commission, Govt. of India for providing research project grant (No. F.PSW-061/15-16 (ERO)). F. Darabi acknowledges financial support of Azarbaijan Shahid Madani University (No. S/5749-ASMU) for the Sabbatical Leave, and thanks the hospitality of ICTP (Trieste) for providing support during the Sabbatical Leave.

References

[1] G. Amelino-Camelia, J.R. Ellis, N.E. Mavromatos, D.V. Nanopoulos, Int. J. Mod. Phys. A 12 (1997) 607.
[2] G. Amelino-Camelia, Living Rev. Relativ. 16 (2013) 5.
[3] G. 't Hooft, Class. Quantum Gravity 13 (1996) 1023.
[4] V.A. Kostelecky, S. Samuel, Phys. Rev. D 39 (1989) 683.
[5] G. Amelino-Camelia, J.R. Ellis, N.E. Mavromatos, D.V. Nanopoulos, S. Sarkar, Nature 393 (1998) 763.
[6] R. Gambini, J. Pullin, Phys. Rev. D 59 (1999) 124021.
[7] S.M. Carroll, J.A. Harvey, V.A. Kostelecky, C.D. Lane, T. Okamoto, Phys. Rev. Lett. 87 (2001) 141601.
[8] G. Amelino-Camelia, J. Lukierski, A. Nowicki, Phys. At. Nucl. 61 (1998) 1811, Yad. Fiz. 61 (1998) 1925.
[9] G. Amelino-Camelia, J. Lukierski, A. Nowicki, Int. J. Mod. Phys. A 14 (1999) 4575.
[10] G. Amelino-Camelia, New J. Phys. 6 (2004) 188.
[11] G. Amelino-Camelia, Int. J. Mod. Phys. D 11 (2002) 35.
[12] J. Magueijo, L. Smolin, Phys. Rev. D 67 (2003) 044017.
[13] J. Magueijo, L. Smolin, Class. Quantum Gravity 21 (2004) 1725.
[14] P. Galan, G.A. Mena Marugan, Phys. Rev. D 70 (2004) 124003.
[15] J. Hackett, Class. Quantum Gravity 23 (2006) 3833.
[16] F. Girelli, S. Liberati, L. Sindoni, Phys. Rev. D 75 (2007) 064015.
[17] C.-Z. Liu, J.-Y. Zhu, Gen. Relativ. Gravit. 40 (2008) 1899.
[18] H. Li, Y. Ling, X. Han, Class. Quantum Gravity 26 (2009) 065004.
[19] R. Garattini, G. Mandanici, Phys. Rev. D 85 (2012) 023507.
[20] R. Garattini, F.S.N. Lobo, Phys. Rev. D 85 (2012) 024043.
[21] R. Garattini, G. Mandanici, Phys. Rev. D 83 (2011) 084021.
[22] J.-J. Peng, S.-Q. Wu, Gen. Relativ. Gravit. 40 (2008) 2619.
[23] J.D. Barrow, J. Magueijo, arXiv:1310.2072 [astro-ph.CO].
[24] G. Amelino-Camelia, M. Arzano, G. Gubitosi, J. Magueijo, Phys. Rev. D 88 (2013) 041303.
[25] A. Awad, A.F. Ali, B. Majumder, J. Cosmol. Astropart. Phys. 1310 (2013) 052.
[26] Y. Ling, J. Cosmol. Astropart. Phys. 0708 (2007) 017.
[27] Y. Ling, Q. Wu, Phys. Lett. B 687 (2010) 103.
[28] P. Horava, Phys. Rev. D 79 (2009) 084008.
[29] P. Horava, Phys. Rev. Lett. 102 (2009) 161301.
[30] R. Garattini, E.N. Saridakis, Eur. Phys. J. C 75 (2015) 343.
[31] M. Weidner, Fortschr. Phys. 55 (2007) 843.
[32] H. Samtleben, Class. Quantum Gravity 25 (2008) 214002.
[33] R.R. Metsaev, A.A. Tseytlin, Nucl. Phys. B 293 (1987) 385.
[34] M.K. Parikh, Phys. Rev. D 84 (2011) 044048.
[35] B. Sahoo, A. Sen, J. High Energy Phys. 0701 (2007) 010.
[36] A. Sen, J. High Energy Phys. 0603 (2006) 008.
[37] O.J. Rosten, Phys. Rep. 511 (2012) 177.
[38] N.P. Warner, Class. Quantum Gravity 17 (2000) 1287.
[39] P.S. Howe, N.D. Lambert, P.C. West, Phys. Lett. B 416 (1998) 303.
[40] M. Cvetic, H. Lu, C.N. Pope, Phys. Rev. Lett. 83 (1999) 5226.
[41] I. Bakhmatov, Eur. Phys. J. C 76 (2016) 174.
[42] M.P. Garcia del Moral, A. Restuccia, Fortschr. Phys. 64 (2016) 398.
[43] I.V. Lavrinenko, H. Lu, C.N. Pope, K.S. Stelle, Nucl. Phys. B 555 (1999) 201.
[44] L.A.P. Zayas, D. Tsimpis, C.A. Whiting, Phys. Rev. D 96 (2017) 046013.
[45] Z. Chacko, M. Graesser, C. Grojean, L. Pilo, Phys. Rev. D 70 (2004) 084028.
[46] J.F. Vazquez-Poritz, J. High Energy Phys. 0112 (2001) 030.
[47] S.H. Hendi, M. Momennia, B. Eslam Panah, S. Panahiyan, Phys. Dark Universe 16 (2017) 26.
[48] S.H. Hendi, S. Panahiyan, S. Upadhyay, B. Eslam Panah, Phys. Rev. D 95 (2017) 084036.
[49] S.H. Hendi, B. Eslam Panah, S. Panahiyan, Phys. Lett. B 769 (2017) 191.
[50] V.B. Bezerra, H.R. Christiansen, M.S. Cunha, C.R. Muniz, Phys. Rev. D 96 (2017) 024018.
[51] A.G. Riess, et al., Astron. J. 116 (1998) 1009.
[52] S. Perlmutter, et al., Nature 391 (1998) 51.
[53] A.G. Riess, et al., Astron. J. 118 (1999) 2668.
[54] S. Perlmutter, et al., Astrophys. J. 517 (1999) 565.
[55] A.G. Riess, et al., Astrophys. J. 560 (2001) 49.
[56] J.L. Tonry, et al., Astrophys. J. 594 (2003) 1.
[57] A.I. Vainshtein, Phys. Lett. B 39 (1972) 393.
[58] E. Babichev, C. De ayet, Class. Quantum Gravity 30 (2013) 184001.
[59] D.G. Boulware, S. Deser, Phys. Rev. D 6 (1972) 3368.
[60] C. de Rham, G. Gabadadze, A.J. Tolley, Phys. Rev. Lett. 106 (2011) 231101.
[61] C. de Rham, G. Gabadadze, Phys. Rev. D 82 (2010) 04402.
[62] C. de Rham, G. Gabadadze, A.J. Tolley, Phys. Lett. B 711 (2012) 190.
[63] S.F. Hassan, R.A. Rosen, A. Schmidt-May, J. High Energy Phys. 1202 (2012) 026.
[64] S.F. Hassan, A. Schmidt-May, M. von Strauss, Phys. Lett. B 715 (2012) 335.
[65] S.F. Hassan, R.A. Rosen, Phys. Rev. Lett. 108 (2012) 041101.
[66] S.F. Hassan, R.A. Rosen, J. High Energy Phys. 1204 (2012) 123.
[67] K. Hinterbichler, Rev. Mod. Phys. 84 (2012) 671.
[68] M.S. Volkov, Phys. Rev. D 90 (2014) 024028.
[69] M. Wyman, W. Hu, P. Gratia, Phys. Rev. D 87 (2013) 084046.
[70] G. D'Amico, C. de Rham, S. Dubovsky, G. Gabadadze, D. Pirtskhalava, A.J. Tolley, Phys. Rev. D 84 (2011) 124046.
[71] A.E. Gumrukcuoglu, C. Lin, S. Mukohyama, J. Cosmol. Astropart. Phys. 11 (2011) 030.
[72] S.H. Hendi, R.B. Mann, S. Panahiyan, B. Eslam Panah, Phys. Rev. D 95 (2017) 021501.
[73] S.H. Hendi, B. Eslam Panah, S. Panahiyan, Class. Quantum Gravity 33 (2016) 235007.
[74] S.H. Hendi, S. Panahiyan, B. Eslam Panah, J. High Energy Phys. 01 (2016) 129.
[75] A. Adams, D.A. Roberts, O. Saremi, Phys. Rev. D 91 (2015) 046003.
[76] S.H. Hendi, B. Eslam Panah, S. Panahiyan, J. High Energy Phys. 05 (2016) 029.
[77] S.H. Hendi, B. Eslam Panah, S. Panahiyan, J. High Energy Phys. 11 (2015) 157.
[78] P. Rudra, R. Biswas, U. Debnath, Astrophys. Space Sci. 335 (2011) 505.
[79] U. Debnath, P. Rudra, R. Biswas, Astrophys. Space Sci. 339 (2012) 135.
[80] P. Rudra, R. Biswas, U. Debnath, Astrophys. Space Sci. 354 (2014) 597.
[81] P. Rudra, U. Debnath, Can. J. Phys. 92 (11) (2014) 1474.
[82] Y.P. Hu, X.M. Wu, H. Zhang, Phys. Rev. D 95 (2017) 084002.
[83] Y.P. Hu, X.X. Zeng, H.Q. Zhang, Phys. Lett. B 765 (2017) 120.
[84] P. Rudra, M. Faizal, A.F. Ali, Nucl. Phys. B 909 (2016) 725.
[85] P. Aniceto, J.V. Rocha, J. High Energy Phys. 2017 (2017) 035.
[86] S.G. Ghosh, S.D. Maharaj, Phys. Rev. D 89 (2014) 084027.
[87] J. Magueijo, L. Smolin, Phys. Rev. Lett. 88 (2002) 190403.
[88] R.G. Cai, Y.P. Hu, Q.Y. Pan, Y.L. Zhang, Phys. Rev. D 91 (2015) 024032.

[89] D.M. Eardley, L. Smar, Phys. Rev. D 19 (1979) 2239.
[90] P.S. Joshi, I.H. Dwivedi, Commun. Math. Phys. 146 (1992) 333.
[91] P.S. Joshi, T.P. Singh, Phys. Rev. D 51 (1995) 6778.
[92] T.P. Singh, P.S. Joshi, Class. Quantum Gravity 13 (1996) 559.
[93] A. Sinha, J. High Energy Phys. 1006 (2010) 061.
[94] V. Niarchos, Fortschr. Phys. 57 (2009) 646.
[95] X.X. Zeng, H. Zhang, L.F. Li, Phys. Lett. B 756 (2016) 170.
[96] W.J. Pan, Y.C. Huang, arXiv:1612.03627.

Black hole solutions surrounded by perfect fluid in Rastall theory

Y. Heydarzade, F. Darabi *

Department of Physics, Azarbaijan Shahid Madani University, Tabriz, Iran

ARTICLE INFO

Editor: J. Hisano

Keywords:
Rastall theory
Kiselev black holes

ABSTRACT

In this work, we obtain uncharged\charged Kiselev-like black holes as a new class of black hole solutions surrounded by perfect fluid in the context of Rastall theory. Then, we study the specific cases of the uncharged\charged black holes surrounded by regular matter like dust and radiation, or exotic matter like quintessence, cosmological constant and phantom fields. By comparing the Kiselev-like black hole solutions in Rastall theory with the Kiselev black hole solutions in GR, we find an effective perfect fluid behavior for the black hole's surrounding field. It is shown that the corresponding effective perfect fluid has interesting characteristic features depending on the different ranges of the parameters in Rastall theory. For instance, Kiselev-like black holes surrounded by regular matter in Rastall theory may be considered as Kiselev black holes surrounded by exotic matter in GR, or Kiselev-like black holes surrounded by exotic matter in Rastall theory may be considered as Kiselev black holes surrounded by regular matter in GR.

1. Introduction

One of the basic elements of Einstein's general theory of relativity (GR) is the so-called covariant conservation of the energy–momentum tensor which via the Noether symmetry theorem leads to the conservation of some globally defined physical quantities. These conserved quantities appear as the integrals of the components of the energy–momentum tensor over appropriate space-like surfaces. These space-like surfaces admit at least one of the Killing vectors of the background spacetime as their normal. By this way, the total rest energy/mass of a physical system is conserved in the context of GR. On the other hand, some GR based new modified theories have been proposed that relax the condition of covariant energy–momentum conservation. One of these possible modification of the general theory of relativity was introduced by P. Rastall in 1972 [1,2]. In this theory, the usual conservation law expressed by the null divergence of the energy–momentum tensor, i.e. $T^{\mu\nu}{}_{;\mu} = 0$, is questioned. Then, a non-minimal coupling of matter fields to geometry is considered where the divergence of $T_{\mu\nu}$ is proportional to the gradient of the Ricci scalar, i.e. $T^{\mu\nu}{}_{;\mu} \propto R^{,\nu}$, such that the usual conservation law is recovered in the flat spacetime. This can be understood as a direct accomplishment of the Mach principle representing that the inertia of a mass distribution is dependent on the mass and energy content of the external spacetime [3]. The main argument in favor of such a proposal is that the usual conservation law on $T_{\mu\nu}$ is tested only in the flat Minkowski space–time or specifically in a gravitational weak field limit. Indeed, this theory reproduces a phenomenological way for distinguishing features of quantum effects in gravitational systems, i.e. the violation of the classical conservation laws [4–6], which is also reported in the $f(R,T)$ [7] and $f(R,\mathcal{L}_m)$ [8] theories, where R, T and $\mathcal{L}_m$ are the Ricci scalar, trace of the energy–momentum tensor and the Lagrangian of the matter sector, respectively. Also, the condition $T^{\mu\nu}{}_{;\mu} \neq 0$ is phenomenologically confirmed by the particle creation process in cosmology [9–16]. In this regard, the Rastall theory can be considered as a good candidate for classical formulation of the particle creation through its non-minimal coupling [12,17]. Moreover, some astrophysical analysis including the evolution of the neutron stars and cosmological data do not reject this modified theory [18–20]. Specially, in [18] it is shown that the restrictions on the Rastall geometric parameters are of the order of $\leq 1\%$ with respect to the corresponding value of the Einstein GR. In other words, the results in [18] confirm that the Rastall theory is a viable theory in the sense that the deviation of any extended theory of gravity from the standard GR must be weak, to pass the solar system tests. Some studies on the various aspects of this theory in the context of current accelerated expansion phase of the universe as well other cosmological problems can be found in [12,21–28]. Also, some research works are dedicated to incorporate this theory with the Brans–Dicke and scalar-

* Corresponding author.
E-mail addresses: heydarzade@azaruniv.edu (Y. Heydarzade), f.darabi@azaruniv.edu (F. Darabi).

tensor theories of gravity [29–31]. A modified Brans–Dicke theory incorporating Rastall's assumption, namely a nonzero divergence of the energy–momentum tensor, is introduced in [32,33] which results in a class of viable theories with consistent field equations and gauge conditions. The implications of Rastall assumption in Kaluza–Klein theory and in inflationary cosmologies have been investigated in [34,35]. It is also shown that this theory regenerates some loop quantum cosmological features of the universe expansion [36]. Apart from the cosmological solutions, any modified theory must also provide the solutions associated to the stellar and black hole configurations. In this line, some neutron star, black hole and wormholes solutions in the context of Rastall theory are obtained in [18,37–41]. Also, a generalized version of Rastall theory is recently proposed which shows an agreement with the cosmic accelerating expansion [42]. In this regard, a dynamical factor for the proportionality of the energy–momentum tensor divergence and Ricci scalar divergence is considered. It is shown that this consideration leads to a transition from the matter dominated era to the current accelerating phase of the universe representing an agreement with some previous observations [43–45]. Finally, it should be mentioned that although Smalley first tried to get a Lagrangian for a prototype Rastall theory of gravity, with a variable gravitational constant [46], but this theory has been suffered from the lack of a consistent Lagrangian structure. This fact is known as the major drawback of this theory. But, recently a Lagrangian formulation for this theory is provided which may motivate the people to consider this theory more serious than before [47]. Besides, this theory possesses a rich structure that may be connected with some fundamental aspects of a complete theory of gravity and there are some points in favor of this theory. First of all, as mentioned before, the usual energy–momentum conservation law of Einstein's special relativity (SR) can be generalized to the curved spacetime in some different ways, including the appropriate geometric terms. Indeed, GR theory is one of the possible extension of SR to the curved spacetime by simply replacing the standard derivative with a covariant derivative, as the minimal generalization. Moreover, the classical form of the energy–momentum tensor must be modified by introducing quantities related to the curvature of the spacetime when the quantum effects are taking into account [4]. Also, due to the chirality of the quantum modes, the propagation of quantum fields in the spacetimes possessing horizons may lead to the violation of the classical conservation law which result in the so-called gravitational anomaly effect [48]. In this regard, Rastall theory can be a good phenomenological candidate in order to take into account the effects of quantum fields in curved spacetime in a covariant approach. Although, there is no action leading to the Rastall equations by implementing the variational principle, but it is possible to obtain such an action by introducing an external field in the Einstein–Hilbert action through a Lagrange multiplier. There are other geometrical models such as the well known Weyl geometry which may result in the field equations similar to the Rastall's field equations [46,49].

On the other hand, the direct local impacts of cosmic backgrounds upon the known black hole solutions have been paid attention recently. It is shown by Babichev et al. [50] that for a universe filled by phantom field, the black hole mass diminishes due to the accreting particles of the phantom scalar field into the central black hole. But this is a global impact indeed. The local changes in the spacetime geometry next to the central black hole can be obtained by a modified metric including the surrounding space time of the black hole. In this regard, an analytical static spherically symmetric solution to Einstein filed equations has been obtained by Kiselev [51]. This solution is characterized by the equation of state parameters of the black hole surrounding fields which generally can be dust, radiation or a dark energy component [51,52]. In [52], a Reissner–Nordström black hole surrounded by radiation and dust and a Schwarzschild black hole surrounded by quintessence, as the special cases of the Kiselev general solution, their phase transitions as well as their thermodynamical properties are investigated. The dynamics of a neutral and a charged particle around the Schwarzschild black surrounded by a quintessence matter have been discussed in [53]. The rotating Kiselev solution and Kerr–Newman Kiselev solution have been also obtained in [54–57]. Phase transition, quasinormal modes and Hawking radiation of Schwarzschild black hole in the quintessence field are studied in [58–60]. Also, one may refer to [61–66] for more detail in thermodynamical analysis of the Schwarzschild, Reissner–Nordström and Reissner–Nordström-AdS black holes in a quintessence background.

The essence of the Rastall theory is associated to the high curvature environments and consequently the black holes physics can provide an appropriate ground in order to investigate this theory in more details. Therefore, in this paper, our aim is to obtain the surrounded Kiselev-like black hole solutions as a new class of non-vacuum black hole solutions of this theory. The organization of the paper is as follows. In section 2, the general analytical static spherical symmetric surrounded black hole solutions in Rastall theory is obtained. Then, in the next five subsections 2.1–2.5, the special cases of the surrounded uncharged\charged black holes by the dust, radiation, quintessence, cosmological constant and phantom fields are addressed. Finally, in section 3, some concluding remarks are represented.

2. Surrounded black hole solutions in Rastall theory

In this section, we are looking for the general non-vacuum spherically symmetric static uncharged\charged black hole solutions in the context of the Rastall theory of gravity. Based on the Rastall's hypothesis [1,2], for a spacetime with Ricci scalar R filled by an energy–momentum source of $T_{\mu\nu}$, we have

$$T^{\mu\nu}{}_{;\mu} = \lambda R^{,\nu}, \tag{1}$$

where λ is the Rastall parameter, a measure for deviation from the standard GR conservation law. Then, the Rastall field equations can be written as

$$G_{\mu\nu} + \kappa\lambda g_{\mu\nu} R = \kappa T_{\mu\nu}, \tag{2}$$

where κ is the Rastall gravitational coupling constant. This field equations reduce to GR field equations in the limit of $\lambda \to 0$ and $\kappa = 8\pi G_N$ where G_N is the Newton gravitational coupling constant.

In order to obtain black hole solutions, we consider the general spherical symmetric spacetime metric in the standard Schwarzschild coordinates as

$$ds^2 = -f(r)dt^2 + \frac{dr^2}{f(r)} + r^2 d\Omega^2, \tag{3}$$

where $f(r)$ is a generic metric function depending on the radial coordinate r and $d\Omega^2 = d\theta^2 + sin^2\theta d\phi^2$ is the two dimensional unit sphere. Using this metric, we obtain nonvanishing components of the Rastall tensor defined as $H_{\mu\nu} = G_{\mu\nu} + \kappa\lambda g_{\mu\nu} R$ as

$$H^0{}_0 = G^0{}_0 + \kappa\lambda R = -\frac{1}{f} G_{00} + \kappa\lambda R = \frac{1}{r^2}\left(f'r - 1 + f\right) + \kappa\lambda R,$$

$$H^1{}_1 = G^1{}_1 + \kappa\lambda R = f G_{11} + \kappa\lambda R = \frac{1}{r^2}\left(f'r - 1 + f\right) + \kappa\lambda R,$$

$$H^2{}_2 = G^2{}_2 + \kappa\lambda R = \frac{1}{r^2} G_{22} + \kappa\lambda R = \frac{1}{r^2}\left(rf' + \frac{1}{2}r^2 f''\right) + \kappa\lambda R,$$

$$H^3{}_3 = G^3{}_3 + \kappa\lambda R = \frac{1}{r^2 sin^2\theta} G_{33} + \kappa\lambda R$$
$$= \frac{1}{r^2}\left(rf' + \frac{1}{2}r^2 f''\right) + \kappa\lambda R, \tag{4}$$

where the Ricci scalar reads as

$$R = -\frac{1}{r^2}\left(r^2 f'' + 4rf' - 2 + 2f\right), \tag{5}$$

in which the prime sign represents the derivative with respect to the radial coordinate r. Then, regarding the nonvanishing components of the Rastall tensor $H^\mu{}_\nu$, the total energy–momentum tensor supporting this spacetime should have the following diagonal form

$$T^\mu{}_\nu = \begin{pmatrix} T^0{}_0 & 0 & 0 & 0 \\ 0 & T^1{}_1 & 0 & 0 \\ 0 & 0 & T^2{}_2 & 0 \\ 0 & 0 & 0 & T^3{}_3 \end{pmatrix}, \tag{6}$$

which must also obey the symmetry properties of the Rastall tensor $H^\mu{}_\nu$. Regarding the equations in (4), the equalities $H^0{}_0 = H^1{}_1$ and $H^2{}_2 = H^3{}_3$ require $T^0{}_0 = T^1{}_1$ and $T^2{}_2 = T^3{}_3$, respectively. Then, one can construct a general total energy–momentum tensor $T^\mu{}_\nu$ possessing these symmetry properties in the following form

$$T^\mu{}_\nu = E^\mu{}_\nu + \mathcal{T}^\mu{}_\nu, \tag{7}$$

where $E^\mu{}_\nu$ is the trace-free Maxwell tensor given by

$$E_{\mu\nu} = \frac{2}{\kappa}\left(F_{\mu\alpha}F_\nu{}^\alpha - \frac{1}{4}g_{\mu\nu}F^{\alpha\beta}F_{\alpha\beta}\right), \tag{8}$$

so that $F_{\mu\nu}$ is the antisymmetric Faraday tensor satisfying the following vacuum Maxwell equations

$$F^{\mu\nu}{}_{;\mu} = 0,$$
$$\partial_{[\sigma}F_{\mu\nu]} = 0. \tag{9}$$

Considering the spherical symmetry existing in the spacetime metric (3) imposes the only non-vanishing components of the Faraday tensor $F^{\mu\nu}$ to be $F^{01} = -F^{10}$. Then, from the equations in (9), one obtains

$$F^{01} = \frac{Q}{r^2}, \tag{10}$$

where Q is an integration constant playing the role of a electrostatic charge. Thus, the equations (3), (8) and (10) give the only non-vanishing components of the Maxwell tensor $E^\mu{}_\nu$ as

$$E^\mu{}_\nu = \frac{Q^2}{\kappa r^4}\, diag(-1,-1,1,1), \tag{11}$$

representing an electrostatic field and clearly possesses the symmetries in $H^\mu{}_\nu$ tensor. On the other hand, $\mathcal{T}^\mu{}_\nu$ describes the energy–momentum tensor of the surrounding field defined as [51]

$$\mathcal{T}^0{}_0 = -\rho_s(r),$$
$$\mathcal{T}^i{}_j = -\rho_s(r)\alpha\left[-(1+3\beta)\frac{r_i r^j}{r_n r^n} + \beta\delta^i{}_j\right]. \tag{12}$$

This form of $\mathcal{T}^\mu{}_\nu$ indicates that the spatial sector is proportional to the time sector, denoting the energy density ρ_s, with the arbitrary parameters α and β related to the internal structure of the black hole surrounding field. Here, we used the subscript “s” for denoting the surrounding field which generally can be a dust, radiation, quintessence, cosmological constant, phantom field or even any combination of them. By taking the isotropic average over the angles we have [51]

$$<\mathcal{T}^i{}_j> = \frac{\alpha}{3}\rho_s\delta^i{}_j = p_s\delta^i{}_j, \tag{13}$$

since it is supposed that $<r^i r_j> = \frac{1}{3}\delta^i{}_j r_n r^n$. Thus, one has the barotropic equation of state for the surrounding field

$$p_s = \omega_s\rho_s, \quad \omega_s = \frac{1}{3}\alpha, \tag{14}$$

where p_s and ω_s are the pressure and equation of state parameter, respectively. Thus, the field equations (4) with respect to the total energy–momentum tensor in (7), (11) and (12) exactly provide the principle of additivity and linearity condition supposed in the reference [51] which was proposed to determine the free parameter β of the energy momentum-tensor of the surrounding field as

$$\beta = -\frac{1+3\omega_s}{6\omega_s}. \tag{15}$$

Then, the non-vanishing components of the $\mathcal{T}^\mu{}_\nu$ tensor can be obtained in the following form

$$\mathcal{T}^0{}_0 = \mathcal{T}^1{}_1 = -\rho_s,$$
$$\mathcal{T}^2{}_2 = \mathcal{T}^3{}_3 = \frac{1}{2}(1+3\omega_s)\rho_s, \tag{16}$$

which also possess the same symmetries in the Rastall tensor $H^\mu{}_\nu$. Consequently, our total constructed energy–momentum tensor in (7) admits all of the symmetry properties of $H^\mu{}_\nu$. One may just consider the $\mathcal{T}^\mu{}_\nu$ as the only supporting energy–momentum tensor of the Rastall field equations. In this way, the obtained solutions will describe the surrounded uncharged black hole solutions in the context of the Rastall theory which differ from the ones in GR, as we see later. Including the Maxwell tensor $E^\mu{}_\nu$ in $T^\mu{}_\nu$ provides the possibility of obtaining most general class of the static surrounded charged black hole solutions in the framework of this theory. In the following, we solve the field equations and obtain its general solution. Then, we address both of the uncharged\charged solutions.

The $H^0{}_0 = T^0{}_0$ and $H^1{}_1 = T^1{}_1$ components of the Rastall field equations give the following differential equation

$$\frac{1}{r^2}\left(rf' - 1 + f\right) - \frac{\kappa\lambda}{r^2}\left(r^2 f'' + 4rf' - 2 + 2f\right) = -\kappa\rho_s - \frac{Q^2}{r^4}, \tag{17}$$

and $H^2{}_2 = T^2{}_2$ and $H^3{}_3 = T^3{}_3$ components read as

$$\frac{1}{r^2}\left(rf' + \frac{1}{2}r^2 f''\right) - \frac{\kappa\lambda}{r^2}\left(r^2 f'' + 4rf' - 2 + 2f\right)$$
$$= \frac{1}{2}(1+3\omega_s)\kappa\rho_s + \frac{Q^2}{r^4}. \tag{18}$$

Thus, we have two unknown functions $f(r)$ and $\rho_s(r)$ which can be determined analytically by the above two differential equations. Now, by solving the set of differential equations (17) and (18),[1] one obtains the following general solution for the metric function

$$f(r) = 1 - \frac{2M}{r} + \frac{Q^2}{r^2} - \frac{N_s}{r^{\frac{1+3\omega_s-6\kappa\lambda(1+\omega_s)}{1-3\kappa\lambda(1+\omega_s)}}}, \tag{19}$$

with the energy density in the form of

$$\rho_s(r) = -\frac{3\mathcal{W}_s N_s}{\kappa r^{\frac{3(1+\omega_s)-12\kappa\lambda(1+\omega_s)}{1-3\kappa\lambda(1+\omega_s)}}}, \tag{20}$$

[1] Substituting $\kappa\rho_s(r)$ from differential equation (17) into (18) gives a differential equation for $f(r)$ leading to the solution (19). Then, by substituting the obtained $f(r)$ into the differential equations (17) or (18), one obtains the appropriate form of $\rho_s(r)$ as given by (20) and (21).

where M and N_s are two integration constants representing the black hole mass and surrounding field structure parameter, respectively in which

$$\mathcal{W}_s = -\frac{(1-4\kappa\lambda)(\kappa\lambda(1+\omega_s)-\omega_s)}{(1-3\kappa\lambda(1+\omega))^2}, \tag{21}$$

is a geometric constant depending on the Rastall geometric parameters κ and λ as well as the equation of state parameter ω_s of the black hole surrounding field. Note that the integration constant N_s represents the characteristic features of the surrounding field. For $\lambda = 0$, i.e. in the GR limit, we have $\rho_s(r) = -\frac{3}{\kappa}\mathcal{W}_s N_s r^{-3(1+\omega_s)}$ where $\mathcal{W}_s = \omega_s$ as in [51]. Note that in [51], the author used the units of $4\pi G_N = 1$ with a metric possessing a negative signature.

Regarding the weak energy condition representing the positivity of any kind of energy density of the surrounding field, i.e. $\rho_s \geq 0$, imposes the following condition on the geometric parameters of the theory

$$\mathcal{W}_s N_s \leq 0. \tag{22}$$

This condition implies that for the surrounding field with geometric parameter $\mathcal{W}_s > 0$, we need $N_s < 0$ and conversely for $\mathcal{W}_s < 0$, we need $N_s > 0$. Then, considering that $\mathcal{W}_s$ is given by (21), the sign of the metric parameter N_s depends on the Rastall geometric parameters κ, λ and the equation of state parameter ω_s of the surrounding field. In this regard, any set of κ, λ and ω_s parameters may admit a different positive or negative N_s values.

Finally, regarding (19), our metric (3) takes the following form

$$ds^2 = -\left(1 - \frac{2M}{r} + \frac{Q^2}{r^2} - \frac{N_s}{r^{\frac{1+3\omega_s-6\kappa\lambda(1+\omega_s)}{1-3\kappa\lambda(1+\omega_s)}}}\right)dt^2 + \frac{dr^2}{1 - \frac{2M}{r} + \frac{Q^2}{r^2} - \frac{N_s}{r^{\frac{1+3\omega_s-6\kappa\lambda(1+\omega_s)}{1-3\kappa\lambda(1+\omega_s)}}}} + r^2 d\Omega^2. \tag{23}$$

In the limit of $\lambda \to 0$ and $\kappa = 8\pi G_N$, we recover the Reissner-Nordström black hole surrounded by a surrounding field in GR which was firstly found by Kiselev [51] as

$$ds^2 = -\left(1 - \frac{2M}{r} + \frac{Q^2}{r^2} - \frac{N_s}{r^{3\omega_s+1}}\right)dt^2 + \frac{dr^2}{1 - \frac{2M}{r} + \frac{Q^2}{r^2} - \frac{N_s}{r^{3\omega_s+1}}} + r^2 d\Omega^2. \tag{24}$$

Our obtained static solution (23) is new and possesses some interesting features. By comparing the metric (23) with the Kiselev metric (24) in GR, we may obtain an effective equation of state parameter ω_{eff} for the modification term resulting from the geometry of the Rastall theory.

The notion of "effective equation of state" in Rastall theory has already been studied in the cosmological context, where a solution for the entropy and age problems of the Standard Cosmological Model were provided [68] by considering Brans–Dicke and Rastall theories of gravity and performing a perturbative analysis. It was shown that by introducing an "effective equation of state", the Rastall theory exhibits satisfactory properties at perturbative level in comparison to the Brans–Dicke theory.

In the next subsections, the surrounded black hole by the dust, radiation, quintessence, cosmological constant and phantom fields, as the subclasses of the general solution (23), as well as their interesting features are studied in detail. At last, we recall that the cases $\kappa\lambda = \frac{1}{4}$ and $\kappa\lambda = \frac{1}{6}$ are generally excluded due to the divergence of the Rastall gravitational coupling constant, as discussed in [1,39].

2.1. The black hole surrounded by the dust field

For the dust surrounding field, we set $\omega_d = 0$ [51,67]. Then, the metric (23) takes the following form

$$ds^2 = -\left(1 - \frac{2M}{r} + \frac{Q^2}{r^2} - \frac{N_d}{r^{\frac{1-6\kappa\lambda}{1-3\kappa\lambda}}}\right)dt^2 + \frac{dr^2}{1 - \frac{2M}{r} + \frac{Q^2}{r^2} - \frac{N_d}{r^{\frac{1-6\kappa\lambda}{1-3\kappa\lambda}}}} + r^2 d\Omega^2. \tag{25}$$

This metric differs from the metric of the surrounded charged black hole by a dust field in GR [51]. One can realize that in GR, i.e. in the limit of $\lambda \to 0$ and $\kappa = 8\pi G_N$, the black hole in the dust background appears as a charged black hole with an effective mass $M_{eff} = 2M + N_d$. Thus, we see that for $\kappa\lambda \neq 0$, the geometric parameters κ and λ of the Rastall theory can play an important role leading to distinct solutions relative to GR. Setting $Q = 0$ or $E^{\mu}{}_{\nu}$ in the total energy-momentum tensor in (7), one arrives at uncharged Kiselev-like black hole solutions in the dust background. One can realize that for $\kappa\lambda \neq 0$ the Rastall correction term never behaves as the mass or charge terms, and introduces a new character to the black hole, not comparable to the mass and charge terms. The presence of such nontrivial character can drastically change the thermodynamics, causal structure and Penrose diagrams, due to the Rastall geometric parameters, with respect to those of GR.

In this case, the geometric parameter $\mathcal{W}_d$ given by the relation (21) reads as

$$\mathcal{W}_d = -\frac{\kappa\lambda(1-4\kappa\lambda)}{(1-3\kappa\lambda)^2}. \tag{26}$$

Then, regarding the weak energy condition represented by the relation (22), for $0 \leq \kappa\lambda < \frac{1}{4}$ it is required that $N_d > 0$, while for $\kappa\lambda < 0 \cup \kappa\lambda > \frac{1}{4}$, we need $N_d < 0$ for the field structure constant. In this case, $\mathcal{W}_d$ and consequently ρ_d are effectively different from their GR counterparts such that $\rho_d = \frac{3\lambda(1-4\kappa\lambda)N_d}{(1-3\kappa\lambda)^2} r^{-\frac{3-12\kappa\lambda}{1-3\kappa\lambda}}$.

By comparing this metric with the Kiselev metric (24) in GR, we may obtain an effective equation of state parameter ω_{eff} for the modification term resulting from the geometry of the Rastall theory as

$$\omega_{eff} = \frac{1}{3}\left(-1 + \frac{1-6\kappa\lambda}{1-3\kappa\lambda}\right). \tag{27}$$

One may realize that ω_{eff} can never be zero (representing a background dust matter) except for the $\kappa\lambda = 0$ corresponding to GR limit. Then, the solutions of this theory are effectively different from those of GR. Regarding (27), two interesting classes are distinguishable as

- $\frac{1}{6} < \kappa\lambda < \frac{1}{3}$ which leads to $\omega_{eff} \leq -\frac{1}{3}$. In this case, we have an effective surrounding fluid with an effective equation of state parameter ω_{eff}, playing the role of dark energy, which leads to an effective repulsive gravitational effect. Then, regarding this range for $\kappa\lambda$, such black holes may contribute to the accelerating expansion of the universe in the Rastall theory of gravity. In the language of Raychaudhuri equation, such an effective surrounding fluid violating the strong energy condition can account for the accelerating expansion of the universe. Some $\kappa\lambda$ values in the range $\frac{1}{6} < \kappa\lambda < \frac{1}{3}$ and their corresponding effective equation of state ω_{eff} parameters accompanied by the geometric parameters $\mathcal{W}_d$ and N_d are listed in Table 1.

Table 1
Some $\kappa\lambda$ values in the range $\frac{1}{6} < \kappa\lambda < \frac{1}{3}$ and their corresponding effective equation of state ω_{eff} parameters accompanied by the geometric parameters $\mathcal{W}_d$ and N_d.

$\kappa\lambda$ value	ω_{eff} value	SEC	$\mathcal{W}_d$ value	N_d value
$\frac{2}{10}$	$-\frac{1}{2}$	violated	$-\frac{5}{4}$	positive
$\frac{2}{9}$	$-\frac{2}{3}$	violated	-1	positive
$\frac{3}{10}$	-3	violated	20	negative

Table 2
Some $\kappa\lambda$ values in the range $\kappa\lambda < \frac{1}{6} \cup \kappa\lambda > \frac{1}{3}$ and their corresponding effective equation of state ω_{eff} parameters accompanied by the geometric parameters $\mathcal{W}_d$ and N_d.

$\kappa\lambda$ value	ω_{eff} value	SEC	$\mathcal{W}_d$ value	N_d value
$\frac{1}{8}$	$-\frac{1}{5}$	respected	$-\frac{32}{25}$	positive
$\frac{1}{9}$	$-\frac{1}{6}$	respected	$-\frac{5}{4}$	positive
$\frac{1}{10}$	$-\frac{1}{7}$	respected	$-\frac{60}{49}$	positive
$\frac{4}{10}$	2	respected	15	negative
$\frac{1}{2}$	1	respected	4	negative
1	$\frac{1}{2}$	respected	$\frac{3}{4}$	negative

Interestingly, for $\kappa\lambda = \frac{2}{10}$ and $\frac{2}{9}$, the effective equation of state ω_{eff} lies in the quintessence range while for $\kappa\lambda = \frac{3}{10}$, it lies in the strong phantom range. This represents the fact that for a given κ, the more large values of λ, namely the more strong coupling $g_{\mu\nu}R$ in Rastall theory, the more strong acceleration phase.

- $\kappa\lambda < \frac{1}{6} \cup \kappa\lambda > \frac{1}{3}$ which leads to $\omega_{eff} \geq -\frac{1}{3}$. In this case, we have an effective surrounding fluid with an equation of state parameter respecting to the strong energy condition possessing the usual attractive gravitational effect. This may contribute to the decelerating expansion or even the contraction of universe depending on the value of the effective equation of state parameter ω_{eff}. In the language of Raychaudhuri equation, such a regular effective matter which respects to the strong energy condition, can justify the deceleration phase. Some $\kappa\lambda$ values in the range $\kappa\lambda < \frac{1}{6} \cup \kappa\lambda > \frac{1}{3}$ and their corresponding effective equation of state ω_{eff} parameters accompanied by the geometric parameters $\mathcal{W}_d$ and N_d are listed in Table 2.
Interestingly, for $\kappa\lambda = \frac{1}{2}$, the effective equation of state $\omega_{eff} = 1$ belongs to the stiff matter possessing very strong attractive gravitational effect.

2.2. The black hole surrounded by the radiation field

For the radiation surrounding field, we set $\omega_r = \frac{1}{3}$ [51,67]. Then, the metric (23) takes the following form

$$ds^2 = -\left(1 - \frac{2M}{r} + \frac{Q^2 - N_r}{r^2}\right)dt^2 + \frac{dr^2}{1 - \frac{2M}{r} + \frac{Q^2 - N_r}{r^2}} + r^2 d\Omega^2. \tag{28}$$

It is interesting that this case is the same as in GR and the geometric effects of the Rastall parameters do not appear for a black hole surrounded by the radiation field [51]. Also, the geometric parameter $\mathcal{W}_r$ given by the relation (21) reads as

$$\mathcal{W}_r = \frac{1}{3}, \tag{29}$$

and consequently with regard to the weak energy condition for this case, represented by the relation (22), it is required that $N_r < 0$ for the radiation field structure parameter. Then, by defining the positive structure parameter $\mathcal{N}_r = -N_r$, we have

$$ds^2 = -\left(1 - \frac{2M}{r} + \frac{Q^2 + \mathcal{N}_r}{r^2}\right)dt^2 + \frac{dr^2}{1 - \frac{2M}{r} + \frac{Q^2 + \mathcal{N}_r}{r^2}} + r^2 d\Omega^2, \tag{30}$$

which is the metric of the well known Reissner–Nordström black hole with an effective charge $Q_{eff} = \sqrt{Q^2 + \mathcal{N}_r}$. This result is interpreted as the positive contribution of the characteristic feature of the surrounding radiation field to the effective charge of the black hole. The appearance of effective charge in the black hole solution cannot change the causal structure and Penrose diagrams of this black hole solution, in comparison to the Reissner–Nordström black hole.

Setting $Q = 0$ or switching off the electrostatic energy-momentum tensor $E^{\mu}{}_{\nu}$ in the total energy-momentum tensor in (7), one arrives at the Kiselev black hole solutions in the radiation background. In that case, the resulting metric will be the Reissner-Nordström black hole with the charge term $\mathcal{N}_r$. Also, note that for a radiation background, not only the metric and the geometric parameter $\mathcal{W}_r$ are the same as in GR but also the energy density ρ_r of the background radiation has the same form in comparison to the GR's as $\rho_r = \frac{\mathcal{N}_r}{\kappa r^4}$. It is seen that the value of radiation energy density ρ_r of the background depends not only on the characteristic feature of the surrounding radiation field $\mathcal{N}_r$, but also it depends on the gravitational constant of the Rastall theory κ. In general, Rastall's gravitational constant may differs from the Newton gravitational constant. However, if one sets $\kappa = 8\pi G_N$ as in GR, the corresponding energy densities in both of these theories will be the same. Such a situation occurs also in the cosmological context of the Rastall theory [68]. In the cosmological setup, the metric solution, i.e. the scale factor, of the universe filled by the radiation fluid is exactly the same as in GR. Then, the evolutions of the universe during the radiation dominated era are the same for both of the GR and Rastall theories. This fact can be understood by inspecting the original field equations of the Rastall theory such that for a radiation fluid, we have $T = 0$ and $R = 0$ indicating that everything should be the same as in GR theory.

2.3. The black hole surrounded by the quintessence field

For the quintessence surrounding field, we set $\omega_q = -\frac{2}{3}$ [51,67]. Then, the metric (23) takes the following form

$$ds^2 = -\left(1 - \frac{2M}{r} + \frac{Q^2}{r^2} - \frac{N_q}{r^{\frac{-1-2\kappa\lambda}{1-\kappa\lambda}}}\right)dt^2 + \frac{dr^2}{1 - \frac{2M}{r} + \frac{Q^2}{r^2} - \frac{N_q}{r^{\frac{-1-2\kappa\lambda}{1-\kappa\lambda}}}} + r^2 d\Omega^2. \tag{31}$$

This metric differs from the metric of the surrounded charged black hole by a quintessence field in GR [51]. Here, it is seen that for $\kappa\lambda \neq 0$, the geometric parameters κ and λ of the Rastall theory can play an important role leading to distinct solutions, in comparison to GR. In this case, setting $Q = 0$ or $E^{\mu}{}_{\nu} = 0$ in the total energy-momentum tensor in (7), one arrives at uncharged Kiselev-like black hole solutions in the quintessence background. Due to the appearance of nontrivial N_q term, the causal structure and Penrose diagram will be different from those of Reissner–Nordström black hole in GR.

In this case, the geometric parameter $\mathcal{W}_q$ given by the relation (21) reads as

$$\mathcal{W}_q = -\frac{(1 - 4\kappa\lambda)(2 + \kappa\lambda)}{3(1 - \kappa\lambda)^2}. \tag{32}$$

Table 3
Some $\kappa\lambda$ values in the range $-\frac{1}{2} \leqslant \kappa\lambda < 1$ and their corresponding effective equation of state ω_{eff} parameters with their behaviors, accompanied by the geometric parameters $\mathcal{W}_q$ and N_q.

$\kappa\lambda$ value	ω_{eff} value	SEC	$\mathcal{W}_q$ value	N_q value
$-\frac{1}{2}$	$-\frac{1}{3}$	violated	$-\frac{12}{25}$	positive
$\frac{4}{10}$	$-\frac{4}{3}$	violated	15	negative
$\frac{1}{2}$	$-\frac{5}{3}$	violated	4	negative

Then, considering the weak energy condition given by the relation (22), we require $N_q > 0$ for $0 \leq \kappa\lambda < \frac{1}{4}$ and $N_q < 0$ for $\kappa\lambda > \frac{1}{4}$. The equation (32) shows that $\mathcal{W}_q$ and consequently the corresponding energy density ρ_q effectively differ from their GR counterparts such that $\rho_q = \frac{(1-4\kappa\lambda)(2+\kappa\lambda)N_q}{\kappa} r^{-\frac{1-4\kappa\lambda}{1-\kappa\lambda}}$.

In this case, by comparing the metric (31) with the original Kiselev metric (24) in GR, one can obtain an effective equation of state parameter ω_{eff} for the modification term resulting by the geometry of the Rastall theory as

$$\omega_{eff} = \frac{1}{3}\left(-1 - \frac{1+2\kappa\lambda}{1-\kappa\lambda}\right). \tag{33}$$

One may realize that ω_{eff} can never be $-\frac{2}{3}$ (the background quintessence filed), except for the $\kappa\lambda = 0$ which corresponds to the GR limit. Then, the solutions of this theory are effectively different from GR's. Regarding (33), two interesting classes are distinguishable as

- $-\frac{1}{2} \leqslant \kappa\lambda < 1$ which leads to $\omega_{eff} \leq -\frac{1}{3}$. In this case, we have an effective surrounding fluid with an equation of state parameter violating the strong energy condition which leads to a repulsive gravitational effect like as the background quintessence field but with a different repulsive strength. This may contribute to the accelerating expansion of the universe. Regarding the appropriate range for $\kappa\lambda$, such black holes may contribute to the accelerating expansion of the universe in the Rastall theory. Using the Raychaudhuri equation, such an effective surrounding quintessence field violating the strong energy condition can justify the acceleration expansion of the universe. Some $\kappa\lambda$ values in the range $-\frac{1}{2} \leqslant \kappa\lambda < 1$ and their corresponding effective equations of state ω_{eff} parameter with its behavior, accompanied by the geometric parameters $\mathcal{W}_q$ and N_q are given in Table 3.
  Interestingly, the case of $\kappa\lambda = -\frac{1}{2}$ leads to $\omega_{eff} = -\frac{1}{3}$ representing an effective surrounding quintessence field weaker than the one with $\omega_q = -\frac{2}{3}$. In the cosmological setup and through the second Friedmann equation, the acceleration equation, $\omega_{eff} = -\frac{1}{3}$ corresponds to a universe with a uniform expanding velocity, i.e. $\ddot{a} = 0$ where a is the scale factor of the ambient FRW universe filled by an effective field with $\omega_{eff} = -\frac{1}{3}$. For, $\kappa\lambda = \frac{4}{10}$ and $\kappa\lambda = \frac{1}{2}$, it is seen that the effective surrounding field possesses a repulsive character stronger than the quintessence with $\omega_{eff} = -\frac{4}{3}$ and $\omega_{eff} = -\frac{5}{3}$ which eventually lie in the phantom regime.
- $\kappa\lambda \leqslant -\frac{1}{2} \cup \kappa\lambda > 1$ which leads to $\omega_{eff} \geq -\frac{1}{3}$. In this case, we have an effective surrounding fluid with an equation of state parameter respecting to the strong energy condition possessing an attractive gravitational effect. This may contribute to the decelerating expansion or even in contraction of the universe. In this case, although the black hole is surrounded by the quintessence field with $\omega_q = -\frac{2}{3}$, however the effective equation of state ω_{eff} regarding the appropriate range for $\kappa\lambda$ does not belong to the quintessence range. For such a regular effective matter which respects to the strong energy condition, the Raychaudhuri equation can justify the deceleration phase or even the contraction of the universe. Some $\kappa\lambda$ values in the range $\kappa\lambda \leqslant -\frac{1}{2} \cup \kappa\lambda > 1$ and their corresponding effective equations of state ω_{eff} parameters with their behaviors, accompanied by the geometric parameters $\mathcal{W}_q$ and N_q are given in Table 4.
  In this case, the Rastall's correction term in metric (31) can never behave as the charge term, i.e. as $\frac{1}{r^2}$, to increases or decrease the charge's effect. But interestingly for $\kappa\lambda = -2$, which leads to the effective equation of state $\omega_{eff} = 0$ representing an effective dust matter, it exactly behaves like the mass term, i.e. $\frac{1}{r}$. The sign of metric parameter N_q for $\kappa\lambda = -2$ is positive and consequently, the correction term contributes to increase the effect of Schwarzschild mass term. A similar but reverse effect is reported in [50] in which for a universe filled by a phantom field, the black hole mass smoothly decreases due to the accreting particles of the phantom scalar field into the central black hole. This fact can be investigated for the case of a universe filled by a quintessence filed, which is out of the scope of the present paper. Also, $\kappa\lambda = \frac{5}{2}$ leads to the equation of state parameter $\omega_{eff} = 1$ denoting a stiff matter. In conclusion, it is seen that although the surrounding field is an essentially quintessence but the effective field is not the quintessence like filed, possessing a negative equation of state parameter, rather it can behave effectively as dust or even stiff matter possessing a zero or positive equation of state parameters, respectively.

Table 4
Some $\kappa\lambda$ values in the range $\kappa\lambda \leqslant -\frac{1}{2} \cup \kappa\lambda > 1$ and their corresponding effective equation of state ω_{eff} parameters with their behaviors, accompanied by the geometric parameters $\mathcal{W}_q$ and N_q.

$\kappa\lambda$ value	ω_{eff} value	SEC	$\mathcal{W}_q$ value	N_q value
-1	$-\frac{1}{6}$	respected	$-\frac{5}{16}$	positive
$-\frac{3}{2}$	$-\frac{1}{15}$	respected	$-\frac{28}{121}$	positive
-2	0	respected	$-\frac{9}{49}$	positive
$\frac{3}{2}$	$\frac{7}{3}$	respected	$\frac{20}{49}$	negative
2	$\frac{4}{3}$	respected	$\frac{7}{25}$	negative
$\frac{5}{2}$	1	respected	$\frac{36}{169}$	negative

2.4. The black hole surrounded by the cosmological constant field

For the cosmological constant surrounding field, we set $\omega_c = -1$ [51,67]. Then, the metric (23) takes the following form

$$ds^2 = -\left(1 - \frac{2M}{r} + \frac{Q^2}{r^2} - N_c r^2\right) du^2 + \frac{dr^2}{1 - \frac{2M}{r} + \frac{Q^2}{r^2} - N_c r^2} + r^2 d\Omega^2. \tag{34}$$

It is interesting that this case is the same as what was already obtained in GR by Kiselev [51]. Then, the Rastall and Einstein theories behave the same in the cosmological constant background. Here, setting $Q = 0$ or switching off $E^{\mu}{}_{\nu}$ in the total energy-momentum tensor in (7), one arrives at uncharged Kiselev-like black hole solutions in the de Sitter or anti-de Sitter background.

In this case, the geometric parameter $\mathcal{W}_c$ given by the relation (21) reads as

$$\mathcal{W}_c = -(1 - 4\kappa\lambda). \tag{35}$$

Then, considering the weak energy condition given by the relation (32), we require $N_c > 0$ for $0 \leq \kappa\lambda < \frac{1}{4}$, and $N_c < 0$ for $\kappa\lambda > 1/4$, corresponding to de Sitter or anti-de Sitter backgrounds, respectively. This shows that the sign of cosmological constant in the Rastall theory depends on its geometric parameters κ and λ. Although the form of metric (34) in this theory is the same as in GR for cosmological constant background, but the energy density of the cosmological constant differs from the GR due to the geometric features of the Rastall theory through the equations (20) and (35). In this case, the energy density of the cosmological constant is given by $\rho_c = \frac{3(1-4\kappa\lambda)N_c}{\kappa}$. A similar situation occurs in the cosmological context of the Rastall theory where the metric solution of the field equations, i.e. the scale factor, for the universe dominated by the cosmological constant has a similar form as in GR, i.e. it has an exponential form. In this case, by comparing the obtained result in [68] as $H \propto \sqrt{1 - \frac{2}{3}(\frac{3-2\lambda}{2\lambda-1})\rho}$ with the GR's as $H \propto \sqrt{\Lambda}$, we see that although the solutions have the same form but the geometric properties of the Rastall theory may affect the energy density of the background cosmological constant.[2]

2.5. The black hole surrounded by the phantom field

For the phantom surrounding field, we set $\omega_p = -\frac{4}{3}$ [67]. Then, the metric (23) takes the following form

$$ds^2 = -\left(1 - \frac{2M}{r} + \frac{Q^2}{r^2} - \frac{N_p}{r^{\frac{-3+2\kappa\lambda}{1+\kappa\lambda}}}\right) dt^2 + \frac{dr^2}{1 - \frac{2M}{r} - \frac{N_p}{r^{\frac{-3+2\kappa\lambda}{1+\kappa\lambda}}}} + r^2 d\Omega^2. \tag{36}$$

This metric differs from the metric of the surrounded charged black hole by a phantom field in GR [51]. For $\kappa\lambda \neq 0$, the geometric parameters κ and λ of the Rastall theory plays an important role leading to distinct solutions in comparison to GR. Also, setting $Q = 0$ or switching of $E^{\mu}{}_{\nu}$ in the total energy-momentum tensor in (7), one arrives at uncharged Kiselev-like black hole solutions in the phantom background. Due to the appearance of nontrivial N_p term, the causal structure and Penrose diagram will be different from those of Reissner-Nordström black hole in GR.

In this case, the geometric parameter $\mathcal{W}_p$ given by the relation (21) reads as

$$\mathcal{W}_p = -\frac{1}{3}\frac{(1-4\kappa\lambda)(4-\kappa\lambda)}{(1+\kappa\lambda)^2}. \tag{37}$$

Then, considering the weak energy condition given by the relation (22), we require $N_p > 0$ for $0 \le \kappa\lambda < \frac{1}{4} \cup \kappa\lambda > 4$ and $N_p < 0$ for $\frac{1}{4} < \kappa\lambda < 4$. The equation (37) shows that $\mathcal{W}_p$ and consequently the corresponding phantom energy density ρ_p effectively differs from their GR counterparts such that $\rho_p = \frac{(1-4\kappa\lambda)(4-\kappa\lambda)}{\kappa(1+\kappa\lambda)^2} N_p\, r^{\frac{1-4\kappa\lambda}{1+\kappa\lambda}}$.

By comparing this metric with the Kiselev metric (24), we may obtain an effective equation of state parameter ω_{eff} for the modification term resulting from the geometry of the Rastall theory as

$$\omega_{eff} = \frac{1}{3}\left(-1 - \frac{3-2\kappa\lambda}{1+\kappa\lambda}\right). \tag{38}$$

One may realize that ω_{eff} never can be $-\frac{4}{3}$ (the background phantom field), except for the $\kappa\lambda = 0$ corresponding to GR limit. Then, two interesting classes are distinguishable as

Table 5
Some $\kappa\lambda$ values in the range $-1 < \kappa\lambda < \frac{3}{2}$ and their associated effective equation of state parameters ω_{eff} parameters with their behaviors, accompanied by the geometric parameters $\mathcal{W}_q$ and N_q.

$\kappa\lambda$ value	ω_{eff} value	SEC	$\mathcal{W}_p$ value	N_p value
$-\frac{1}{2}$	-3	violated	$-\frac{12}{25}$	positive
$\frac{1}{2}$	$-\frac{7}{9}$	violated	4	negative
1	$-\frac{1}{2}$	violated	$\frac{3}{4}$	negative

Table 6
Some $\kappa\lambda$ values in the range $\kappa\lambda < -1 \cup \kappa\lambda \ge \frac{3}{2}$ and their corresponding effective equation of state ω_{eff} parameters with their behaviors, accompanied by the geometric parameters $\mathcal{W}_p$ and N_p.

$\kappa\lambda$ value	ω_{eff} value	SEC	$\mathcal{W}_p$ value	N_p value
$-\frac{3}{2}$	$\frac{11}{3}$	respected	$-\frac{28}{121}$	positive
-2	2	respected	$-\frac{9}{49}$	positive
$-\frac{5}{2}$	$\frac{13}{9}$	respected	$-\frac{44}{289}$	positive
2	$-\frac{2}{9}$	respected	$\frac{7}{25}$	negative
$\frac{5}{2}$	$-\frac{1}{7}$	respected	$\frac{36}{169}$	negative
3	$-\frac{1}{12}$	respected	$\frac{11}{64}$	negative
4	0	respected	$\frac{15}{121}$	negative

- $-1 < \kappa\lambda < \frac{3}{2}$ leading to $\omega_{eff} \le -\frac{1}{3}$. Then, we have a surrounding fluid with an effective equation of state parameter ω_{eff} which violates the strong energy condition resulting in a repulsive gravitational force. Then, in this range of $\kappa\lambda$, these black holes may contribute to the accelerating expansion of the universe in the Rastall theory. For such an effective surrounding quintessence field violating the strong energy condition, the Raychaudhuri equation can account for the acceleration expansion of the universe. In Table 5, some $\kappa\lambda$ values in the range $-1 < \kappa\lambda < \frac{3}{2}$ and their associated effective equations of state parameters ω_{eff} parameter with their behaviors, accompanied by the geometric parameters $\mathcal{W}_q$ and N_q are given. Interestingly, for $\kappa\lambda = -\frac{1}{2}$, we have $\omega_{eff} = -3$ which has a repulsive character stronger than the background phantom field with $\omega_p = -\frac{4}{3}$ while for $\kappa\lambda = \frac{1}{2}$ and $\kappa\lambda = 1$, we have an effective field with a repulsive character weaker than the background phantom field with $\omega_p = -\frac{4}{3}$ but still lying in the quintessence range.
- $\kappa\lambda < -1 \cup \kappa\lambda \ge \frac{3}{2}$ which leads to $\omega_{eff} \ge -\frac{1}{3}$. In this case, we have an effective surrounding fluid with an equation of state parameter respecting to the strong energy condition which leads to a attractive gravitational effect. This may contribute to the decelerating expansion or even in contraction of the universe. In this case, although the black hole is surrounded by the phantom field with $\omega_p = -\frac{4}{3}$, but the effective equation of state ω_{eff} regarding the appropriate range of $\kappa\lambda$ does not belong to the phantom range. This effect may cause the contraction of universe filled by such a black holes in the Rastall theory of gravity. For such a regular effective matter which respects to the strong energy condition, the Raychaudhuri equation can justify the deceleration phase or even the contraction of the universe. Some $\kappa\lambda$ values in the range $\kappa\lambda < -1 \cup \kappa\lambda \ge \frac{3}{2}$ and their corresponding effective equations of state ω_{eff} parameter with their behaviors, accompanied by the geometric parameters $\mathcal{W}_p$ and N_p are given in Table 6. In this case, the Rastall's correction term in metric (36) can never behave like the charge term, i.e. as $\frac{1}{r^2}$, but interestingly for $\kappa\lambda = 4$, which leads to the effective equation of state $\omega_{eff} = 0$ representing an effective dust matter, it exactly behaves like the mass term, i.e. $\frac{1}{r}$. For this case, the sign of

[2] One should note to a little different notation for the field equations in our work and [68], where the field equations are defined as $R_{\mu\nu} - \frac{\lambda}{2}Rg_{\mu\nu} = \kappa T_{\mu\nu}$ and $T^{\mu\nu}{}_{;\mu} = \frac{1-\lambda}{2\kappa}T^{;\nu}$.

N_q for $\kappa\lambda = 4$ is negative and the correction term contributes to decreases the effect of Schwarzschild mass term. Such an effect is reported in [50] in which for a universe filled by a phantom field approaching to the Big Rip, the black hole mass gradually decreases due to the accreting particles of the phantom scalar field into the central black hole. In conclusion, it is seen that although the surrounding field is an essentially phantom field but the effective surrounding field is not the phantom field, rather it can be effectively a quintessence, dust or even stiff matter.

3. Conclusion

We have obtained general uncharged\charged Kiselev-like black hole solutions surrounded by perfect fluid in the context of Rastall theory. Then, we have investigated in more detail the specific cases of the black holes surrounded by dust, radiation, quintessence, cosmological constant and phantom fields. In each case, the weak energy condition, representing a positive energy density, is applied to put constraint on the physical parameters of this modified theory. By comparing the new term in the metric, resulted from the Rastall theory, with the Kiselev solution in GR, an effective behavior for the black hole surrounding field is realized. It is shown that the effective fluid has different characteristics through its effective equation of state parameter ω_{eff} depending on the $\kappa\lambda$ values. In the case of black hole in a dust background with $\omega_d = 0$, for $\frac{1}{6} < \kappa\lambda < \frac{1}{3}$, we have $\omega_{eff} \leq -\frac{1}{3}$ violating the strong energy (SEC) condition, while for $\kappa\lambda < \frac{1}{6} \cup \kappa\lambda > \frac{1}{3}$, we have $\omega_{eff} \geq -\frac{1}{3}$ respecting to strong energy condition. For a black hole in a quintessence background with $\omega_q = -2/3$, we have $\omega_{eff} \leq -\frac{1}{3}$ for $-\frac{1}{2} \leqslant \kappa\lambda < 1$ violating the strong energy condition, while $\omega_{eff} \geq -\frac{1}{3}$ for $\kappa\lambda \leqslant -\frac{1}{2} \cup \kappa\lambda > 1$ respecting to strong energy condition. In the case of a black hole in phantom background with $\omega_p = -4/3$, for $-1 < \kappa\lambda < \frac{3}{2}$, we have $\omega_{eff} \leq -\frac{1}{3}$, while for $\kappa\lambda < -1 \cup \kappa\lambda \geq \frac{3}{2}$, we have $\omega_{eff} \geq -\frac{1}{3}$. For such an effective surrounding fluid violating/respecting the strong energy condition, the Raychaudhuri equation can account for the accelerating/decelerating expansion of the universe, respectively. For each of these special classes, some interesting $\kappa\lambda$ values and their corresponding ω_{eff} as well as the defined Rastall geometric parameters $\mathcal{W}_p$ and N_p are given in the Tables 1 to 6. For example, for the black hole in dust background, for $\kappa\lambda = \frac{2}{10}$ and $\frac{2}{9}$, the effective equation of state ω_{eff} lies in the quintessence range while for $\kappa\lambda = \frac{3}{10}$, it lies in the strong phantom regime possessing repulsive gravitational effect. For $\kappa\lambda = \frac{1}{2}$, we have $\omega_{eff} = 1$ which belongs to the stiff matter with stronger gravitational attraction than the background dust. In the case of a black hole in a quintessence background with $\omega_q = -\frac{2}{3}$, the case of $\kappa\lambda = -\frac{1}{2}$ leads to $\omega_{eff} = -\frac{1}{3}$ representing an effective surrounding quintessence field weaker than the background. For, $\kappa\lambda = \frac{4}{10}$ and $\kappa\lambda = \frac{1}{2}$, it is seen that the effective surrounding field possesses a repulsive character stronger than the quintessence with $\omega_{eff} = -\frac{4}{3}$ and $\omega_{eff} = -\frac{5}{3}$, respectively, which lie in the phantom regime. Also, for $\kappa\lambda = -2$, we have $\omega_{eff} = 0$ representing an effective dust field while $\kappa\lambda = \frac{5}{2}$ leads to the equation of state parameter $\omega_{eff} = 1$ denoting a stiff matter. In latter cases, it is seen that although the surrounding field is an essentially quintessence but the effective field is not the quintessence like filed, possessing a negative equation of state parameter, rather it can behave effectively as dust or even stiff matter possessing a zero or positive equation of state parameters, respectively. Finally, for a black hole in a phantom background with $\omega_p = -4/3$, for $\kappa\lambda = -\frac{1}{2}$, we have $\omega_{eff} = -3$ which has a repulsive character stronger than the background phantom field, while for $\kappa\lambda = \frac{1}{2}$ and $\kappa\lambda = 1$, we have effective fields with repulsive character weaker than the background phantom field, still lying in the quintessence range. For $\kappa\lambda = 4$, we have $\omega_{eff} = 0$ representing an effective dust field. Then, it is seen that for the latter cases, although the surrounding field is an essentially phantom field but the effective surrounding field is not the phantom field, rather it can be effectively a quintessence, dust or even stiff matter. It is predicted that the new terms appearing in the Kiselev-like black holes may cause for some drastic changes in their horizons, causal structures and thermodynamical aspects, in comparison to the Kiselev black holes in GR. Such study is under work by the authors and will be reported, elsewhere.

References

[1] P. Rastall, Phys. Rev. D 6 (1972) 3357.
[2] P. Rastall, Can. J. Phys. 54 (1976) 66.
[3] Vladimir Majernik, Lukas Richterek, arXiv:gr-qc/0610070.
[4] N.D. Birrell, P.C.W. Davies, Quantum Fields in Curved Space, Cambridge University Press, Cambridge, 1982.
[5] T. Koivisto, Class. Quantum Gravity 23 (2006) 4289.
[6] O. Minazzoli, Phys. Rev. D 88 (2013) 027506.
[7] T. Harko, F.S. Lobo, S. Nojiri, S.D. Odintsov, Phys. Rev. D 84 (2011) 024020.
[8] T. Harko, F.S. Lobo, Galaxies 2 (2014) 410.
[9] G.W. Gibbons, S.W. Hawking, Phys. Rev. D 15 (1977) 2738.
[10] L. Parker, Phys. Rev. D 3 (1971) 346;
L. Parker, Phys. Rev. D 3 (1971) 2546.
[11] L.H. Ford, Phys. Rev. D 35 (1987) 2955.
[12] C.E.M. Batista, M.H. Daouda, J.C. Fabris, O.F. Piattella, D.C. Rodrigues, Phys. Rev. D 85 (2012) 084008.
[13] S.H. Pereira, C.H.G. Bessa, J.A.S. Lima, Phys. Lett. B 690 (2010) 103.
[14] S. Calogero, J. Cosmol. Astropart. Phys. 11 (2011) 016.
[15] S. Calogero, H. Velten, J. Cosmol. Astropart. Phys. 11 (2013) 025.
[16] H. Velten, S. Calogero, arXiv:1407.4306.
[17] E.R. Bezerra de Mello, J.C. Fabris, B. Hartmann, Class. Quantum Gravity 32 (2015) 085009.
[18] A.M. Oliveira, H.E.S. Velten, J.C. Fabris, L. Casarini, Phys. Rev. D 92 (2015) 044020.
[19] C.E.M. Batista, J.C. Fabris, O.F. Piattella, A.M. Velasquez-Toribio, Eur. Phys. J. C 73 (2013) 2425.
[20] J.C. Fabris, O.F. Piattella, D.C. Rodrigues, M.H. Daouda, AIP Conf. Proc. 1647 (2015) 50, arXiv:1403.5669v1.
[21] M. Capone, V.F. Cardone, M.L. Ruggiero, J. Phys. Conf. Ser. 222 (2010) 012012.
[22] J.C. Fabris, O.F. Piattella, D.C. Rodrigues, C.E.M. Batista, M.H. Daouda, Int. J. Mod. Phys. Conf. Ser. 18 (2012) 67.
[23] J.P. Campos, J.C. Fabris, R. Perez, O.F. Piattella, H. Velten, Eur. Phys. J. C 73 (2013) 2357.
[24] J.C. Fabris, M.H. Daouda, O.F. Piattella, Phys. Lett. B 711 (2012) 232.
[25] J.C. Fabris, arXiv:1208.4649v1.
[26] H. Moradpour, Phys. Lett. B 757 (2016) 187.
[27] A.S. Al-Rawaf, M.O. Taha, Phys. Lett. B 366 (1996) 69.
[28] A.S. Al-Rawaf, M.O. Taha, Gen. Relativ. Gravit. 28 (1996) 935.
[29] T.R.P. Caramês, M.H. Daouda, J.C. Fabris, A.M. de Oliveira, O.F. Piattella, V. Strokov, Eur. Phys. J. C 74 (2014) 3145.
[30] I.G. Salako, M.J.S. Houndjo, A. Jawad, Int. J. Mod. Phys. D 25 (7) (2016) 1650076.
[31] T.R.P. Caramês, M.H. Daouda, J.C. Fabris, A.M. Oliveira, O.F. Piattella, V. Strokov, arXiv:1503.04882.
[32] L.L. Smalley, Phys. Rev. D 9 (1974) 1635.
[33] L.L. Smalley, Phys. Rev. D 1 (2) (1975) 376.
[34] C. Wolf, Phys. Scr. 34 (1986) 193.
[35] C. Wolf, Phys. Scr. 38 (1988) 129.
[36] G.F. Silva, O.F. Piattella, J.C. Fabris, L. Casarini, T.O. Barbosa, Gravit. Cosmol. 19 (2013) 156.
[37] Y. Heydarzade, H. Moradpour, F. Darabi, arXiv:1610.03881.
[38] H. Moradpour, N. Sadeghnezhad, arXiv:1606.00846.
[39] H. Moradpour, I.G. Salako, Adv. High Energy Phys. 2016 (2016) 3492796.
[40] A.M. Oliveira, H.E.S. Velten, J.C. Fabris, L. Casarini, Phys. Rev. D 93 (2016) 124020.
[41] K.A. Bronnikov, J.C. Fabris, O.F. Piattella, et al., Gen. Relativ. Gravit. 48 (2016) 162.
[42] H. Moradpour, Y. Heydarzade, F. Darabi, Ines G. Salako, Eur. Phys. J. C 77 (2017) 259, http://dx.doi.org/10.1140/epjc/s10052-017-4811-z.
[43] R.A. Daly, et al., Astrophys. J. 677 (2008) 1.
[44] E. Komatsu, et al., WMAP Collaboration, Astrophys. J. Suppl. 192 (2011) 18.
[45] V. Salvatelli, A. Marchini, L.L. Honorez, O. Mena, Phys. Rev. D 88 (2013) 023531.
[46] L.L. Smalley, Il Nuovo Cimento B 80 (1) (1984) 42.

[47] R.V. dos Santos, J.A.C. Nogales, arXiv:1701.08203.
[48] R. Bertlmann, Anomalies in Quantum Field Theory, Oxford University Press, Oxford, 2000.
[49] T.S. Almeida, M.L. Pucheu, C. Romero, J.B. Formiga, Phys. Rev. D 89 (2014) 064047.
[50] E. Babichev, V. Dokuchaev, Yu. Eroshenko, Phys. Rev. Lett. 93 (2004) 021102.
[51] V.V. Kiselev, Class. Quantum Gravity 20 (2003) 1187.
[52] B. Majeed, M. Jamil, P. Pradhan, Adv. High Energy Phys. 2015 (2015) 124910.
[53] M. Jamil, S. Hussain, B. Majeed, Eur. Phys. J. C 75 (2015) 24.
[54] S.G. Ghosh, Eur. Phys. J. C 76 (2016) 222.
[55] T. Oteev, A. Abdujabbarov, Z. Stuchl'ik, B. Ahmedov, Astrophys. Space Sci. 361 (2016) 269.
[56] B. Toshmatov, Z. Stuchl'ik, B. Ahmedov, arXiv:1512.01498.
[57] Z. Xu, J. Wang, arXiv:1609.02045.
[58] R. Tharanath, N. Varghese, V.C. Kuriakose, Mod. Phys. Lett. A 29 (2014) 1450057.
[59] Y. Zhang, E.K. Li, J.L. Geng, Gen. Relativ. Gravit. 46 (2014) 1728.
[60] S. Chen, J. Jing, Class. Quantum Gravity 22 (2005) 4651.
[61] K. Ghaderi, B. Malakolkalami, Nucl. Phys. B 903 (2016) 10.
[62] K. Ghaderi, B. Malakolkalami, Astrophys. Space Sci. 361 (2016) 161.
[63] Y.H. Wei, Z.H. Chu, Chin. Phys. Lett. 28 (2011) 100403.
[64] Y.H. Wei, J. Ren, Chin. Phys. B 22 (2013) 030402.
[65] B.B. Thomas, M. Saleh, T.C. Kofane, Gen. Relativ. Gravit. 44 (2012) 2181.
[66] P. Pradhan, Int. J. Mod. Phys. D 26 (2017) 1750010.
[67] A. Vikman, Phys. Rev. D 71 (2005) 023515.
[68] J.C. Fabris, R. Kerner, J. Tossa, Int. J. Mod. Phys. D 9 (2000) 111.

Baryogenesis in Lorentz-violating gravity theories

Jeremy Sakstein *, Adam R. Solomon

Center for Particle Cosmology, Department of Physics and Astronomy, University of Pennsylvania, 209 S. 33rd St., Philadelphia, PA 19104, USA

ARTICLE INFO

Editor: S. Dodelson

ABSTRACT

Lorentz-violating theories of gravity typically contain constrained vector fields. We show that the lowest-order coupling of such vectors to U(1)-symmetric scalars can naturally give rise to baryogenesis in a manner akin to the Affleck–Dine mechanism. We calculate the cosmology of this new mechanism, demonstrating that a net $B-L$ can be generated in the early Universe, and that the resulting baryon-to-photon ratio matches that which is presently observed. We discuss constraints on the model using solar system and astrophysical tests of Lorentz violation in the gravity sector. Generic Lorentz-violating theories can give rise to the observed matter–antimatter asymmetry without violating any current bounds.

1. Introduction

Why is there so much more matter than antimatter? One most likely cannot appeal to initial conditions, as these would be washed away by inflation. The standard model can provide such an asymmetry during the electroweak phase transition, but cannot produce enough to accommodate observations [1]. It seems probable, then, that a dynamical generation mechanism, or *baryogenesis*, arises from new physics beyond the standard model.[1]

In this paper we point out that Lorentz violation might play a key role in this new physics. While Lorentz invariance is extraordinarily well tested in the matter sector, the possibility of gravitational Lorentz violation remains relatively unconstrained. If boosts are broken but rotational invariance is maintained—i.e., if gravity picks out a preferred rest frame—then the low-energy physics is described by Einstein-æther theory, a vector–tensor theory in which the vector field is constrained to have a fixed, timelike norm. This is the general effective field theory when boosts are broken [8]; for example, a special case of Einstein-æther arises in the low-energy limit of Hořava–Lifschitz gravity [9,10], a putative UV completion of general relativity which relies on the existence of a preferred foliation.

We demonstrate that if a $U(1)_{B-L}$ scalar couples to the vector of Einstein-æther theory, then the lowest-order interactions between the two can lead to baryogenesis. This operates in a manner qualitatively similar to Affleck–Dine baryogenesis [11] or the recent model of Ref. [12], in which the $U(1)_{B-L}$ symmetry is broken at early times due to a tachyonic mass proportional the Hubble parameter H appearing in the effective scalar potential.[2] In purely metric theories this is difficult to achieve, as H is not a spacetime scalar. Breaking boosts cures this difficulty, and indeed in Einstein-æther H is simply proportional to the divergence of the timelike vector field. As Einstein-æther is the most general low-energy effective theory for broken boosts, our conclusion can be stated as follows: if the Universe contains a $U(1)_{B-L}$ scalar with softly broken symmetry and spontaneous Lorentz violation, a working baryogenesis mechanism comes for free.[3]

This paper is organized as follows. In section 2 we introduce the model of scalar-æther baryogenesis and discuss known constraints on the theory. In section 3 we derive the cosmology and verify that this model can yield the observed baryon-to-photon ratio with sensible parameters, and we conclude in section 4.

2. Model

The model we will consider is the constrained vector (or "æther") u^{μ} of Einstein-æther theory coupled to a new $U(1)_{B-L}$ scalar ϕ. At leading order, the most general action we can write down is

* Corresponding author.
E-mail addresses: sakstein@physics.upenn.edu (J. Sakstein), adamsol@physics.upenn.edu (A.R. Solomon).

[1] For reviews of baryogenesis, see, e.g., Refs. [1–7].

[2] In the former model, this arises due to a coupling of the scalar to the inflaton, while in the latter, it arises from a Weyl coupling to dark matter.

[3] Baryogenesis in Lorentz-violating theories has also been studied in various other contexts in Refs. [13–15]. Our method is distinct from these. In particular, we do not consider a coupling of the vector to the baryon current of the form $u_{\mu}j^{\mu}_{B}$, which would give rise to spontaneous baryogenesis.

$$S = \int d^4x\sqrt{-g}\left[\frac{M_{\rm Pl}^2}{2}R - \mathcal{K}^{\mu\nu}_{\alpha\beta}\nabla_\mu u^\alpha \nabla_\nu u^\beta + \kappa\left(u^\mu u_\mu + m^2\right)\right.$$
$$- \partial_\mu\phi\partial^\mu\phi^\dagger - m_\phi^2|\phi|^2 - \frac{\lambda}{2}|\phi|^4 - \frac{\varepsilon}{4}\phi^4 - \frac{\varepsilon^\dagger}{4}\phi^{\dagger 4}$$
$$\left. + \frac{\alpha}{3}|\phi|^2\nabla_\mu u^\mu\right], \quad (1)$$

where

$$\mathcal{K}^{\mu\nu}_{\alpha\beta} = c_1 g^{\mu\nu}g_{\alpha\beta} + c_2\delta^\mu_\alpha\delta^\nu_\beta + c_3\delta^\mu_\beta\delta^\nu_\alpha + c_4 u^\mu u^\nu g_{\alpha\beta} \quad (2)$$

is the most general kinetic term for the Lorentz-violating vector u^μ, and κ is a Lagrange multiplier that ensures that the vector is timelike and of fixed norm, $u^\mu u_\mu = -m^2$. We have included some $U(1)_{B-L}$-violating terms proportional to ε in order to generate a net $B-L$. The last line is the leading-order interaction one can write down between ϕ and u^μ given the symmetries.[4] This is the general low-energy effective theory with ϕ, broken boosts, and softly-broken $U(1)_{B-L}$.

The coupling between u^μ and ϕ can give rise to baryogenesis using a mechanism akin to (but distinct from) that of Affleck and Dine [11] or similar generalizations [12]. In particular, the effective potential for ϕ is

$$V_{\rm eff}(\phi,\phi^\dagger) = \left(m_\phi^2 - \frac{1}{3}\alpha\nabla_\mu u^\mu\right)|\phi|^2 + \frac{\lambda}{2}|\phi|^4 + \frac{\varepsilon}{4}\phi^4 + \frac{\varepsilon^\dagger}{4}\phi^{\dagger 4} \quad (3)$$

so that $\nabla_\mu u^\mu$ acts as a tachyonic mass term for ϕ. We can see this explicitly by considering a homogeneous and isotropic cosmological setting, so that the Universe is described by a Freidmann–Lemaître–Robertson–Walker metric (in cosmic time)

$$ds^2 = -dt^2 + a^2(t)d\vec{x}^2. \quad (4)$$

The on-shell condition $u_\mu u^\mu = -m^2$ and symmetry imply that the vector must be of the form [22–24]

$$u^\mu = (m, 0, 0, 0). \quad (5)$$

In this case, the divergence of u^μ is $\nabla_\mu u^\mu = m\partial_t \ln\sqrt{-g} = 3mH$, so that the effective potential becomes

$$V_{\rm eff}(\phi,\phi^\dagger) = \left(m_\phi^2 - \alpha m H\right)|\phi|^2 + \frac{\lambda}{2}|\phi|^4 + \frac{\varepsilon}{4}\phi^4 + \frac{\varepsilon^\dagger}{4}\phi^{\dagger 4}. \quad (6)$$

This potential leads to baryogenesis similarly to the well-known Affleck–Dine mechanism. In the early Universe, the $U(1)_{B-L}$ symmetry is broken as the tachyonic mass term is more important than the bare mass, while at late times the symmetry is restored. During the broken-symmetry phase, the motion of the angular component of the scalar generates a net $B-L$ due to the symmetry-breaking terms $\varepsilon\phi^4 + \varepsilon^\dagger\phi^{\dagger 4}$. When the symmetry is restored, the $B-L$ is stored in the field, and can be transferred to the standard model through sphaleron processes [25], although one must first transfer the asymmetry to left-handed standard model particles. The details of the transfer were discussed for models such as ours in Ref. [12], where the neutrino portal [26,27] was identified as one promising mechanism. We note that our model differs quantitatively from Affleck–Dine: the tachyonic mass scales like mH rather than H^2, which can lead to novel and interesting new features. In what follows, we will calculate the cosmology of this model, paying special attention to the generation of a net $B-L$.

2.1. Constraints

In this subsection, we briefly summarize observational and theoretical constraints on the parameters in our model. Most of these will apply to Einstein-æther theory or to its coupling to a real scalar. We will use the notation $c_{12} = c_1 + c_2$, $c_{123} = c_1 + c_2 + c_3$, etc.

Experimental constraints on Einstein-æther theory tend to place upper bounds on the æther vacuum expectation value (VEV) m, with the result $c_i m^2 \ll M_{\rm Pl}^2$ for generic values of the c_i parameters. We note that any of these constraints can be weakened or removed entirely by tuning the c_i parameters, although these tunings cannot all be done simultaneously. We refer the reader to Sec. V.D of Ref. [19] for a more comprehensive summary of constraints on the æther.

The strongest constraints come from gravitational Čerenkov radiation: high-energy cosmic rays could lose energy to subluminal æther-graviton modes, leading to a degradation in cosmic ray propagation which has not been observed, constraining $m/M_{\rm Pl} < 3\times 10^{-8}$ [28]. These constraints can be avoided by tuning the c_i or allowing for superluminal propagation in æther-graviton modes; since this is an explicitly Lorentz-breaking theory, superluminality may not be as deadly as one normally expects. The preferred-frame parameters $\alpha_{1,2}$ in the parametrized post-Newtonian formalism are modified by the æther, constraining $m/M_{\rm Pl} < 6\times 10^{-4}$ in the absence of tuning c_i [29,30]. Note that the tuning which eliminates gravitational Čerenkov radiation ($c_3 = -c_1$, $c_2 = c_1/(1-2c_1)$) also sets $\alpha_1 = 0$, so that the dominant constraint comes from α_2, in which case the strongest constraint on m is rather mild, $m/M_{\rm Pl} \lesssim 10^{-2}$.

Under the assumption that $c_i m^2 \ll M_{\rm Pl}^2$, we are justified in ignoring the mixing with gravity [23], in which case there are a few constraints on the c_i from flat space perturbation theory. In the vector sector, the absence of ghosts requires $c_1 > 0$ [23]; coupling a scalar to $\nabla_\mu u^\mu$, as in this paper, does not modify the vector perturbations around flat space [19]. The no-ghost condition for the spin-0 piece of u^μ is the same, while gradient stability requires $c_{123} > 0$. Some authors require the spin-0 æther mode to propagate subluminally, which would imply $c_{123} < c_1$ [23], although the scalar coupling relaxes this bound to $c_{123} < c_1 + x$ for some $x > 0$ [19]. If we require the sound speed of tensors to be subluminal then we would require $c_{13} > 0$ [23].

The æther-scalar coupling can lead to a gradient instability, placing an upper bound on α [19]. Writing $\phi(x) = \frac{1}{\sqrt{2}}\rho(x)e^{i\theta(x)}$, the real scalar ρ interacts with the æther through a potential $V(\theta,\rho) = \frac{1}{2}m_\phi^2\rho^2 + \frac{1}{4}\lambda\rho^4 - \frac{1}{6}\alpha\Theta\rho^2$, where $\Theta \equiv \nabla_\mu u^\mu$. Gradient stability around flat space requires[5]

$$V_{\Theta\rho}^2 \le 2c_{123}\left(V_{\rho\rho} + k^2\right), \quad (7)$$

where $V_{\Theta\rho} = \partial_\Theta\partial_\rho V$ and $V_{\rho\rho} = \partial_\rho^2 V$ are evaluated at the background values of Θ and ρ. Applying this to our potential, we find that the constraint is trivially satisfied in the unbroken symmetry

[4] Couplings between a scalar and $\nabla_\mu u^\mu$ were first introduced in Ref. [16], and have also been considered in, e.g., Refs. [17–21]. One could also consider a term $|\phi|^2 u^\mu u_\mu$, but this can always be absorbed into the mass and Lagrange multiplier when the vector is on-shell.

[5] While the analysis of Ref. [19], in contrast to our model, assumed a single real scalar, the field θ decouples from ρ and u^μ around the background $\partial_\mu\bar{\theta} = 0$. In principle a non-zero $\partial_\mu\bar{\theta} = c_\mu$ could modify the constraint by shifting the effective mass for ρ fluctuations, $m_\phi^2 \to m_\phi^2 + c^2$.

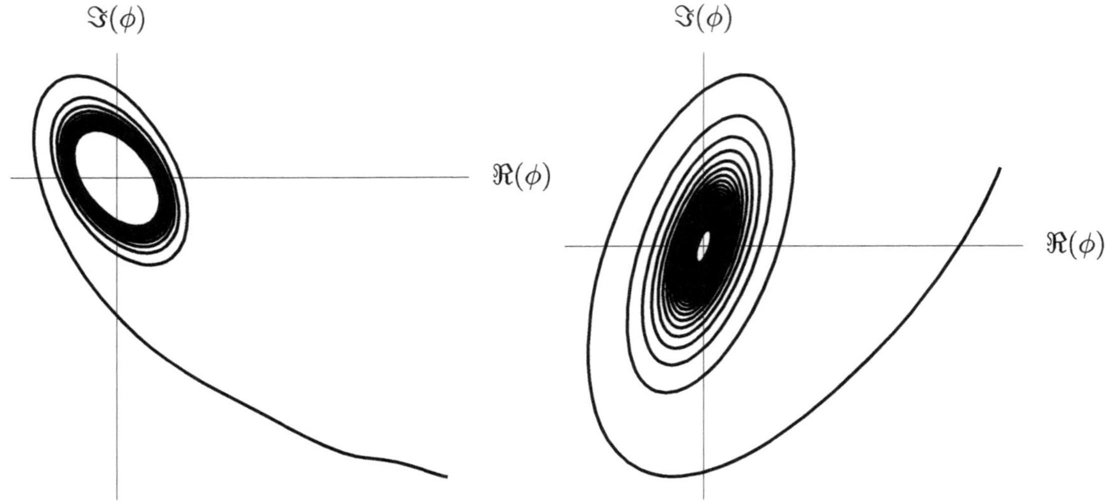

Fig. 1. The motion of the complex scalar in field space. The left panel corresponds to a model with $m_\phi/\alpha m = 10^{-4}$, $\alpha m/M_{\rm Pl} = 10^{-2}$, and $T_R = 10^{12}$ GeV. The right panel has $m_\phi/\alpha m = 7\times 10^{-3}$, $\alpha m/M_{\rm Pl} = 2\times 10^{-6}$, and $T_R = 10^{13}$ GeV.

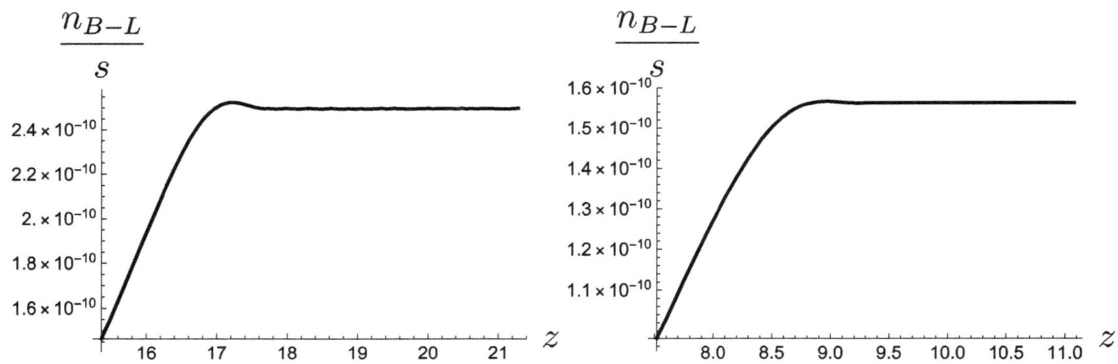

Fig. 2. n_{B-L}/s as a function of $z = \ln(\alpha m t)$. The left panel corresponds to a model with $m_\phi/\alpha m = 10^{-4}$, $\alpha m/M_{\rm Pl} = 10^{-2}$, and $T_R = 10^{12}$ GeV. The right panel has $m_\phi/\alpha m = 7\times 10^{-3}$, $\alpha m/M_{\rm Pl} = 2\times 10^{-6}$, and $T_R = 10^{13}$ GeV.

phase ($\rho = 0$), while in the broken symmetry phase ($\rho = \bar\rho$) we have, in the $k \to 0$ limit,[6]

$$\alpha^2 \leq 54 c_{123}\lambda, \tag{8}$$

where we have assumed $m_\phi \ll \bar\rho$, as we will throughout this paper. This constraint places a mild upper bound on the coupling, $\alpha \lesssim \sqrt{\lambda}$. We expect $\lambda \lesssim \mathcal{O}(1)$, otherwise the theory is strongly coupled. Only the combination αm is relevant for baryogenesis, and this constraint then implies that we cannot simultaneously take $m < M_{\rm Pl}$ to satisfy the constraints above whilst having αm parametrically larger.

3. Cosmology

In this section we will assume a homogeneous and isotopic cosmological background and derive a simple estimate (19) for the baryon-to-photon ratio generated by our model. In order to verify that the approximations we make are valid, we also solve the equations of motion numerically, with the results plotted in Figs. 1 and 2.

We can see from the potential (6) that the U(1) symmetry is broken at early times and restored at late times, when

$$H < \bar H \equiv \frac{m_\phi^2}{\alpha m}. \tag{9}$$

When $H > \bar H$ there is a time-dependent symmetry-breaking minimum at

$$|\phi_{\rm min}|^2 = \frac{\alpha m H - m_\phi^2}{\lambda} \approx \frac{\alpha m H}{\lambda}, \tag{10}$$

where we have assumed that the small symmetry-violating terms $(\varepsilon\phi^4 + \varepsilon^\dagger\phi^{\dagger 4})$ are negligible, or, equivalently, have chosen the coefficients ε so that this is the case at early times.

The field tracks this minimum nearly adiabatically until $H \approx \bar H$, at which point the symmetry is restored and the field begins to oscillate around the symmetry-restored minimum at $\phi = 0$. When this occurs, the $U(1)_{B-L}$-violating terms play an important role. We would like these to become important around the time that the symmetry is restored. Expanding $\varepsilon = \bar\varepsilon e^{i\psi}$, the correction to the potential is

$$\Delta V_{\rm eff} = \frac{1}{2}\bar\varepsilon|\phi|^4\cos(4\theta + \psi), \tag{11}$$

where $\phi = |\phi|e^{i\theta}$. The $B-L$ charge density is

$$n_{B-L} = J^0 = i(B-L)(\phi^\dagger \overleftrightarrow{\partial^0}\phi) = 2(B-L)|\phi|^2\dot\theta, \tag{12}$$

[6] Strictly speaking the $k \to 0$ limit is not physical, as we will be dealing with cosmological spacetimes, which only resemble flat space for $k \gg H$. A more exact condition is obtained by setting $k = \bar H$, where $\bar H$ is the value of the Hubble parameter below which the symmetry is restored, as discussed in the next section. This only significantly relaxes the constraint (8) if $m_\phi \gg \alpha m$, in which case the constraint becomes $\alpha^2 \leq \sqrt{18 c_{123}\lambda}\, m_\phi/m$.

so we see that the motion of the angular field is responsible for generating a net $B-L$. This means that θ should not sit at its minimum in the early Universe, and, indeed, one expects it to be frozen at some initial value due to Hubble damping. We would like it to begin rolling around the time of symmetry restoration in order to generate a net $B-L$ before the field settles into the new minimum at $\phi = 0$, which will be the case if the canonically-normalized field's mass[7] $m_\theta^2 = \Delta V_{\mathrm{eff}\,\theta\theta}/|\phi|^2$ is of order $\bar{H}^2$. If $m_\theta < \bar{H}$ the angular field will not roll after symmetry restoration and no $B-L$ will be generated. Similarly, if $m_\theta > \bar{H}$ the field starts rolling long before symmetry restoration, and the value of $B-L$ is set by tuning the initial conditions. Setting $m_\theta \sim \bar{H}$ implies that

$$\bar{\varepsilon} \sim \lambda \left(\frac{m_\phi}{\alpha m}\right)^2, \tag{13}$$

where we have used $|\phi| = |\phi_{\min}|$ at the time of symmetry restoration.

Using the angular field's equation of motion,

$$|\phi|^2\left(\ddot{\theta} + 3H\dot{\theta}\right) + 2|\phi||\dot{\phi}|\dot{\theta} = \bar{\varepsilon}|\phi|^4 \sin(4\theta + \psi), \tag{14}$$

one has

$$\dot{n}_{B-L} + 3Hn_{B-L} = 2(B-L)\bar{\varepsilon}|\phi|^4 \sin(4\theta + \psi). \tag{15}$$

Making the approximation $\dot{n}_{B-L} \approx Hn_{B-L}$ [31,32], which we will verify numerically later, one finds

$$n_{B-L} \sim \bar{\varepsilon}\frac{|\phi_{\min}|^4}{\bar{H}}, \tag{16}$$

where we have omitted factors of order unity. Using equations (9), (10), and (13) we can estimate the $B-L$ conserved charge density as

$$n_{B-L} \sim \frac{m_\phi^4}{\lambda \alpha m}. \tag{17}$$

We do not directly observe n_{B-L}, but rather the baryon-to-photon ratio $n_b = n_{B-L}/s$, where s is the entropy density. This introduces some model dependence; for concreteness, and to minimize the number of free parameters, we will focus on a minimal model in which the Universe reheats instantaneously after inflation, and the $U(1)_{B-L}$ symmetry is restored shortly thereafter. The assumption of instantaneous reheating yields

$$s = \frac{4\rho_I}{3T_R} \sim \bar{H}^2 M_{\mathrm{Pl}}^2 T_R, \tag{18}$$

where T_R is the reheat temperature. Combining this with equation (17) we find

$$n_b = \frac{n_{B-L}}{s} \sim \frac{10^{-10}}{\lambda}\left(\frac{T_R}{10^8\ \mathrm{GeV}}\right)\left(\frac{\alpha m}{M_{\mathrm{Pl}}}\right). \tag{19}$$

We see that this new mechanism can produce the observed baryon-to-photon ratio, $n_b \sim 10^{-10}$, with sensible choices for the reheat temperature and model parameters. Note that the parameters α and m only appear in the combination αm, while it is m^2 (in combination with the c_i) which is constrained by experimental tests of Lorentz violation, as discussed in section 2.1. We expect $\lambda \lesssim \mathcal{O}(1)$, otherwise the theory is strongly coupled, and as discussed above, we should have $\alpha^2 \lesssim \mathcal{O}(10)\lambda$ to ensure gradient stability around flat space.

[7] Recall that $\mathcal{L}_\theta/\sqrt{-g} \supset |\phi|^2(\partial\theta)^2$, necessitating the factor of $|\phi|^{-2}$ in the canonical normalization.

In order to verify the approximations we have made above, we have numerically integrated the scalar field equations assuming a radiation-dominated Universe. In Fig. 1 we plot the motion of the complex scalar for two different models. One can see the behavior we predicted qualitatively above: the field tracks its time-dependent minimum at early times before the angular field begins to roll when the symmetry is restored, giving rise to a spiral trajectory. In Fig. 2 we plot the baryon-to-photon ratio n_{B-L}/s for the same models. One can see that our numerical results agree well with our prediction (19).

4. Conclusions

In this paper we have studied baryogenesis in Lorentz-violating theories of gravity, which, at low energies, are naturally described by a constrained vector so that there is a preferred frame. Baryogenesis requires a field charged under $U(1)_{B-L}$, the simplest choice being a complex scalar. We have demonstrated here that the lowest-order interaction between the scalar and vector can give rise to a tachyonic mass term for the scalar proportional to the Hubble parameter so that the $U(1)_{B-L}$ symmetry is broken at early times. Inverse phase transitions such as these can generate a net $B-L$ through the coherent motion of the scalar when the symmetry is restored at late times, and we have shown here that Lorentz-violating theories can successfully generate the observed baryon-to-photon ratio using this phenomenon. Furthermore, this can be achieved for parameter choices that are not ruled out by current constraints. Our theory differs from the quintessential paradigm—the Affleck–Dine mechanism—in that the tachyonic mass is proportional to H rather than H^2, which gives rise to new features and a qualitatively different cosmology, which we have calculated in detail. Lorentz-violating gravity theories continue to be important in the study of dark energy and quantum gravity. Here, we have shown that they may also shed light on the origin of the matter–antimatter asymmetry.

Acknowledgements

We are grateful to Mark Trodden for enlightening discussions. JS and ARS are supported by funds provided to the Center for Particle Cosmology by the University of Pennsylvania. This paper is dedicated to the memory of Little Pete's diner, 1978–2017. *Benedictio caro cum caseo sit.*

References

[1] D.E. Morrissey, M.J. Ramsey-Musolf, New J. Phys. 14 (2012) 125003, arXiv:1206.2942 [hep-ph].
[2] M. Trodden, Rev. Mod. Phys. 71 (1999) 1463, arXiv:hep-ph/9803479.
[3] M. Trodden, in: Proceedings, 32nd SLAC Summer Institute on Particle Physics: Cosmic Connections, SSI 2004, Menlo Park, California, August 2–13, 2004, eConf C040802 (2004) L018, arXiv:hep-ph/0411301.
[4] A. Riotto, M. Trodden, Annu. Rev. Nucl. Part. Sci. 49 (1999) 35, arXiv:hep-ph/9901362.
[5] M. Dine, A. Kusenko, Rev. Mod. Phys. 76 (2003) 1, arXiv:hep-ph/0303065.
[6] J.M. Cline, in: Les Houches Summer School – Session 86: Particle Physics and Cosmology: The Fabric of Spacetime, Les Houches, France, July 31–August 25, 2006, 2006, arXiv:hep-ph/0609145.
[7] R. Allahverdi, A. Mazumdar, New J. Phys. 14 (2012) 125013.
[8] C. Armendariz-Picon, A. Diez-Tejedor, R. Penco, J. High Energy Phys. 10 (2010) 079, arXiv:1004.5596 [hep-ph].
[9] P. Horava, Phys. Rev. D 79 (2009) 084008, arXiv:0901.3775 [hep-th].
[10] D. Blas, O. Pujolas, S. Sibiryakov, J. High Energy Phys. 04 (2011) 018, arXiv:1007.3503 [hep-th].
[11] I. Affleck, M. Dine, Nucl. Phys. B 249 (1985) 361.
[12] J. Sakstein, M. Trodden, arXiv:1703.10103 [hep-ph], 2017.
[13] O. Bertolami, D. Colladay, V.A. Kostelecky, R. Potting, Phys. Lett. B 395 (1997) 178, arXiv:hep-ph/9612437.
[14] S.M. Carroll, J. Shu, Phys. Rev. D 73 (2006) 103515, arXiv:hep-ph/0510081.

[15] M. de Cesare, N.E. Mavromatos, S. Sarkar, Eur. Phys. J. C 75 (2015) 514, arXiv:1412.7077 [hep-ph].
[16] W. Donnelly, T. Jacobson, Phys. Rev. D 82 (2010) 064032, arXiv:1007.2594 [gr-qc].
[17] D. Blas, S. Sibiryakov, J. Cosmol. Astropart. Phys. 1107 (2011) 026, arXiv:1104.3579 [hep-th].
[18] J.D. Barrow, Phys. Rev. D 85 (2012) 047503, arXiv:1201.2882 [gr-qc].
[19] A.R. Solomon, J.D. Barrow, Phys. Rev. D 89 (2014) 024001, arXiv:1309.4778 [astro-ph.CO].
[20] P. Sandin, B. Alhulaimi, A. Coley, Phys. Rev. D 87 (2013) 044031, arXiv:1211.4402 [gr-qc].
[21] M.M. Ivanov, S. Sibiryakov, J. Cosmol. Astropart. Phys. 1405 (2014) 045, arXiv:1402.4964 [astro-ph.CO].
[22] S.M. Carroll, E.A. Lim, Phys. Rev. D 70 (2004) 123525, arXiv:hep-th/0407149.
[23] E.A. Lim, Phys. Rev. D 71 (2005) 063504, arXiv:astro-ph/0407437.
[24] I. Carruthers, T. Jacobson, Phys. Rev. D 83 (2011) 024034, arXiv:1011.6466 [gr-qc].
[25] J.A. Harvey, M.S. Turner, Phys. Rev. D 42 (1990) 3344.
[26] A. Falkowski, J. Juknevich, J. Shelton, arXiv:0908.1790 [hep-ph], 2009.
[27] V. Gonzalez Macias, J. Wudka, J. High Energy Phys. 07 (2015) 161, arXiv:1506.03825 [hep-ph].
[28] J.W. Elliott, G.D. Moore, H. Stoica, J. High Energy Phys. 08 (2005) 066, arXiv:hep-ph/0505211.
[29] B.Z. Foster, T. Jacobson, Phys. Rev. D 73 (2006) 064015, arXiv:gr-qc/0509083.
[30] T. Jacobson, in: Proceedings, Workshop on From Quantum to Emergent Gravity: Theory and Phenomenology, QG-Ph, Trieste, Italy, June 11–15, 2007, PoS QG-PH (2007) 020, arXiv:0801.1547 [gr-qc].
[31] T. Asaka, M. Fujii, K. Hamaguchi, T. Yanagida, Phys. Rev. D 62 (2000) 123514, arXiv:hep-ph/0008041.
[32] B. von Harling, K. Petraki, R.R. Volkas, J. Cosmol. Astropart. Phys. 1205 (2012) 021, arXiv:1201.2200 [hep-ph].

A dark matter model that reconciles tensions between the cosmic-ray $e^{\pm}$ excess and the gamma-ray and CMB constraints

Qian-Fei Xiang [a,b], Xiao-Jun Bi [a], Su-Jie Lin [a], Peng-Fei Yin [a]

[a] Key Laboratory of Particle Astrophysics, Institute of High Energy Physics, Chinese Academy of Sciences, Beijing 100049, China
[b] School of Physical Sciences, University of Chinese Academy of Sciences, Beijing 100049, China

ARTICLE INFO

Editor: J. Hisano

ABSTRACT

The cosmic-ray (CR) $e^{\pm}$ excess observed by AMS-02 can be explained by dark matter (DM) annihilation. However, the DM explanation requires a large annihilation cross section which is strongly disfavored by other observations, such as the Fermi-LAT gamma-ray observation of dwarf galaxies and the Planck observation of the cosmic microwave background (CMB). Moreover, the DM annihilation cross section required by the CR $e^{\pm}$ excess is also too large to generate the correct DM relic density with thermal production. In this work we use the Breit–Wigner mechanism with a velocity dependent DM annihilation cross section to reconcile these tensions. If DM particles accounting for the CR $e^{\pm}$ excess with $v \sim \mathcal{O}(10^{-3})$ are very close to a resonance in the physical pole case, their annihilation cross section in the Galaxy reaches a maximal value. On the other hand, the annihilation cross section would be suppressed for DM particles with smaller relative velocities in dwarf galaxies and at recombination, which may affect the gamma-ray and CMB observations, respectively. We find a proper parameter region that can simultaneously explain the AMS-02 results and the thermal relic density, while satisfying the Fermi-LAT and Planck constraints.

1. Introduction

Astrophysics and cosmology observations reveal that the dominant matter component in the universe is dark matter (DM), but the particle nature of DM remains unknown [1,2]. The existence of DM cannot be explained within the framework of the standard model (SM), and thus provides a hint of the physics beyond the SM. Great efforts have been devoted to DM researches, including collider detection, direct detection, and indirect detection experiments.

DM particles can be traced by cosmic ray (CR) experiments through their annihilation products from the Galaxy halo. The Alpha Magnetic Spectrometer (AMS-02), launched in 2011, is able to measure CR spectra with an unprecedented precision [3]. The precise results released by AMS-02 have confirmed the CR $e^{\pm}$ excess above ~ 10 GeV, which indicates the existence of exotic $e^{\pm}$ sources. Many astrophysical explanations have been proposed for this excess, such as primary sources like pulsars [4–6], or the CR interactions occurring around CR acceleration sources [7–10]. Interestingly, this excess can also be explained by DM annihilations/decays to charged leptons [11–17].

On the other hand, DM particles would also generate high energy photons associated with charged leptons. The related gamma-ray signatures can be significant in systems with high DM densities and low baryon densities, such as dwarf galaxies. However, the Fermi-LAT observations do not find such signatures, and set strong constraints on the DM annihilation cross section [18–20]. Since the large annihilation cross section required by the CR $e^{\pm}$ excess seems not to be allowed by the Fermi-LAT constraints [17], the DM annihilation explanation is strongly disfavored.

Moreover, the electromagnetically interacting particles generated by DM annihilations at recombination could affect cosmic microwave background (CMB) [21–27]. Precise measurements performed by WMAP [1] and recently by Planck [2] have been used to set constraints on the DM energy injections and the DM annihilation cross sections for specified final states. Compared to the results from CR and gamma-ray observations, these constraints are more stringent, and are free of some astrophysical uncertainties, which arise from the large-scale structure formation, DM density files and so on [26].

E-mail address: yinpf@ihep.ac.cn (P.-F. Yin).

Apparently, the results from the Fermi-LAT and Planck observations strongly disfavor the large DM annihilation cross sections required by the CR $e^{\pm}$ excess [17]. However, note that DM particles have very different relative velocities in different circumstances. For the DM particles potentially impacting on the CR $e^{\pm}$, dwarf galaxy gamma-ray, and CMB observations, the typical relative velocities are $v \sim 10^{-3}$, 10^{-4}, and $\ll 10^{-6}$, respectively. Therefore, the inconsistence between the DM explanations for different experimental results can be relaxed or even avoided by a velocity dependent annihilation cross section. In fact, the velocity dependent DM annihilation models, such as the Sommerfeld [28–35] and Breit–Wigner mechanisms [36–42], have been widely used to simultaneously explain the thermal DM relic density and the CR $e^{\pm}$ excess. In these models, DM particles have a much larger annihilation cross section in the Galaxy with $v \sim 10^{-3}$ than that in the early Universe for explaining the relic density with $v \sim 10^{-1}$.

In this paper, we explain the AMS-02 $e^{\pm}$ excess in an annihilating DM scenario with the Breit–Wigner mechanism. The DM relic density and the constraints from the Fermi-LAT and Planck observations are also taken into account. In this scenario, two DM particles resonantly annihilate via the s-channel exchange of a heavy mediator. The typical form of the DM annihilation cross section is characterized by two parameters, namely $\gamma \equiv \Gamma_{Z'}/m_{Z'}$ and $\delta \equiv 1 - m_{Z'}^2/4m_\chi^2$, where $\Gamma_{Z'}$, $m_{Z'}$, and m_χ are the mediator decay width, the mediator mass, and the DM mass, respectively. The assumptions of $\delta > 0$ and $\delta < 0$ correspond to the cases with an unphysical pole and a physical pole, respectively. As shown in Ref. [36–38], both these two cases can simultaneously explain the high energy positron excess observed by PAMELA and the DM relic density. In our analysis, we perform a fitting to the AMS-02 $e^{\pm}$ data with the DM contribution, and derive the corresponding DM annihilation cross sections for $\mu^+\mu^-$ and $\tau^+\tau^-$ final states. Then we adjust the parameters γ and δ to obtain suitable DM annihilation cross sections with different relative velocities. We find that there exists a parameter region with $\delta < 0$, simultaneously accounting for the AMS-02 $e^{\pm}$ excess and DM relic density, which is also allowed by the Fermi-LAT dwarf galaxy gamma-ray and the Planck CMB observations (for other studies of the similar topic, see also Ref. [43,44,42]).

This paper is organized as follows. In Sec. 2 we perform a fitting to the AMS-02 data, and derive the corresponding DM annihilation cross sections for $\mu^+\mu^-$ and $\tau^+\tau^-$ final states. In Sec. 3 we briefly introduce the Breit–Wigner scenario. In Sec. 4 we show how to relax the tension between DM explanations for the AMS-02, Fermi, and Planck observations, and obtain the correct DM relic density. Sec. 5 is our conclusions and discussions.

2. Fit to the AMS-02 data

The complicated CR propagation process can be described by a propagation equation involving some free parameters. In order to predict the CR $e^{\pm}$ background, some additional parameters describing the primary and secondary CR injections are needed. In principle, these parameters are determined by available CR observations. In this work, we use the package GALPROP [45,46] to resolve the propagation equation, and perform a Markov chain Monte Carlo fitting to the AMS-02 data in the high dimensional parameter space.

The propagation parameters are dominantly determined by a fitting to the measured secondary-to-primary ratios [17], including the B/C data from ACE [47] and AMS-02 [48], and the $^{10}Be/^{9}Be$ data from several experiments. Two kinds of propagation models, namely the diffusion–convection (DC) model and the diffusion–reacceleration (DR) model, are taken into account in [17]. The injection spectrum of the primary electron background is assumed to be a three-piece broken power law with two breaks. Comparing to the spectrum with only one break at the low energy, we find that the spectrum with an additional break around 60 GeV can provide a better fit to the AMS-02 data. The nucleon injection parameters are constrained by fitting the proton flux of AMS-02 [48]. After deriving the propagated proton spectrum, the injection of the secondary $e^{\pm}$ backgrounds is calculated by using the parameterized cross section presented in Ref. [49].

For the DM signature, we assume that DM particles purely annihilate to $\mu^+\mu^-$ or $\tau^+\tau^-$. The initial $e^{\pm}$ spectra from DM annihilation are calculated by PPPC 4 DM ID [50], which includes the electroweak corrections [51]. The DM density profile is taken to be the NFW profile [52] defined by $\rho(r) = \rho_s r_s / r(1 + r/r_s)^2$, with a characteristic halo radius $\rho_s = 20$ kpc and a characteristic halo density $\rho_s = 0.26$ GeV cm^{-3}. Since the dominant contributions to the observed high-energy CR $e^{\pm}$ are provided by DM particles located in a range of ~ 1 kpc around the Solar system due to the CR propagation effect, the different choices of the DM density profile would not significantly modify the prediction for CR $e^{\pm}$.

Combining the contributions of primary CR e^-, secondary CR $e^{\pm}$, and $e^{\pm}$ from DM annihilation, we perform a fit to the latest AMS-02 $e^{\pm}$ data, including the positron fraction $\frac{e^+}{e^+ + e^-}$ and the fluxes of e^+, e^-, and $e^+ + e^-$ [53–55]. We provide the fitting results to the observed e^+ flux with the DR propagation model in Fig. 1; the bands representing 2σ uncertainties are also shown. The best-fit values of m_χ and related 2σ regions of $\langle\sigma_{\mathrm{ann}} v\rangle$ (in cm^{3}s^{-1}) are listed in Table 1. The corresponding exclusion limits derived from the Fermi-LAT dwarf galaxy gamma-ray [19] and Planck CMB [26] observations are also given. It is obvious that the parameter regions of $\langle\sigma_{\mathrm{ann}} v\rangle$ for explaining the AMS-02 $e^{\pm}$ excess are excluded by other two kinds of observations. Compared to the $\mu^+\mu^-$ channel, the tension in the $\tau^+\tau^-$ channel is severer due to tremendous photons from the hadronic decays of τ.

3. Breit–Wigner enhancement

In the Breit–Wigner scenario, the DM annihilation cross section has a typical form of

$$\sigma v \propto \frac{1}{16\pi m_\chi^2} \frac{1}{(\delta + v^2/4)^2 + \gamma^2}. \tag{1}$$

This form is valid in the non-relativistic limit with $v^2 << 1$ and $\delta << 1$ at the center-of-mass energy $\sqrt{s} \sim \sqrt{4m_\chi^2 + m_\chi^2 v^2}$.

As an example, we consider a simple leptophilic fermionic DM model, where DM particles interact with charged leptons through a vector mediator Z' [39]. The corresponding lagrangian is

$$\mathcal{L}_{int} \supset -g(a\bar{\chi}\gamma^\mu\chi + \bar{l}_i\gamma^\mu l_i)Z'_\mu, \tag{2}$$

where l_i represents the species of leptons, g and ag are the couplings of Z' to the leptons and DM particles, respectively. This model can easily avoid the constraints from DM direct detection and collider experiments due to its leptophilic property.

The DM annihilation cross section in this model is given by

$$\sigma v = \frac{1}{6\pi} \frac{a^2 g^4 s}{(s - m_{Z'}^2)^2 + m_{Z'}^2 \Gamma_{Z'}^2} (1 + \frac{2m_\chi^2}{s}), \tag{3}$$

where m_χ, $m_{Z'}$ and $\Gamma_{Z'}$ are the DM mass, the Z' mass, and the decay width of Z', respectively, v is the relative velocity between two incident DM particles. Note that the lepton mass has been neglected in Eq. (3) due to the large $\sqrt{s}$ considered in our analysis. The decay width of Z' can be expressed as

$$\Gamma_{Z'} = \frac{m_{Z'}}{12\pi} a^2 g^2 \xi_\chi^3 \Theta(m_{Z'} - 2m_\chi) + \frac{m_{Z'}}{12\pi} g^2 \xi_{l_i}^3, \tag{4}$$

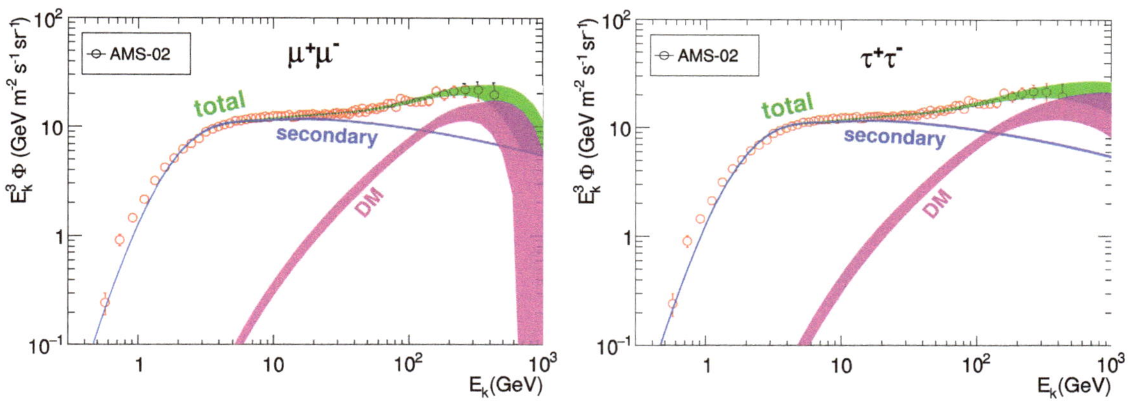

(a)$\mu^+\mu^-$ channel. (b)$\tau^+\tau^-$ channel.

Fig. 1. Fittings to the positron flux measured by AMS-02 for DM annihilations to $\mu^+\mu^-$ (left panel) and $\tau^+\tau^-$ (right panel), respectively. The pink bands indicate the contributions from DM annihilation within 2σ uncertainty. The blue lines represent the secondary CR positron flux. Total positron fluxes are shown as green bands. (For interpretation of the references to color in this figure legend, the reader is referred to the web version of this article.)

Table 1
The best-fit values of DM masses m_χ and corresponding thermally averaged annihilation cross sections $\langle\sigma_{\rm ann}v\rangle$ (in units of cm^{3}s^{-1}) given by the fitting to the AMS-02 data with the DR propagation model. The corresponding limits from the Fermi-LAT and Planck observations are also shown.

Channels	m_χ (TeV)	AMS-02 (2σ)	Fermi limits	Planck limits
$\mu^+\mu^-$	0.89	$3.79\times10^{-24}<\langle\sigma_{\rm ann}v\rangle<6.48\times10^{-24}$	2.95×10^{-24}	2.58×10^{-24}
$\tau^+\tau-$	3.89	$5.29\times10^{-23}<\langle\sigma_{\rm ann}v\rangle<1.06\times10^{-22}$	1.25×10^{-23}	1.06×10^{-23}

where $\xi_\chi\equiv\sqrt{1-4m_\chi^2/m_{Z'}^2}$, $\xi_{l_i}\equiv\sqrt{1-4m_{l_i}^2/m_{Z'}^2}$, and $\Theta(x)$ is the unit step function. For $m_{Z'}\sim 2m_\chi$, Z' dominantly decays to leptons with the decay width given by $\sim g^2 m_{Z'}/12\pi^2$.

Then we calculate the thermally averaged DM annihilation cross section through the formula of [56]

$$\langle\sigma_{\rm ann}v\rangle=\frac{1}{n_{EQ}^2}\frac{m_\chi}{64\pi^4 x}\int_{4m_\chi^2}^{\infty}\hat\sigma(s)\sqrt{s}K_1\left(\frac{x\sqrt{s}}{m_\chi}\right)ds, \tag{5}$$

with

$$\begin{aligned} n_{EQ}&=\frac{g_i}{2\pi^2}\frac{m_\chi^3}{x}K_2(x),\\ \hat\sigma(s)&=2g_i^2 m_\chi\sqrt{s-4m_\chi^2}\,\sigma v, \end{aligned} \tag{6}$$

where $K_i(x)$ is the modified Bessel function of order i, g_i is the internal degree of freedom of the DM particle, which equals 4 in this model. In this work, we assume that DM particles have the same temperature as the thermal radiation with a Maxwell–Boltzmann distribution above a low kinetic decoupling temperature, e.g. $x_{kd}\gg 10^6$. The discussion of the kinetic decoupling effect is given in the next section.

The evolution of the DM density is determined by numerically solving the Boltzmann equation

$$\frac{dY}{dx}=-\frac{s(x)}{Hx}\langle\sigma_{\rm ann}v\rangle(Y^2-Y_{eq}^2), \tag{7}$$

where $Y\equiv n/s$, n is the DM number density, $s=\frac{2\pi^2}{45}g_{*s}\frac{m^3}{x^3}$ is the Universe entropy density, $H=\sqrt{\frac{4\pi^3 g_*}{45m_{pl}^2}\frac{m^2}{x^2}}$ is the Hubble parameter, and g_{*s} and g_* are the effective degrees of freedom defined by the entropy density and the radiation density, respectively.

4. Results

In principle, we can accommodate the DM explanations for observations with different DM relative velocities. Only the DM particles located in the Galaxy within a range of ~ 1 kpc around the Solar system could provide significant contributions to the observed high energy CR $e^\pm$, because of the propagation effects. The typical relative velocities of these particles are $\sim 10^{-3}$, while the typical relative velocities of DM particles in dwarf galaxies are $\sim 10^{-4}$. Their annihilation cross sections may be very different in the velocity dependent annihilation models. In order to obtain the constraints on $\langle\sigma_{\rm ann}v\rangle_{\rm G}$, the constraints on $\langle\sigma_{\rm ann}v\rangle_{\rm D}$ from the Fermi-LAT observation should be rescaled by a factor of $1/S\equiv\langle\sigma_{\rm ann}v\rangle_{\rm G}/\langle\sigma_{\rm ann}v\rangle_{\rm D}$, where $\langle\sigma_{\rm ann}v\rangle_{\rm D}$ and $\langle\sigma_{\rm ann}v\rangle_{\rm G}$ are the thermally averaged DM annihilation cross sections in dwarf galaxies and near the solar system in the Galaxy, respectively. In order to relax the tension between the DM explanations for the Fermi-LAT and AMS-02 observations, the S factor should be smaller than 1.

We show the S factor in Fig. 2, and find that a parameter region with $10^{-8}<\gamma<10^{-6}$ and $-4\times10^{-6}<\delta<-10^{-7}$ can satisfy our requirement with $S\ll 1$. For the cases of $\delta>0$ corresponding to an unphysical pole, there is no parameter region with $S<1$ as shown in Fig. 2(b). This can be understood by Eq. (1): the DM annihilation cross section always increases with decreasing relative velocity for $\delta>0$. Therefore, only the cases of $\delta<0$ can be used to relax the tension between different observations.

In Fig. 3, we compare the parameter regions accounting for the AMS-02 $e^\pm$ excess with the dwarf galaxy gamma-ray constraints, which are obtained by rescaling the limits given by the Fermi-LAT collaboration [19]. It is shown that the cases with a negative tiny $-\delta\lesssim 10^{-6}$ can evade the dwarf galaxy constraints. Since σv is proportional to $1/m_\chi^2$ as can be seen from Eq. (1), the ratio of $1/S=\langle\sigma_{\rm ann}v\rangle_{\rm G}/\langle\sigma_{\rm ann}v\rangle_{\rm D}$ is almost independent of the DM mass. Note that the limit on $\langle\sigma_{\rm ann}v\rangle_{\rm G}$ is obtained by $\langle\sigma_{\rm ann}v\rangle_{\rm lim,G}=\langle\sigma_{\rm ann}v\rangle_{\rm lim,D}/S$. Therefore, the modified dwarf galaxy gamma-ray constraints for different sets of δ and γ are almost par-

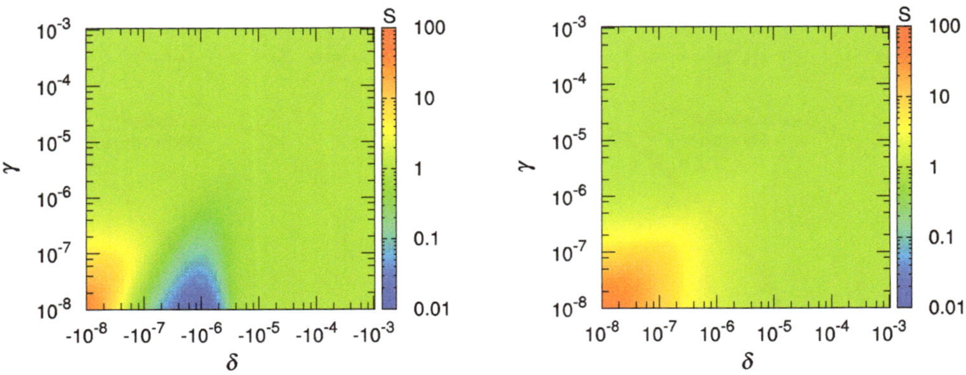

(a)$\delta < 0$. (b)$\delta > 0$.

Fig. 2. The scaling factor $S \equiv \langle\sigma_{\rm ann} v\rangle_{\rm D}/\langle\sigma_{\rm ann} v\rangle_{\rm G}$ in the $\delta - \gamma$ plane, where $\sigma_{\rm D}$ and $\sigma_{\rm G}$ denote the annihilation cross sections in dwarf galaxies with $v = 10^{-4}$ and near the solar system in the Galaxy with $v = 10^{-3}$, respectively. The left and right panels represent the physical pole case with $\delta < 0$ and unphysical pole case with $\delta > 0$, respectively.

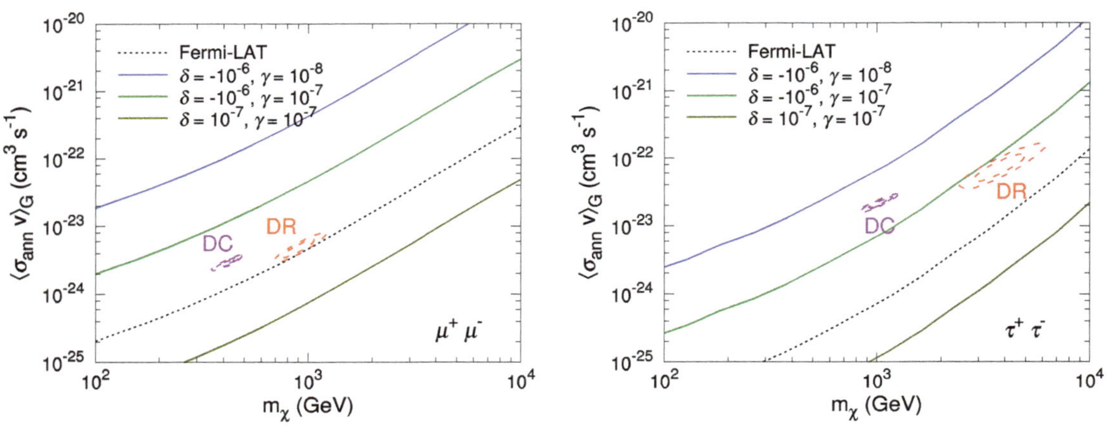

(a)$\mu^+\mu^-$ channel. (b)$\tau^+\tau^-$ channel.

Fig. 3. Contour regions represent the parameter regions accounting for the AMS-02 results in the DC and DR propagation models. Solid lines are the constraints on $\langle\sigma_{\rm ann} v\rangle_{\rm G}$ from the Fermi-LAT dwarf galaxy gamma-ray observation for different parameter sets of δ and γ. The original the Fermi-LAT limits on $\langle\sigma_{\rm ann} v\rangle_{\rm D}$ are also shown. The left and right panels represent the cases of DM annihilation to $\mu^+\mu^-$ and $\tau^+\tau^-$, respectively.

allel to the limits given by the Fermi-LAT collaboration in Fig. 3. In the above analysis, we fix the DM relative velocity in dwarf galaxies to be $v = 10^{-4}$. Strictly speaking, since DM particles in dwarf galaxies have different typical relative velocities with an order of $\mathcal{O}(10^{-4})$, the total constraint should be obtained by combining the individual constraints specified for each dwarf galaxy with a large J factor. A detailed discussion can be found in Ref. [57].

The above analysis can be directly applied to reconcile the tension between the DM explanations for the AMS-02 $e^{\pm}$ and Planck CMB observations (see Fig. 4). In order to derive the constraints on $\langle\sigma_{\rm ann} v\rangle_{\rm G}$ from CMB observations, we define a rescaling factor of $1/S' \equiv \langle\sigma_{\rm ann} v\rangle_{\rm G}/\langle\sigma_{\rm ann} v\rangle_{z_{\rm r}}$, where $\langle\sigma_{\rm ann} v\rangle_{z_{\rm r}}$ is the thermally averaged annihilation cross section of DM particles affecting CMB at recombination with $v \ll 10^{-6}$. In fact, the Breit–Wigner effect would saturate for DM particles with such a small v. In Fig. 3, we compare the parameter regions accounting for the AMS-02 $e^{\pm}$ excess with the CMB constraints, which are obtained by rescaling the limits given by Ref. [27]. We find that the cases with a negative tiny $\delta \sim -10^{-6}$ can also evade the CMB constraints.

For each point in the $\delta - \gamma$ plane with $\delta < 0$, we determine ag^2 and a through the correct relic density $\Omega h^2 = 0.1188$ [2] by resolving the Boltzmann equation, and derive corresponding $\langle\sigma_{\rm ann} v\rangle_{\rm G}$, $\langle\sigma_{\rm ann} v\rangle_{\rm D}$, and $\langle\sigma_{\rm ann} v\rangle_{z_{\rm r}}$. In Fig. 5, the red bands represent the parameter regions simultaneously accounting for the AMS-02 CR $e^{\pm}$ excess and the correct relic density. Here we only consider m_χ and $\langle\sigma_{\rm ann} v\rangle_{\rm G}$ derived with the DR propagation model as given in Table 1. The parameter regions excluded by the Fermi-LAT and Planck limits are also shown in Fig. 5. We find that there exists a parameter region with $\gamma \lesssim 10^{-7}$ and $\delta \sim -10^{-6}$, which can accommodate all the observations.

We also show the isolines of ag^2 and a satisfying the correct DM relic density in Fig. 5. The behavior of these lines can be understood as follows. Roughly speaking, the thermal relic density Ωh^2 in the usual DM models is determined by the freeze-out temperature $x_f \sim \mathcal{O}(10)$ (corresponding to $v^2 \sim 10^{-1}$) and $\langle\sigma_{\rm ann} v\rangle_{\rm f}$ as $\Omega h^2 \propto x_f/\langle\sigma_{\rm ann} v\rangle_{\rm f}$. For the resonant case, since the annihilation cross section would increase with dropping temperature, the annihilation process may be significant until the Breit–Wigner effect almost saturates at a temperature of x_b. x_b can be roughly determined by $|\delta|^{-1}$ for $\delta < 0$. This is because that there are not enough DM particles with velocities of $v \sim |\delta|^{\frac{1}{2}}$ for sufficient resonant annihilation when $x \gg 1/|\delta|$. Using the approximated form of $\langle\sigma_{\rm ann} v\rangle_{\rm b} \propto a^2 g^4 |\delta|^{\frac{1}{2}} x_b^{\frac{3}{2}}/\gamma$ by integrating out the pole [42], we get $\Omega h^2 \propto x_b/\langle\sigma_{\rm ann} v\rangle_{\rm b} \propto \gamma/a^2 g^4$. Therefore, the correct relic den-

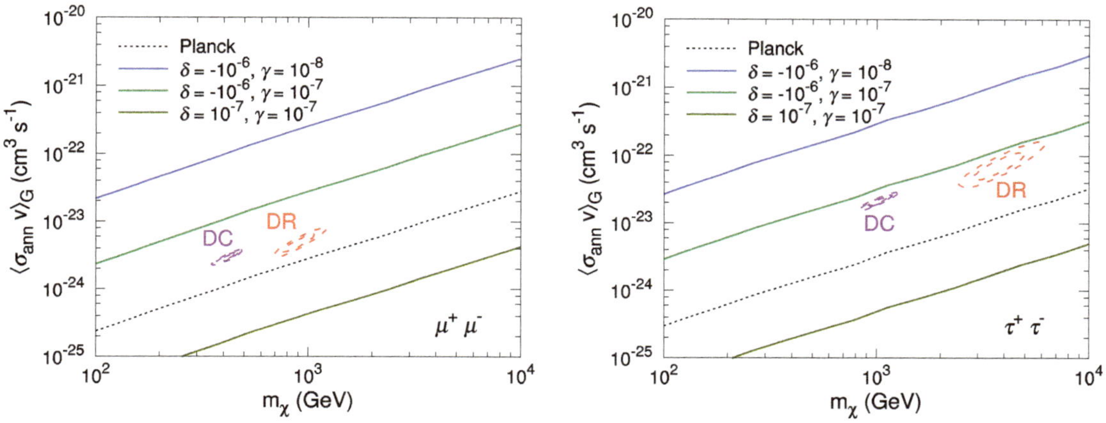

(a)$\mu^+\mu^-$ channel.

(b)$\tau^+\tau^-$ channel.

Fig. 4. The same as Fig. 3 but the constraints are derived from the Planck CMB observation.

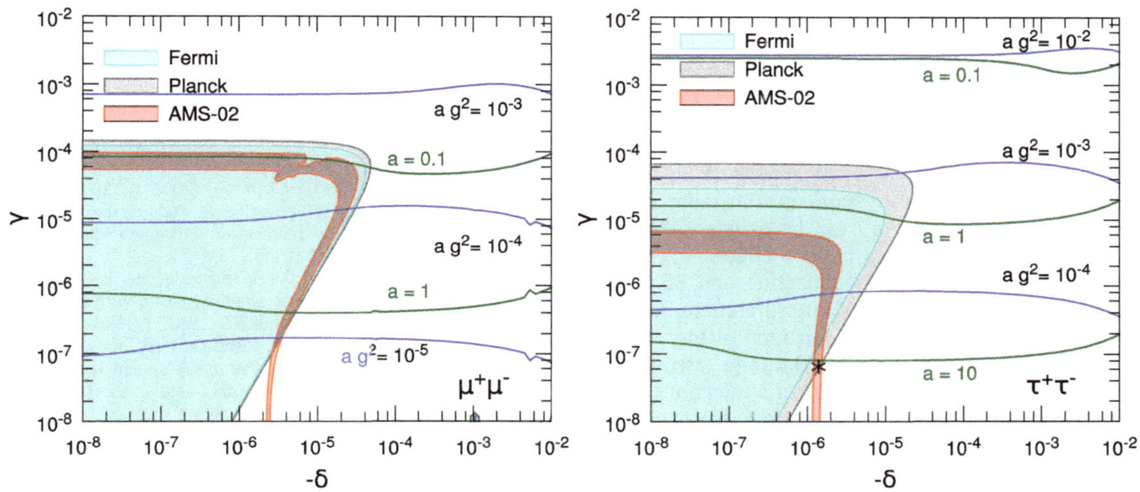

(a)$\mu^+\mu^-$ channel.

(b)$\tau^+\tau^-$ channel.

Fig. 5. Parameter regions accounting for various observations in the $\delta - \gamma$ plane with $\delta < 0$ for DM annihilation to $\mu^+\mu^-$ (left panel) and $\tau^+\tau^-$ (right panel), respectively. The DM mass is taken to be the value given in Table 1. In each parameter point, a and ag^2 are derived by requiring the correct relic density $\Omega h^2 = 0.1188$; then $\langle\sigma_{\rm ann} v\rangle_{\rm G}$, $\langle\sigma_{\rm ann} v\rangle_{\rm D}$, and $\langle\sigma_{\rm ann} v\rangle_{z_r}$ can also be obtained. Red shaded region are the parameter regions corresponding to $\langle\sigma_{\rm ann} v\rangle_{\rm G}$ given in Table 1 which can explain the AMS-02 results. The gray and cyan regions denote parameter regions excluded by the Planck and Fermi-LAT observations, respectively. The green and blue solid lines are the isolines of a and ag^2, respectively. (For interpretation of the references to color in this figure legend, the reader is referred to the web version of this article.)

sity can be easily obtained by adjusting ag^2 with $\gamma^{\frac{1}{2}}$ as shown in Fig. 5.

An important issue that should be addressed is the kinetic decoupling effect. In the parameter regions discussed above, since the scatterings between DM particles and SM radiations are not sufficient due to the t-channel exchange of a heavy mediator, the kinetic decoupling would occur at a high temperature of $T > \mathcal{O}(1)$ GeV. The velocities of DM particles drop as $\sim R^{-1}$ after the kinetic decoupling rather than $\sim R^{-\frac{1}{2}}$ before the kinetic decoupling, where R is the scale factor of the Universe. Then the Breit–Wigner mechanism would significantly enhance the DM annihilation cross section at the freeze-out epoch and drastically reduce the DM relic density. As discussed in Ref. [40], it is difficult to simultaneously explain the CR $e^{\pm}$ excess and the relic density with such a significant kinetic decoupling effect. Moreover, after the kinetic decoupling, the velocity distribution of DM particles would depart from the thermal distribution and is difficult to deal with in the calculation of the relic density. A solution is introducing some additional mediators, which can enhance the scattering rate between DM particles and SM radiations and/or the DM self scattering rate. The detailed discussions can be found in Ref. [34,42].

5. Conclusions and discussions

In this work we show that the DM annihilation through the Breit–Wigner mechanism can reconcile the tension between the DM explanation for the AMS-02 CR $e^{\pm}$ excess and the constraints from Fermi-LAT dwarf galaxy gamma-ray and Planck CMB observations. Since DM particles affecting these observations have different relative velocities, their annihilation cross sections are different for interpretating the experimental results. In order to check whether the DM explanation for the AMS02 results is excluded by other observations, we should translate all the limits into those on $\langle\sigma_{\rm ann} v\rangle_{\rm G}$ for DM particles with a typical relative velocity $v \sim 10^{-3}$.

We take a leptophilic Z' model as a benchmark model. This kind of leptophilic model is not constrained by the results of current direct detection and collider experiments. For the tiny values of the mediator decay width and the mass deviation from the pole, $\langle\sigma_{\rm ann}v\rangle$ would be sensitive to the relative velocity. For the unphysical pole case with $\delta > 0$, $\langle\sigma_{\rm ann}v\rangle$ increases with deceasing velocity. Thus the enhanced constraints on $\langle\sigma_{\rm ann}v\rangle_{\rm G}$ from the dwarf galaxy gamma-ray and CMB observations exclude the explanation for the CR $e^{\pm}$ excess in this case.

For the physical pole case with $\delta \sim -10^{-6}$, DM particles accounting for the CR $e^{\pm}$ excess with $v \sim \mathcal{O}(10^{-3})$ have the largest annihilation cross section close to the pole. On the other hand, the DM annihilation cross section is suppressed for DM particles with smaller relative velocities in dwarf galaxies and at recombination, which may impact on the gamma-ray and CMB observations, respectively. Therefore, the constraints on $\langle\sigma_{\rm ann}v\rangle_{\rm G}$ from these observations are weaken. We find that a parameter region with $\delta \sim -10^{-6}$ and $\gamma \lesssim 10^{-7}$ can simultaneously account for the AMS-02, Fermi-LAT dwarf galaxy gamma-ray, and Planck CMB observations, and the relic density.

From the perspective of model building, a question is how to naturally realize the tiny values of δ and γ derived in above analysis. Here we consider the benchmark point with $\gamma = 7.1 \times 10^{-8}$ and $\delta = -1.5 \times 10^{-6}$ marked in the right panel of Fig. 5. For the small decay width of the mediator, we get $g \sim 1.8 \times 10^{-3}$ and $ag \sim 1.8 \times 10^{-2}$. These values are easy to realize in a realistic model. The problem is how to achieve a tiny $\delta \sim -10^{-6}$, which seems to require a significant fine-tuning. A solution is given by Ref. [42] through the nontrivial flavor symmetry-breaking in the dark sector. By assigning a particular symmetry-breaking mode, a resonance with a mass of almost $2m_\chi$ can be realized. The tiny mass deviation of δ is naturally induced by loop effects.

Finally, it is worth noting that the reionization and BBN histories can also set constraints on the DM interpretations for the CR $e^{\pm}$ excess. The constraints from the reionization history would not exclude the DM interpretations due to the large astrophysical uncertainties. In the BBN epoch, the energetic injections from velocity-dependent DM annihilation with a large cross section would significantly modify the light element abundances [25]. We find that the available BBN constraints on the velocity-independent DM annihilation cross section from the purely electromagnetic or hadronic channels [58,59] are not stricter than the CMB constraints. Therefore, these analyses would not set more stringent constraints on the parameter region of interest with $|\delta| \sim 10^{-6}$ in our paper, where the Breit–Wigner effect saturates at the BBN epoch with $x \gg 10^6$. Note that the injections of pions from the $\tau^+\tau^-$ channel would induce an overproduction of ^{4}He via the pn conversion. The study of this effect can set the constraint on the velocity-dependent DM annihilation cross section.

Acknowledgement

This work is supported by the National Key Program for Research and Development (No. 2016YFA0400200) and by the National Natural Science Foundation of China under Grants No. 11475189 and 11475191.

References

[1] WMAP collaboration, E. Komatsu, et al., Five-year Wilkinson Microwave Anisotropy Probe (WMAP) observations: cosmological interpretation, Astrophys. J. Suppl. 180 (2009) 330–376, arXiv:0803.0547.

[2] Planck collaboration, P.A.R. Ade, et al., Planck 2015 results. XIII. Cosmological parameters, Astron. Astrophys. 594 (2016) A13, arXiv:1502.01589.

[3] AMS collaboration, M. Aguilar, et al., First result from the Alpha Magnetic Spectrometer on the International Space Station: precision measurement of the positron fraction in primary cosmic rays of 0.5–50 GeV, Phys. Rev. Lett. 110 (2013) 141102.

[4] D. Hooper, P. Blasi, P.D. Serpico, Pulsars as the sources of high energy cosmic ray positrons, J. Cosmol. Astropart. Phys. 0901 (2009) 025, arXiv:0810.1527.

[5] H. Yuksel, M.D. Kistler, T. Stanev, TeV gamma rays from Geminga and the origin of the GeV positron excess, Phys. Rev. Lett. 103 (2009) 051101, arXiv:0810.2784.

[6] S. Profumo, Dissecting cosmic-ray electron–positron data with Occam's Razor: the role of known pulsars, Cent. Eur. J. Phys. 10 (2011) 1–31, arXiv:0812.4457.

[7] P. Blasi, The origin of the positron excess in cosmic rays, Phys. Rev. Lett. 103 (2009) 051104, arXiv:0903.2794.

[8] H.-B. Hu, Q. Yuan, B. Wang, C. Fan, J.-L. Zhang, et al., On the e+ e- excesses and the knee of the cosmic ray spectra—hints of cosmic rays acceleration in young supernova remnants, Astrophys. J. 700 (2009) L170–L173.

[9] Y. Fujita, K. Kohri, R. Yamazaki, K. Ioka, Is the PAMELA anomaly caused by the supernova explosions near the Earth?, Phys. Rev. D 80 (2009) 063003, arXiv:0903.5298.

[10] K. Kohri, K. Ioka, Y. Fujita, R. Yamazaki, Can we explain AMS-02 antiproton and positron excesses simultaneously by nearby supernovae without pulsars or dark matter? PTEP 2016 (2016) 021E01, arXiv:1505.01236.

[11] L. Bergstrom, T. Bringmann, J. Edsjo, New positron spectral features from supersymmetric dark matter—a way to explain the PAMELA data? Phys. Rev. D 78 (2008) 103520, arXiv:0808.3725.

[12] V. Barger, W.Y. Keung, D. Marfatia, G. Shaughnessy, PAMELA and dark matter, Phys. Lett. B 672 (2009) 141–146, arXiv:0809.0162.

[13] M. Cirelli, M. Kadastik, M. Raidal, A. Strumia, Model-independent implications of the e+-, anti-proton cosmic ray spectra on properties of dark matter, Nucl. Phys. B 813 (2009) 1–21, arXiv:0809.2409.

[14] P.-f. Yin, Q. Yuan, J. Liu, J. Zhang, X.-j. Bi, et al., PAMELA data and leptonically decaying dark matter, Phys. Rev. D 79 (2009) 023512, arXiv:0811.0176.

[15] J. Zhang, X.-J. Bi, J. Liu, S.-M. Liu, P.-F. Yin, et al., Discriminating different scenarios to account for the cosmic e + /- excess by synchrotron and inverse Compton radiation, Phys. Rev. D 80 (2009) 023007, arXiv:0812.0522.

[16] L. Bergstrom, J. Edsjo, G. Zaharijas, Dark matter interpretation of recent electron and positron data, Phys. Rev. Lett. 103 (2009) 031103, arXiv:0905.0333.

[17] S.-J. Lin, Q. Yuan, X.-J. Bi, Quantitative study of the AMS-02 electron/positron spectra: implications for pulsars and dark matter properties, Phys. Rev. D 91 (2015) 063508, arXiv:1409.6248.

[18] Fermi-LAT collaboration, M. Ackermann, et al., Dark matter constraints from observations of 25 Milky Way satellite galaxies with the Fermi Large Area Telescope, Phys. Rev. D 89 (2014) 042001, arXiv:1310.0828.

[19] Fermi-LAT collaboration, M. Ackermann, et al., Searching for dark matter annihilation from Milky Way dwarf spheroidal galaxies with six years of Fermi Large Area Telescope data, Phys. Rev. Lett. 115 (2015) 231301, arXiv:1503.02641.

[20] Fermi-LAT collaboration DES, A. Albert, et al., Searching for dark matter annihilation in recently discovered Milky Way satellites with Fermi-LAT, Astrophys. J. 834 (2017) 110, arXiv:1611.03184.

[21] J.A. Adams, S. Sarkar, D.W. Sciama, CMB anisotropy in the decaying neutrino cosmology, Mon. Not. R. Astron. Soc. 301 (1998) 210–214, arXiv:astro-ph/9805108.

[22] X.-L. Chen, M. Kamionkowski, Particle decays during the cosmic dark ages, Phys. Rev. D 70 (2004) 043502, arXiv:astro-ph/0310473.

[23] N. Padmanabhan, D.P. Finkbeiner, Detecting dark matter annihilation with CMB polarization: signatures and experimental prospects, Phys. Rev. D 72 (2005) 023508, arXiv:astro-ph/0503486.

[24] S. Galli, F. Iocco, G. Bertone, A. Melchiorri, CMB constraints on dark matter models with large annihilation cross-section, Phys. Rev. D 80 (2009) 023505, arXiv:0905.0003.

[25] J. Hisano, M. Kawasaki, K. Kohri, T. Moroi, K. Nakayama, T. Sekiguchi, Cosmological constraints on dark matter models with velocity-dependent annihilation cross section, Phys. Rev. D 83 (2011) 123511, arXiv:1102.4658.

[26] T.R. Slatyer, Indirect dark matter signatures in the cosmic dark ages. I. Generalizing the bound on s-wave dark matter annihilation from Planck results, Phys. Rev. D 93 (2016) 023527, arXiv:1506.03811.

[27] T.R. Slatyer, Indirect dark matter signatures in the cosmic dark ages II. Ionization, heating and photon production from arbitrary energy injections, Phys. Rev. D 93 (2016) 023521, arXiv:1506.03812.

[28] J. Hisano, S. Matsumoto, M.M. Nojiri, Explosive dark matter annihilation, Phys. Rev. Lett. 92 (2004) 031303, arXiv:hep-ph/0307216.

[29] K. Belotsky, D. Fargion, M. Khlopov, R.V. Konoplich, May heavy neutrinos solve underground and cosmic ray puzzles?, Phys. At. Nucl. 71 (2008) 147–161, arXiv:hep-ph/0411093.

[30] J. Hisano, S. Matsumoto, M. Nagai, O. Saito, M. Senami, Non-perturbative effect on thermal relic abundance of dark matter, Phys. Lett. B 646 (2007) 34–38, arXiv:hep-ph/0610249.

[31] M. Cirelli, A. Strumia, M. Tamburini, Cosmology and astrophysics of minimal dark matter, Nucl. Phys. B 787 (2007) 152–175, arXiv:0706.4071.

[32] N. Arkani-Hamed, D.P. Finkbeiner, T.R. Slatyer, N. Weiner, A theory of dark matter, Phys. Rev. D 79 (2009) 015014, arXiv:0810.0713.

[33] J.L. Feng, M. Kaplinghat, H.-B. Yu, Halo shape and relic density exclusions of Sommerfeld-enhanced dark matter explanations of cosmic ray excesses, Phys. Rev. Lett. 104 (2010) 151301, arXiv:0911.0422.
[34] J.L. Feng, M. Kaplinghat, H.-B. Yu, Sommerfeld enhancements for thermal relic dark matter, Phys. Rev. D 82 (2010) 083525, arXiv:1005.4678.
[35] M. Cirelli, P. Panci, K. Petraki, F. Sala, M. Taoso, Dark matter's secret liaisons: phenomenology of a dark u(1) sector with bound states, J. Cosmol. Astropart. Phys. 1705 (2017) 036, arXiv:1612.07295.
[36] D. Feldman, Z. Liu, P. Nath, PAMELA positron excess as a signal from the hidden sector, Phys. Rev. D 79 (2009) 063509, arXiv:0810.5762.
[37] M. Ibe, H. Murayama, T. Yanagida, Breit–Wigner enhancement of dark matter annihilation, Phys. Rev. D 79 (2009) 095009, arXiv:0812.0072.
[38] W.-L. Guo, Y.-L. Wu, Enhancement of dark matter annihilation via Breit–Wigner resonance, Phys. Rev. D 79 (2009) 055012, arXiv:0901.1450.
[39] X.-J. Bi, X.-G. He, Q. Yuan, Parameters in a class of leptophilic models from PAMELA, ATIC and Fermi, Phys. Lett. B 678 (2009) 168–173, arXiv:0903.0122.
[40] X.-J. Bi, P.-F. Yin, Q. Yuan, Breit–Wigner enhancement considering the dark matter kinetic decoupling, Phys. Rev. D 85 (2012) 043526, arXiv:1106.6027.
[41] P.S.B. Dev, D.K. Ghosh, N. Okada, I. Saha, Neutrino mass and dark matter in light of recent AMS-02 results, Phys. Rev. D 89 (2014) 095001, arXiv:1307.6204.
[42] Y. Bai, J. Berger, S. Lu, Supersymmetric resonant dark matter: a thermal model for the AMS-02 positron excess, arXiv:1706.09974.
[43] K. Belotsky, R. Budaev, A. Kirillov, M. Laletin, Fermi-LAT kills dark matter interpretations of AMS-02 data. Or not? J. Cosmol. Astropart. Phys. 1701 (2017) 021, arXiv:1606.01271.
[44] D. Kim, J.-C. Park, S. Shin, Dark matter "transporting" mechanism explaining positron excesses, arXiv:1702.02944.
[45] A.W. Strong, I.V. Moskalenko, Propagation of cosmic-ray nucleons in the galaxy, Astrophys. J. 509 (1998) 212–228, arXiv:astro-ph/9807150.
[46] I.V. Moskalenko, A.W. Strong, Production and propagation of cosmic ray positrons and electrons, Astrophys. J. 493 (1998) 694–707, arXiv:astro-ph/9710124.
[47] R.A. Mewaldt, J.R. Jokipii, M.A. Lee, E. Möbius, T.H. Zurbuchen (Eds.), Acceleration and Transport of Energetic Particles Observed in the Heliosphere, American Institute of Physics Conference Series, vol. 528, Sept. 2000.
[48] AMS-02 collaboration, in: International Cosmic Ray Conference, 2013, http://www.ams02.org/2013/07/new-results-from-ams-presented-at-icrc-2013/.
[49] T. Kamae, N. Karlsson, T. Mizuno, T. Abe, T. Koi, Parameterization of gamma, e+/- and neutrino spectra produced by p-p interaction in astronomical environment, Astrophys. J. 647 (2006) 692–708, arXiv:astro-ph/0605581.
[50] M. Cirelli, G. Corcella, A. Hektor, G. Hutsi, M. Kadastik, P. Panci, et al., PPPC 4 DM ID: a poor particle physicist cookbook for dark matter indirect detection, J. Cosmol. Astropart. Phys. 1103 (2011) 051, arXiv:1012.4515.
[51] P. Ciafaloni, D. Comelli, A. Riotto, F. Sala, A. Strumia, A. Urbano, Weak corrections are relevant for dark matter indirect detection, J. Cosmol. Astropart. Phys. 1103 (2011) 019, arXiv:1009.0224.
[52] J.F. Navarro, C.S. Frenk, S.D. White, A universal density profile from hierarchical clustering, Astrophys. J. 490 (1997) 493–508, arXiv:astro-ph/9611107.
[53] AMS Collaboration, L. Accardo, et al., High statistics measurement of the positron fraction in primary cosmic rays of 0.5–00 GeV with the alpha magnetic spectrometer on the International Space Station, Phys. Rev. Lett. 113 (2014) 121101.
[54] AMS Collaboration, M. Aguilar, et al., Electron and positron fluxes in primary cosmic rays measured with the alpha magnetic spectrometer on the International Space Statio, Phys. Rev. Lett. 113 (2014) 121102.
[55] AMS Collaboration, M. Aguilar, et al., Precision measurement of the $(e^+ + e^-)$ flux in primary cosmic rays from 0.5 gev to 1 TeV with the alpha magnetic spectrometer on the International Space Station, Phys. Rev. Lett. 113 (2014) 221102.
[56] P. Gondolo, G. Gelmini, Cosmic abundances of stable particles: improved analysis, Nucl. Phys. B 360 (1991) 145–179.
[57] Y. Zhao, X.-J. Bi, H.-Y. Jia, P.-F. Yin, F.-R. Zhu, Constraint on the velocity dependent dark matter annihilation cross section from Fermi-LAT observations of dwarf galaxies, Phys. Rev. D 93 (2016) 083513, arXiv:1601.02181.
[58] J. Hisano, M. Kawasaki, K. Kohri, T. Moroi, K. Nakayama, Cosmic rays from dark matter annihilation and Big-Bang nucleosynthesis, Phys. Rev. D 79 (2009) 083522, arXiv:0901.3582.
[59] M. Kawasaki, K. Kohri, T. Moroi, Y. Takaesu, Revisiting Big-Bang nucleosynthesis constraints on dark-matter annihilation, Phys. Lett. B 751 (2015) 246–250, arXiv:1509.03665.

Permissions

All chapters in this book were first published in PLB, by Elsevier B.V.; hereby published with permission under the Creative Commons Attribution License or equivalent. Every chapter published in this book has been scrutinized by our experts. Their significance has been extensively debated. The topics covered herein carry significant findings which will fuel the growth of the discipline. They may even be implemented as practical applications or may be referred to as a beginning point for another development.

The contributors of this book come from diverse backgrounds, making this book a truly international effort. This book will bring forth new frontiers with its revolutionizing research information and detailed analysis of the nascent developments around the world.

We would like to thank all the contributing authors for lending their expertise to make the book truly unique. They have played a crucial role in the development of this book. Without their invaluable contributions this book wouldn't have been possible. They have made vital efforts to compile up to date information on the varied aspects of this subject to make this book a valuable addition to the collection of many professionals and students.

This book was conceptualized with the vision of imparting up-to-date information and advanced data in this field. To ensure the same, a matchless editorial board was set up. Every individual on the board went through rigorous rounds of assessment to prove their worth. After which they invested a large part of their time researching and compiling the most relevant data for our readers.

The editorial board has been involved in producing this book since its inception. They have spent rigorous hours researching and exploring the diverse topics which have resulted in the successful publishing of this book. They have passed on their knowledge of decades through this book. To expedite this challenging task, the publisher supported the team at every step. A small team of assistant editors was also appointed to further simplify the editing procedure and attain best results for the readers.

Apart from the editorial board, the designing team has also invested a significant amount of their time in understanding the subject and creating the most relevant covers. They scrutinized every image to scout for the most suitable representation of the subject and create an appropriate cover for the book.

The publishing team has been an ardent support to the editorial, designing and production team. Their endless efforts to recruit the best for this project, has resulted in the accomplishment of this book. They are a veteran in the field of academics and their pool of knowledge is as vast as their experience in printing. Their expertise and guidance has proved useful at every step. Their uncompromising quality standards have made this book an exceptional effort. Their encouragement from time to time has been an inspiration for everyone.

The publisher and the editorial board hope that this book will prove to be a valuable piece of knowledge for researchers, students, practitioners and scholars across the globe.

List of Contributors

Luís C. B. Crispino
Faculdade de Física, Universidade Federal do Pará, 66075-110, Belém, Pará, Brazil

Sam R. Dolan
Consortium for Fundamental Physics, School of Mathematics and Statistics, University of Sheffield, Hicks Building, Hounsfield Road, Sheffield S3 7RH, United Kingdom

Luiz C. S. Leite
Faculdade de Física, Universidade Federal do Pará, 66075-110, Belém, Pará, Brazil
Consortium for Fundamental Physics, School of Mathematics and Statistics, University of Sheffield, Hicks Building, Hounsfield Road, Sheffield S3 7RH, United Kingdom

Aurélien Barraua and Boris Bolliet
Laboratoire de Physique Subatomique et de Cosmologie, Université Grenoble-Alpes, CNRS-IN2P3, 53, avenue des Martyrs, 38026 Grenoble cedex, France

Francesca Vidotto
Radboud University, Institute for Mathematics, Astrophysics and Particle Physics, Mailbox 79, 6500 GL Nijmegen, The Netherlands

Marrit Schutten
Laboratoire de Physique Subatomique et de Cosmologie, Université Grenoble-Alpes, CNRS-IN2P3, 53, avenue des Martyrs, 38026 Grenoble cedex, France
Radboud University, Institute for Mathematics, Astrophysics and Particle Physics, Mailbox 79, 6500 GL Nijmegen, The Netherlands

Chong Oh Lee
Department of Physics, Kunsan National University, Kunsan 573-701, Republic of Korea

Antonella Garzilli and Alexey Boyarsky
Lorentz Institute, Leiden University, Niels Bohrweg 2, Leiden, NL-2333 CA, The Netherlands

Oleg Ruchayskiy
Discovery Center, Niels Bohr Institute, Blegdamsvej 17, DK-2100 Copenhagen, Denmark

A. O. Barvinsky
Theory Department, Lebedev Physics Institute, Leninsky Prospect 53, Moscow 119991, Russia
Tomsk State University, Department of Physics, Lenin Ave. 36, Tomsk 634050, Russia

A. Yu. Kamenshchik
Dipartimento di Fisica e Astronomia, Università di Bologna and INFN, Via Irnerio 46, 40126 Bologna, Italy
L. D. Landau Institute for Theoretical Physics of the Russian Academy of Sciences, Kosygin str. 2, 119334 Moscow, Russia

Imanol Albarran
Departamento de Física, Universidade da Beira Interior, Rua Marquês D'Ávila e Bolama, 6201-001 Covilhã, Portugal
Centro de Matemática e Aplicações da Universidade da Beira Interior (CMA-UBI), Rua Marquês D'Ávila e Bolama, 6201-001 Covilhã, Portugal

Mariam Bouhmadi-López
Department of Theoretical Physics, University of the Basque Country UPV/EHU, 48080 Bilbao, Spain
IKERBASQUE, Basque Foundation for Science, 48011, Bilbao, Spain

Che-Yu Chen and Pisin Chen
Department of Physics and Center for Theoretical Sciences, National Taiwan University, Taipei, 10617, Taiwan
LeCosPA, National Taiwan University, Taipei, 10617, Taiwan
Kavli Institute for Particle Astrophysics and Cosmology, SLAC National Accelerator Laboratory, Stanford University, Stanford, CA 94305, USA

Y. Bisabr, F. Ahmadi
Department of Physics, Shahid Rajaee Teacher Training University, Lavizan, Tehran 16788, Iran

J. G. FerreiraJr., C. A. de S. Pires, J. G. Rodrigues and P. S. Rodrigues da Silva
Departamento de Física, Universidade Federal da Paraíba, Caixa Postal 5008, 58051-970, João Pessoa, PB, Brazil

M. R. Setare
Department of Science, Campus of Bijar, University of Kurdistan, Bijar, Iran

F. Felegary and F. Darabi
Department of Physics, Azarbaijan Shahid Madani University, Tabriz, 53714-161, Iran

P. Michel L. T. da Silva, A. de Souza Dutra and J. M. Hoff da Silva
Departamento de Física e Química, Universidade Estadual Paulista, Av. Dr. Ariberto Pereira da Cunha, 333, Guaratinguetá, SP, Brazil

Alan A. Coley
Department of Mathematics and Statistics, Dalhousie University, Halifax, Nova Scotia, B3H 3J5, Canada

David D. McNutt
Faculty of Science and Technology, University of Stavanger, N-4036 Stavanger, Norway

Andrey A. Shoom
Department of Mathematics and Statistics, Memorial University, St. John's, New found land and Labrador, A1C 5S7, Canada

Neil D. Barrie and Archil Kobakhidze
ARC Centre of Excellence for Particle Physics at the Terascale, School of Physics, The University of Sydney, NSW 2006, Australia

Xue-Mei Deng
Purple Mountain Observatory, Chinese Academy of Sciences, Nanjing 210008, China

Yi Xie
School of Astronomy and Space Science, Nanjing University, Nanjing 210093, China
Key Laboratory of Modern Astronomy and Astrophysics, Nanjing University, Ministry of Education, Nanjing 210093, China

D. J. Brooker and R. P. Woodard
Department of Physics, University of Florida, Gainesville, FL 32611, United States

N. C. Tsamis
Institute of Theoretical Physics & Computational Physics, Department of Physics, University of Crete, GR-710 03 Heraklion, Greece

Ernest Ma
Physics and Astronomy Department, University of California, Riverside, CA92521, USA

Nayem Sk
Dept. of Physics, University of Kalyani, West Bengal, 741235, India

Mauricio Bellini
Departamento de Física, Facultad de Ciencias Exactas y Naturales, Universidad Nacional de Mar del Plata, Funes 3350, C.P. 7600, Mar del Plata, Argentina
Instituto de Investigaciones Físicas de Mar del Plata (IFIMAR), Consejo Nacional de Investigaciones Científicas y Técnicas (CONICET), Mar del Plata, Argentina

Asuka Ito and Jiro Soda
Department of Physics, Kobe University, Kobe 657-8501, Japan

Wei Liao
Institute of Modern Physics, School of Science, East China University of Science and Technology, 130 Meilong Road, Shanghai 200237, PR China

Huibert het Lam and Tomislav Prokopec
Institute for Theoretical Physics, Spinoza Institute and EMME, Utrecht University, Postbus 80.195, 3508 TD Utrecht, The Netherlands

Shan-Shan Zhao and Yi Xie
School of Astronomy and Space Science, Nanjing University, Nanjing 210023, China
Key Laboratory of Modern Astronomy and Astrophysics, Nanjing University, Ministry of Education, Nanjing 210093, China

Ganim Gecim and Yusuf Sucu
Department of Physics, Faculty of Science, Akdeniz University, 07058 Antalya, Turkey

Joan Solà, Adrià Gómez-Valent and Javier de Cruz Pérez
Departament de Física Quàntica i Astrofísica, and Institute of Cosmos Sciences, Universitat de Barcelona, Av. Diagonal 647, E-08028 Barcelona, Catalonia, Spain

Konstantinos Dimopoulos
Consortium for Fundamental Physics, Physics Department, Lancaster University, Lancaster LA1 4YB, UK

Yaghoub Heydarzade and FarhadDarabi
Department of Physics, Azarbaijan Shahid Madani University, Tabriz, 53714-161, Iran

Prabir Rudra
Department of Mathematics, Asutosh College, Kolkata700 026, India

Ahmed Farag Ali
Netherlands Institute for Advanced Study, Korte Spinhuissteeg 3, 1012 CG Amsterdam, Netherlands
Department of Physics, Faculty of Science, Benha University, Benha, 13518, Egypt

Mir Faizal
Irving K. Barber School of Arts and Sciences, University of British Columbia-Okanagan, Kelowna, BC V1V 1V7, Canada
Department of Physics and Astronomy, University of Lethbridge, Lethbridge, AB T1K 3M4, Canada

Y. Heydarzade and F. Darabi
Department of Physics, Azarbaijan Shahid Madani University, Tabriz, Iran

Jeremy Sakstein and Adam R. Solomon
Center for Particle Cosmology, Department of Physics and Astronomy, University of Pennsylvania, 209 S. 33rd St., Philadelphia, PA 19104, USA

Xiao-Jun Bi, Su-Jie Lin and Peng-Fei Yin
a Key Laboratory of Particle Astrophysics, Institute of High Energy Physics, Chinese Academy of Sciences, Beijing 100049, China

Qian-Fei Xiang
Key Laboratory of Particle Astrophysics, Institute of High Energy Physics, Chinese Academy of Sciences, Beijing 100049, China
School of Physical Sciences, University of Chinese Academy of Sciences, Beijing 100049, China

Index

Q

R

S

T

U

V

W

X

CPSIA information can be obtained
at www.ICGtesting.com
Printed in the USA
LVHW062007100122
708212LV00002B/76

9 781647 283551